AF334012

Knots and Applications

K&E Series on Knots and Everything — Vol. 6

KNOTS AND APPLICATIONS

Editor

Louis H. Kauffman
Department of Mathematics,
Statistics and Computer Science
University of Illinois at Chicago

World Scientific
Singapore • New Jersey • London • Hong Kong

Published by

World Scientific Publishing Co. Pte. Ltd.

P O Box 128, Farrer Road, Singapore 9128

USA office: Suite 1B, 1060 Main Street, River Edge, NJ 07661

UK office: 57 Shelton Street, Covent Garden, London WC2H 9HE

Library of Congress Cataloging-in-Publication Data

Knots and applications / edited by Louis H. Kauffman.
       p.   cm. -- (Series on knots and everything ; vol. 6)
     Includes bibliographical references.
     ISBN 9810220049   ISBN 9810220308 (pbk)
     1. Knot polynomials.   2. Mathematical physics.   I. Kauffman,
Louis H., 1945–   . II. Series: K & E series on knots and
everything ; vol. 6
QC20.7.K56K56   1995
501'.514224--dc20   94-30312
                                            CIP

Printed in Singapore by Uto-Print

Preface

This volume of papers is devoted to knots and their applications. From the outset, knot theory has been an applied science with roots in weaving and in practical knot tying. The theory of knots and links in three-dimensional space is a branch of pure mathematics that is a good model of the behaviour of actual rope in the three space of our direct experience.

This theory received its inception in the 19th century with the help of Lord Kelvin (William Thomson) who, inspired by his theory of vortex atoms, asked the mathematicians Kirkwood, Little and Tait to compile tables of topologically distinct knots. These tables were an impetus to the newly growing field of topology, and by the 1920's there was a genuine mathematical theory of knots with its own special characteristics.

Along with this initial mathematical theory came an isolation of the mathematics of knots from potential applications. This isolation was partly due to the collapse of the ether theory upon which Lord Kelvin's vortices were based, and partly upon the new and abstract nature of the topological methods that were being applied to knots. Nevertheless, as time went on applications and ideas for applications came forth. In 1984 Vaughan Jones discovered a new and very simply defined invariant of knots and links. The Jones polynomial is intimately related to mathematical physics, and this new circumstance changed the entire landscape of both knot theory and its applications. This book is a collection of research articles on knots and their applications that span the entire time period from Lord Kelvin to the present day.

Here is a brief description of the contents of this volume.

The first article, "Knot Logic" by Louis Kauffman, is a research article written for this volume. Here the application of knot theory is to the foundations of mathematics. The technique of knot diagrams is used to give non-standard models for set theory and this is generalized to include a diagrammatic version of the Church-Curry lambda calculus and its formal relationship with invariants of knots and links (the quandle, crystal and rack). One advantage of this approach via the lambda calculus is that it allows a generalization of the rack to an interlock algebra that is a complete knot invariant. Many examples are given that relate paradoxes of self-reference and puzzles (such as the "Chinese rings" — see the example of recursive unlinking in section 2) to the theory of knots. Other topics in this essay include an exposition of recent work of Laver and DeHornoy applying the Artin braid group to the solution of the word problem for certain algebras arising in set theory, relations between switching circuit theory, electrical networks and knot theory and a look at the formalism of self-reproducing machines in knot diagrammatic terms. The essay ends with a discussion of the idea of knots and links as the "pregeometry" desired by John Wheeler [CTW] for foundations of physics. An appendix outlines the bracket construction for the Jones polynomial.

Other than brief mention of the bracket polynomial and related matters in the Knot Logic paper, this volume does not directly address the interplay between knot theory and statistical mechanics that has ensued since the discovery of the Jones polynomial. The interested reader should consult [K1],[K2],[K3],[J],[W],[Wu] for further information on this topic.

Next comes a reprinting of three articles by Lord Kelvin on the theory of vortex atoms. These are entitled
1. "On Vortex Atoms"
2. "On Vortex Motion"
3. "Vortex Statics".
The first and third articles are qualitative in nature, and show very clearly Kelvin's physical mode of thinking. The middle article is more mathematical, and will give the reader a good flavor of his mode of analysis in relation to these problems. The idea of a knotted vortex is very attractive and we shall see it recur in later papers in this book.

The papers by Kelvin were written between 1867 and 1880. We skip, for the next paper, to a reprinting of the 1968 paper by David Finkelstein and Julio Rubenstein entitled "Connection between Spin, Statistics, and Kinks". This paper uses knots (and a trick with a rubber band that you can perform on your fingers) to analyse a modern problem involving spin in quantum mechanical systems.

Next we reprint a remarkable 1972 paper by Herbert Jehle entitled "Flux Quantization and Particle Physics". Jehle made the hypothesis that elementary physical particles are quantized electromagnetic flux loops. In his theory a particle can be identified with a knot or link representing this flux. The antiparticle for a given knot or link is its mirror image. This vision of particulate matter as trapped loops of quantum electromagnetic energy is a direct descendant of Lord Kelvin's vortices. Jehle's theory is incomplete, but a fascinating signpost in the history of attempts to apply knot theoretic ideas to basic physics.

The next article is a reprint of the 1977 paper by Eckehard W. Mielke entitled "Knot Wormholes in Geometrodynamics?". It has been suggested by John Wheeler that a particle may "really" be the entrance to a long tube leading out of observable space and returning elsewhere (possibly in another universe) as a paired anti-particle.   Such an entity is called a wormhole. Mielke considers knotted wormholes in the context of a theory of general relativity with two spatial dimensions and one temporal dimension. Here again the idea of the knotted vortex or flux tube has taken hold in a new way in a modern physical context.

We follow with a recent (1992) article by H.K. Moffatt and Renzo L. Ricca entitled "Helicity and the Calugareanu Invariant". This article is a rigorous application of topological ideas to flux tubes in the context of fluid dynamics. It also contains interesting comments about the history of a famous relationship about twist, writhe and linking number for space curves. As Moffatt and Ricca point out, the original source for the idea of "self-linking" or writhe of a space curve is the (1959) paper of Calugareanu [C]. On the other hand, it is certainly the case that the modern point of view on this relationship should be credited to James White in his paper [Wh] of 1969. The relationship in question states that the linking number (Lk) of the two edges of a closed oriented band embedded in three-dimensional space can be decomposed as a sum of the Twist (Tw) and the Writhe (Wr) of the band. Twist measures how the band twists around its core and Writhe measures the "self-linking" of the core. The core is the space curve described by the central curve in the band. This equation, Lk = Tw + Wr, has been applied by White and his co-workers to a myriad of problems in molecular biology. See [K3], Part II - Section 15 for a concise description of this work.

Next is an article written for this volume by Lev Rozansky entitled "Witten's Invariant of 3-Dimensional Manifolds: Loop Expansion and Surgery Calculus". This article describes Rozansky's research in the quantum field theoretic interpretation of invariants of links and three manifolds. This approach was pioneered by Edward Witten in his 1989 paper "Quantum Field Theory and the Jones Polynomial" [W]. Rozansky's article can serve as an introduction to the intricate methods of quantum field theory — applied to the study of knots and three manifolds. Here there is an extraordinary interplay between methods of mathematical physics and problems in topology.

The paper by Maurizio Martellini and Mario Rasetti entitled "2+1 Dimensional Quantum Gravity as a Gaussian Fermionic System and the 3D-Ising Model" turns the tables, using the quantum field theoretic interpretation of link invariants to apply them to a problem in mathematical physics.

We turn to molecular biology and chemistry with three papers:

1. L. Kauffman and Y.B. Magarshak, "Vassiliev Knot Invariants and the Structure of RNA Folding ".

2. Alison MacArthur, "The Entanglement Structures of Polymers".

3. David W. Walba, Timothy C. Homan, Rodney M. Richards and R. Curtis Haltiwanger, "Synthesis and Cutting "In Half" of a Molecular Mobius Strip — Applications of Low Dimensional Topology in Chemistry".

The paper by Kauffman and Magarshak describes how to create invariants of graph embeddings that correspond to molecules, such as RNA, that fold on themselves. The paper can also be read for an introduction to the concept of a Vassiliev invariant of graphs and links and its relationship with the structure of Lie algebras.

The paper by Alison MacArthur describes the use of knot theoretic techniques in modelling the properties of long chain molecules including rubber and other materials.

The paper by David Walba and his collaborators describes not only the synthesis of molecular Mobius strips, but also of knotted molecules by the "hook and ladder" technique described in the text of the paper. These are excursions in the synthetic chemistry of molecules with particular topological structures. In the case of the knotted molecule, the editor of this book had an input that bore fruit in the laboratory. I suggested the hook and ladder technique to David Walba in 1986 while visiting the University of Iowa under Jon Simon's project on Stereochemical topology. David took up the idea with great enthusiasm and it eventually worked! It is not easy to synthesize small molecules with specified topology. Walba is the original discoverer of a molecular Mobius strip. His work gave a new impetus to the study of graph embeddings as a generalization of knot theory.

The last paper, "Turning a Penrose Triangle Inside Out" by Thaddeus M. Cowan, is an application of topology to the psychology of visual perception. Cowan uses the descriptive language of braids to coordinate and invent the structure of visual illusions, "impossible figures" and their properties in the eye of the beholder.

I would like to take this opportunity to thank the many people who have contributed to this volume by their direct contributions and many pleasing discussions. In particular I would like to thank Dr. K.K. Phua of World Scientific Publishing Company for suggesting the possibility of a book of papers on knots and their applications, David Tranha, Publishing Director, Mathematical Sciences, Cambridge University Press for the permission to reproduce the papers by Lord Kelvin, Arthur Greenspoon of Mathematical Reviews for great help in proof reading the papers and my wife Diane for many discussions, including the one in 1986 when we tried out the hook and ladder with rope and tape.

Louis H. Kauffman
Chicago, July 4, 1994

References

[C] G. Calugareanu, Sur les classes d'isotopie des noeds tridimensionale et leurs invariants, *Czechoslovak Math. J.* **11** (1959), 5-20.

[CTW] C.W. Misner, K.S. Thorne and J.A. Wheeler, *Gravitation*, W.H. Freeman and Co. (1971).

[J] V.F.R. Jones, On knot invariants related to some statistical mechanics models, *Pacific J. Math.* **137**, no. 2 (1989), 311-334.

[K1] Louis H. Kauffman. State models and the Jones polynomial, *Topology* **26** (1987), 395-407.

[K2] Louis H. Kauffman, Statistical mechanics and the Jones polynomial, Proceedings of the Santa Cruz Conference on the Artin Braid Group - Summer 1986 -AMS Contemporary Math. Series Vol. 78 (1989), 263-297.

[K3] Louis H. Kauffman, *Knots and Physics*, (Second Edition 1993), World Scientific Pub. Co.

[W] E. Witten. Quantum field theory and the Jones polynomial, *Commun. Math. Phys.* **121** (1989), 351-399.

[Wh] J. White. Self-linking and the Gauss integral in higher dimensions, *Amer. J. Math.* **91** (1969), 693-728.

[Wu] F.Y.Wu. Knot theory and statistical mechanics, *Rev. Modern Phys.* **64**, no. 4 (October 1992), 1099-1131.

References

Table of Contents

Knot Logic

Louis H. Kauffman

Department of Mathematics, Statistics and Computer Science
851 South Morgan Street
University of Illinois at Chicago, Chicago, Illinois 60607-7045
e-mail: U10451@UICVM.BITNET

Abstract.   Knot and link diagrams are used to represent nonstandard sets, and to represent the formalism of combinatory logic (lambda calculus). These diagrammatics create a two-way street between the topology of knots and links in three dimensional space and key considerations in the foundations of mathematics.

Key Words. knot, knot logic, topology, combinatory logic, quandle, crystal, rack, interlock algebra, LD-magma, quantum link invariants, circuit logic, topological quantum field theory, replication,
self-replication.

Contents

I. Introduction

This paper introduces the use of knot and link diagrams for representing nonstandard sets and also for representing the formalism of combinatory logic (lambda calculus). These diagrammatics create a two-way street between the topology of knots and links in three dimensional space and key considerations in the foundations of mathematics. The paper explores the relationship of this foundational study with the structure of quantum link invariants and with applications of knot theory to biological structure.

Section II reviews concepts of set theory from an original point of view, emphasizing the relative consistency of sets that do not satisfy the axiom of foundation — by constructing models in terms of notations, graphs and subsets of the plane. Section II also introduces ideas from knot theory and shows how to prove that you cannot cancel knots , just skirting paradox in the process. Section II includes a discussion of reentry and recursive forms in relation to knots, wild embeddings and fractals. An example is given of a sequence of graph embeddings whose complexity increases linearly, while an associated unlinking number is conjectured to increase exponentially. The section ends with a discussion of indicational calculi, non-standard logic, quantum logic and boundary logic.

Section III introduces knot set theory, a set theory whose membership relation is represented by one arc underpassing another. Knot set theory accomodates sets that are members of themselves and sets whose members are defined mutually. The diagrammatic representation of knot sets is so constructed that topologically equivalent diagrams represent the same set. One of the consequences in involving the topology in this way is that knot sets use a "fermionic" convention for the treatment of lists of identicals. The fermionic convention is that identicals cancel in pairs. Thus in the fermionic convention the set {a,a} is equivalent to the empty set. Ordinary set theory uses the "bosonic" convention that identicals condense in pairs (so that {a,a} = {a} in standard sets).

Section IV discusses concepts of reference in relation to knot set theory.

Section V gives a construction that translates between knot diagrams and combinatory logic. In this formalism the broken arcs of the diagram are used to represent different elements in a lambda calculus, and the diagrams themselves naturally represent non-associative compositions of these elements. We show how to write key constructions in the lambda calculus such as the Church-Curry fixed point theorem in terms on these diagrams. We then investigate the relationship of this formalism with the topology and show how it is intimately related to the algebraic concepts of quandle, crystal and rack (see [J], [K6], [RF]) as used by knot theorists. The quandle, crystal and rack are non-associative algebras that derive from a diagram of the knot and are topological invariants of it. In section VI we take this correspondence further by defining an extension of the crystal, the *interlock algebra* of a knot.

The interlock algebra is an algebra of lambda operators associated with the knot diagram. It is a topological invariant of the diagram and it contains complete information about the topology of the knot. Two knots are isotopic in three space if and only if their interlock algebras are isomorphic. The interlock algebra of a knot contains two types of lambda elements — those with no free variable and those with one free variable (multiple variables will occur in the case of a link). This presence of operators with free variables in the interlock algebra allows an intrinsic identification of subalgebras that are needed for the topology. The construction of the interlock algebra is an application of combinatory logic to topology. Section VI ends with a brief discussion of the classical Alexander polynomial.

Section VII discusses a problem in universal algebra — the structure of non-associative systems with a single non-commutative binary operation that admits a left-distributive law over itself: $a(bc) = (ab)(ac)$. These algebras are called *LD-magmas*. We have already met this condition in studying quandles in section 4. Here the left-distributive law is studied for its own sake. The word problem for free magmas was solved by Patrick DeHornoy in a beautiful and startling application of the Artin braid group. We sketch his method.

Section VIII sketches how the fixed point theorem for the lambda calculus is related to recursive forms, self-reference and Gödel's incompleteness theorem. This section contains a digression on forms of self-replication, including DNA, the Building Machine, the Mighty Simple Self-Rep and the Knot Logical Self-Rep (which turns out to be a picture of the syntax of the Building Machine). The self-replication of a knot is accomplished by a slide equivalence more drastic than the handle-sliding of Kirby calculus. The section ends with a description of Kirby calculus in this context.

Section IX is an introduction to the logic of Dirac brackets in the context of topological invariants. Section X discusses relations between knot theory, electricity and switching circuit theory. Section XI, on asynchronous automata, is a description of a domain in circuit design that has analogies with knot theory. In this context we see that quandles, crystals and racks (Sections 5 and 6) implicate a concept of knot automata.

Section XII explores pregeometry in the sense of John Wheeler. We make the case that knot and link diagrammatics are central to an appropriate conception of pregeometry. An appendix discusses the bracket model for the Jones polynomial.

The author would like to express his thanks to Louis Crane, Lee Smolin, Carlo Rovelli, Julian Barbour and John Wheeler for helpful conversations. Research for this paper was partially supported by the Program for Mathematics and Molecular Biology, University of California at Berkeley and by NSF Grant No. DMS 9205277.

II. Sets, Knots, Recursions

It is customary either to build the theory of sets axiomatically, or to construct it from the intuitive concepts of membership and collection. It is well-known that a naive approach leads to paradoxes.

For example, the Russell set R is defined to be the set of all sets that are not members of themselves. X is a member of R exactly when X is not a member of X. On substitution of R for X, we find that R is a member of R exactly when R is not a member of R.

Initially, it is not clear whether the difficulty with the Russell set is in the notion of set formation, the idea of self-membership, the use of the word "not" , the use of the word "all" or elsewhere. The Theory of Types [WhR] due to Russell and Whitehead placed the difficulty in the use of self-membership, and solved the paradox by prohibiting this and other ways of mixing different levels of discourse.

The Gödel-Bernays set theory (see [K], Appendix on Elementary Set Theory) creates a different solution to the Russell paradox by making one large distinction between *set* and *class* . Of two sets A and B it can be said without ambiguity that A is a member of B, or B is a member of A, or neither is a member of the other. A class is a *set* if it is a member of another class. Classes are determined by their members, and classes can be defined in terms of properties: Given a property P, there exists a class C(P) equal to the class of all x such that *P(x) is true and x is a set.*

In this system, the Russell class is
R = {x | x is not a member of x and x is a set}.
Thus R is a class, but R is not a set.

In a system of the Gödel-Bernays type, there is nothing inherently wrong with self-membership. In fact, self-membership and other forms of contradiction of the "axiom of foundation" (which disallows infinite descending chains of membership) are very interesting to explore using geometry, topology and diagrams. To this end, let us start from the beginning and construct some sets.

The empty set is commonly denoted by empty brackets: { }.
Notationally, sets indicated only through brackets are a subcollection of all the ways of making well-formed brackets:

A finite expression E in brackets is well-formed if
1. E is empty
or
2. E = { F } G where F and G are well-formed.

These two rules give a complete characterization of the well-formed bracket expressions. A *finite ordered multi-set S* is an expression in the form S = {T} where T is any well-formed expression. It follows that T = A$_1$ A$_2$ ... A$_n$ where n is a positive integer, and each A$_i$ is a finite ordered multi-set. The A$_i$'s are the *members* of S.

We write the members of S without commas between them.

For example, if S= { { } { { } } }, then the members of S are

{ } and {{ }}.

A multi-set may have a multiplicity of identical members as in

X = { {   } {   } {   } }.

Ordered multi-sets are equal exactly when they have identical sequences of brackets. To emphasize this point, let L denote a left bracket , { , and R denote a right bracket , } . Then the set X above is encoded by the sequence LLRLRLRR.

To obtain the usual category of finite sets, factor the ordered multi-sets by the equivalence relation generated by XY = YX and XX = X where X and Y are well-formed expressions. It then follows from our definitions that two finite sets are equal exactly when they have the same members.

It is easy to see that the class of ordered finite multi-sets is isomorphic to the class of rooted planar trees - by graphical duality as indicated below.

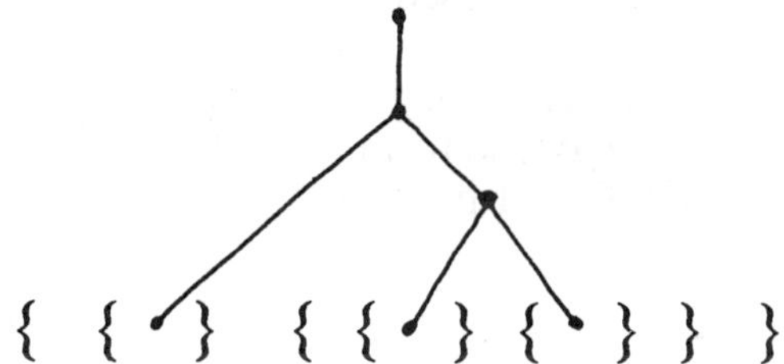

$$\{ \ \{ \ \} \ \{ \ \{ \ \} \ \{ \ \} \ \} \ \}$$

Another way to think of these sets is to replace each pair of brackets by a rectangle in the plane. Then any set is a collection of disjoint rectangles, with a single outermost rectangle — the set boundary. The members of the set are delineated by the rectangles inside this outermost rectangle that are outermost or at the same level as all other rectangles in the pattern. The tree is still obtained by graphical duality as shown below.

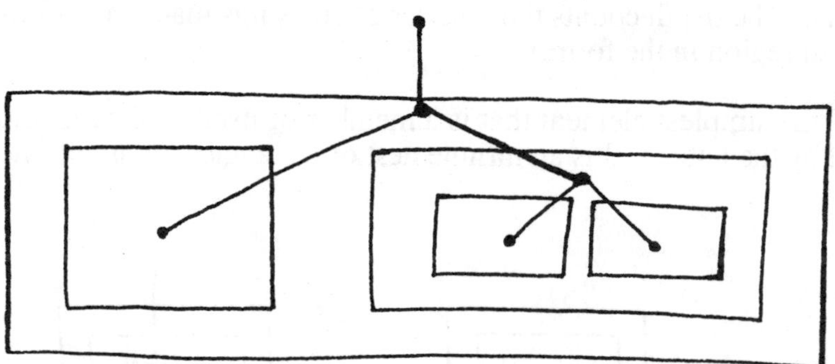

In both cases there is a natural notion of depth obtained by counting crossings inward from the outermost rectangle, or by counting nodes from the root of the tree. The equivalence relation on rectangles that generates finite sets is: *take the collections of rectangles up to homeomorphisms of the plane.* Here we use a sophisticated concept to define an elementary one. The use of this will become apparent at once when we enlarge the category and obtain a model of non-standard sets.

Let FIST (First Infinite Sets) denote the class of (not necessarily finite) disjoint collections of rectangles in the plane such that each collection S has a single outermost rectangle, and the collection of rectangles inside that outermost rectangle is a disjoint union of elements of FIST. (These are the members of S.) If A and B are in FIST, then we shall say that A=B if there is a homeomorphism of the plane that carries A to B.

Call a collection of rectangles in the plane, taken up to homeomorphism of the plane, a *form.* Thus, finite (and some infinite) sets can be interpreted as forms, but not all forms are sets. In any form we can say unambiguously of two rectangles whether one is inside or outside of the other.

Forms can be framed and juxtaposed.

Let {X} denote the result of putting a rectangle around the form X. Call this operation the *framing* of the form X. Let XY denote the *juxtaposition* of the forms X and Y. To get multi-sets from forms, consider forms that are framed.
For example,

0. { }
1. { { } }
2. { { } { } }
3. { { } { } { } }
    ...

can be regarded as a list of multisets, with 0,1,2,3,... members. No commas are needed in the internal list of a set represented in this fashion. One simply searches for the different frames at depth 1, to get the list of members. (The depth counts the number of crossings made inward from the outermost region in the form.)

In FIST the simplest element that is a member of itself is shown below and denoted by the letter J. J is an infinite nest of rectangles, or an infinite linear tree.

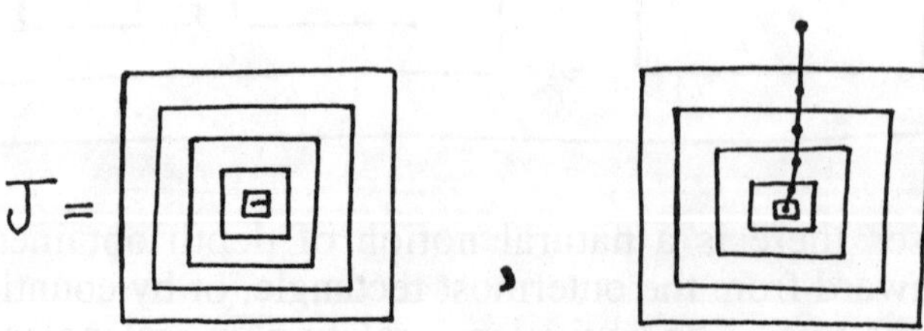

Note that J = { J } where we interpret the brackets as the addition of the outer rectangle. There is nothing inherently infinite about the description J= { J }, but its recursive unfolding   leads to an infinite construction corresponding to an infinity of nested brackets:

$$J = \{ \quad \{ \quad \{ \ \{ \ \{ \{\{...\}\} \} \ \} \ \} \ \} \quad \}$$

With this rectangle model in mind, consider elements of FIST that are defined by systems of equations. For example, A = {{} B}, B = {A} yields

$$A = \{\{\} \ B\} = \{\{\} \ \{A\}\}$$

$$= \{ \ \{\} \ \{\{ \ \{\} \ \{A\} \ \}\} \ \}$$

$$= \{ \ \{\} \ \{\{ \ \{\} \ \{\{ \ \{\} \ \{\{ \ \{\} \ \{...\} \ \}\} \ \}\} \ \}\} \ \}.$$

A and B   correspond to non-homeomorphic systems of rectangles, and so give a pair of distinct but entangled sets in FIST.

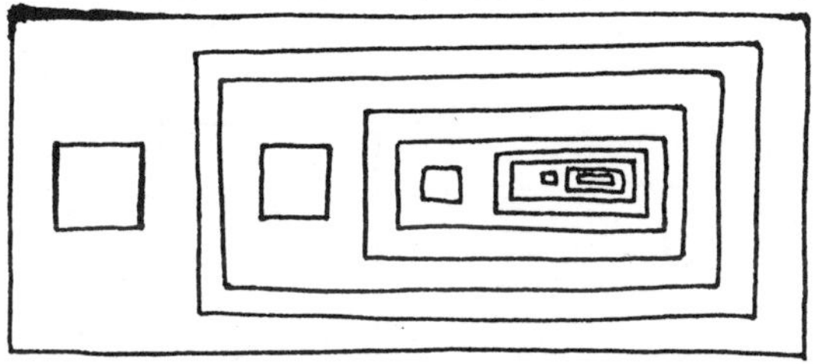

Reentry Notation, Recursive Forms and Infinite Regress
A set that is a member of itself can be diagrammed as a set with an arrow pointing into the inside of the set where the self inclusion occurs (compare [K16]).

$$M = \{ \ M \ \} = \{ \quad \}$$

In this form, one tends to take a model of infinite regress or recursion as in

Similarly, in the case of interlock (a={b}, b={a}) we have a= {{a}} and the reentry description

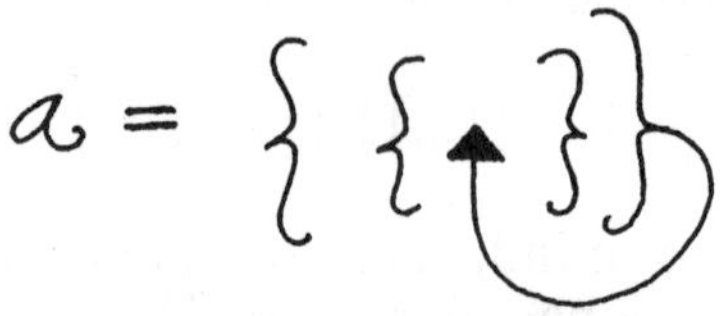

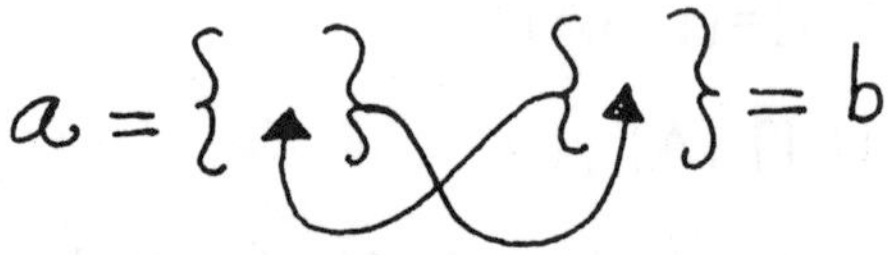

$$a = \{ \{ \blacktriangle \} \}$$

$$a = \{ \blacktriangle \} \{ \blacktriangle \} = b$$

The reentry concept goes beyond set formation to a domain of recursive forms. To indicate recursive forms that are not necessarily interpreted as sets it is convenient to use a rectangular box notation. Thus we write

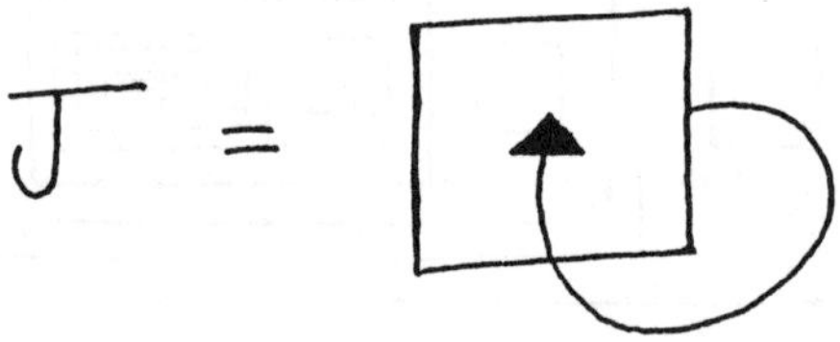

$$J = \boxed{\blacktriangle}$$

and

$$F = \boxed{\boxed{F}\,F} = \boxed{\boxed{\blacktriangle}\,\blacktriangle}$$

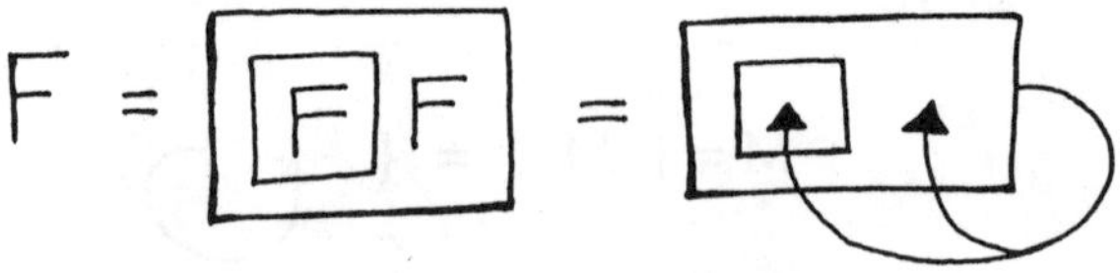

The second recursive form, F, can be called the *Fibonacci Form* since the number of divisions of this form at depth n is the nth Fibonacci number.

(The form divides the plane into disjoint connected regions. These are the *divisions* of the form. A division is said to have *depth n* if it requires n inward crossings of rectangle boundaries to reach that region from the outermost region in the plane. Each rectangle divides the plane into a bounded region and an unbounded region. A crossing of the boundary of a given rectangle is said to be an *inward crossing* if it goes from the unbounded region to the bounded region.)

To see this and other facts about the divisions of a form, *let F_n denote the number of divisions of an arbitrary form F at depth n.*

Then, for any forms X and Y,
1. $(XY)_n = X_n + Y_n$
2. $\{X\}_n = X_{n-1}$.

In the case of the Fibonnaci form, we have   **F={{F}F}**.   Hence $F_n = F_{n-2} + F_{n-1}$. Since $F_0 = F_1 = 1$, this proves our assertion about the Fibonacci series as the depth counts of the Fibonacci form.

From here it is quite natural to define the *growth rate*, **m(F)**, of a form F as the limit of F_{n+1}/F_n as n goes to infinity.

$$\mathbf{m(F) = \lim_{n \to \infty} (F_{n+1}/F_n)}.$$

The growth rate of the Fibonacci form is the golden ratio, $(1+ \sqrt{5})/2$.

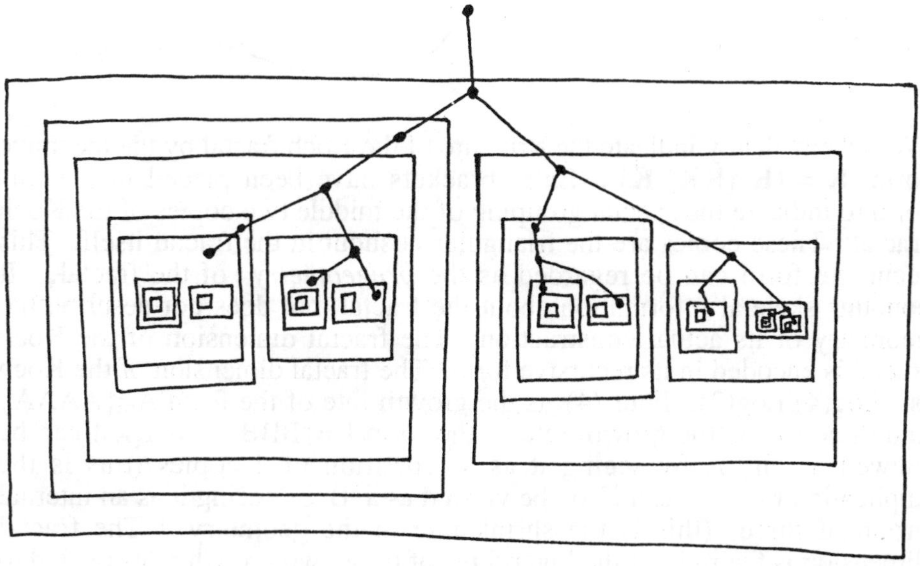

Recursive forms and their growth rates are intimately related to fractals. For example, the Koch fractal reenters its own indicational space in four major places as shown below.

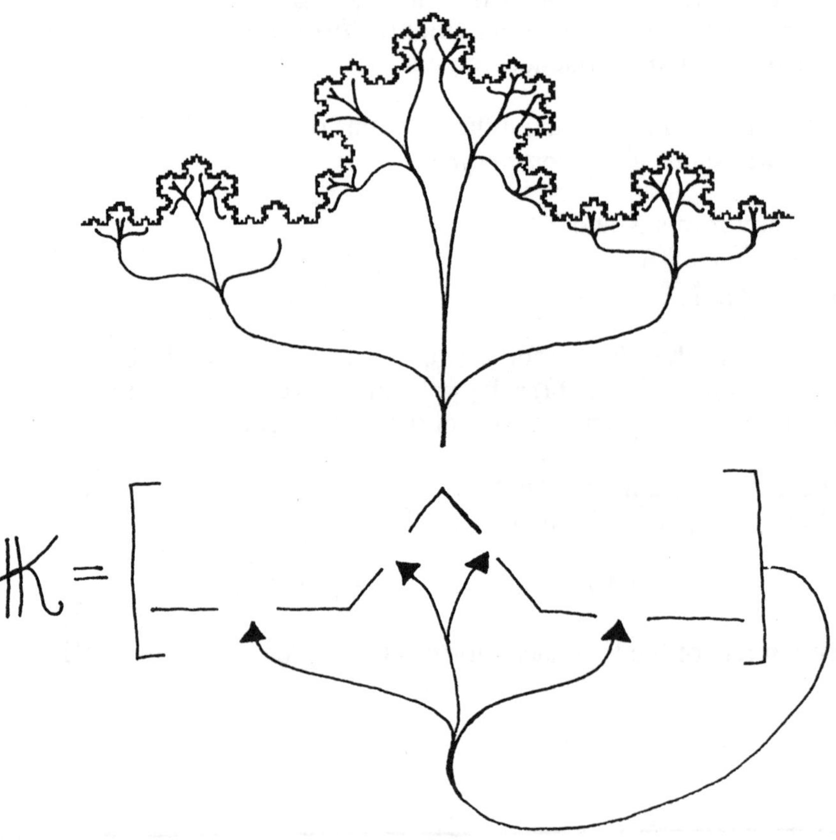

We schematically indicate the structure of the Koch fractal by the recursive form **K = {K {KK} K}** . Extra brackets have been placed inside this form to indicate the special grouping of the middle two copies of the Koch fractal. These copies are the triangular pushout in the fractal itself. This recursive form can be regarded as the *pregeometry* of the fractal. It contains skeletal information about the fractal, but does not describe the geometry of its actual construction. The fractal dimension of the Koch fractal is encoded in its recursive form. The fractal dimension of the Koch is Log(4)/Log(3). Four (4) is the growth rate of the form A={AAAA} and three (3) is the growth rate of the form B={BBB}. K itself can be viewed as an A by seeing it as a repetition of 4 copies (this is the duplication rate). K can also be viewed as a B by seeing it as an internal group of three (this is the shrink rate in the geometry). The fractal dimension is the ratio of the logarithms of these two growth rates related to the recursive form.

Alexander's Horned Sphere

Now we go to topology and look at the reentry form associated with the famous Alexander Horned Sphere [HY]. The schematic for this construction is illustrated below.

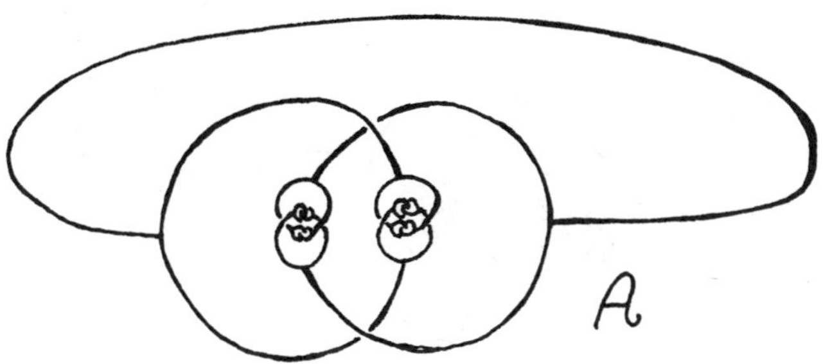

This infinite graph can be described as a reentry form as shown below.

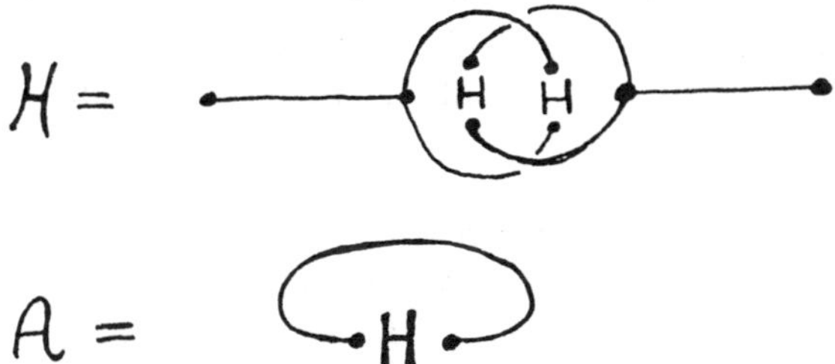

The limit of this construction produces a wildly embedded tree in three-space that is self-linked (i.e. the fundamental group of the complement of this tree is non-trivial). Any finite stage of the construction produces an unlinked embedding of a tree. The Alexander Horned Sphere is obtained by taking a limit of the boundaries of tubular neighborhoods of the finite trees in this construction. It is an embedding of a two dimensional sphere into three dimensional space such that the inside of the sphere is simply connected, but the outside is not simply connected.

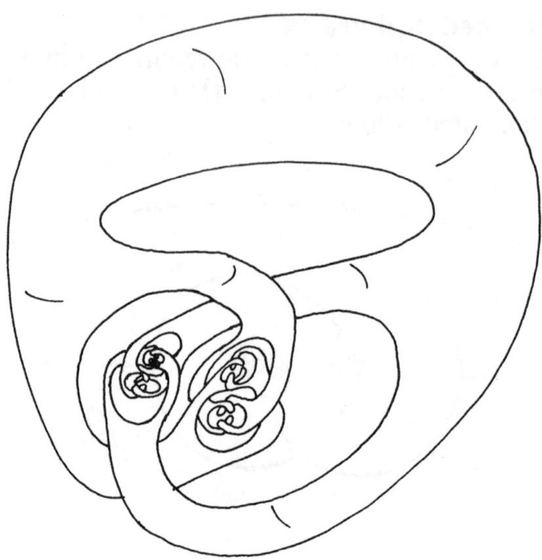

The most remarkable thing about the horned sphere is that it is a sphere. The limit construction does not touch itself anywhere. There is a Cantor's set worth of wild points on this embedded sphere such that any neighborhood of a wild point contains infinitely many branches of the structure.

An example of recursive unlinking
Consider the graph embedding shown below.

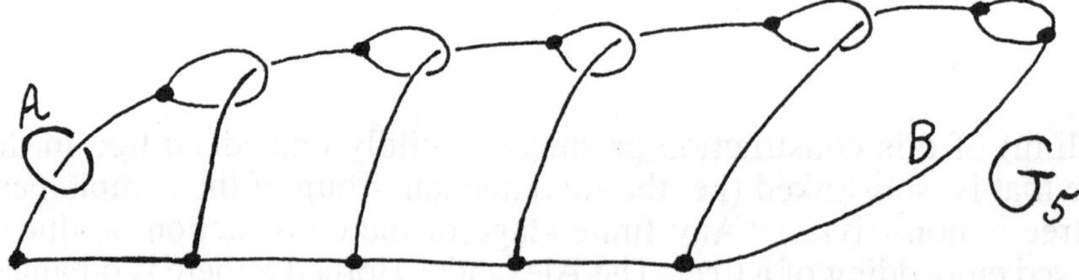

This is a special case of the graph embedding G_n where n is equal to 5. In G_n there is a series of n hoops, each one successively slipped through the previous one, all tied together at their bases, and so that the arc B is attached from the last hoop to its own base. Suppose that it is desired to unlink the circle labelled A from this graph under the stipulation that A is allowed to make crossing exchanges only with the arc labelled B. One can perform any isotopy of the embedding coupled with these allowed crossing changes. Then *I conjecture that G_n requires at least 2^{n-1} crossing exchanges with B in order to become unlinked.* If this conjecture is true, then we have an unlinking problem whose complexity goes up exponentially, while the complexity of the underlying graph embeddings that support it goes up linearly. This example shows how the sort of recursive construction associated with an object like the horned sphere can pose an actual complexity problem in topology for the finite stages of the recursion.

The isotopy shown below of G_5 to a graph G' with the hoops unentangled, should give the reader a glimpse of evidence for this conjecture. It is clear that A can be unlinked from G' by 2^4 exchanges. Hence, up to isotopy, A can be unlinked in G_5 by 2^4 exchanges. A similar construction shows that A can be unlinked in G_n with 2^{n-1} exchanges. We conjecture that this procedure is minimal.

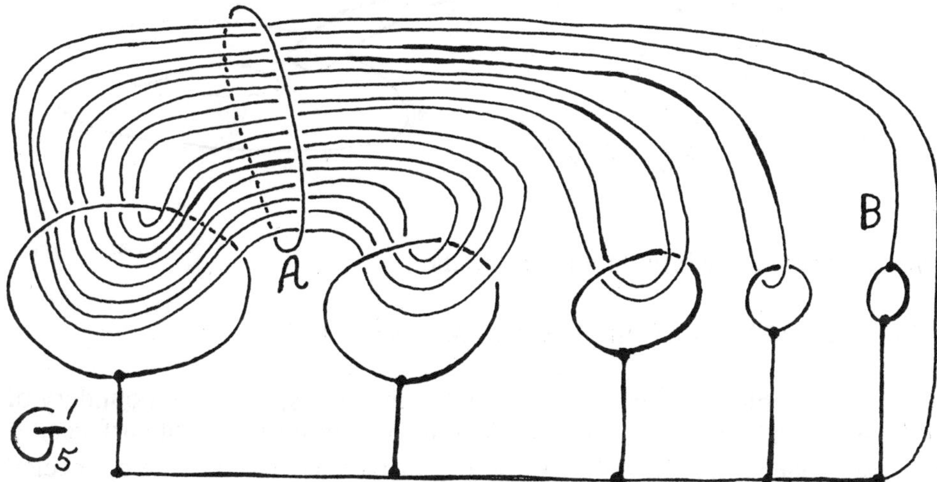

The Method of Infinite Repetition
There is a technique in topology called *the method of infinite repetition* . It begins with the paradox:

$$
\begin{aligned}
0 &= (1\text{-}1)+ (1\text{-}1)+ (1\text{-}1)+ (1\text{-}1)+ \; ... \\
&= \;\; 1\text{-}1+1\text{-}1+1\text{-}1+1\text{-}1+... \\
&= 1 + (\text{-}1+1) + (\text{-}1+1) + (\text{-}1 +1) + \; ... \\
&= 1 + 0 \\
&= 1.
\end{aligned}
$$

Theorem. Let it be assumed that infinite sums make sense and that $a+b = b+a$ and $x+(y+z) = (x+y)+z$, $0+x =0$ for all $a,b,x,y,z.$ Then $a+b = 0$ implies that $a=0$ and $b=0.$
Proof:
$$
\begin{aligned}
0 &= 0 + 0 + 0 + 0 + ... \\
&= (a+b) + (a+b) + (a+b) + ... \\
&= a + b + a + b + a + b + ... \\
&= a + (b+a) + (b+a) + (b+a) + ... \\
&= a + 0 + 0 + 0 + ... \\
&= a.
\end{aligned}
$$
Similarly, $b=0.$ This completes the proof.//

Of course, for numbers, infinite sums do not necessarily make sense, and so we have not proved that zero equals one. There are, however, topological applications to this formalism. Here is an example: Let M and M' be (compact orientable) surfaces. The connected sum of M and M', M#M', is obtained by excising a disk from each surface and connecting them to each other by a tube whose ends are glued to the circular boundaries of the two regions left by the excision in each surface.

We shall prove, by infinite repetition, the

Theorem. $M\#M' = S^2$ implies that $M = S^2$ and $M' = S^2$.

Here S^2 denotes the surface of a two dimensional sphere (the boundary of a three dimensional ball.) and $=$ denotes homeomorphism of surfaces. It is easy to check that $M\#S^2 = M$ for any surface M and that the connected sum operation is well-defined for finite sums, and that it is commutative and associative. Can we make sense of an infinite sum? The answer is yes, but one leaves the category of compact surfaces: Put the surfaces $M_1, M_2, M_3, \ldots$ in a row extending to the (viewer's) right. Form $M_\infty = M_1\#M_2\#M_3\#\ldots$ by connecting them together by straight tubes between adjacent surfaces. The resulting surface M_∞ is well-defined but no longer compact. For example $S_\infty = S^2\#S^2\#S^2\#\ldots$ is homeomorphic to the plane R^2.

In this case an infinite sum of "zeroes" is not zero! However, for any surface M, $M\#S_\infty = M-\{pt\}$, since removing a point is equivalent to the connected sum with R^2. Thus:
If $M\#M' = S^2$, then

$$S_\infty$$
$$= (M\#M')\#(M\#M')\# \ldots$$
$$= M\#(M'\#M)\#(M'\#M)\#\ldots$$
$$= M\#(S_\infty)$$
$$= M-\{pt\}.$$

Now form the one-point compactification of both sides and conclude that $S^2 = M$.

Because S_∞ is not the 2-sphere, we cannot use this argument to conclude that if M#M' is smoothly homeomorphic to the 2-sphere, then M is smoothly homeomorphic to the 2-sphere. Differentiability may fail in the neighborhood of the missing point. In fact, for surfaces the theorem still holds in the smooth category, but the same argument transposed to higher dimensions has this limitation. For example in dimension 7, there are manifolds M and M' homeomorphic to spheres but not diffeomorphic to spheres such that M#M' is diffeomorphic to the standard 7 sphere (See [KM]).

You Can't Cancel Knot

Tie a knot in a piece of rope and then tie another knot adjacent to it. (In this picture of knots, one is *not allowed* to move any rope past the end points. Think of the end-points as attached to opposite walls of a room. With the ends attached to the wall, the rope can be moved so long as it is not removed from the wall or torn apart.)

Is it possible that the two knots taken together can undo one another even though they are individually knotted?   The answer is NO. The proof is by infinite repetition [F]: Let O denote the unknot. Let K#K' denote the connected sum of knots obtained by adjacent tying. Instantiate $K_\infty = K\#K'\#K\#K'\#K\#\ldots$ as an infinite weave in a compact space by introducing a limit point as shown below.

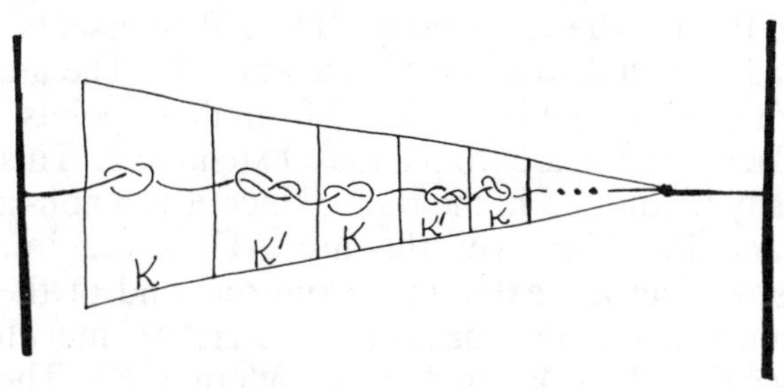

Then K_∞ is, by the method of infinite repetition, equal to both K and to O. Hence K must be unknotted.

This argument goes into the larger category of knots with infinite amounts of weave to make its conclusions. In order to show that the conclusion holds in the usual category of finite weaves, a topological theorem is needed stating that if finitely woven knots are equivalent in the larger category of infinite weaves then they are equivalent in the category of finite weaves. The result that supports this conclusion is found in [MO].

The Conway Proof
There is a very beautiful proof of the impossibility of knot cancellation due to John Conway (see [G]). His proof does not go off into infinite weave. Here is a sketch of Conway's proof:

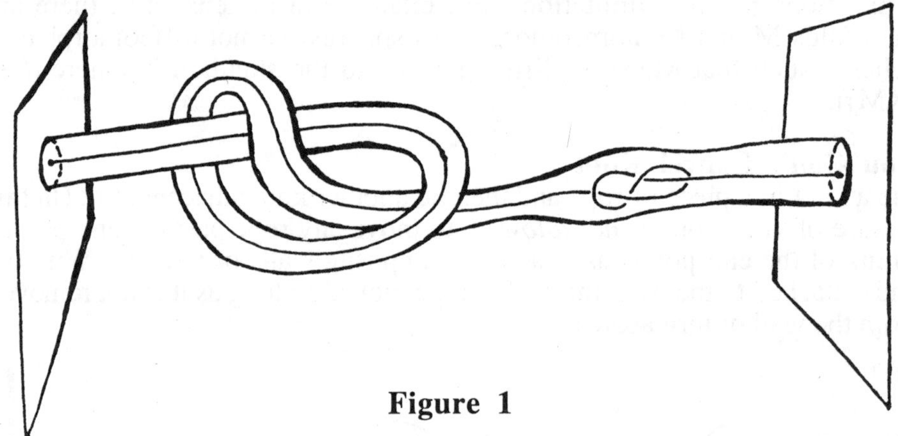

Figure 1

Put a tube T around K#K' (as shown in Figure 1 above) so that the tube is a tubular neighborhood of K and so that the tube engulfs K'. If K#K' = O, then there is a homeomorphism of the room to whose walls K#K' is attached that leaves the walls of the room fixed, and straightens K#K' to a straight line L extending from the left wall to the right wall. The tube T will be deformed by this homeomorphism to a new tube T' that does not intersect the line L. Let P be plane in the room containing L. Then P intersects the left and right walls of the room in the endpoints of L and in four points of the tube T' (two on each wall). Let a and b denote the intersection of P with T' on the left wall and let c and d denote the intersection of P with the right wall. Then P intersects T' in arcs that emanate from a,b,c,d and some closed curves in P. The arc from a cannot reach either b or d because it is separated from these points by the line L in the plane P. Therefore the arc from a must extend to c. This arc AC from a to c is necessarily unknotted in the room, since it is a non-self-intersecting arc in the plane P. However the arc AC is the image under the homeomorphism of an arc extending from one end of the tube T to the other, and by construction, this means that the arc AC must be equivalent to the knot K (since the tube is knotted in the pattern of K). Therefore we have shown that in the course of unknotting K#K' we have necessarily unknotted K itself! Therefore you cannot cancel knots.//

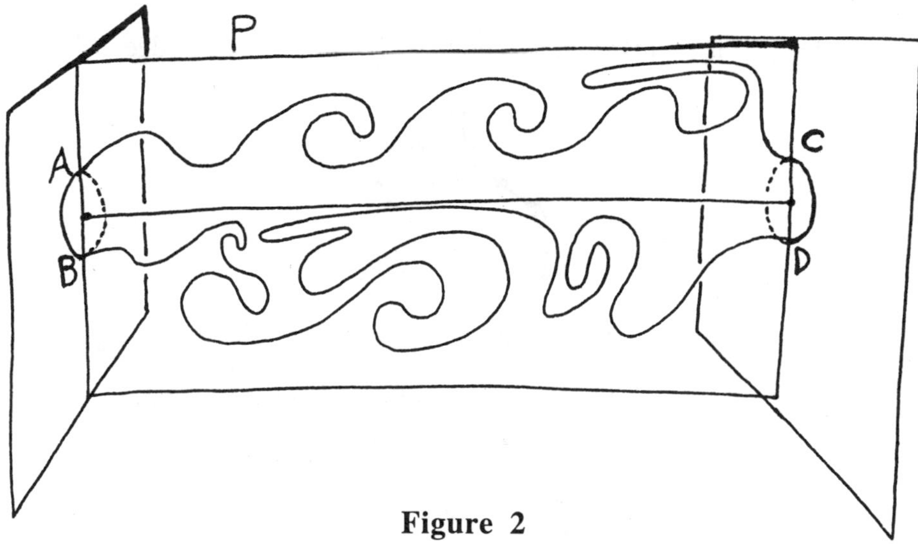

Figure 2

Graphs that Encapsulate Infinity
There is a very elegant way to represent sets in FIST that are described by systems of equations: *Any* directed graph represents such a set.

Each node in the graph represents a set. An edge directed from node A to node B encodes the relation that *B is a member of A.*

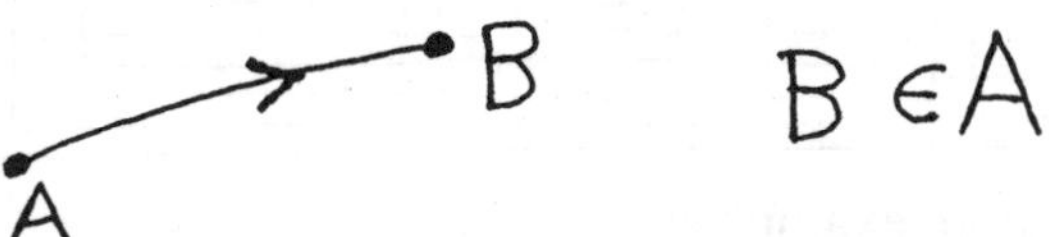

(This method of representation is used by Aczel [AC].)
A single finite set is a rooted tree where all the edges are directed away from the root as in the examples preceding this discussion. Nevertheless, any directed graph yields a set, or sets. For example,

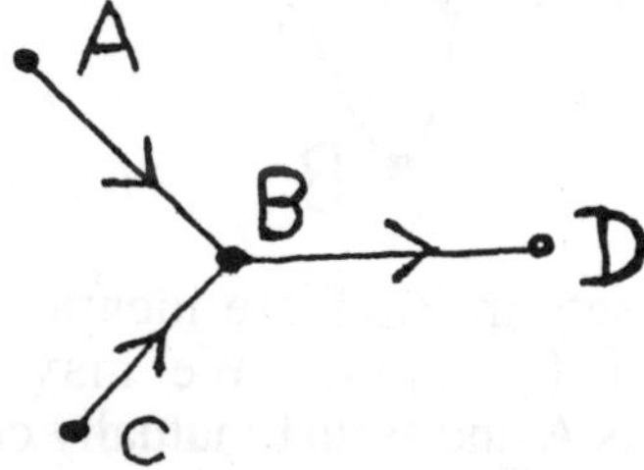

Here A = { B }, B = { D }, C = { B }, D = {   }. (A node with no outwardly directed edges connotes the empty set.) In this case, we see that

18

A = {{{ }}}, B = {{ }}, C = A, D = { }. The symmetry of the graph with respect to the nodes A and C corresponds to the equality of the corresponding sets.

The set J = {J} is represented as a node with a self-directed edge.

The category of sets in FIST that are represented by finite directed graphs is pleasant to contemplate, but it only scratches the surface of FIST. For example, the following infinite tree has no finite graph description:

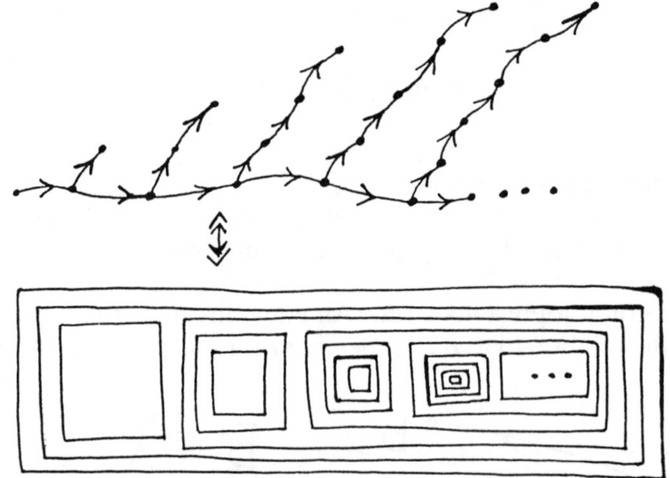

Here are a few more examples:

1. A = {B}   and   B={A}.

Here the corresponding sets in FIST are identical since we obtain A = {{{{....}}}}   and B = {{{{...}}}}. We may wish that this graph represented two distinct sets A and B that mutually create one another. This end can be achieved by taking the graphs at face value, rather than accepting the model involving these recursive limits as the end of the story. In the next section we shall do just this in the context of knot sets. In the FIST context, one can obtain the effect of distinguishing A and B by giving one a different membership structure from the other via a "label" as in A = {B, { } } and B = {A}.

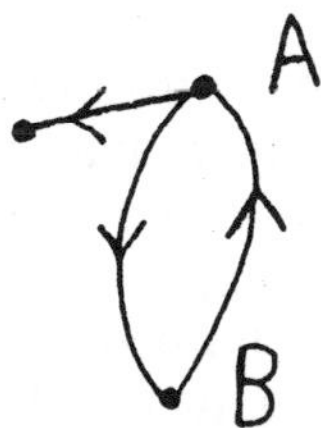

2. F= {{F} F}.

The solution in FIST is F = { { {{ ... } ... } } { { ... } ... } }.   This is the Fibonacci form (considered earlier in this section).

3. Consider the set in FIST specified by the graph shown below

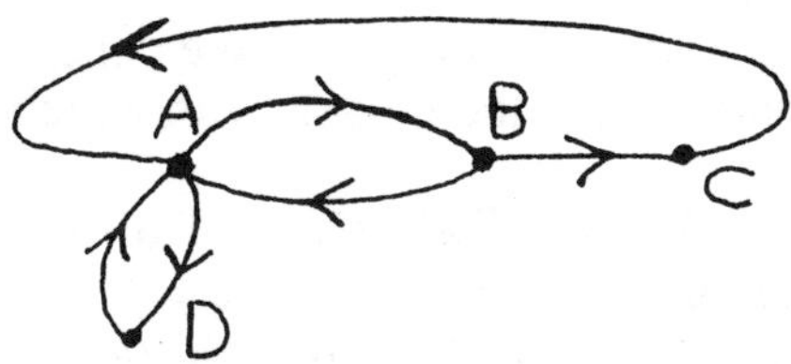

The corresponding system of equations is
A={B,D}
B={A,C}
C={A}
D={A} .
The last two equations force C=D, and these then force A=B. Thus the system is equivalent to the system A={A,D} and D= {A} or to the graph

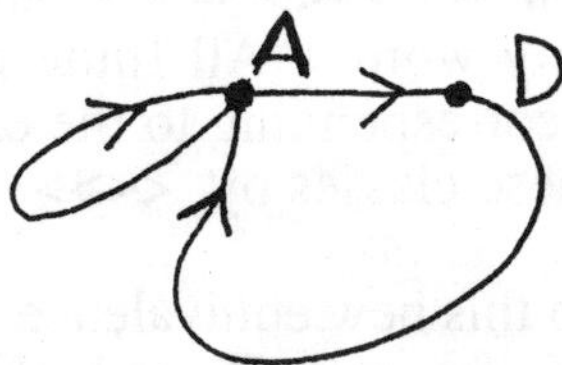

This example shows how different graphs can lead to the same elements of FIST.   It is an interesting question to determine the minimal graphs that represent a given system of mutually defined sets in FIST. The nodes of such a directed graph are mutually distinguished from one another in terms of the mutual membership relations. An analogy to this situation for undirected graphs is found in the extremal variety graphs of Barbour and Smolin [BaS]. In an extremal variety graph, all points are distinct due to the presence of distinguishing neighborhood structures in the unoriented graph. Thus, the extremal variety graph represents a space in which the points are distinguished from one another due entirely to their mutual relationships.

Minimal directed graphs for sets in FIST are an oriented analog of the extremal varieties.

Pregeometry

These remarks look forward to the discussion of pregeometry in section 10. A minimal directed graph or a maximal variety graph can be regarded as a miniature world in which the nodes are the observers. Each observer obtains its identity from its relations with the other observers. In the case of directed graphs, each observer's immediate perception is of its members (the nodes that are one directed edge away). Further reports yield the members of members and eventually the full system of relationships that constitute this world. The problem of pregeometry is how it can come to pass that such worlds acquire geometry and topology that is natural with respect to the structure of relations, and giving rise to known physical law. It is our contention (see Section 10) that knot theory gives a new way to consider the question of pregeometry.

In the next section, we discuss a representation of sets that interfaces with knots and links in three dimensional space. We conclude the present section with two general remarks about the models with which this section began.

Remark 1. Indicational Calculus, Boolean logic and the Calculus of Indications

We have seen that the full set of well-formed parenthesis structures is a background of the theory of finite sets. Let us denote these structures modulo the relations XY=YX and XX=X by parentheses written in angle-bracket form. Thus $< \diamond <<\diamond>> <\diamond> > = < \diamond <\diamond> <\diamond> <<\diamond>> > = < \diamond <<\diamond>> >$ denotes the set whose members are an empty set and a set consisting of an empty set. The expression $<<\diamond>><\diamond>$ is a *form* but not a set in the terminology used earlier in this section. Now consider the quotient of the class of forms generated by the extra relation $<<\diamond>> = \mathbf{e}$ where $\mathbf{e}$ denotes the empty word. Let $=$ continue to denote this equivalence relation. Then $<<\diamond>><\diamond> = <\diamond>$ and $<<\diamond>> =$ where the blank space is the empty word. All finite forms fall into the two distinct equivalence classes corresponding to the empty word and the mark $<\diamond>$. We represent these classes by $<<\diamond>>$ and $<\diamond>$.

The collection of forms up to this new equivalence satisfy many equations. For example, $<<X>> = X$ for any X and $<X>X = <\diamond>$ for any X. By interpreting

$<X>$ as the *negation* of **X**,

XY as **X or Y,**

$<<X><Y>>$ as **X and Y,**

$<<\diamond>>$ as **False,** $<\diamond>$ as **True,**

one recovers the full structure of Boolean algebra. This is the calculus of indications of G. Spencer-Brown [S-B] expressed in parenthesis notation. Boolean algebra arises from the boundary structure of finite set theory. The calculus of indications begins with well-formed parenthetical expressions modulo the equivalence generated by

$$\langle\rangle\langle\rangle \;=\; \langle\rangle \quad \text{and} \quad \langle\langle\rangle\rangle = \qquad .$$

These equivalences can be performed within otherwise identical larger expressions.

Imaginary Boolean Values
Infinite expressions in the context of the calculus of indications, give non-Boolean values. For example, if **P** = <<<...>>>, then **P** = **<P>**. Infinite expressions are not necessarily reducible to one of the two states **<<>>** or **<>.** It is an interesting problem to enlarge the context of Boolean algebra to handle such values. See [K15], [K16], [K17], [KV] for a discussion of solutions to this problem. Spencer-Brown [S-B] makes the perspicuous observation that there is a direct analogy between the imaginary Boolean value **P = <P>** and **i**, the square root of minus one: **i** is the solution to **i = -1/i**. If we ask to solve $x = F(x)$ with $F(x) = -1/x$, then x=1 implies x = -1 and x=-1 implies that x = 1. The problem of finding a square root of minus one is analogous to the liar paradox. Complex numbers provide a solution to this paradox in the numerical domain. Just so one can consider imaginary values in logical domains.

The solution **P** = <<<...>>> to **P = <P>** is the analog of the solution x = a + b/(a +b/(a + b/(a + ...))) for x = a + b/x. In the case where a = 0, b = -1 there is no real numerical value for this continued fraction. When x^2 = ax +b has a real root, then the continued fraction converges and gives a real answer. When x^2 = ax +b does not have a real root then the continued fraction does not converge, but the recursion x ----> a + b/x is quite interesting to study in its own right, producing an intriguing class of oscillations of the form x_{n+1} = a + b/x_n. (*Exercise*: show that these oscillations all take the form x_n = tan(nq +F) for appropriate choice of theta and phi depending upon a and b.) In Figure 3 we show a typical plot of x_n (vertical axis) against n (horizontal axis) in the case where x^2 = ax +b has no real root. (Here the starting value for x is 1 and a = 1, b = -6.)

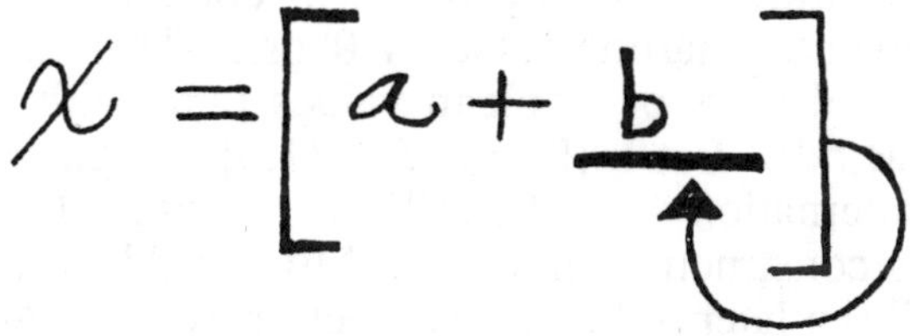

$$\chi = \left[a + \frac{b}{} \right]$$

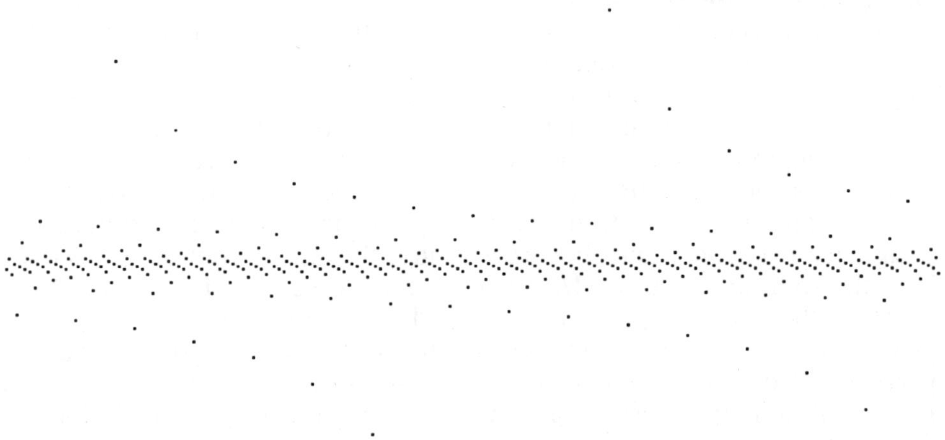

Figure 3

Paradox can be studied through the recursive process inherent in its syntactic form. (See [K16], [K17], [KV], [H1], [H2].)   In the case of the complex numbers it is interesting to point out that the view of the square roots of minus one as oscillations between 1 and -1 is mirrored in the matrix representation of these roots by the matrices whose squares are minus the identity.

$$\begin{matrix} 0 & 1 \\ -1 & 0 \end{matrix} \qquad \text{and} \qquad \begin{matrix} 0 & -1 \\ 1 & 0 \end{matrix} .$$

In thinking about the square root of minus one, one must ask which one (i or -i)? Similarly, in regarding the imaginary value   **P = <P>**, one encounters *two* oscillations. There are two corresponding sequences, depending on whether the starting value is 0 or 1. These solutions can be formalized as ordered pairs of Boolean values [a,b] with [a,b]' = [b',a'], and [a,b][c,d] = [ac,bd]. Let I=[0,1] and J=[1,0]. Then I and J are the two views of the alternation   ...0101010101...   with   I' = I , J' = J and IJ = [0,0]=0. This construction gives a DeMorgan Algebra [K15],[K16], [KV]. As we shall see later in this essay (section 10) an entirely different world opens up if we ask for the same conditions, but IJ = 0.

Remark 2. Quantum Logic

Recall the simplest form of quantum logic (See [F1] ,[F2],[F3], [O]) based on a vector space V with a notion of orthogonal complement for subspaces (W' is the orthocomplement of W). Elements in the algebra of this logic are subspaces of V. The negation of W is its orthocomplement W'. The sum of subspaces A and B (A+B) is the subspace spanned by A and B in V. The product of A and B (A*B) is their set theoretic intersection.   Let 1 denote V and 0 denote the zero subspace.

In this logic, we have A+A' = 1, A*A' = 0   for any A.   The law of the excluded middle still holds, and there is no element J in the logic such that J' = J. On the other hand, if V is two dimensional, and P and Q represent perpendicular lines in V, while R represents a line independent from both P and Q then we have 1= 1*R = (P + Q)*R while P*R + Q*R = 0 + 0 = 0. The distributive law does not hold in the quantum logic.

Such a non-Boolean logic is called a quantum logic because it models the operations of states and projections in a quantum mechanical system. Addition of vectors corresponds to the superposition of states. Here we are concerned not with the naturality of this structure with respect to quantum mechanics, but rather with its naturality in respect to mathematical foundational ideas. Vector spaces are a rather late development in the hierarchy of mathematical constructions. Can one encounter quantum logic nearer to the bottom? One answer is an appeal to geometry. If we describe in notation this move to quantum logic it becomes: Let (for three dimensions) the whole space, a plane, a line or a point indicate a given proposition. Let the negation of this proposition be indicated by a linear space that is perpendicular to the indicator for a given proposition. Thus, in a plane, if we diagram P by a line

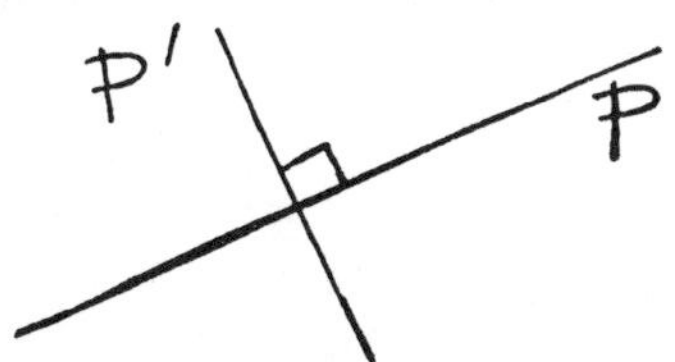

then P' is a line perpendicular to P.

$$\left\{ \begin{array}{l} R*(P+P')=R \\ R*P+R*P'=0 \end{array} \right\}$$

At once there arises the infinite multiplicity of lines in between P and P'. If the plane itself is all (1) and a point the void (0), then we can only save the law of the excluded middle by letting P+P' indicate the plane spanned by these two lines. It is nevertheless this very existence of intermediates that makes the logic non-distributive. For we take R to be a line going straight between P and P', and we find that R*(P+P') is not equal to R*P + R*P'. The quantum logic is the logic of the first movement of notation into geometry.

Quantum logic is the pre-geometry of notation. Boolean logic is obtained *in notation* by ignoring the existence of intermediate states.

This discussion makes no claim that its remarks about notation and quantum logic have a direct bearing on quantum mechanics. Such issues deserve more exploration.

Remark 3. Ordered Parentheses, Boundary Logic and the Temperley Lieb Algebra

In this section we have taken the point of view that ordered parenthetical expressions in brackets (finite ordered multi-sets) are precursors to finite set theory. In examining the structure of such expressions it is useful to tie left and right ends of the parenthesis into a single form that shall be called a *cap*. This notational device is indicated below.

Call parenthetical expressions written in this notation *capforms*. The capforms are intimately related to a number of topological problems. One way to see this is to draw a simple closed curve (i.e. a curve with no self-intersections) in the plane and slice it with a straight line. the line cuts the curve into two capforms such that the feet of each cap are on the line.

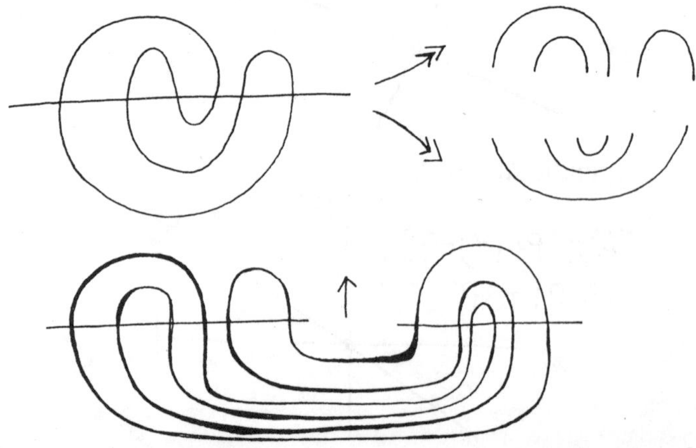

The interaction of these two capforms produces the single simple closed curve. In fact, we formalize the interaction of the two capforms as a cancellation (or connection) of nearby boundaries. We indicate nearby interacting boundaries by an arrow.

This gives rise to the following rules in a calculus of capform boundaries that we call *boundary logic* . (See [BRI] for a distinct but related use of this term.)

To determine whether two capforms interact to produce a single simple closed curve, one can either calculate in boundary logic or draw geometric connections and trace the resulting plane curves:

Remark. By using the boundary logic in parenthetical form, we can formalize it with rules for string replacements. Then the equivalent of the above graphical calculations can be performed by a digital computer. (See [K6], Appendix to Second Edition, pp. 605-608.)

If C_n denotes the capforms with n caps, define a binary operation C_n x C_n -----> C_n by $X \# Y = X \mid^n Y$ where $\mid^n$ denotes the n-fold iteration of the boundary joining operation. This product operation can be described quite explicitly by regarding a capform in C_n as having n left legs and n right legs. $X \# Y$ is the result of joining the right legs of X to the left legs of Y as shown below.

The structure of this product on C_n is better understood by rewriting the elements of C_n so that the left legs appear at the top of a box, and the right legs appear at the bottom.

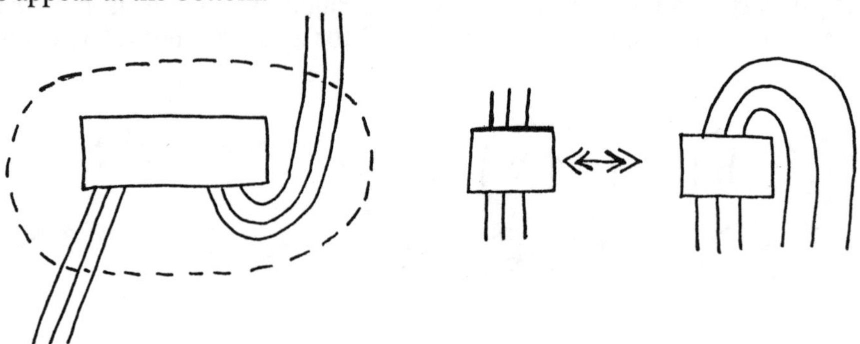

Then one can verify that every capform is a product of the elementary capforms shown below. These forms are the generators of the (diagrammatic) Temperley-Lieb algebra [K3], [K6].

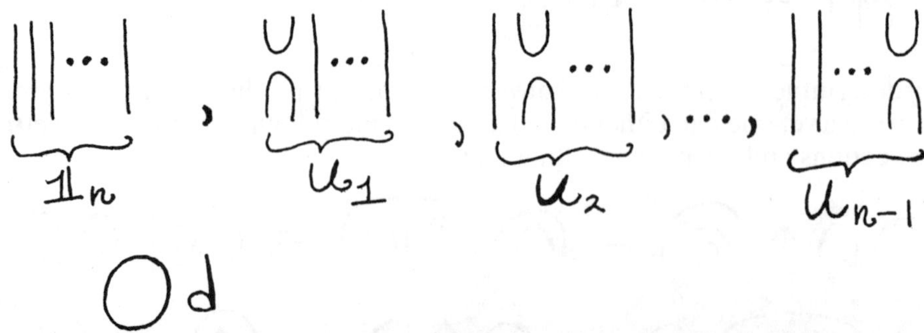

The following relations describe this algebra

$$U_iU_{i+1}U_i = U_i,$$
$$U_iU_{i-1}U_i = U_i,$$
$$U_iU_j = U_jU_i \text{ if } |i-j|>2,$$
$$(U_i)^2 = dU_i.$$

Here d denotes the value assigned to a single free loop (the loop is taken to commute with other elements of the algebra.)

The last relation is illustrated below.

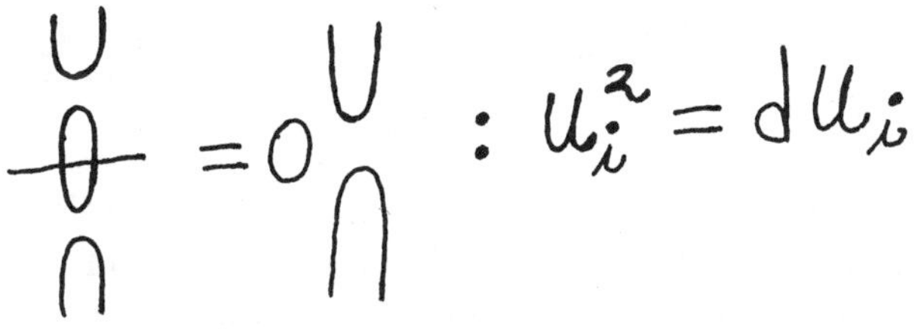

The Temperley-Lieb algebra originated in certain problems in statistical mechanics (see [BX]), and it has a very strong influence on many problems in the theory of knots and links.

The fundamentals of set theory are intimately connected, through combinatorial structures and the theme of boundaries, with logic, topology and mathematical physics.
All this from framing nothing!

III. Knot Set Theory

A diagrammatic alternative to Venn diagrams can model a non-standard set theory.

This section describes such a diagrammatic model and explains its relationship with the theory of knots and links in three dimensional space.

We begin with undefined objects denoted by letters a,b,c,... and a notion of membership denoted $a \in b$ (a "belongs" to b). It will be possible for a to belong to itself ($a \in a$) or for a to belong to b while b belongs to a. In the model there is no infinite regress and the system, a formal diagrammatic theory, is consistent relative to standard discrete mathematics.

Here is a description of the model. Objects will be indicated by non- self intersecting arcs in the plane. A given object may correspond to a multiplicity of arcs. This is indicated by labelling the arcs with the label corresponding to the object. Thus the arc below corresponds to the label a.

Membership is indicated by the diagram shown below.

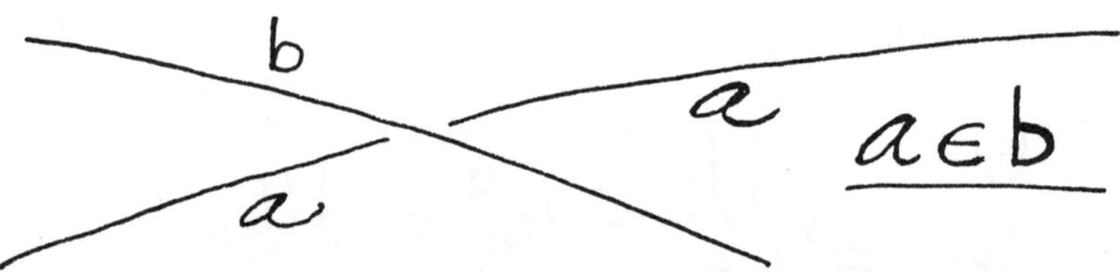

Here we have shown $a \, \varepsilon \, b$. The arc b is unbroken, while a labels two arcs that meet on opposite sides of b. Following the pictorial convention of illustrating one arc passing behind another by putting a break in the arc that passes behind, one says that *a passes under b*. The pictorial convention is important both for the logic and for the deeper relationship with three dimensional space that we shall elucidate shortly.

It is an easy matter to illustrate certain basic constructions in set theory. For example, the von Neumann construction of sets of arbitrary finite cardinality is traditionally done by starting with the empty set $\phi = \{ \ \}$, and building a sequence of sets X_n with $X_0 = \{ \ \}$, $X_1 = \{\{ \ \}\}$, $X_2 = \{\{ \ \}, \{\{ \ \}\}\}$.
Here $X_{n+1} = X_n \, U \, \{ X_n \}$ where U denotes the operation of union. The diagrams below show how to implement this construction using the overcrossing convention for membership.

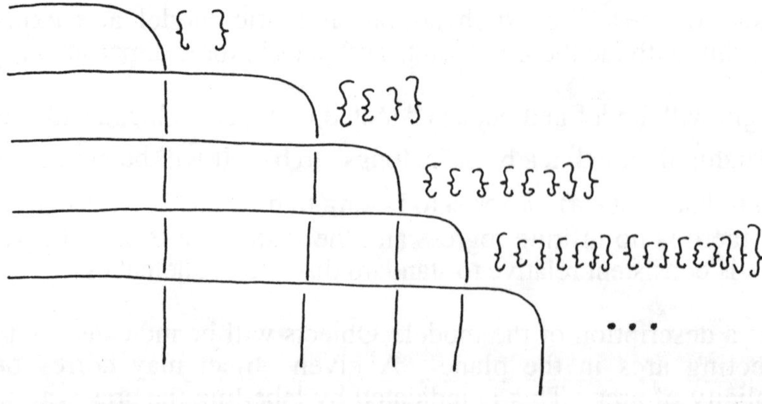

With these same diagrams it is possible to indicate sets that are members of themselves

$$a = \{a\}$$

and sets that are members of each other

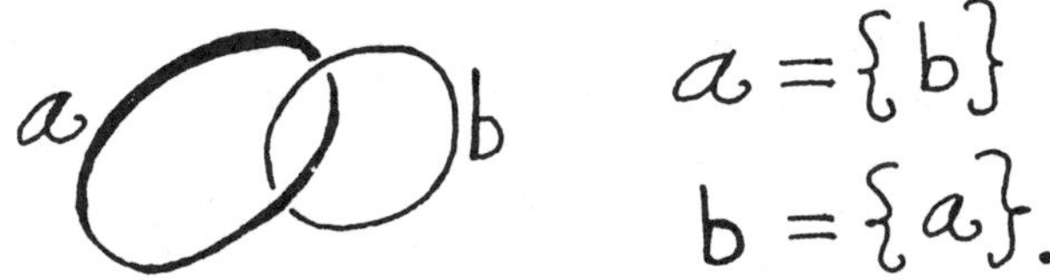

$$a = \{b\}$$
$$b = \{a\}.$$

As they stand, these diagrams   indicate sets that may have a multiplicity of identical members. Thus

Here b = {a,a}   and a = { }.

The traditional way to condense multiplicities of identicals is to regard them as all equivalent to one another. This amounts to the condensation rule { ... a,a ...} = { ...a... }. In the case of our diagrams another solution is suggested. In this solution, *identicals cancel in pairs* and we have { ... a,a ...} = { ... ... }. Thus {a, a} = { }. This is diagrammed as shown below:

It is easy to remember this diagrammatic transformation, since it can be interpreted as a drawing of one strand of rope being slipped out from under another. We shall accordingly adopt the rule of cancellation of identicals as fundamental to knot set theory.

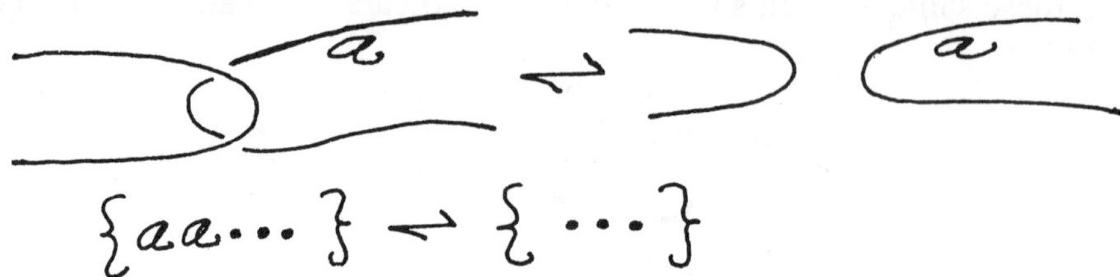

Digression on Knots

The diagrams that we are drawing have a well-known interpretation as diagrams of knots, links and tangles in three dimensional space. By convention, a knot consists in a single closed curve, a link may have many closed curves and a tangle has arcs with free ends. Also by convention, topological changes in a tangle do not involve moving the free ends or in passing strands over the free ends.

There is a direct relationship between the topology of these knots, links and tangles and the properties of the knot set theory.

Reidemeister [R] proved that any knot or link in three dimensional space can be represented by a diagram containing only crossings of the type indicated below,

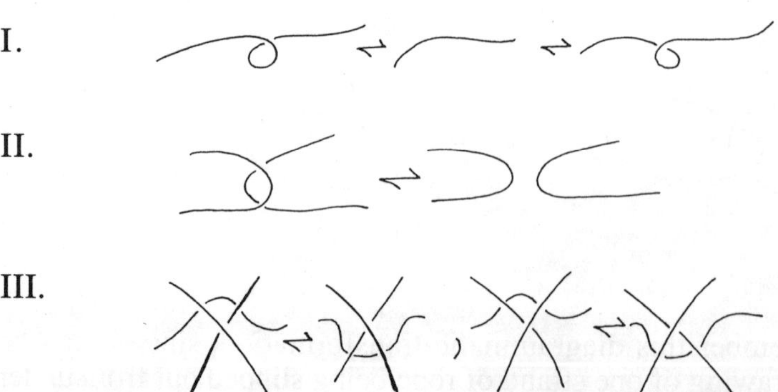

and that two knots or links (A knot is an embedding of a single closed curve into three space. A link is an embedding of a collection of curves into three space.) are isotopic in three space if and only if their diagrams are equivalent to one another under a finite sequence of transformations of the types I, II, and III as indicated below. (Isotopy corresponds directly to the physical picture of transforming one rope to another by pushing, pulling , stretching but no tearing.)

I.

II.

III.

The same theorems apply to tangles, with the caveat that the free ends of the tangles remain fixed during the applications of the moves, and that strands are not allowed to pass over the ends of the tangles.

Here is a simple example of unknotting via the Reidemeister moves.

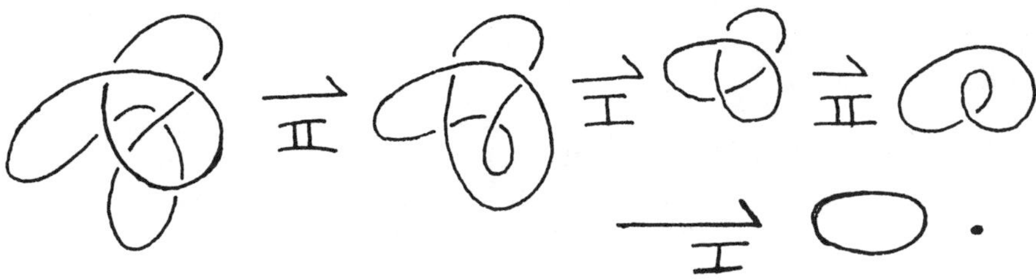

Here is a subtler example, turning the figure eight knot into its mirror image.

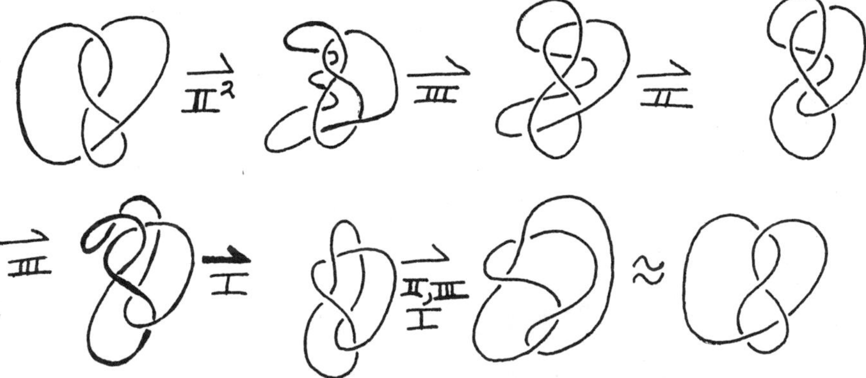

It is a very tricky matter to extract topological data about knots and links from their diagrams. We shall have more to say about this later.

The Triangle Move. The Reidemeister moves derive from properties of the projection of a curve from three-dimensional space to a plane or to the surface of a sphere. In fact Reidemeister had a *single* move for knots and links in three space. This single move, the *triangle move* , generates the three Reidemeister moves. The triangle move is defined for piecewise linear knots and links in three-space. A piecewise linear link is made up from finitely many straight line segments. Any link represented by a differentiable embedding, or any link that can be drawn by hand in a finite amount of time, can be approximated by a piecewise linear link. Given a pl (short for piecewise linear) link, a triangle move is performed by the following prescription:

Perform one of the following two types of operations.
1. Mark a straight segment A on the link K. Let r and s denote the endpoints of A. This segment A can be a proper subsegment or an entire segment of K. Let p be a point in the complement of the link K such that

the triangle with vertices r,s,p intersects K only along A. Let B denote the segment rp and C the segment sp. Cut the segment A from the link and replace it by the union of the segments B and C.

2. Let B and C be consecutive segments marked on the link K. (By consecutive I mean that they share a single endpoint.) Let A be the segment determined by the endpoints of B and C that are not shared between them. Let ABC denote the triangle (surface) determined by the segments A, B and C. Assume that ABC intersects K in exactly B and C. Then cut A and B from K and paste in C.

The diagrams below illustrate how projections of triangle moves generate the three Reidemeister moves. Two pl links in three dimensional space are ambient isotopic if and only if they can be related by a finite sequence of triangle moves.   Careful consideration of the projections shows that sequences of Reidemeister moves on diagrams capture the content of an ambient isotopy.

I.

II.

III.

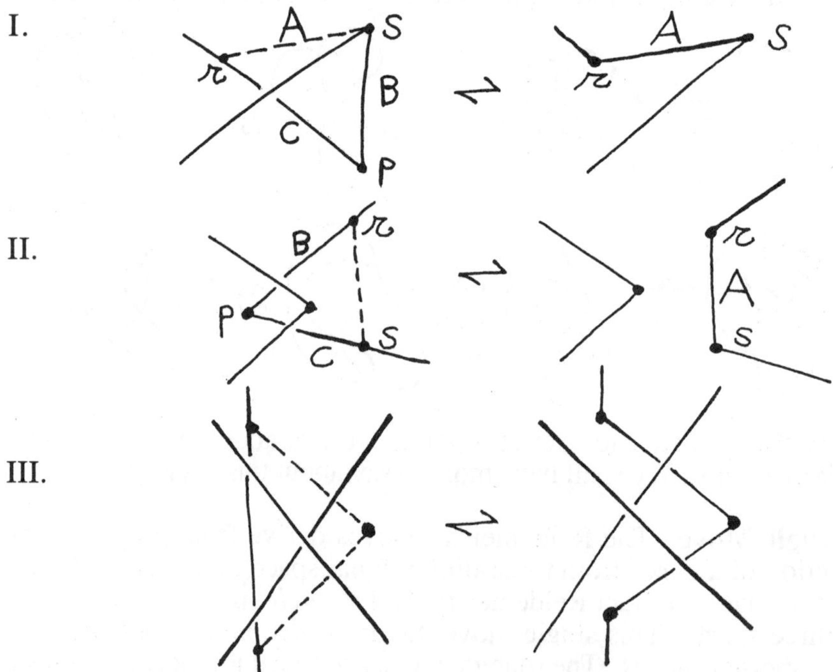

It is worth considering how the first Reidemeister move is generated by a simple triangle move.   This shows clearly the illusory nature of self-membership from the point of view of three dimensional space *if* we stick to pure topology.

On the other hand, if the loop is actually a physical loop in a rope, then the cancellation of the loop shown in the the first move must be paid for by a corresponding twist in the rope. This is most easily illustrated by replacing the line drawing by a drawing of a twisted band as shown below.

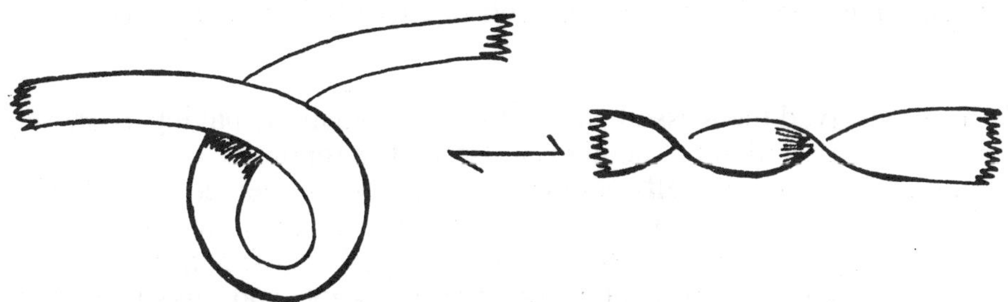

This band picture of the first Reidemeister move shows that we can regard it as an exchange rather than an elimination or creation of the loop.

The reason for dwelling on the first Reidemeister move in our context is that this move allows the creation or cancellation of self-membership in the corresponding knot set. If we take the point of view that the diagrams represent twisted bands (called framed knots and links), then the self-membership is not lost as we go to the topology. A corresponding equivalence relation on links is called *regular isotopy*. Regular isotopy is generated by the second and third Reidemeister moves. We shall return to this idea later in the discussion.

End of Digression.

Note that by the cancellation of identicals, diagrams related by the second Reidemeister move represent the same knot set. The third move does not change any membership relations. Finally, invariance of a knot set under the first Reidemeister move would entail quotienting the theory by self membership. As we have remarked above, it is natural to consider only equivalence of knots and links up to regular isotopy — the equivalence relation generated by the second and third Reidemeister moves — or to regard the diagrams as representative of embedded bands in space. In the latter case, self membership is catalogued by the twists in a thickened arc, as well as loops in that arc.

If we maintain the distinction of self-membership by using only regular isotopy on the diagrams, then the Russell paradox becomes meaningful in the knot set domain, but there is still a strange twist about self-membership. By the convention of cancellation of identicals we have the equivalence, $X = \{X,X, C\}$ where $X = \{C\}$ is the reduced form of the knot set X, and C denotes the contents of X. *Any knot set has a representative that is a member of itself. It is only of the reduced forms for the knot sets that we can speak of a set that is or is not a member of itself.*

The most radical interpretation is: Use diagrams with free ends (tangles) and allow the first Reidemeister move on knot sets. This means that any knot set has representatives that are members of themselves and it has representatives that are not members of themselves. The states of self-membership and non-self membership are equivalent. Up to representation, a (radical) knot set is a member of itself if and only if it is not a member of itself!

We have resolved the Russell paradox in this domain by having every set a member of itself and not a member of itself. The topological interpretation of knot sets shows that self-membership can be quotiented from the set theory (so that a given set has representatives that are members of themselves and representatives that are not members of themselves). The quotient theory is as consistent as the theory of knots and links in 3-space. Since this theory can be expressed in terms of ordinary set theory, this provides a relative consistency proof for radical knot sets.

Mutuality such as   a = {b} , b={a} is another matter. Here there is no reduction to anything simpler, and topologically, mutuality corresponds to nothing more paradoxical than the linking of two space curves.

In this version of knot sets one can make a diagram of a given knot set , and then use this diagram as a weaving pattern for a physical weave. Throw that weave into three space. Flatten the weave back onto a plane.   The result is an equivalent knot set.   The information in a knot set is encoded into the topology.

Knot Sets Avoid Infinite Regress
The knot set gives a way to conceptualize nonstandard sets without recourse to infinite regress. Infinity has been transposed into topology where inside and outside can equivocate through a twist in the boundary. In knot sets we obtain the multiple levels of ordinary set theory without the seemingly necessary hierarchy. This is nowhere more evident than in the self membering set represented by a curl.

Here an observer on the curl itself will go continuously from being container to being member as he walks along the ramp. Membership becomes topological relationship.

Remark. The reader may be familiar with other non-standard models for set theory such as those in the book by Peter Aczel [AC]. The constructions given here are very close in spirit to those of Aczel. There are two major differences. The first difference is in our choice to handle identicals via cancellation rather than condensation. The second is in the background use of reentering forms to indicate recursively defined constructions. We do not utilize the same demand for uniqueness of labelling as in [AC]. This is a technical matter and will be discussed elsewhere. The surprise in our construction is that the theory has a topological interpretation.

The version of knot sets discussed herein has a precursor in the work of the Swedish logician Stig Kanger in the early 1940's ([P], pp. 13-14.). Kanger represented sets as cords — with a cord tied around another cord representing a set with the other cord as a member. A cord tied around itself becomes a set that is a member of itself. Our knot sets, based on the diagrams for knots, turn out to have a deeper relationship to the topology of knots than the Kanger system. Kanger's idea is very significant, and it is interesting to compare it to the earlier systems of numeration (Quipu) that are based upon tying knots in a rope.

IV. Arrow Epistemology
An arrow points.

The arrow accomplishes its pointing via the distinction between inside and outside (convex side versus concave side) made by the arrow head. The body of the arrow extends the domain of the concave side into a flexible arm that can reach outward from the base of the arrow.

Once the body of the arrow becomes flexible, then elementary notational topology makes it possible for the arrow to point to itself. At this point, two forms of self pointing arise - pointing to the base (origin) of the body of the arrow and pointing to the interior of the body of the arrow. The former is the simplest form of self-reference, and leads via the unfolding shown below to a direct relationship with fixed points and recursion. The unfolding corresponds to describing all the trips that one can make around the circle formed by the self-pointing arrow. Thus **a = ----->** denotes one trip while **aa = ----->----->** denotes two trips and **A = aaa... = ----->----->----->** .... denotes infinitely many trips. Note that **A = aA.** In this way the unfolding **A** of the self-pointing arrow is a a fixed point for the operation of "affixing an arrow on the left".

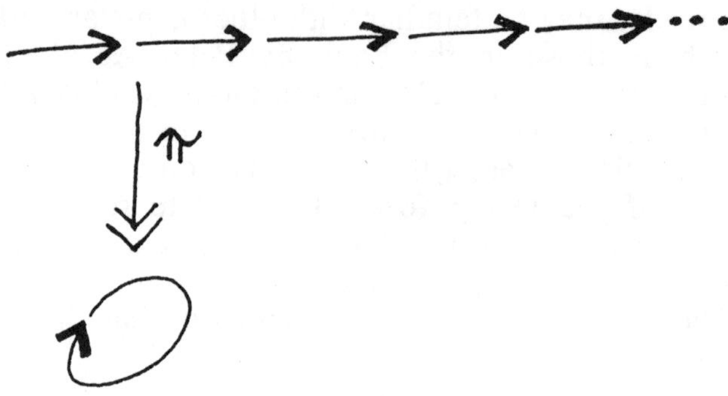

In the second alternative, the arrow points to its own body.

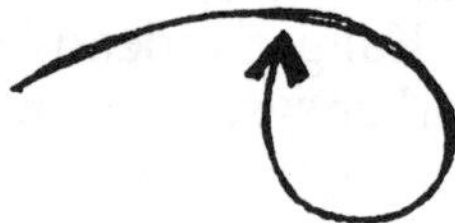

We have seen that this alternative can be extended to a notation for self-membership or reference of the body of the arrow to itself in the form in which an undercrossing points to (is a member of) the overcrossing line.

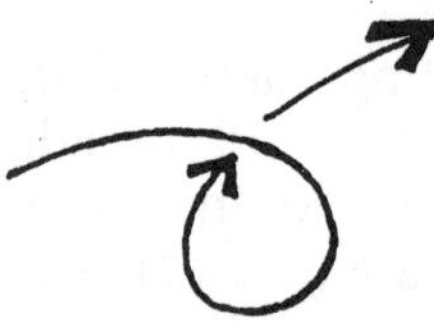

Self-pointing of an arrow or a line bifurcates into two interpretations depending upon whether the end of segment is seen as a pointer or whether an interior point of a segment is seen in relation to another interior point. In projection these two points of view come together through the convention of the cut segment at a crossing.

Any reference is a distinction. The notation adopted for a fundamental distinction has a remarkable influence on the way we think about it. In standard set theory a set is indicated by a pair of curly brackets: { }. (This is the empty set.)

A Story

The brackets themselves indicate bifurcation from a point.

Each bracket instantiates the growth of a distinction from a state of unity (the point of the cusp).

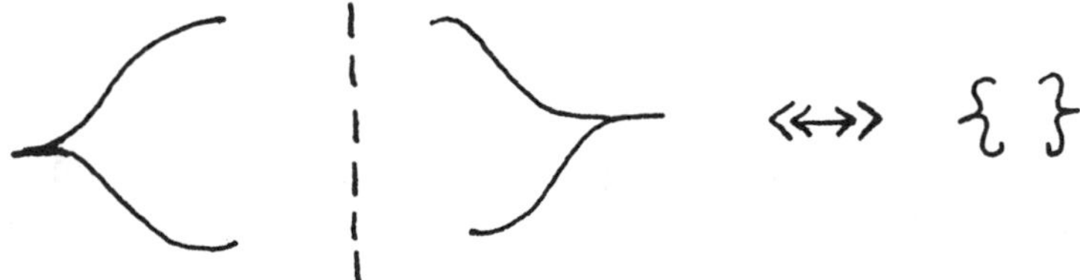

A further operation beyond bifurcation is necessary for the formation of a set. The bifurcation that is the (left) bracket is copied and mirrored to form the (right) bracket. A left and right bracket taken together become a container.   Once we have reached the level of being able to make a distinction, and to make a copy of that distinction intrinsically distinct (the mirror imaging) from the original, then we are prepared to form a new distinction (the container). The new distinction occurs at a different level from the original distinctions. This allows the hierarchy that is set theory.

Knot Structure

The self-pointing arrow is not a knot. The circle diagram for an unknot does not point to itself at all, but is simply a closed circular form. Examine the trefoil.

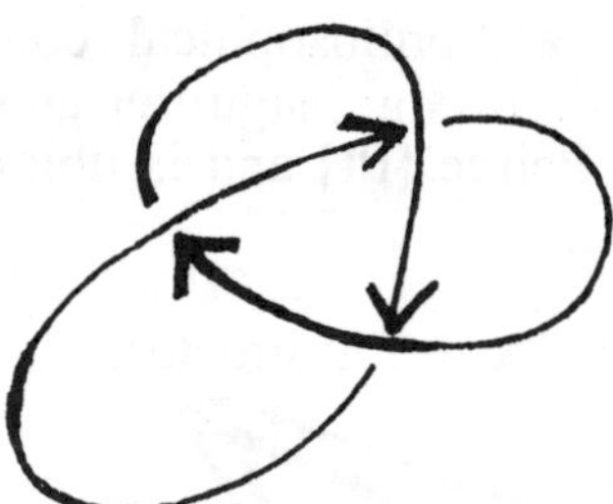

The diagram consists in three arrows, each one pointing to the body of the next.

The extra convention that the base of one arrow is always correlated with the tip of another is special to the knot theory.

It allows the interpretation of the two arrows taken together as part of an undercrossing line, and hence the set theoretic and geometric interpretations that we have already discussed. If we contravene this convention, then we obtain diagrams such as the one below, where the base of an arrow simply begins from some point on an arc.

This gives us a set of planar diagrams that can be studied on their own terms. Self-pointing can take the form shown below.

The triplet structure of the trefoil is still present in diagrams such as this one.

One reason for considering such a wider class of diagrams it that it enables us to draw connections with the kind of diagrammatics that occurs in artistic, linguistic, physical and philosophical contexts. For example, the irreducible tripartite relation of sign, signifier and signified occurs in the work of Charles Sanders Peirce [PI] and is ubiquitous in semiotics and linguistics.

A symbol almost identical to the trefoil structure

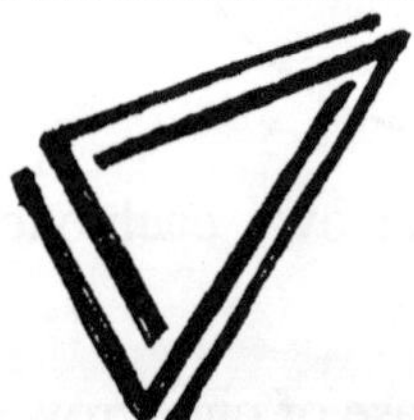

occurs in the stylized form shown above in the the work of Annetta Pedretti [P] on language. Here the three parts represent the distinguishing subject, the that which is distinguished and the   background binding the distinguished and the one who distinguishes.

A similar tripartite structure arises as soon as one includes the boundary in any distinction. Two sides and the boundary joining them form a tripartite structure where each part is determined by the other two parts.   No boundary exists without the two sides. No side exists without the potential to cross the boundary from the other side. Frederick Joseph Staley [STA] calls such triplets triadic dualisms.

A triadic dualism need not have the appearance of either a trefoil or a distinction. The most striking topological example of a triadic relation is the link shown below. This link, the Borommean rings, is topologically linked, consisting of three unknotted circles. The rings fall apart upon the removal of any one of the triplet.

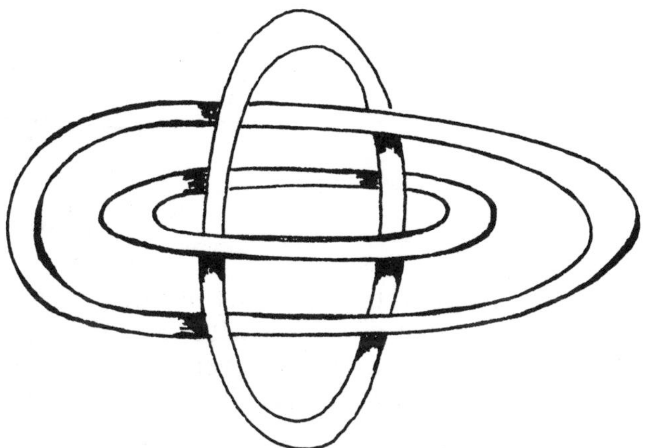

V. Lambda Calculus and Topology
It is natural to enquire whether the knot sets shed light on the topological structure of knots and links themselves.
Consider a trefoil knot:

The set is just the self-membering a={a,a,a}, and hence equivalent to the empty set in the radical theory and to one twist (a={a}) in the regular theory (regular   theory   uses regular isotopy).   Many topologically distinct diagrams correspond to a given knot set.

It is tempting to consider the possibility that the knots and links can be viewed in terms of a subtle kind of logic. This is in fact the case.

Non-Associative Formalism in Knot Diagrams

Label the arcs in a link diagram. Regard the label on the arc c obtained by underpassing b from a as a product of a and b : c= ab.

Here we abandon the notion of membership at a crossing and replace it with an algebraic product. Think of the overcrossing line as acting on the undercrossing line to produce the label for the continuation of the undercrossing. This is an inherently non-associative formalism, as the diagrams below demonstrate.

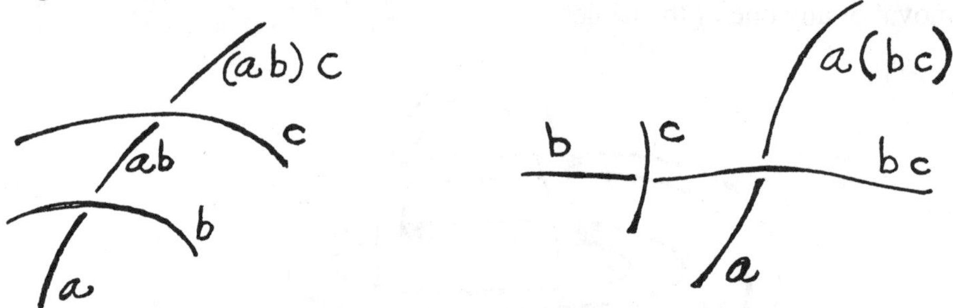

In this mode we can diagram the constructions of the lambda calculus of Church and Curry [B]. No direct knowledge of the lambda calculus is needed for the discussion to follow. However, the last part of this section is a discussion of the lambda calculus in relation to knots.

Consider $Gx = F(xx)$. If we substitute G for x, we obtain $GG = F(GG)$. At this level of formalism, every F has a fixed point GG where $Gx = F(xx)$. Diagramming the nonassociative algebra inherent in this discussion we have:

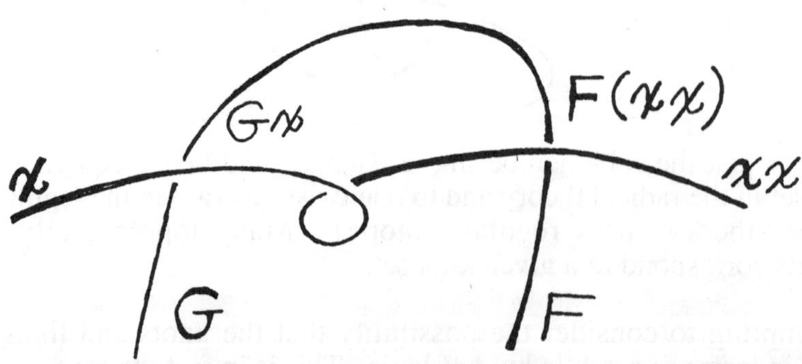

Taking G=x in the above diagram, by tying together the lines, we obtain GG = F(GG):

In this way, we obtain a knot diagrammatic interpretation of the basic fixed point construction of the lambda calculus. The analogy with our previous construction of self membering knot sets is striking, but these lambda calculus constructions use much more of the structure of the knot and link diagrams.

Here is a knot diagrammatic interpretation of the equation **Ya = a(Ya)**. It is a double leveled twist.

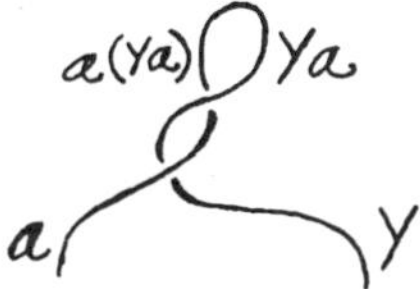

At this point we must take a more careful look at our conventions for handling diagrammatic non-associative products. If we take the convention for multiplication literally, then it can be read in two ways at a given crossing as shown below.

$$\left.\begin{array}{c} c = ab \\ a = cb \end{array}\right\} \Rightarrow c = (cb)b.$$

Thus $c = ab$ and $a = cb$. Hence $c = (cb)b$. For consistency, we demand that $c = (cb)b$ for all elements b and c.   Look at the diagrammatic consequences of taking the axiom $c = (cb)b$. We have the following diagram.

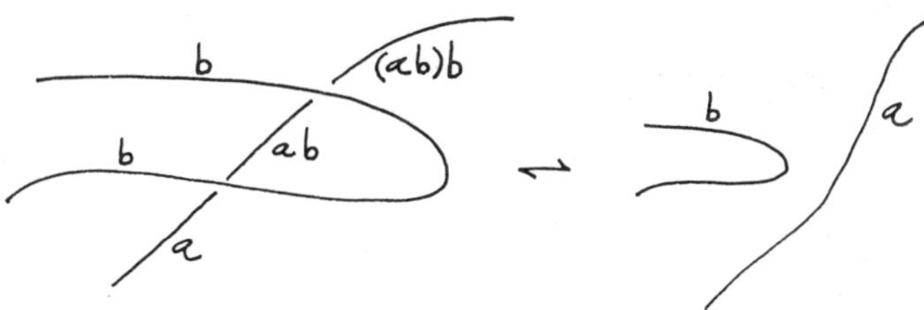

Under the axiom $c = (cb)b$, the algebra cannot see the second Reidemeister move.   The demand for invariance under the third Reidemeister move leads to yet another axiom: $(ab)c = (ac)(bc)$. This states that the algebra is right-distributive over itself.

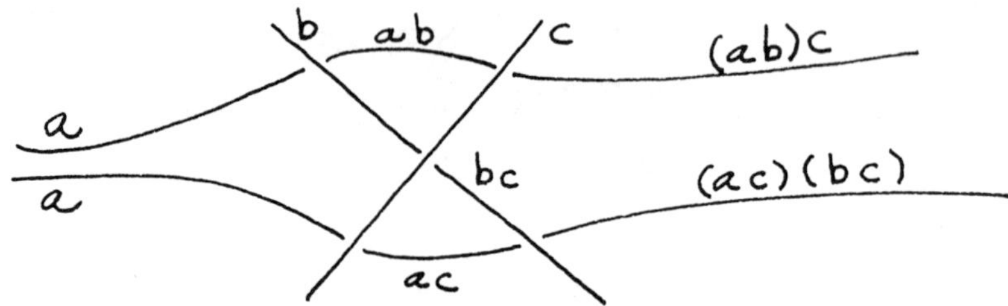

Finally for the type I move we need $aa = a$ for all a.

Thus we need an algebraic system with one binary operation and satisfying the axioms:

1. $aa = a$

2. $c = (cb)b$

3. $(ab)c = (ac)(bc)$

An algebra satisfying these axioms is called an *involutory quandle* [J]. If we eliminate the first axiom it is called a *light crystal* [K2], [K6].

The simplest example of an involutory quandle is as follows: Let R be a commutative ring, and define $a*b = 2b-a$. Then the operation * satisfies the axioms for an involutory quandle. That is

1. $a*a = a$

2. $c = (c*b)*b$

3. $(a*b)*c = (a*c)*(b*c)$

The knot theory associated with this algebra is non-trivial. For example, label the edges of the trefoil knot with elements a,b,c as shown below:

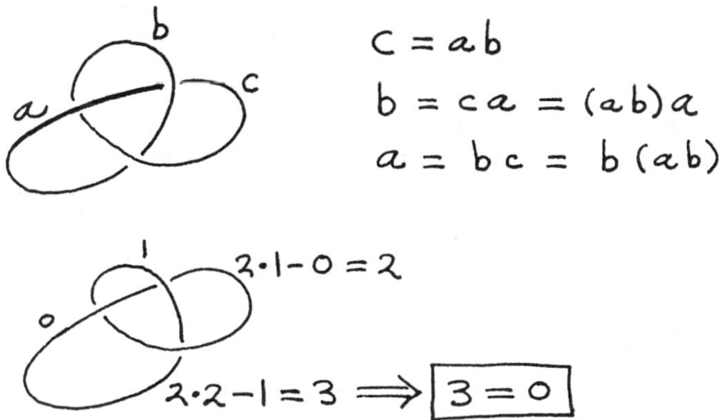

We see that it is necessary that a=b(ab) and (ab)a =b.   In the specific representation we require that   a = b*(a*b) = 2(a*b) -b = 2(2b-a) - b = 3b-2a, whence 3(b-a) =0.   Similarly, b=(a*b)*a = 2a- (2b-a) = 3a-2b and 3(b-a) =0. This is satisfied in R= Z/3Z (Z denotes the integers) and so we can consistently label the knot from this involutory quandle. The number 3 is a topological characteristic of the trefoil knot. The fact that this modulus is non-zero proves that the trefoil is in fact knotted.

By representing this fixed point (a = b(ab), b= (ab)a) into the structure of the involutory quandle , we prove that the trefoil knot is in fact knotted. (Subtler methods are required to distinguish the trefoil from its mirror image.) These methods in fact show that the fixed point pair

$$a=b(ab) \; , \; b=(ab)a$$

is non-trivial in the lambda calculus associated with the involutory quandle axioms. Knot theory provides a rich domain for studying combinatory fixed points and their properties. The topology provides an expansion of the original context for lambda calculus. Insights from lambda calculus inform the theory of knots and links.

Lambda Calculus
Lambda calculus is concerned with the formalism of composition of functions in an arbitrary non-associative algebra with one binary operation.

The basic notation is illustrated by the forms A= λx.xx, B= λxy. (xy)x. The prefix on the form tells the variables that are free to accept substitution, and the order in which this substitution is to take place. Thus

$$A= \lambda x.xx \quad \text{means} \quad Aq = qq \quad \text{and}$$

$$B= \lambda xy. (xy)x \quad \text{means} \quad (Bp)q = (pq)p.$$

A completely left associated expression is written (by convention) without parentheses. Note that the variable that appears first in the list $(\lambda xy...)$ is the primary acceptor for substitution. Thus, given $C = \lambda xyz.F(x,y,z)$, and any other expression D, we can write $CD = \lambda yz.F(D,y,z)$ where in CD we have substituted D for every appearance of x in F, and we have removed x from the list of free variables in CD. This means that expressions in the lambda calculus have a well-defined binary law of composition. This composition is not associative. Note that the composition of expressions without free variables is just their formal juxtaposition in the free non-associative algebra (on one operation) generated by these expressions.

Consider $G = \lambda x.F(xx)$ where F is an element in the lambda calculus, and xx denotes the composition of an expression with itself. Then $GG = F(GG)$ and hence *the function F(z) has a fixed point in the context of the lambda calculus.* This is the well- known fixed point theorem for the untyped lambda calculus [B].

This fixed point is as mysterious as the set that is its own member. We have produced it without any use of infinity, and yet the substitution process does not stop inside F. We get the sequence

$$GG = F(GG) = F(F(GG)) = ...$$

In this sense GG may be regarded as the creator of the limiting formal fixed point

$$L = F(F(F(F(F...)))).$$

That is, by allowing infinite expressions we have the identity $F(L) = L$.

Note the striking difference between the application of G to itself and the application of G to any expression p. $Gp = F(pp)$. If pp does not involve further substitutions then the process stops, while GG goes on forever. The key to this formation of recursion and fixed point lies in the duplication (pp) involved in the definition of G.

G is an operator on x that inserts x and duplicates it in the process of insertion. The application of a duplicating operator to itself results in an interior application of a copy of the duplicating operator to itself, and this process goes on forever. The pattern fits in a myriad of contexts. It is the basis of jokes, paradoxes, the theory of self-reproducing machines and even Gödel's incompleteness theorem. See section 8 of this paper for a discussion of these connections.

One can do ordinary logic in the lambda calculus by the simple expedient of identifying true (T) and false (F) with the following elementary lambda expressions: $T = \lambda xy.x,$ $F = \lambda xy.y.$

$$(Tx)y = x$$
$$(Fx)y = y \ .$$

It is then easy to see that

xyz = (xy)z means exactly "If x then y , else z".

(For example, let x=T. Then Tyz = (Ty)z = y. Thus Tyz is true if y is true and Tyz is false exactly when y is false.)

It is easy to define all the other logical operations in terms of the if-then-else. For example, let ~ a = aFT. Then ~ F = FFT = T and ~ T = TFT = F. Thus ~ a denotes the negation of a.

Now consider the following construction: Qx = ~ (xx). Let P = QQ. Then P = QQ = ~ (QQ) = ~ P.   Thus P = ~ P.   P is a paradoxical combinator, the direct consequence of the fixed point theorem in the domain of lambda calculus.   The Russell paradox itself appears if we interpret AB   as   "B is a member of A". Then RX = ~(XX) defines the set R of all X that are not members of themselves, and the substitution of R for X gives the paradoxical value: RR = ~(RR).

In terms of our conventions for non-associative algebra and link diagrams, the interpretation of AB as B ε A   is backwards unless we work with the opposite algebra where A represents an overcrossing line and B and undercrossing line. Just for the rest of this section, lets do that. Thus, the operation AB will be represented in link diagrams as shown below.

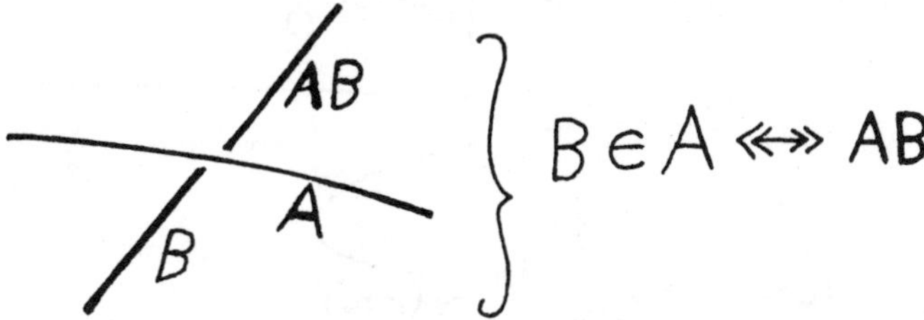

In this form we see that the non-associative algebraic interpretation of knot diagrammatic formalism is a generalization of knot set formalism where membership is indicated not by the full undercrossing line, but by one arc B that approaches the overpass A. The outgoing line, labelled AB, expresses the relation B ε A.

Just so, the lambda calculus for sets (with AB denoting B as a member of A) involves generalized sets corresponding to the membership relations themselves. Thus A(BC) says that "C ε B" is a member of A. While the entity "C ε B" is not defined as a collection, it is defined as a new arc in the diagrammatics. In this diagrammatic system, each arc stands for a statement of membership relative to the other arcs in the diagram. Self-membership is diagrammed via

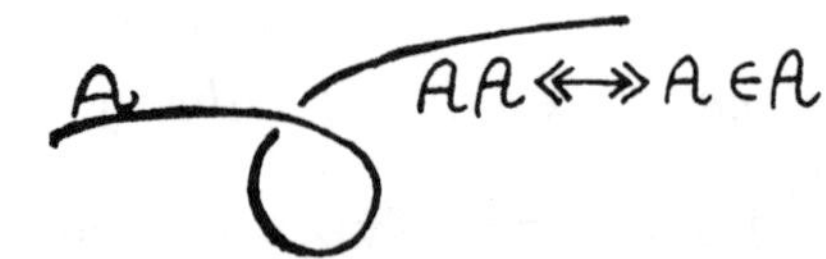

and it is indicated by the exit line AA:

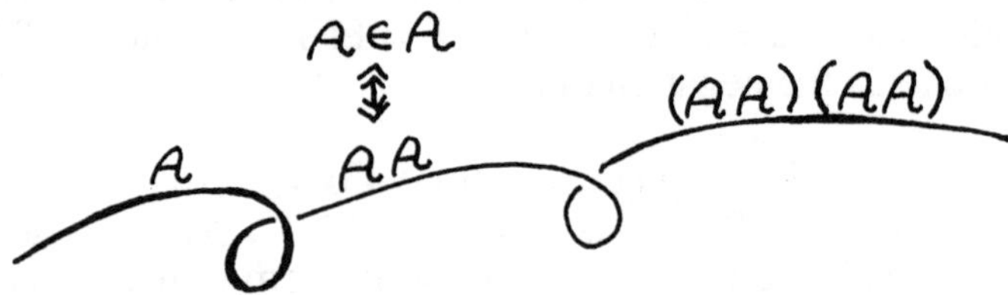

In this pictorial language, statements (such as A) and meta-statements (such as A ε A) are handled at the same level of the formalism.

Diagramming the Russell paradox we find:

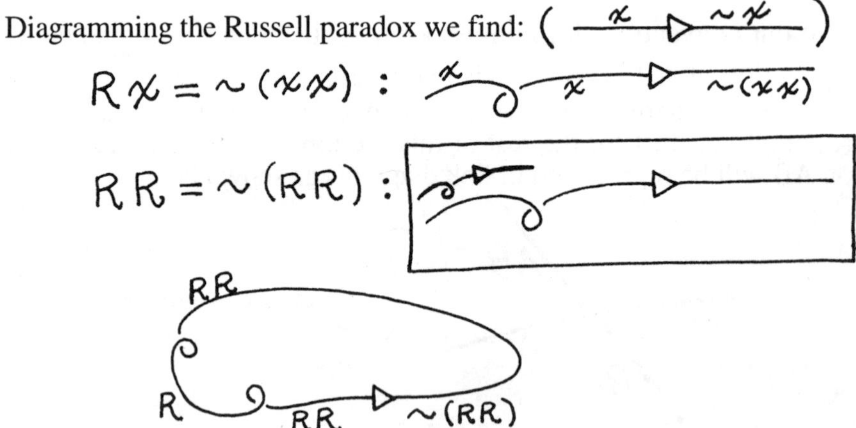

The final circuit corresponding to RR = ~(RR)   is balanced (labelled consistently) if there is a value for RR such that RR = ~(RR). Otherwise, it can be regarded as a runaway feedback system related to the recursion $X_{n+1} = \sim X_n$. For the input values T or F, the system oscillates between T and F. For an appropriate imaginary value for RR it is balanced.

Remark on Insertion and Reentry.   The lambda calculus shares a structural component with the reentry notation of section 2. A lambda expression is equipped with pointers into itself. We devise a notation that makes these pointers explicit.

Let $\quad \lambda x.F(x) = F(\ \underset{\smile}{\ }\ )\qquad$ so that $\qquad F(\ \underset{\smile}{\ }\ )\ a\ =\ F(a)$.

Similarly, $\quad F(\ \underset{\smile}{\ }\ )\ a\ =\ F(aa)$.

Thus, if $\quad G = F(\ \underset{\smile}{\ }\ )\quad$ then $\quad GG = F(\ \underset{\smile}{\ }\ )\ G\ = F(GG)$.

In reentry notation, $GG = F(GG)$ is denoted by an arrow pointing into the place where the expression reenters its own indicational space:

$$GG = F(\ \underset{\uparrow}{\ }\ )\ \Longleftrightarrow\ GG = F(GG)$$

Thus

$$F(\ \underset{\smile}{\ }\ )\ F(\ \underset{\smile}{\ }\ )\ =\ F(\ \underset{\uparrow}{\ }\ )$$

In this formalism the simplest instance of the fixed point theorem is the statement

$$\bigcirc\ \bigcirc\ =\ \bigcirc$$

The notation is useful for the construction of specific fixed points.

Lemma. There is an expression Y in the lambda calculus such that for any a, $Ya = a(Ya)$. Thus Ya is a fixed point for a.

Proof 1.

$$Ya = a(Ya) \Rightarrow Ya = a(\ \underset{\uparrow}{\ }\ )$$

$$\Rightarrow Ya = a(\ \underset{\smile}{\ }\ )\ a(\ \underset{\smile}{\ }\ )$$

$$\Rightarrow Y = (\ \underset{\smile}{\ }\ )(\ \underset{\smile}{\ }\ )$$

Thus $Y = \lambda a.(\lambda x.a(xx))(\lambda x.a(xx))$. //

Proof 2.

$$Ya = a(Ya)$$

$$\Rightarrow Y = \overline{(\,Y\,)} = \overline{(\,\uparrow\,)}$$

$$= \,\overline{(\,)} \; \overline{(\,)}$$

Thus $Y = (\lambda xa.(a((xx)a))(\lambda xa.(a(xx)a))$. //

The two solutions to the assertion in the Lemma are (in standard notation)
due respectively to Church and to Turing. Without the reentry and insertion
notations and their interrelationship expressed by the fixed point theorem, it
would be hard to see that these two solutions express the very same
process.

Dana Scott's Tower
For many years there was a question about the relative consistency of
lambda calculus. This was particularly serious since the paradoxical
combinators leap directly out of the mouth of the fixed point theorem. In [S]
Dana Scott produced a model by using a "hierarchy of languages".
<<Recall Bertrand Russell's comments in the introduction to Wittgenstein's
Tractatus [WITT] about the possibility of an exit from the "Wovon man
nicht sprechen kann, darüber muß man schweigen." through a hierarchy of
languages.>>   Scott's hierarchy is a tower construction in the form

$$X_0 \subset X_1 \subset X_2 \subset X_3 \subset \;\cdots\; \subset X_\infty$$

Each X_n is a topological space with a weak (non-Hausdorff) topology and
such that it contains the autohomeomorphisms of the previous level. There
is a projection from each level to the previous level.   The next higher level
is produced by adjoining this projection.   The direct limit X_∞ exists and is
equal to its own set of autohomeomorphisms. Every point in the limit space
is also a homeomorphism of that space. Enough properties are obtained so
that X_∞ becomes a model for the lambda calculus.

It is an interesting coincidence that a tower construction (due to Vaughan
Jones [JO]) of great formal similarity to the Scott tower was the motivating

force behind the discovery of the Jones polynomial in knot theory. Jones' construction creates a tower of von Neumann algebras and it is central to problems involving the classification of these algebras. It is worthwhile exploring the parallels between the Scott tower and the Jones tower. This will be the subject of another paper.

This ends our sketch of the lambda calculus.

VI. Interlock Algebra

It is the purpose of this section to explain how one further generalization of the methods of section 5 leads directly to the main considerations of classical knot theory and to the problems of the relationship of classical knot theory with the theory of quantum invariants of knots and links [see e.g. K6].

The generalization assumes that the knot or link diagram is oriented. An orientation consists in a choice of direction for each component — indicated by an arrow drawn on the component. With orientations indicated, we can define two binary operations corresponding to the two possible orientations at a crossing:

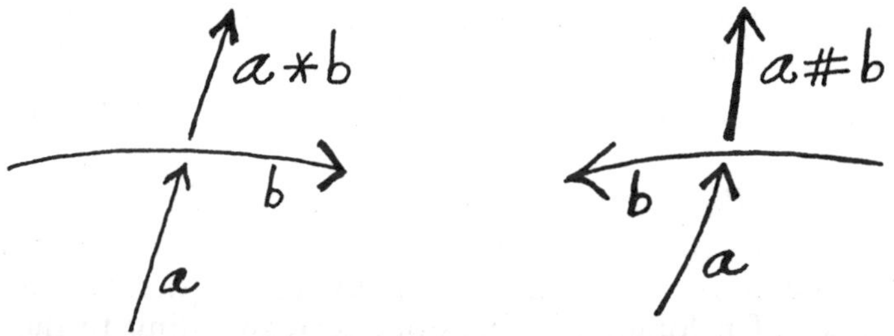

These operations are denoted by a*b for the right handed crossing and a#b for the left handed crossing as illustrated above. Note that we in fact need only indicate the orientation of the overcrossing line and take the convention that if it goes to the right as one approaches it along the under crossing line, then the operation is *, while if it goes to the left as one approaches it along the undercrossing line, then the operation is #. This means that we can once again take the reverse view of a given crossing and conclude that (a*b)#b = a   and that   (a#b)*b = a   for all a and b. This means that the algebra is automatically invariant under the type two move.

Invariance under the type three move demands adding self distributivity for each of the operations. Thus we arrive at the axiomatic definition of a *quandle* [J]:

1. a*a = a,   a#a=a

2. (a*b)#b = a,   (a#b)*b = a

3. $(a*b)*c = (a*c)*(b*c)$, $(a\#b)\#c = (a\#c)\#(b\#c)$

A *crystal* is the algebraic structure that results from dropping the first axiom (see [K6]). (In [RF] a crystal written in exponential notation — $a*b = a^b$ is called a *rack*.) The crystal is an invariant of framed links (see section 2). A quandle or crystal is associated to any oriented knot or link by taking the free (non- associative) algebra (in the sense of universal algebra) on the arcs of the diagram (one label for each arc) modulo one relation for each crossing as indicated above ($c=a*b$ or $c=a\#b$) and the axioms for the quandle, crystal or rack.

Regularly isotopic links have isomorphic crystals, and isotopic links have isomorphic quandles. The quandle completely classifies knots up to mirror images [J]. That is , if two oriented knots have isomorphic quandles, then they are either isotopic, or one is isotopic to the (orientation reversed) mirror image of the other.

The crystal can be used to completely classify knots, but at the cost of adding a subalgebra generated by special elements called longitudes. In keeping with our lambda formulation of these matters, a *longitude* L is a certain element in the form $\Lambda = \lambda x.xP_1P_2P_3...P_n$ where we use the convention that a product that is not explicitly associated is put in left associated form.   Thus if

$\Lambda = \lambda x.xABC$ then $\Lambda = \lambda x.(((xA)B)C)$, whence $\Lambda x = (((xA)B)C)$.

Here $P_1, P_2, P_3, ..., P_n$ *are the operators (overcrossing lines) met in order as one traverses the knot diagram from some given point on the diagram.* These operators are well defined up to cyclic order, and there are, accordingly, a set of n longitude operators corresponding to the cyclic permutations of the Pi's. We add these longitude operators to the crystal, and take the algebra so generated, calling it the *interlock algebra* of the knot.

Theorem. *The interlock algebra is a complete classifier for the knot* in the following sense: Diagrams for two knots can be adjusted so that they both have writhe zero. (The writhe is the sum of the signs of the crossings. See the example below.) If these diagrams have isomorphic interlock algebras, then the knots are isotopic in three dimensional space.

Proof. This follows directly from known theorems [Wald] about the classification of knots. The knot is completely classified by the fundamental group of the complement plus the peripheral subgroup generated by meridians and a standard longitude on the tubular neighborhood of the knot. This information can be read from the interlock algebra. By writing the interlock algebra as a lambda algebra, we are able to include the longitude in the algebra. *//*

Given a knot diagram K, let **I**(K) denote its interlock algebra and let $\Lambda(K)$ denote the set of longitudes in **I**(K). Note that the longitudes themselves are elements at a different level than the elements of the underlying crystal **C**(K). Elements of the crystal have no free variables. Thus the algebra of longitudes and their compositions can be directly identified in any version of the interlock algebra. This is a case where combinatory logic impinges directly on topological applications.

Example. The writhe [K2] is the sum of the crossing signs

Thus the standard right-handed trefoil has writhe 3.

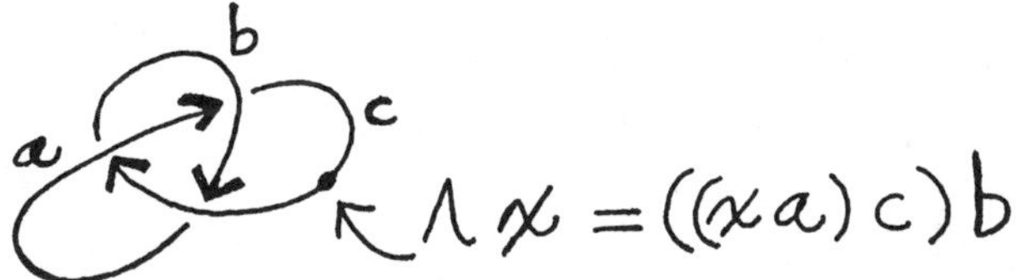

In this trefoil, the crystal is generated by a,b,c with the relations a*b=c, b*c=a, c*a=b. A representative longitude (starting at the segment labelled c) for the interlock algebra of this projection is $\Lambda = \lambda x . x * a * c * b = \lambda x . ((x*a)*c)*b$.

A trefoil projection of writhe zero is shown below.

The crystal of this projection is generated by a,b,c with relations a*b=c, c#c=d,d#d=e,e#e=f,f*a=b,b*f=a. A representative longitude Λ is defined by the operation $\Lambda x = x*b\#c\#d\#e*a*f$ (left associated). The interlock algebra contains complete information about the trefoil knot.

Classical Knots

The fact that the interlock algebra is a complete classifier for knots follows from known results about the fundamental group and peripheral subgroup. It is an open problem to give a purely diagrammatic or knot logical proof of this theorem.

One illustration in relation to the Alexander polynomial is useful. Consider the trefoil knot indicated below.

In the crystal, we have the equation $a*(x*a) = x$, derived from this diagram.
The reader will easily prove that the definitions
$x*y = tx + (1-t)y$
$x\#y = (1/t)x + (1-(1/t))y$
satisfy the crystal axioms. Here x, and y belong to a module M over the ring of Laurent polynomials in t.

The trefoil equation $a*(x*a) = x$ makes an extra demand on this module structure. In particular, the specific equation $a*(0*a) = 0$ becomes $(t^2-t+1)a=0$.

$$0 = a*(0*a)$$
$$= ta + (1-t)(t0 + (1-t)a)$$
$$= ta + (1-t)^2a$$
$$= (t^2-t+1)a.$$

It turns out that the polynomial (t^2-t+1) is *itself* an invariant of the trefoil (up to multiplication by units in the ring of Laurent polynomials in t). This is the Alexander polynomial of the trefoil. Thus the Alexander polynomial is the annihilator of a module that is associated with the knot.
This completes our discussion of the relationship of the lambda calculus to classical knot theory and the theory of link diagrams.

Operator Notation

Another notation is worth mentioning in this context. Rather than work in a non-associative algebra we can use an operator notation for a*b as follows:

Let $a*b = a\,\overline{b|}$ and $a\#b = a\,\overline{|b}$

We regard a, $\overline{b|}$ and $\overline{|b}$ as elements in a non-commutative algebra.

Thus $(a*b)*c = a\,\overline{b|}\,\overline{c|}$ while $a*(b*c)=a\,\overline{\overline{b\ c|}\,|}$.

A formalism of this type is equivalent to exponential notation ($a*b = a^b$), but can be handled more systematically. Since in the crystal we have two operations a*b and a#b, two operator notations are also required.

Then inversion $(a*b)\#b = a$ becomes the equation $a\,\overline{b|}\,\overline{|b} = a$. We can therefore regard $\overline{b|}$ and $\overline{|b}$ as elements of a group of automorphisms of the underlying set S of the crystal, and b is the inverse of b . If we isolate this associated group structure G, then the longitudes of the interlock algebra find a natural home as special elements of G. With this viewpoint, we can eliminate the lambda notations and use the pair consisting of the crystal and special longitudinal elements in G to form a knot classifier.

Remark. One way to handle the operator notation in ordinary typography is to write $A\{B\}$ for $A\,\overline{B|}$. We will use this convention in the next section.

Remark on Terminology

Brieskorn [BR] called structures such as quandles, crystals and racks *automorphic sets*. The term quandle came first in David Joyce's 1979 thesis [J]. Crystals and racks mean essentially the same thing, and crystals first appeared in [K2],[K6]. Fenn and Rourke independently invented the concept and called it the rack in [RF]. Their concept includes new points of view about the homotopy and topology related to these structures.

VII. The LD-Magma

We now turn to a remarkable application of the theory of braids to a problem in the borderline between universal algebra, set theory and logic. It is a problem that fits naturally into this discussion of knot logic. It is the problem of understanding a non-associative algebra with one binary operation that distributes over itself. In the notation of the previous sections we have seen that right-distributivity is an expression of the third Reidemeister move in the theory of knots. An equivalent convention will give similar pictures for left-distributivity, and we shall here discuss the structure of algebras that are *left-distributive*: $a*(b*c) = (a*b)*(a*c)$.

In [L] Laver raised the question of the word problem for *free left distributive algebras* (called *LD-magmas*). He solved the word problem under the assumption of the existence of certain types of infinite cardinal numbers. Dehornoy [DH] discovered a direct solution to the word problem that was purely combinatorial. In the process Dehornoy showed how to embed an LD-magma into the Artin braid group and to thereby reduce the word problem to the topological and already-solved word problem for the braid group. In this section we discuss some of the properties of the LD-magma and how Dehornoy puts it inside the Artin braid group.

Here we work with a single binary operation a*b. Let a*b = a{b} in operator notation as explained at the end of the last section. Let us assume left distributivity so that a*(b*c) = (a*b)*(a*c). In operator notation,

$$a*(b*c) = a\{b\{c\}\}$$
$$(a*b)*(a*c) = a\{b\}\{a\{c\}\}$$

Thus, for left distributivity, we assume the equation below

$$\mathbf{a\{b\{c\}\}} \quad = \quad \mathbf{a\{b\}\{a\{c\}\}}$$

for any a, b and c in the algebra.   In order to get an intuition for this structure it is useful to do a few computations. We restrict ourselves to the case of the free algebra that is generated by one element a with left distributivity. Call this the *LD-Magma.*
Lets begin by listing some elements of the magma:

a
a{a}
a{a{a}}.

The elements **a** and **a{a}** are not subject to the distributive law. But **a{a{a}}** is subject to this law and it is in this way infinitely productive:

a{a{a}}
=   a{a}{a{a}}
=   a{a}{a}{a{a}{a}}
= a{a}{a}{a{a}}{a{a}{a}{a}}   **= ...**

At the last stage shown we meet the possibility of distributing in two ways and enter a branching infinity of expressions derived from **a{a{a}}**. Note that all expressions in the LD-Magma are of the form **a{A}{B}{C}...** where **A,B,C, ...** are themselves expressions in the Magma. The set of *truncations* of a given expression **a{A}{B}{C}...**   are the expressions

a, a{A}, a{A}{B}, ... We shall say that **X** is *less than* **Y** (denoted **X** < **Y**) if **X** is a truncation of any expression that is equivalent to **Y**. Thus we have (by continuing the computation started above) that **a < a{a} < a{a}{a} < a{a}{a}{a} < ... < a{a{a}}.**
Dehornoy proves that, with this notion of inequality, the LD-Magma is a linearly ordered set. *Given any two elements of the magma, either they are equal, or one is a truncation of an equivalent version of the other.* There is great subtlety in this ordering.

Just to give the flavor of this enterprise, consider the elements $a^{[n]}$ defined inductively by

$$a^{[1]} = a \quad \text{and} \quad a^{[n]} = a\{a^{[n]}\}.$$

Thus $a^{[1]}$ = a, $a^{[2]}$ = a{a}, $a^{[3]}$ = a{a{a}}, $a^{[4]}$ = a{a{a{a}}}, ...

Proposition. Let P be any element of the LD-Magma.
Then $P < a^{[n]}$ for some natural number n. In fact, for any P there is a natural number r such that $P\{a^{[n-1]}\} = a^{[n]}$ for all n > r.

Proof. We take the second sentence of the proposition as an inductive hypothesis, and proceed by mathematical induction. The simplest element of the magma is a, and we have $a\{a^{[n-1]}\} = a^{[n]}$ for all n by the definition of $a^{[n]}$. Thus r=1 for the element a. This establishes the base for the induction argument. Now suppose that $R\{a^{[n-1]}\} = a^{[n]}$ for all n > r and that $S\{a^{[m-1]}\} = a^{[m]}$ for all m > s. Let n > r+s. Let P = R{S}. Then

$$
\begin{aligned}
&P\{a^{[n]}\} \\
&= R\{S\}\{a^{[n]}\} \\
&= R\{S\}\ \{R\{a^{[n-1]}\}\} \\
&= R\{S\{a^{[n-1]}\}\} \\
&= R\{a^{[n]}\} \\
&= a^{[n+1]}.
\end{aligned}
$$

Since any expression in the magma can be built in the form R{S}, this completes the inductive step and hence the proof of the theorem.//

Dehornoy gives an inductive construction that embeds the LD-Magma into the Artin braid group. Equivalent expressions in the Magma go to topologically equivalent braids. In order to describe Dehornoy's construction, we must first recall the structure of the braid group. We regard the Artin braid group $B\infty$ as the union of the braid groups B_n on n strands where B_n is embedded in B_{n+1} by adding a trivial n+1st strand on the right.

Then B_n is generated by the elementary braids $\sigma_1, \sigma_2, \ldots, \sigma_{n-1}$ and their inverses where σ_i is a braid where only the ith and i+1th strands cross as shown below.

$$\sigma_1, \quad \sigma_2, \quad \ldots, \quad \sigma_{n-1}$$

In general, a braid in B_n is a configuration of n strands in a plane crossed with the unit interval, so that the strands have a specific row of starting points in the top plane and a corresponding row of ending points in the bottom plane. Each planar cross section of the strands consists in n points. Thus each strand descends from the top plane to the bottom plane, possibly winding about its neighbors. B_n becomes a group through the composition induced by attaching the bottom points of one braid to the top points of the other. The inverse of a braid is its mirror image obtained by reversing all the crossings in a planar projection of the braid. The group B_n is generated by $\sigma_1, \sigma_2, \ldots, \sigma_{n-1}$ and has a complete list of relations: $\sigma_i \sigma_{i+1} \sigma_i = \sigma_{i+1} \sigma_i \sigma_{i+1}$ and $\sigma_i \sigma_j = \sigma_j \sigma_i$ for $|i-j|>1$. The first relation is a version of the third Reidemeister move. The fact that $\sigma_i \sigma_i^{-1} = 1$ is an expression of the second Reidemeister move.

Dehornoy's construction takes elements of the magma into B_∞. Let X be an element of the magma and $b(X)$ its corresponding braid. Then $b(X)$ is defined inductively by the formulas

$$b(a) = 1 \quad \text{and} \quad b(X\{Y\}) = b(X)s(b(Y))\sigma_1 s(b(X)^\wedge).$$

Here $s(b)$ is the braid obtained from the braid b by shifting all its strands to the right by one strand (add one straight strand to the left of b.) and $b(X)^\wedge$ denotes the inverse of $b(X)$ in the braid group.

Lemma. $b(A*(B*C)) = b((A*B)*(A*C))$ where the equality on the right is equality of braids in the braid group.
Proof. It is convenient to prove this result by checking that the corresponding operation on braids is left distributive. That is, we define on braids the operation $X*Y = X\{Y\} = Xs(Y)\sigma_1 s(X)^\wedge$. We shall demonstrate that for braids A,B,C we have the equality in the braid group $A*(B*C) = (A*B)*(A*C)$.

A*(B*C)
= A{B{C}}
= A s(B{C}) σ_1 s(A^)
= A s(B s(C) σ_1 s(B^)) σ_1 s(A^)
= A s(B) ss(C) σ_2 ss(B^) σ_1 s(A^)

(A*B)*(A*C)
= A{B}{A{C}}
= A{B} s(A{C}) σ_1 s(A{B})^
= A s(B) σ_1 s(A)^ s(A s(C) σ_1 s(A)^) σ_1 s(A s(B) σ_1 s(A)^)^
= A s(B) σ_1 s(A)^ s(A) ss(C) σ_2 ss(A)^ σ_1 [s(A) ss(B) σ_2 ss(A)^]^
= A s(B) σ_1 **s(A)^ s(A)** ss(C) σ_2 ss(A)^ σ_1 ss(A) σ_2^{-1}ss(B)^ s(A)^
= A s(B) σ_1 ss(C) σ_2 ss(A)^ σ_1 **ss(A)** σ_2^{-1}ss(B)^ s(A)^
= A s(B) σ_1 ss(C) σ_2 **ss(A)^ ss(A)** σ 1 σ_2^{-1}ss(B)^ s(A)^
= A s(B) σ_1 **ss(C)** σ_2 σ_1 σ_2^{-1}ss(B)^ s(A)^
= A s(B) ss(C) **σ_1 σ_2 σ_1** σ_2^{-1}ss(B)^ s(A)^
= A s(B) ss(C) σ_2 σ_1 **σ_2 σ_2^{-1}**ss(B)^ s(A)^
= A s(B) ss(C) σ_2 **σ_1 ss(B)^** s(A)^
= A s(B) ss(C) σ_2 ss(B)^ σ_1 s(A)^
= A*(B*C)

This completes the proof of the Lemma. //
See Figure 4 for a diagrammatic illustration of this proof.
Dehornoy proves that b: LD-Magma -----> B∞ is injective. Thus the word problem in the magma is reduced to the already solved word problem in the braid group. He uses his results to prove that certain classes of braids are non-trivial, and raises the question of further interactions between this theory of magmas and the topology associated with the braid group.

In the spirit of this quest I suggest an investigation of the composition **LD-Magma -----> B∞ -----> TL** where TL denotes the Temperley Lieb algebra (See Remark 2 at the end of section 2.) with generators 1, $U_1, U_2, U_3,...$ and relations $U_iU_{i+1}U_i = U_i$, $U_iU_{i-1}U_i = U_i$, $U_iU_j = U_jU_i$ if li-jl>2, $(U_i)^2 = dU_i$. Here TL is regarded with coefficients in $Z[A, A^{-1}]$ and $d = -A^2 - A^{-2}$. We then have the Jones representation [JO] $\rho: B∞ -----> TL$ of the braid group to the Temperley Lieb algebra given by the formulas $\rho(s_i) = AU_i + A^{-1}$ and $\rho(s_{i-1}) = A^{-1}U_i + A$. It is possible that the Jones representation is faithful. If this is so, then the composition $\rho b: LD\text{-}Magma -----> TL$ is an embedding of the magma into the Temperley Lieb algebra.

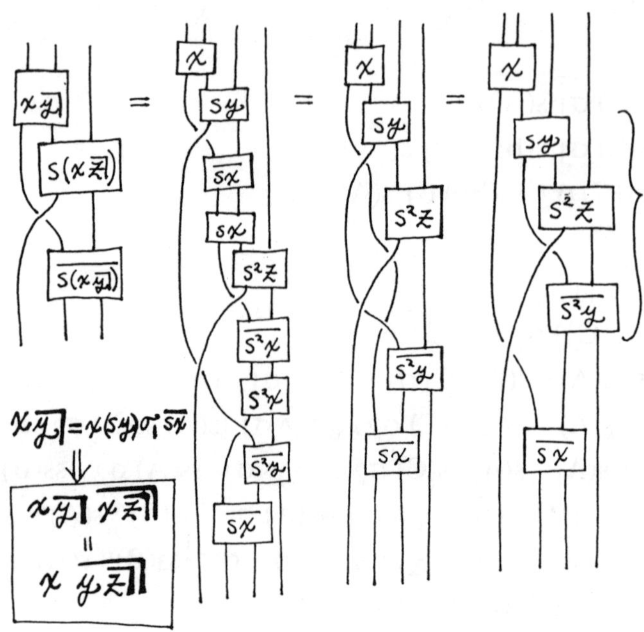

Figure 4

VIII. On Gödel's Theorem, Self-Reproducing Machines, Knots and the Lambda Calculus

We point out the pattern that relates the fixed point theorem in lambda calculus to Gödel's Incompleteness Theorem, and to other issues about reference. Then we discuss mechanisms of replication, and even how to make knots reproduce themselves.

The pattern: Let **G** be defined by the equation **Gx = F(xx)**.
Then, by substituting **G** for **x**, **GG = F(GG)**.

In one step, we have gone from finity to infinity, for **GG** demands to be substituted again and again in the form

$$\begin{aligned}
\mathbf{GG} &= \mathbf{F(GG)} \\
&= \mathbf{F(F(GG))} \\
&= \mathbf{F(F(F(GG)))} \\
&= \ldots\ .
\end{aligned}$$

In the limit,   **J = F(F(F(F(F(...)))))** and **F(J) = J**.

The Building Machine

Suppose **B** *is a universal building machine* . Give **B** a description **x**; then **B** builds the machine **X** described by **x**. **B** sends along the description **x** with **X**:

$$\mathbf{Bx} \ \text{------->} \ \mathbf{Xx}.$$

Let **b** be the description of the Building machine itself. Then

$$\textbf{Bb} \;\text{-------}\!> \;\textbf{Bb.}$$

This process will not stop. An endless cycle of self-reproduction ensues.

Gödel's Theorem, Indicative Shift, Quine and Knot
It is best to understand that **Gx=F(xx)** means that **G** is an entity that will create two copies of x for any x that it meets, and tuck these two copies inside the parentheses of F. In the Lambda Calculus of Church and Curry [B] one writes **G = λx.F(xx).** The lambda just indicates explicitly what variables are free for substitution and *in what order* the substitutions are to take place. With this notation in mind, let **#X = XX.** We then have

$$\textbf{G} = \lambda\textbf{x.F(xx)} = \lambda\textbf{x.F(\#x)}$$

Whence

$$\textbf{\#G = GG} = \lambda\textbf{x.F(\#x)G} = \textbf{F(\#G).}$$

Now replace the equality sign by a sign of reference as we did in the case of the building machine. We obtain [K17]

The Indicative Shift: G -----> H   then   #G -----> HG.

In this context the line above becomes the *definition* of **#G.** **#G** no longer connotes direct repetition of **G,** rather it connotes a new reference to G's referent and to G. As such *#G is a symbolic description of the movement to a meta-level where sign and referent are together in one setting.*

Self-reference occurs when the meta-operator becomes a referent.

 "I am the observed relation between myself and observing myself." [VF]

Let **I -------> #,**   then substituting into the indicative shift, we have

$$\textbf{G -----> H   then   \#G -----> HG.}$$

$$\textbf{I -----> \#   then   \#I -----> \#I.}$$

The simplest case of this self-reference occurs when we take "#" to refer to #. Then   "#" -----> # shifts to #"#" -----> #"#".   This is a syntactic analogue of the famous Quine sentence: "Refers to itself when appended to its own quotation" refers to itself when appended to its own quotation.

A student at the School of the Art Institute of Chicago once remarked to the author of this paper that the trefoil knot was rather like the Quine sentence, if we regard "quoting" as "putting a loop around it". Then the trefoil has the structure "A"A where A is a loop looping about itself.

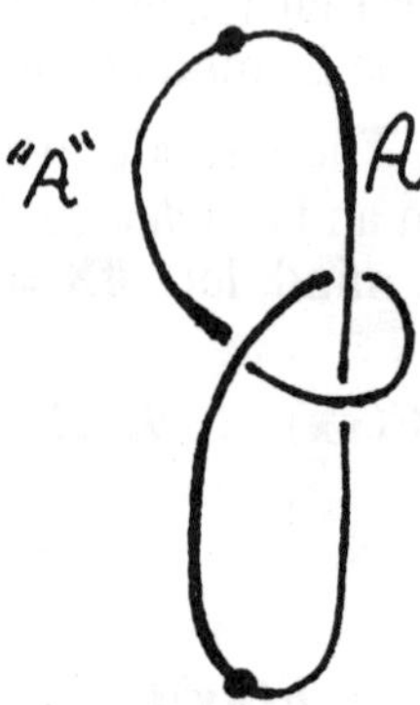

Now return to original format, with a lambda expression:

$$\mathbf{g} \text{-----}> \lambda\mathbf{x}.\mathbf{NE}(\#\mathbf{x}) \quad \text{then} \quad \#\mathbf{g} \text{-----}> \mathbf{NE}(\#\mathbf{g}).$$

The reference of **g** to $\lambda\mathbf{x}.\mathbf{NE}(\#\mathbf{x})$ is *shifted* into a reference of **#g** to **NE(#g)**. The statement **NE(#g)** speaks about its own indicator, **#g**. This is the core of the logic of the Gödel Theorem on the incompleteness of formal systems. In that context, **NE(Y)** is the statement that there is no proof of the statement indicated by **Y**. **Y** is an integer. Every statement is indicated by a specific integer.

With **g** the indicator of $\lambda\mathbf{x}.\mathbf{NE}(\#\mathbf{x})$, **#g** is the indicator of **NE(#g)**. **NE(#g)** asserts its own unprovability, and thereby becomes a theorem that cannot be proved within the system without an inconsistency. If the system is consistent, then the theorem **NE(#g)** is in fact true, and proved by a meta-argument outside the system. The meta-argument is precisely that **NE(#g)** cannot be proved within the system, coupled with its interpretation stating that **NE(#g)** asserts its own unprovability within the system.

Can a knot deny its own detectability? This is the curious speculation that emerges from the present line of thought. Such a Godelian knot would have to occur within the context of a given method of knot detection. In the case of the Alexander polynomial there are knots that avoid detection. For example, the Kinoshita-Terasaka knot shown below has Alexander polynomial equal to 1, the same value as the unknot, and it is indeed knotted.

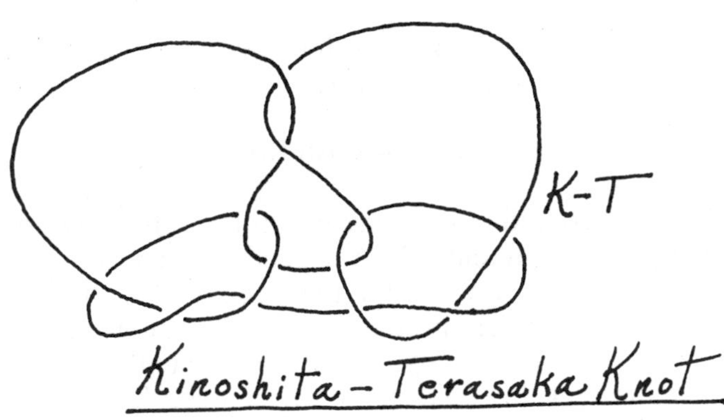

Alexander cannot detect the K-T knot, even though his predecessor slew the Gordian Knot.   Is the K-T knot a Godelian knot with respect to the Alexander polynomial? Does the structure of the K-T knot "say" that the Alexander polynomial is incapable of detecting it? We cannot answer this question with the present tools. It is tantalizing to imagine that the K-T knot encodes such a denial in its diagrammatic surface structure. If this could be understood, then it might be possible to generalize it and locate knots that are Godelian with respect to the Jones polynomial and other invariants. To this date, no one has found a knot that the Jones polynomial can not detect. Is there a knot that says
"I am (k)not detectable by the Jones polynomial"?

The Diagrammatics of Self-Reproduction

1. DNA
DNA self-rep [RZvL] (*self-rep* is short for self-reproduction) is based on the fact that DNA =WC, a joining of base-paired Watson (W) and Crick (C) strands. Here WC denotes the joined strands.
Let   (WC)E -----> (WE) (CE)   denote the process of strand separation (The Watson and Crick strands pull apart into two separate strands during mitosis.) in an appropriate environment E. The remaining abstract rules for this self-rep are:
WE -----> (WC)E   and CE-----> (WC)E. These rules are instantiated in the living medium, as free floating molecules attach themselves to the bare Watson or Crick strands. Thus the schema of DNA self-rep is as follows

DNA = (WC)E -----> (WE) (CE) -----> (WC)E   (WC)E = DNA DNA.

2. The Building Machine (again)

Placed in this same format the Building Machine self-rep takes the form

$$Bx \longrightarrow Bx \; Xx$$

Here we have indicated a building machine that produces its work without destroying itself. Applied to its own description, the Machine appears to reproduce itself.

$$Bb \longrightarrow Bb \; Bb.$$

The scenario for the building machine is similar to the scenario for DNA, but conceptually quite different. DNA depends upon a form that can divide into complementary parts (W and C) that each rebuild their missing mate from the materials in the environment. The building machine follows arbitrary instructions to build "anything". Consequently, it can build itself. The formalisms reflect this conceptual difference. The DNA formalism accomplishes repetition through matching complementarities. The Building Machine formalism has repetition built into its structure.

3. Mighty Simple Self-Rep

Here is indeed a mighty simple set of formal rules for self-rep.

$$C \longrightarrow A \; A$$
$$A \longrightarrow C$$

The "cell" C splits apart into intermediate entities A and A.
Each A becomes a copy of C.

$$C \longrightarrow A \; A \longrightarrow C \; C.$$

Of course one might wonder why we don't go directly for the *very simple* self-rep C $\longrightarrow$ C C, repetition pure and simple. But after all, *Mighty Simple Self-Rep* has the advantage that it has a referent to an intermediate stage. Thus we can diagram *Mitosis* as a mighty simple self-rep:

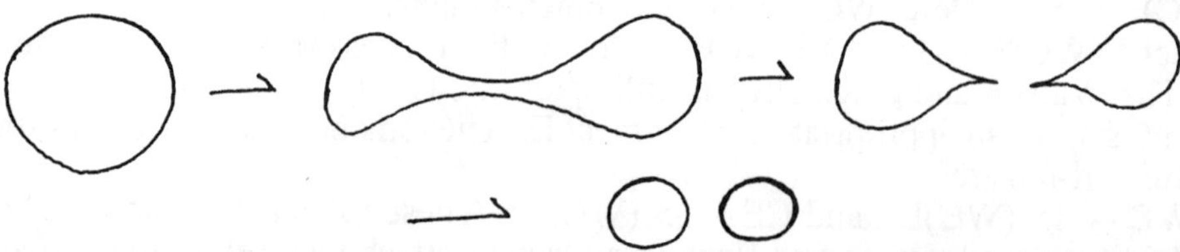

A **symbolic schema** for this mitosis might be

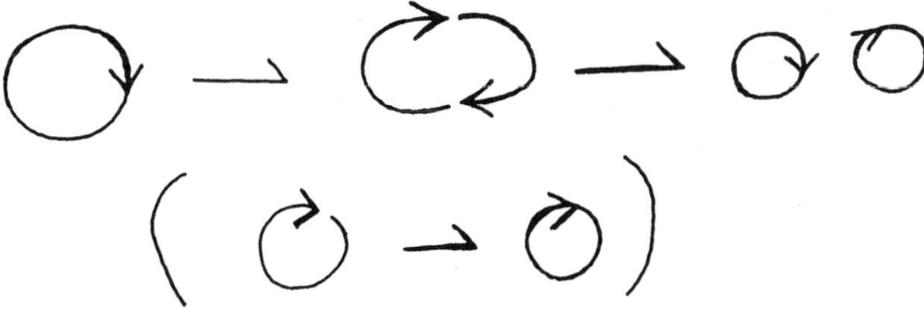

We are thrown into the arena of arrow epistemology (see section 4 of the present essay). The entity C that self-replicates is a circularity, hence a self-pointing arrow. The arrow duplicates into two non-self-pointing arrows (C -----> A A), but arrows being what they are tend to curl up and become self-pointing (A -----> C). Cut a circularity, and it is no longer circular. Cutting the circularity is essential for its duplication. The intermediate stage is syntactically and semantically inevitable.

These structures for self-rep can also be seen as precursors for the topological complexities inherent in the self-rep for circularly closed DNA embedded in three-dimensional space. Here the individual strands are linked and a more complex environment (topological strand switching enzymes and more [SU]) is needed to navigate the appropriate intermediate stages. At this point the whole structure of knot theory comes to bear upon molecular biology.

4. A Knot-Logical Self-Rep

Here is a self-rep of a topological circle that is based upon letting go of part of the formal structure of knot diagrams. As we have discussed, the standard representation of a crossing involves breaking the undercrossing line into two local arcs at the crossing site. In the usual convention the end points of these arcs are paired across the boundary formed by the overcrossing line. It is as though the endpoints were magnetically attracted to one another so that if you move one of them, then the other one follows. Let us let loose of this convention. Let the arcs move independently of one another, but keep the convention that the endpoint of the arc must hug the overcrossing line. We then have

Call the equivalence relation generated by this slide together with the Reidemeister moves *slide equivalence* of knot and link diagrams. It is not

hard to see that slide equivalence does away with most of the interesting topology, but it is quite fascinating to see *how* it does this. In particular, a single circle in the plane is slide equivalent to the disjoint union of *two* circles. This is our knot logical self-rep, as shown below.

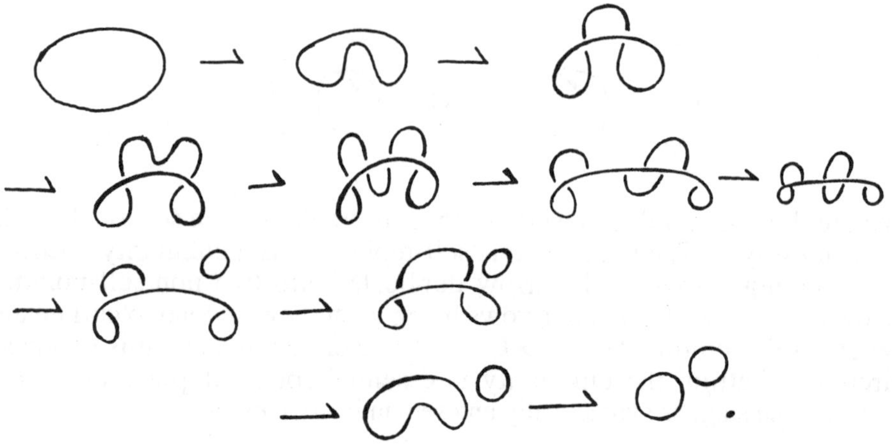

Notice how this process goes. First, by an ordinary Reidemeister move, the circle bifurcates into two arcs.

One arc stands as a platform for the other arc, so we shall label the platform B, and the arc standing upon it x. Then by another Reidemeister move we see that x can now triple into X (below the platform) and two copies of x.

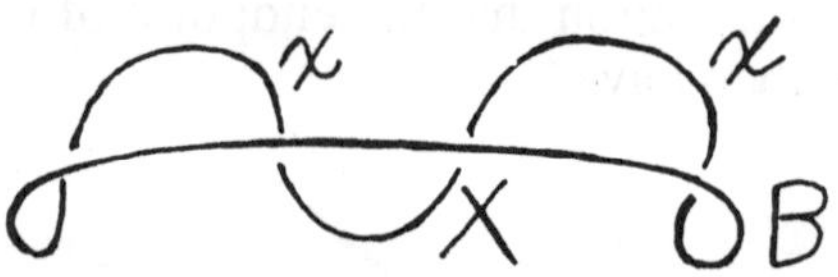

Sliding rearrangements let us pair X and x to form a new circle, Xx, that disengages from the platform to form a separate circle C'. This leaves a single x on the platform, and a slide plus Reidemeister move reassembles

this back to the original circle C. We have accomplished the self-rep
C -----> C C'.

C -----> Bx

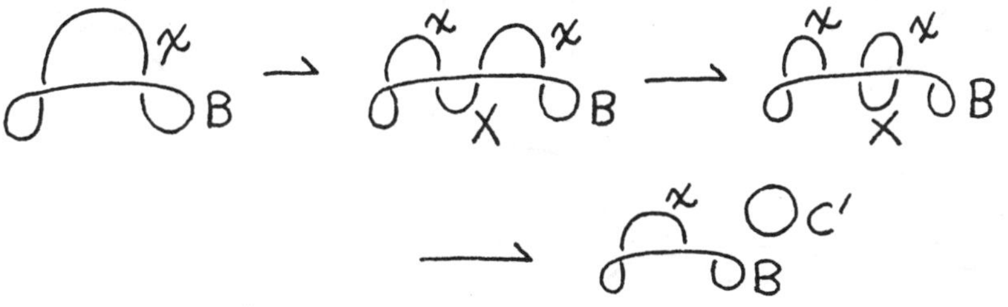

Bx -----> Bx Xx -----> Bx C'

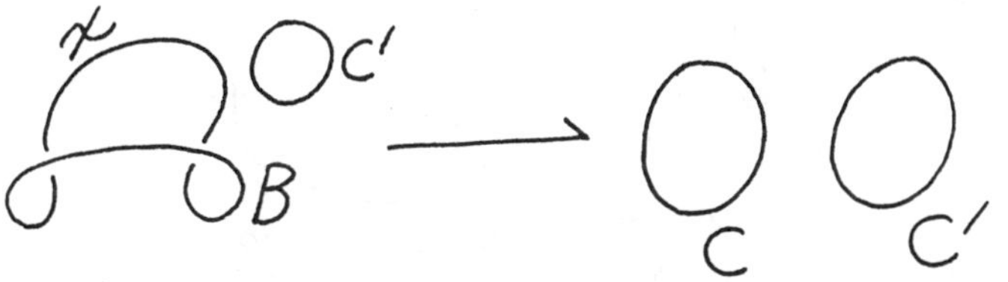

Bx C' -----> C C'

This self-reproduction via sliding is a way to make the formalism of the
Building Machine into diagrams. It corresponds to the syntax of building
machine formalism, but gives us a topological picture of this syntax. The
mighty simple self-rep and the building machine self-rep are seen to be
gremlins of the same clan.

End of Section on Self-Reproduction

Remark on Slide Equivalence. The slide equivalence used in knot
logical self reproduction is actually a way to build any knot or link from the
unknot. Any two knots or links are related by this equivalence relation.
From the point of view of a topologist, this renders the equivalence relation
trivial. From the point of view of biological analogy and epistemology it is
intriguing to find a domain of forms just below the more rigid domain that
holds the topology of knots and links. Slide equivalence is pre-topological
(see the discussions of pre-geometry in sections 2 and 10). For the reader
interested in how to build any knot or link by slide equivalence, we include

here the appropriate sequence of Lemmas in diagrammatic form in Figure 5 For a related discussion, see [KH].

i)   (self-rep).

ii)

iii)

iv)

v)

vi)

vii)

viii)

Figure 5

A modified version of slide equivalence is central to the study of three dimensional manifolds. Knots and links classify 3-manifolds when taken up to regular isotopy (see section 2) and the handle-sliding equivalence illustrated below together with creation and annihilation of singly framed circles. [KI]

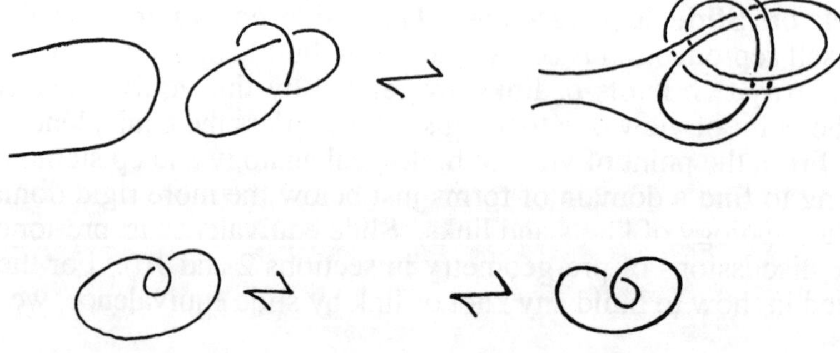

In this context, the link diagram is a code for the construction of the associated 3-manifold via surgery. The surgery process is accomplished by regarding the three dimensional space that contains the knots as the boundary of a four dimensional ball. To this ball, a thickened disk (thickened to a family of disks parametrized by another disk) is attached along its boundary to each knot component in the diagram. Each such attachment is called a "handle". The boundary of the resulting 4-manifold is the 3-manifold obtained by surgery along the given link. The handle-siding move indicated above is exactly the result of sliding one of the four-dimensional handles over another so that at the end of the process the boundaries of both handles are in the original 3-dimensional space.

We can tell the story of the handle-sliding more slowly by moves on the links as shown below.

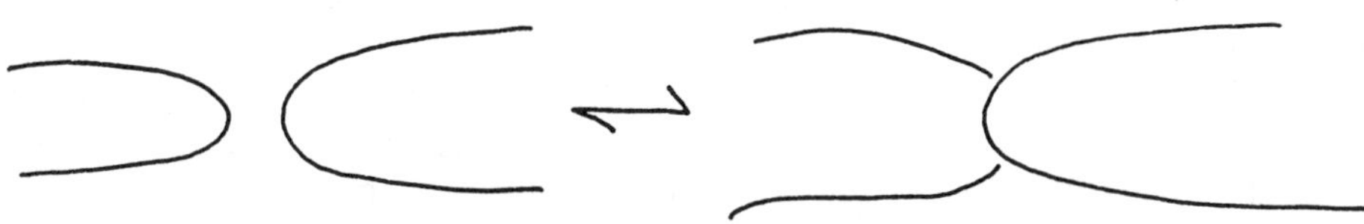

In the diagram shown above the handle on the left has just begun to slide up on the handle on the right. Part of the left diagram disappears into 4-space. The process continues, mediated by the rules

Thus

These local sliding rules resurrect the usual theory of handle sliding for link diagrams. It is understood, however, that free-wheeling recombinations of arcs are not allowed so that the self-rep is not part of this formal system. This modified version of slide equivalence for the study of 3-manifolds embodies a significant part of the relationship of knots with the theory of spin networks [K21], [K22] , [K19] and will be discussed at greater length in another paper.

IX. Quantum Knots and Topological Quantum Field Theory

Topological quantum field theory is a generalization of quantum mechanics. In quantum mechanics one computes amplitudes <a|b> where <a| denotes preparation and |b> denotes detection.

In computing such an amplitude we cut the world via the distinction preparation/detection. It is the rules of quantum mechanical amplitudes that tell us how to combine networks of such cuts to form more complex amplitudes. These rules are utterly categorical, and they fit into topology in the following way: Imagine a topological space M, and a direction associated with that space that we can call "time". In this time direction there is an evolution of the slices of that space perpendicular to time. Thus the space is seen as a process that goes from vacuum to vacuum.

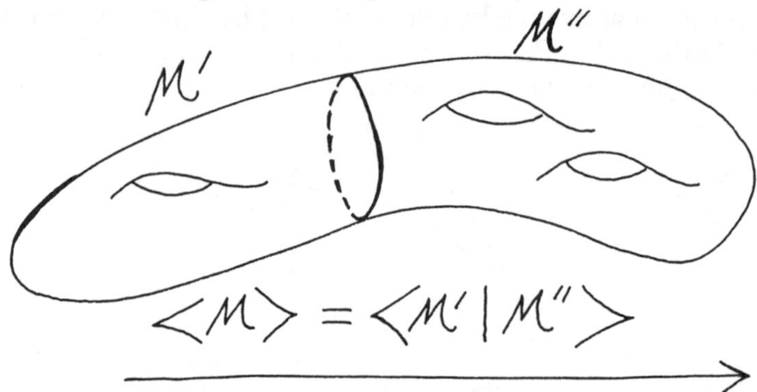

Let <M> denote the vacuum to vacuum amplitude for this time evolution. Let us cut M in two pieces M' and M" so that M is the union of M' and M" along their common boundary S.   Then we can think of <M> as the amplitude <M'|M''> , and the two halves <M'| and |M''> become preparation and detection. In order for this to make physical sense it must be the case that the amplitude <M> does not depend upon the way in which we cut M into two pieces, and it must not depend upon the direction of time either.   It is out of these very stringent conditions that one comes upon <M> as a topological invariant of the space M! In the process, the notion of bras (<a|), kets (|b>) and amplitudes <a|b> has been generalized far beyond the confines of standard quantum mechanics.

The simplest case of bras and kets must be considered first. A bra , <a| , is an element of a Hilbert space H. If everything is taken over the complex numbers C, then an element of H can be regarded as a mapping C ----->

H   (that is, the element itself is the image of the unit element 1). The ket, lb>,   is an element of the dual space and hence it is a map   H -----> C.

The bracket is the composition   <alb>: C -----> H -----> C, and the amplitude itself is the image of 1 under this composite.   More complex spatial decompositions lead to more complex compositions, but the basic pattern is the same.

The Logic of Dirac Brackets

This bracket notation of Dirac is subtle. If we write P=lb><al   then the square of P is a multiple of P.   This is embodied directly in the formalism:

$$\mathbf{P^2} = \mathbf{PP} = \text{lb><allb><al} = \text{lb><alb><al} = \text{<alb> lb><al} = \mathbf{dP}$$

where δ = <alb>. Note that we have implicitly adopted the notation $\| = \|$ in order to indicate that the result of the composition of a bra with a ket is a scalar: <allb> = <alb>. The Boolean law of idempotencey (xx=x) underlies the structure of the Dirac bracket.

We can abstract the notation to the formal ket bra **Q= ><** with

$$\mathbf{QQ} = \text{><<>< = <> >< = } \delta \text{ >< = } \delta\mathbf{Q}$$

where <>=δ is regarded as a scalar. Amplifying this notation slightly, we have

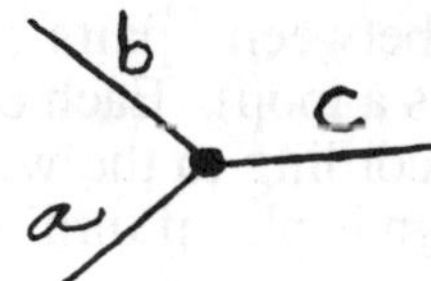

This is the simplest glyph in the diagrammatic interpretation of the Temperley Lieb algebra. (See the remarks on the Temperley Lieb algebra and boundary logic at the end of section 2 of this essay.) Thus the Temperley Lieb algebra can be seen as a generalization of the formalism of Dirac brackets. In this way the brackets are directly related to topology.

Recoupling

When the link diagrams are interpreted as toy models of particle interactions then it is natural to color the lines according to the particle states and to include trivalent interaction vertices that can be interpreted as the emission of a third particle under the interaction of the other two lines.

In the classical theory of quantum angular momentum there are many recoupling formulas such as

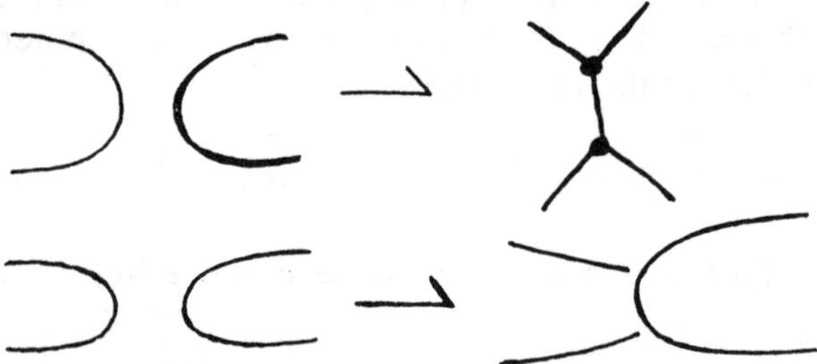

A formula like this matches the pattern of the handleslide as described at the end of the last section.

The marriage of these formalisms in the context of the Temperley Lieb algebra or the context of q-deformed spin networks results in invariants of 3-manifolds and in reconstruction of the invariants of Witten-Reshetikhin-Turaev in a combinatorial form [K19].

X. Knots and Circuits

It is perhaps not surprising that knots and electrical circuits should have some relationship with each other. However, it came as a distinct surprise [GK]   to discover a way to get topological information about a knot by measuring the conductivity of an associated electrical circuit! This section will sketch this method and relate it to the context of logic and switching circuits.

Every knot or link diagram, K, implicates a planar graph, G(K), by the checkerboard construction illustrated below. In this construction, one shades the regions of the diagram, leaving the outer region unshaded. Each shaded region is then taken as a vertex of the graph. The edges correspond to   crossings in common between   pairs of regions (or between a region and itself for an edge that is a loop). Each edge is labelled with a plus sign (+) or a minus sign (-) according to the way the crossing is situated with respect to the edge. The sign is plus if turning the overcrossing line through the shaded regions is a counterclockwise turn. Otherwise, the edge is negative. This convention is illustrated below.

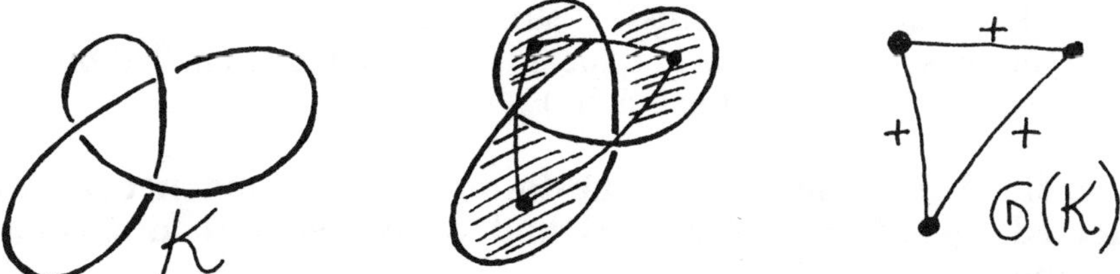

Given a planar graph with signed edges, we can construct a corresponding link diagram by the *medial construction*. The medial construction is inverse to the process that associates the graph G(K) to a link diagram K. Given G, we form the medial, M(G), as follows. First draw simple closed curves each describing the border of one of the regions of the planar graph G. Each curve is drawn just inside the region near the border actually described by the graph.

The result of this process will be that to each edge of G there are associated two parallel arcs from the curves drawn on either side of that edge in the plane. Insert in each such pair of arcs (one pair for each edge of G) a crossing of the type indicated by the sign of the edge. (When the edge has a positive label, turning the overcrossing line counter clockwise sweeps it across the edge.)

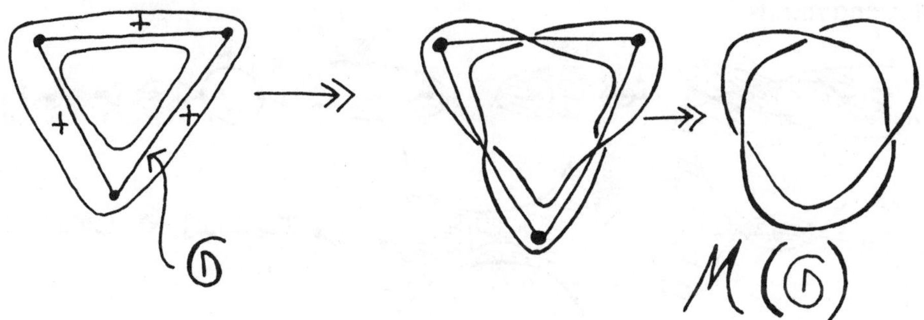

The reader can easily verify that
1. If K is a connected link diagram (i.e. the projected 4-valent graph in the plane associated with K is connected), then M(G(K))=K.
2. If H is a connected planar graph with signed edges, then G(M(H))=H.

(Equality in the first case is up to graphical isotopy of the link diagram in the plane. Graphical isotopy of the link diagram does not allow Reidemeister moves. It is just planar isotopy of the projected 4-valent graph that preserves the over and under crossing structure.
Equality in the second case is graphical isotopy in the plane that preserves the signed edge structure of the graph.)

We conclude that **there is a one-to-one correspondence between connected signed planar graphs and connected link diagrams in the plane.**

Consequently, it is possible to translate knot theory to a theory about signed planar graphs. We now make this translation for the Reidemeister moves. *From now on we shall regard the edge labels + and - as the integers +1 and -1.* This convention is crucial to everything that follows.

The first Reidemeister move becomes the addition or removal of a pendant edge or loop in the signed graph:

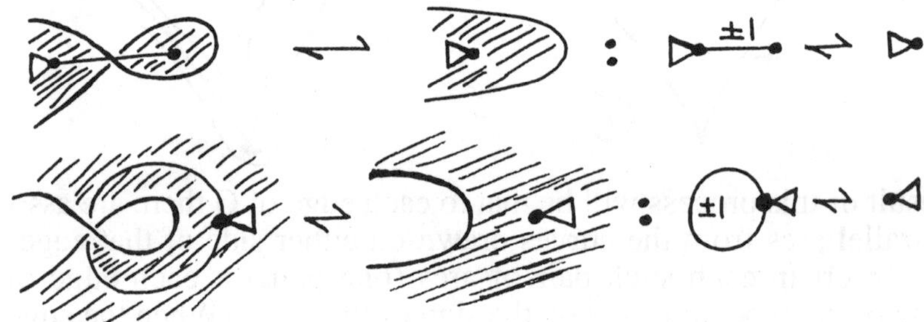

The second Reidemeister move becomes the contraction of a series connection of plus and minus or the deletion of a parallel connection of a plus and minus.

The third Reidemeister move is a replacement of a triangle by a star (or vice versa), with corresponding sign changes. In each of these local configurations there are two signs of one type and one sign of the opposite type. With the edges labelled a,b,c in the triangle and A,B,C in the star we have $X = - x$. (Edges in the triangle are matched with edges in the star so that a superposition of star and triangle has three distinct alphabetical labels at each vertex.)

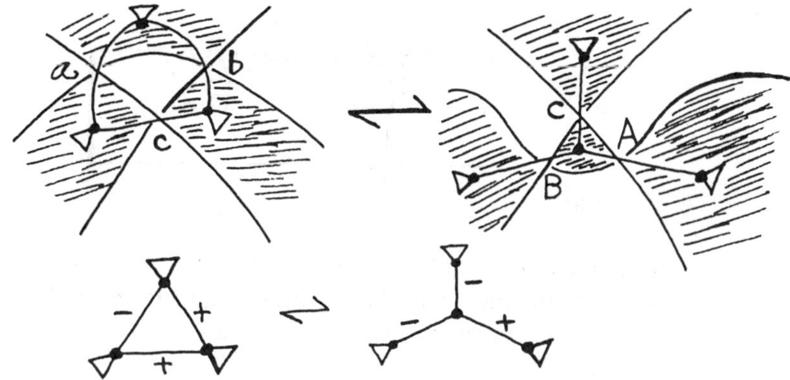

Just for practice, here is the reduction of a graph to a point (corresponding to an isotopy of a diagram that represents the unknot).

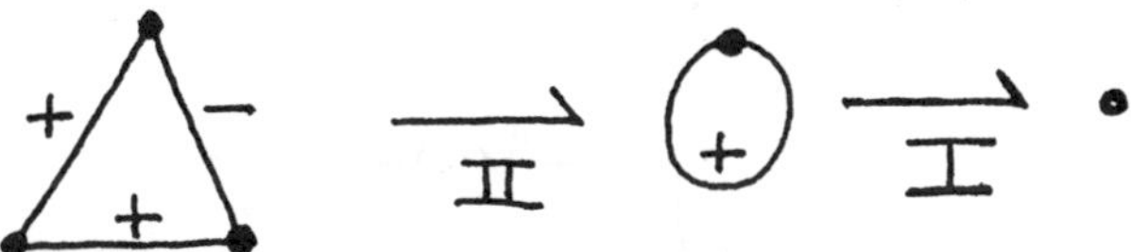

All of knot theory can be done in this category of graphs. In the original setting of the knot theory the Reidemeister moves have a definite topological meaning. One can wonder whether there is a natural interpretation of these moves for signed planar graphs.One answer to this question will emerge as soon as we recall a little elementary electrical theory.

Recalling Electricity

It is most common to consider circuits composed of elements with different values of resistance. Two resistors connected in series have the sum of the resistances of the individuals. Two resistors connected in parallel have the reciprocal of the sum of the reciprocals of the individual resistors.

> These rules follow from the relation $E=IR$ of voltage (E), current (I) and resistance (R) in conjunction with the *Kirchhoff laws* that the sum of the currents at a junction in a circuit is zero, and that the sum of the voltage drops around a closed loop is zero.

> In the case of the series connection of resistances R_1 and R_2, we have a constant current I in the wire and voltage drops E_1 and E_2 across R_1 and R_2 respectively. Thus $E_1=IR_1$, $E_2 = IR_2$ and $E=E_1 + E_2 = IR$ where R is the resistance of the

series connection of R_1 and R_2. Therefore, $IR_1 + IR_2 = IR$ and so $R_1 + R_2 = R$.

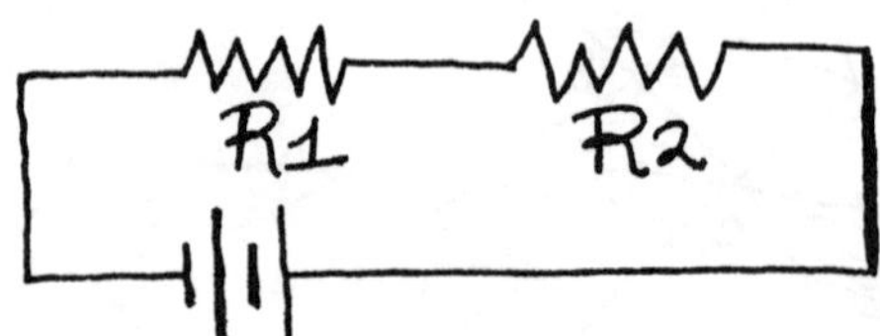

In the case of a parallel connection of resistances R_1 and R_2 we have currents I_1 in the R_1 branch and I_2 in the R_2 branch and a total current of $I = I_1 + I_2$ flowing out of the junction of the two branches.

The voltage drop across the parallel connection of the two branches being E, we have $E = IR$, $I = I_1 + I_2$, $E = I_1 R_1$, $E = I_2 R_2$. Thus $R = E/(I_1 + I_2) = E/(E/R_1 + E/R_2) = 1/(1/R_1 + 1/R_2)$.

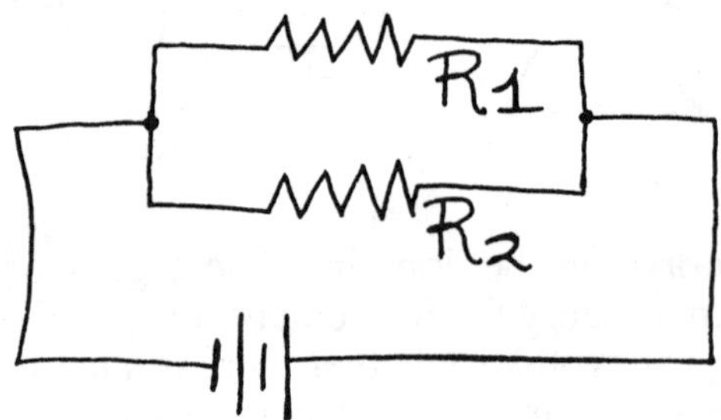

Conductance is the reciprocal of resistance. Thus an open circuit has zero conductance and infinite resistance. A closed circuit with no resistance has infinite conductance. It is convenient in our context to work with conductance. Thus the parallel and series rules for conductance are

$$a\{p\}b = a + b$$

$$a\{s\}b = a\#b = 1/((1/a) + (1/b)) = ab/(a+b)$$

where $\{s\}$ stands for a series connection and $\{p\}$ stands for a parallel connection.

There is a *star-triangle relation* in electrical theory. For corresponding edges x and X from triangle a,b,c to star A,B,C the transformation is $X = S/x$ where $S = ab + ac + bc$. With these assignments of local conductance, stars and triangles can be interchanged in circuits that are otherwise identical without changing any global conductance calculation. (Edges in the triangle are matched with edges in the star so that a superposition of star and triangle has three distinct alphabetical labels at each vertex.)

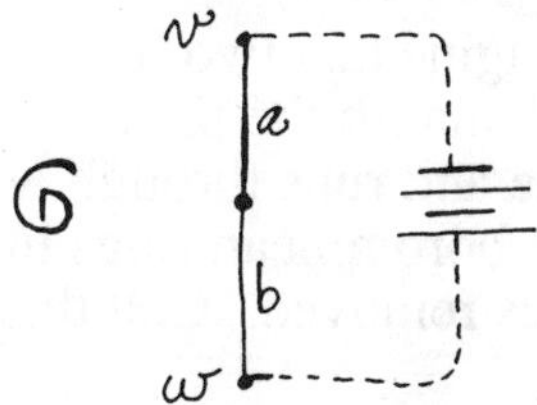

The linear algebra in back of conductance leads to a global formula for the conductance of a circuit between two chosen points v and w (the input and output respectively). Let G be the underlying graph of the circuit. G is a graph with labels on its edges corresponding to the conductance of each edge. Let $\Sigma(G)$ denote the sum over all maximal trees in G of the products of the labels on the edges of each tree. Let G(v,w) be the graph obtained from G by identifying v with w. Let C(G,v,w) denote the conductance of G from v to w. Then (see [GK]) we have the formula

$$(*) \qquad C(G,v,w) = \Sigma(G)/\Sigma(G(v,w)).$$

For example, consider G as shown below. Then G has one tree and $\Sigma(G) = ab$. G(v,w) has two trees, and $\Sigma((G(v,w)) = a + b$. Thus C(G,v,w) = ab/(a+b). This corresponds to the rule for the conductance of a series connection.

By using the formula (*), we can define conductivity for arbitrary algebraic labellings of graphs and the resulting theory will still be invariant under the series, parallel and star-triangle transformations. In particular, one can consider (formally) conductances that take negative or imaginary values. In such cases the quotient $\Sigma(G)/\Sigma(G(v,w))$ may take undefined values (0/0) when the denominator is zero. As long as the denominator is not zero, the transformations remain valid.

Return to Knots
Since we can consider generalized conductances where the labels take negative values, it is now possible to return to the link diagrams, translated

into graphs, and examine the relationship of the conductance with the performance of the Reidemeister moves.

Using the graphical form we see at once from the previous discussion that *each Reidemeister move is a transformation that leaves the conductance invariant.* (It is assumed that the input and output terminals are not involved in the given Reidemeister move.) Therefore the conductance measures a topological property of the original link diagram and the Reidemeister moves have a (generalized) electrical interpretation.

For example, consider the Borommean rings as shown below. These rings are linked as a triple, but any two of them (in the absence of the third) are unlinked. We have drawn the corresponding graph, and see at once that the conductance will be non-zero for any two terminals since all the edges have weight +1. This means that we have proved that the Borommean rings cannot unlinked by any isotopy that does not pass the diagram over the points v and w.

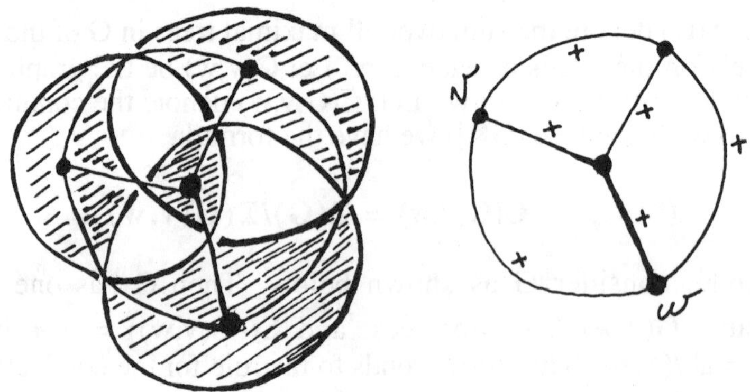

One way to put this is to imagine that two lines have been removed from 3-space. One line is perpendicular to the plane and runs through v, the other is perpendicular to the plane and runs through w. C(G,v,w) is an invariant of the link M(G) (equal the borommean rings in this case) as embedded in 3-space with these two lines removed. Call this space the *tunnel 3-space*, T[v,w].

It is easy to see that if you replace a link by its mirror image (by reversing all the crossings), then G is replaced by G* where all the edges change their signs. From this it follows that C(G*,v,w) = - C(G,v,w) whenever C(G,v,w) is not equal to 0/1, 0/0 or 1/0 = ∞.

A nice application of this result ensues for alternating links. In an alternating link diagram a walker moving along one the strands of the link will alternately go over and under on successive crossings.

A link is said to be alternating if it has an alternating diagram. Furthermore, the graph of an alternating diagram has all positive or all negative signs, just as in the case of the Borommean rings. Thus C(G,v,w) is non-zero (determinate and not infinite) if $G = G(L)$ for L a connected alternating diagram. Therefore we have shown

Theorem[GK]. For any choice of terminals (hence any choice of a pair of shaded regions) in a connected alternating diagram L, the corresponding embedding of L in the tunnel space T[v,w] is not ambient isotopic to its mirror image.

This result is interesting because many alternating knots and links are equivalent to their mirror images in ordinary 3-space. None of this achirality can prolong to the tunnel spaces associated with the diagram. For example, the figure eight knot shown below is equivalent to its mirror image, but it is not equivalent to its mirror image in the tunnel space shown below.

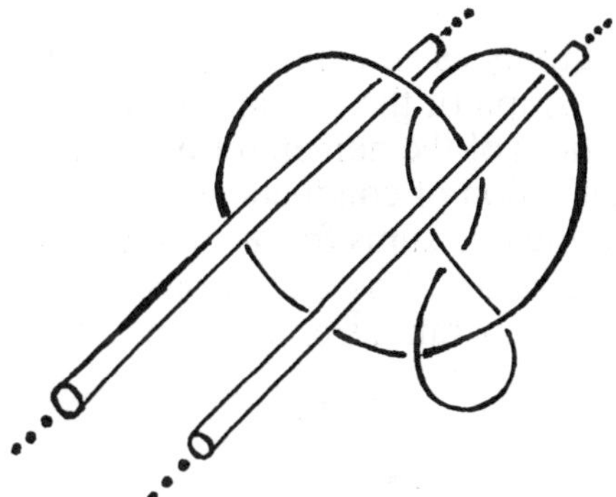

In [GK] we show how the conductance invariant can be expressed in terms of the Alexander-Conway polynomial in the special case where the terminals define a *tangle*. This case of tangles is of interest in its own right. We consider boxes with two inputs (top) and outputs (bottom) where the input and output lines are part of a weave or linkage inside the box. Equivalence of tangles is topological equivalence of these weaves restricted to motion inside the box that leaves the inputs and outputs fixed. It is natural to associate a shaded graph to a tangle as shown below.

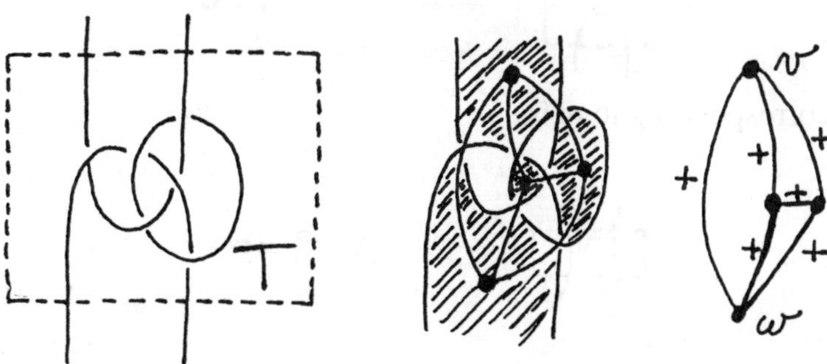

With this association we have natural choices for input and output vertices of the graph corresponding to the input and output lines of the tangle. Therefore, we can directly define the conductance of a tangle, and it is equal to a conductance for the *numerator* of the tangle where the numerator is obtained by tying the two input lines together and the two output lines together as indicated below.

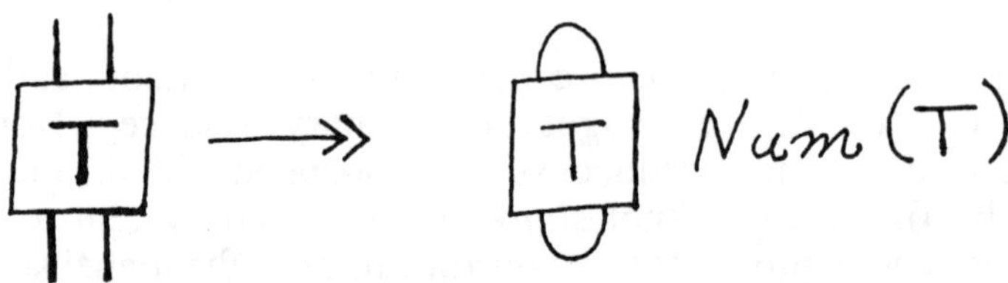

If we tie the inputs to the outputs as shown here, we get the *denominator* of the tangle. In the signed graph of the denominator of the tangle we see that the input and output vertices of the numerator have been identified with each other. Thus, by our definition of conductance, the conductance of the tangle is equal to the ratio of the tree sums for the numerator (graph) and the denominator (graph). Furthermore, the operations of series and parallel combination for graphs become operations of addition and multiplication for tangles. See below.

$$C(T) = \sum G(Num(T)) \Big/ \sum G(Den(T)).$$

With these pictures, we can look again at the properties of conductance in terms of the calculus of tangles. The tangle corresponding to 0 is

$$C = 0$$

The tangle corresponding to ∞ is

$$C = \infty$$

Series (#) and parallel (+) combination of circuits correspond to the following tangle operations — also denoted # and +. The operation # is called tangle multiplication and the operation + is called tangle addition.

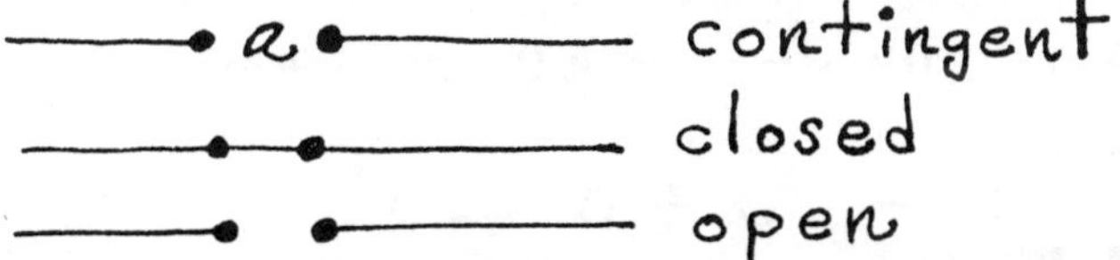

The equations $1 + (-1) = 0$ and $1 \#(-1) = \infty$ correspond to the tangle deformations

The tangle calculus is quite useful because many classes of knots and links can be built from elementary tangle operations.

From Electricity to Circuit Logic
We now turn to a relationship among elementary electricity, topology and switching logic. In studying the logic of switching circuits, one is concerned with networks of elementary switches.
An elementary switch is a circuit element with one input lead, one output lead and two states. The states of a switch are *closed* and *open*.   In the closed state the switch presents infinite conductivity. In the open state the switch presents zero conductivity. Thus we can represent the binary values for switching circuits by the values zero (0) and infinity (∞).   In this way we   view switching circuits as special cases of circuits composed of conductances that vary between 0 and ∞, or as special cases of circuits composed of arbitrary conductance values.

A switch will be denoted as shown below, with a labelled arc to indicate that no state has been chosen. A solid arc indicates that the switch is closed and the absence of an arc indicates that the switch is open.

When we refer to a switching circuit we mean one with a designated input and a designated output line. Such circuits can be combined in series and in parallel. A given circuit may, however, not be obtained by series and parallel combination from elementary switches. For example, the switching circuit below is not so obtained.

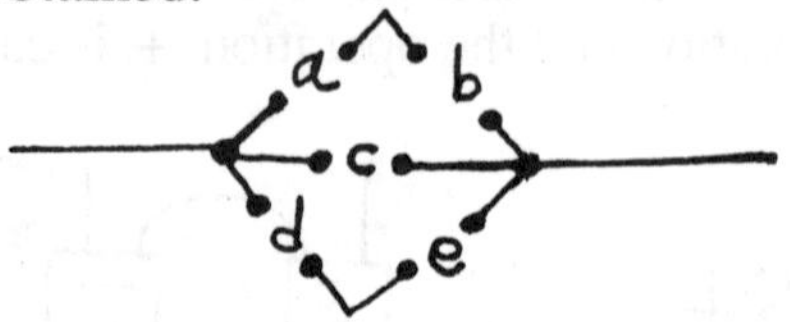

Equality of switching circuits (A=B) means that they are both open or they are both closed. This makes sense if all the switches in a and all the switches in b have been set to specific states. We shall extend this notion of equality to circuits with labelled, but undetermined switches. Once a circuit has labelled switches it is possible for more than one switch to have the same label. The convention is that this is a *ganged* switch: All occurrences of a given label are either open or they are closed. Mechanical examples of switches that control a multiplicity of contacts are quite common.   Now suppose that A and B are switching circuits with switches labelled a,b,c,... . Call the set of labels L(A&B) = {a,b,c,...}. Each label appears in A or in B or in both A and B. A *state of A&B* is a choice of values for the elements of L(A&B). For each state of A&B there is a specific choice of contacts in each circuit and hence each circuit is either open or it is closed. We say that A=B if this equality is true for each state of A&B.

For example, if A and B are as shown below, then A=B. Note that the switch c in A is ganged. With c open both A and B are closed only if a and b are closed. With c closed, both A and B are closed.

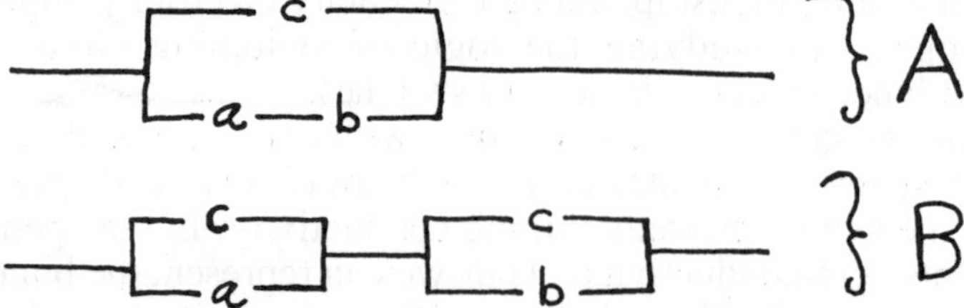

Let a+b denote parallel combination of switching circuits

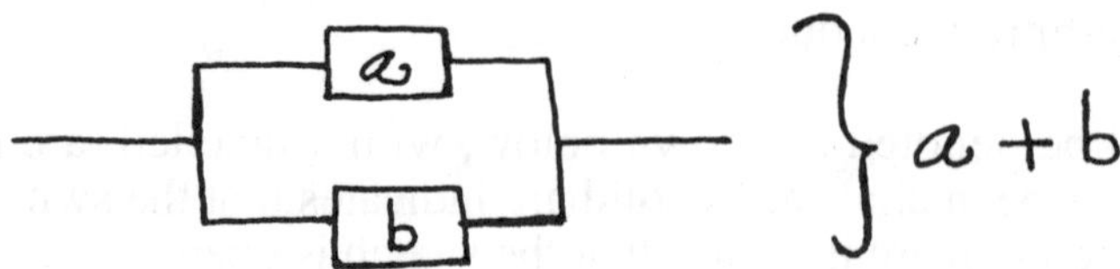

Note that $0+0 = 0, 0 + \infty = \infty = \infty + 0$ and $\infty + \infty = \infty$.

Let a#b denote series combination of switching circuits.

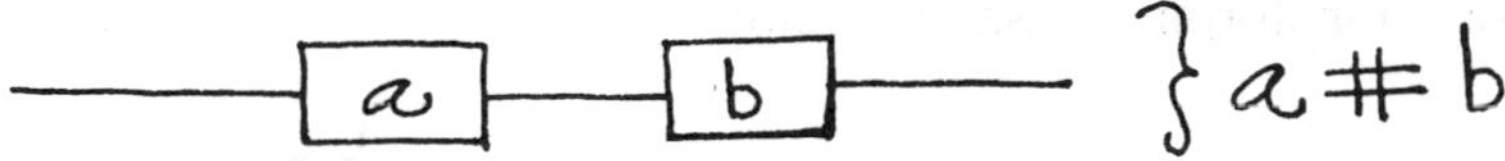

Note that 0#0 = 0, 0#∞ = 0 = ∞#0   and ∞#∞ = ∞.

$$0 \# 0 = 0$$

$$0 \# \infty = 0$$

$$\infty \# \infty = \infty$$

Since each switch has the two states (zero and infinity) , we can define a
unary operation , a -----> a', on switching circuits (with one input and one
output designated) with $0' = \infty$ and $\infty' = 0$.
The definition is:
a' is the circuit obtained by flipping all the switches in a.

It was Claude Shannon [SH] who observed that one input, one output
switching circuits have the structure (as we have just described it) of a
Boolean algebra. That is, Shannon observed that the operations + and # are
associative and commutative, that each distributes over the other: a#(b+c) =
a#b + a#c   and a+(b#c) = (a+b)#(a+c). Furthermore, (a+b)' = a' # b' ,
a+ a'=∞,   a#a' = 0,   and 0+a = a, ∞+a=∞, 0#a= 0 ∞#a=a   for any
circuits a and b.   Here we have written the underlying two-valued Boolean
arithmetic with the symbols 0 and ∞.   The infinity symbol is usually
denoted by the notation "1" in   presentations of Boolean algebra.

Shannon also proved a star-triangle relation in switching theory. The
Shannon relation is shown below.

$$b = a' \# c'$$
$$a = b' \# c'$$
$$c = a' \# b'$$

$$\left\{ \begin{array}{l} b' = a+c \\ a' = b+c \\ c' = a+b \end{array} \right\}$$

This Boolean star-triangle relation is a direct consequence of the distributive law in Boolean algebra. (Each of the operations + and # distributes over the other.)   Thus a#b + c = (a+c)#(b +c), and this equation corresponds to the network transformation shown below.

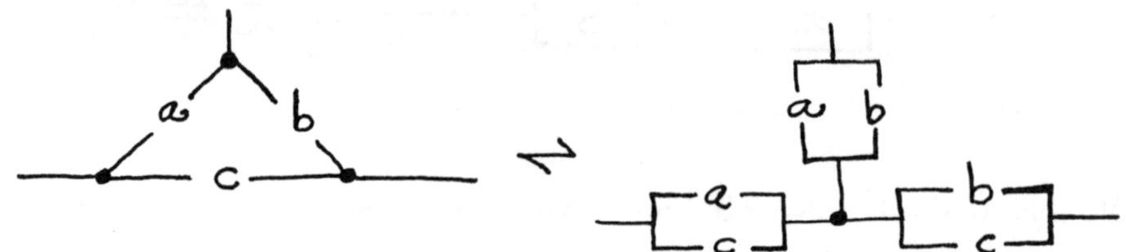

Note that the Boolean star-triangle relation is exactly the limiting case of the electrical one. One way to keep track of this is to use formal fractions involving 0 and ∞ with $\infty = 1/0$ so that $0\infty = 1$. Keep track of orders of zero and orders of infinity to the end of the calculation. In the example below, the triangle has labels $0, \infty, \infty$ and so by Boolean rules the star has labels $0 + \infty = \infty$, $\infty + \infty = \infty$ and $\infty + 0 = \infty$. Here $S = 0\infty + \infty\infty + \infty 0 = 2 + \infty^2$. Thus in the limit calculation for electricity, the star is labelled with $S/0 = S\infty = 2\infty + \infty^3 = \infty$ (final reduction of orders), $S/\infty = S0 = 2x0 + \infty\infty 0 = 0 + \infty = \infty$.

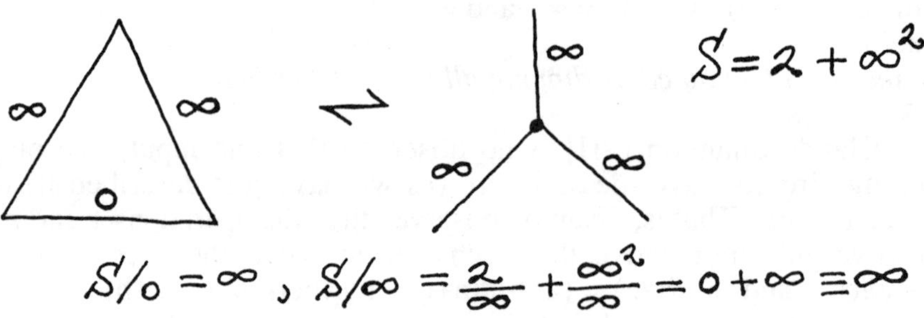

In this sense we see that the extended real numbers (including ∞ and $-\infty$) with operations + (ordinary addition) and # ($a\#b = 1/((1/a) + (1/b)) = ab/(a+b)$) form a natural extension of Boolean algebra. Of course the Boolean rules no longer hold in this larger system. The operations + and # fail to distribute over one another and the values $+1$ and -1 satisfy $x' = x$, paradoxical elements if restricted to a Boolean algebra. Note that it is exactly these paradoxical values that serve to label the two crossings in the knot theory.

Note that a universe (a link diagram with no indication of over or under crossings) can be regarded as a Boolean switching network in the plane. Each crossing is a switch with the two settings

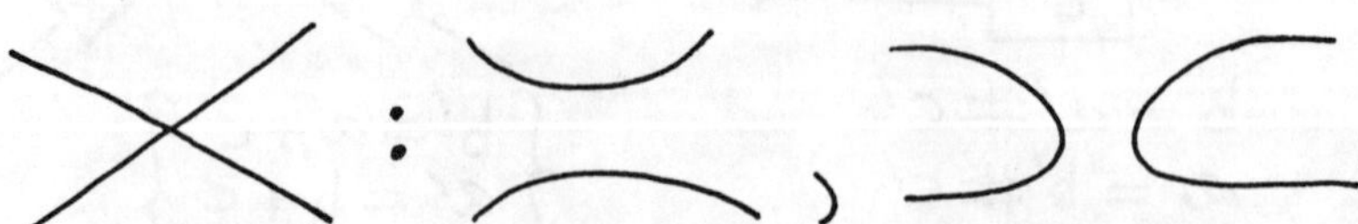

In order to make input and output specific, lets take a flattened tangle T for our Boolean net. Once the switches are set, we get a state of our tangle, and its (Boolean) value is 0 if it is possible to walk between the input lines and end up at the output lines without crossing any arcs. Otherwise the value is ∞.

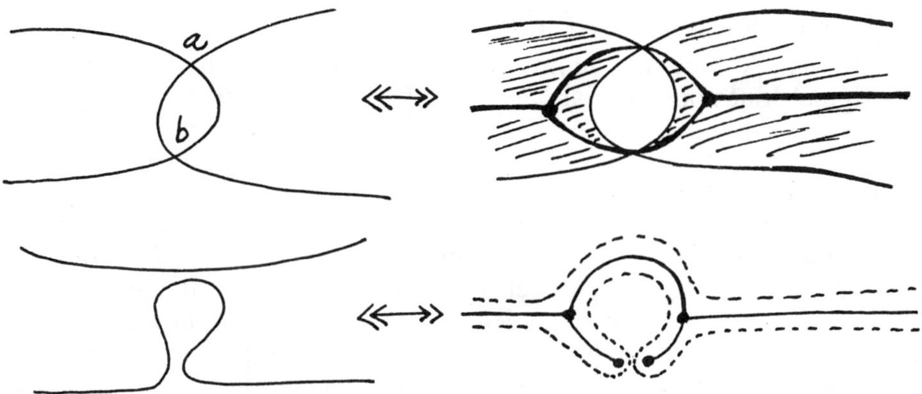

Let S be such a state for T. Let ‖ num(S)‖ denote the number of simple closed curves in the numerator of S, and let ‖den(S)‖ denote the number of simple closed curves in the denominator of S. Let $D(S) = ‖num(S)‖ - ‖den(S)‖$. Then it is easy to see that S has Boolean value 0 if and only if $D(S) = 1$ and that S has Boolean value ∞ if and only if $D(S) = -1$. Thus we have (in our formal conventions for orders of 0 and infinity) the equation $VAL(S) = 0^{D(S)}$, where $VAL(S)$ is the Boolean value of the state S.

From Circuit Logic to Electricity

This formulation for the Boolean case shows that the states of the knot theoretic switching net are in fact the states of the summation for the bracket polynomial [K2],[K3],[K4]. This suggests that the bracket polynomial at loop value zero may have something to do with conductance. This is correct, and gives the following formula in the full case of tangles T with arbitrary over and undercrossings: Let BR(K) denote the bracket polynomial for a link K evaluated at $A = \sqrt{i}$ where $i^2 = -1$. (Note that the loop value is zero in this case.) Then the conductance of a tangle T is given by the formula **C(T) = -i BR(num(T))/BR(den(T))**. This formula follows from the results in [GK].

Let K* be the mirror image of the link K (obtained by reversing all the crossings of T). Then $BR(K^*) = BR(K)^*$ where z* denotes the complex conjugate of the complex number z. It follows directly from this formula that $C(T^*) = - C(T)$, a fact that we know directly from conductance calculations.

Since the bracket evaluation is based on the expansion formula
BR(✕) = √i BR(✕) + (1/√i)BR(⊃ ⊂), we see that *the conductance of an arbitrary tangle is the ratio of weighted averages of Boolean evaluations for states of the switching network that underlie the tangle.*

[In general BR(✕) = ABR(✕) + A⁻¹BR(⊃ ⊂) with loop value equal to -A² - A⁻² defines an invariant of regular isotopy that is a version of the original Jones polynomial. See [JO], [K3], [K5], [K6].]

Abstract Tangle Calculus

We have just seen a natural evolution of definitions for the multiplication (#) and addition (+) of tangles.   In the electrical theory the analogs of these operations (series (#) and parallel (+) combination) satisfy the relation a#b = (a' + b')' where x' = 1/x. This leads to the question: Can we define an "inverse" operation on tangles T -----> T' such that for tangles S and T the equation S#T = (S' + T')' is a topological identity? The answer is that we can come very close.

The difference between the topology and the algebra is that topologically neither addition nor multiplication of tangles is commutative. Furthermore, the tangle obtained by turning a tangle upside down (exchanging inputs and outputs) is not necessarily the same as the original tangle. Lets define this operation by T -----> T^ and call T^ the *flip* of the tangle T. The flip of T is obtained by turning T around in the plane by 180 degrees.

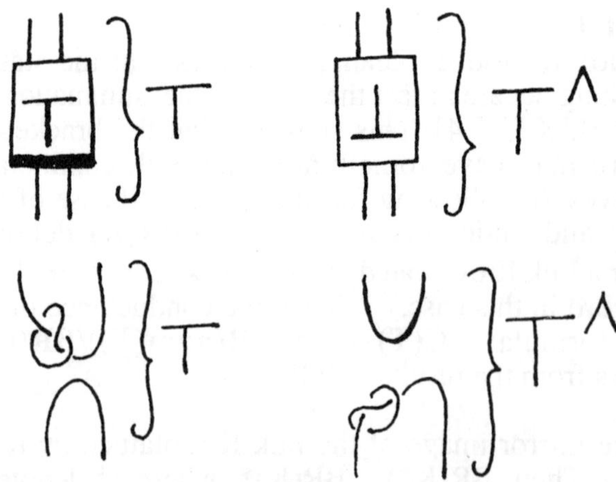

The diagrams above give an example of a tangle where T and T^ are topologically distinct. Note that we know that the conductance invariant cannot tell the difference between T and T^.

That is, $C(T) = C(T^\wedge)$ for any tangle T.

We now define the *inverse* T' of a tangle T by the formula $T' = t(T^*)$ where T* denotes the mirror image tangle obtained by reversing all the crossings of T, and t(T*) is the *twist* of T* obtained by making the left input the left output and the right output the right input, as shown below.

$$T' : \quad O' = \ \simeq \ || \ = \ \infty$$

Lemma. For any tangles S and T, the following equation is valid topologically: $(S' + T')' = (S\#T)^\wedge$. Thus for any evaluation on tangles $T \longrightarrow v(T)$ satisfying $v(A^\wedge) = v(A)$ and $v(A\#B) = v(B\#A)$ it follows that $v(\ (S' + T')') = v(S\#T)$.

Proof.

$$(S'+T')' \quad \simeq \quad (S\#T)^\wedge$$

//

Lemma. Let $C(T)$ be the conductance for tangles as defined previously in this section. Then the conductance of the inverse of a tangle is the inverse of its conductance. That is, $C(T') = 1/C(T)$.

Proof. Note that $num(T') = den(T^*)$ and $den(T') = num(T^*)$. Since $BR(K^*) =$ the complex conjugate of $BR(K)$ for any link K, the result then follows from the formula $C(T) = -i\ BR(num(T))/BR(den(T))$.
This completes the proof.//

Example.

$$T = \quad : \quad C(T) = 3.$$

$$T' = \quad \simeq \quad : \quad C(T') = \frac{1}{3}.$$

In the case of conductance we can say more about the relationship of the evaluations and the tangle calculus. Let $N(T) = BR(num(T)$ and $D(T) = BR(den(T)$ so that $C(T) = -i F(T)$ where $F(T) = N(T)/D(T)$. We shall refer to $F(T)$ as the *fraction* of the tangle T. Given fractions a/b and c/d their formal sum is $(ad+bc)/bd$. We know that $C(S + T) = C(S) + C(T)$ and hence $F(S + T) = F(S) + F(T)$. In fact, this is true formally in the sense that $N(S + T) = N(S)D(T) + D(S)N(T)$ and $D(S + T) = D(S)D(T)$. The proof (as in [C]) depends upon reducing everything to the case of the tangles $0=0/1$ and $\infty=1/0$. Here there is one case that must be mentioned: $\infty + \infty = 1/0 + 1/0 = (1.0 + 0.1)/ 0.0 = 0/0.0$.
This tells us correctly that

$$C[\infty + \infty] = C\left[\ \right) \cup \left(\ \right] = \leq[\boxed{0}]\bigg/\leq[\boxed{\infty}]$$
$$= 0/0^{2}.$$

We may still wish to regard the sum of infinity and infinity as infinity rather than 0/0. This works if one is willing to take the different powers of zero into account (as we did in the evaluation of a Boolean network). Formally, we take $0^a . 0^b = 0^{(a+b)}$ and $0^a + 0^b = 0^{\min(a,b)}$. This suggests working in a realm of formal fractions [a, b] where [a, b]+[c, d] = [ad+bc, bd] and powers of zero are handled according to the above rules.

In the tangle category, this brings us right back to the phenomenon that the powers of the zero tangle act like the Dirac bracket: Let G denote this tangle. Identify G with the formal fraction of 0/1 : G=[0,1]. Then G#G = 0G where 0 denotes a single loop and is taken to be the value zero in our conductance calculations. Since $G\#G = (G' + G')' = (\infty + \infty)^{-1} = [0,0.0]^{-1} = [0.0, 0] = 0[0, 1] = 0G$ this multiplicative phenomenon is identical to the matter of adding the infinity tangle to itself.

$$G = \bigcup_{\bigcap} \ , \quad G\#G = \overset{\cup}{\underset{\bigcap}{0}} \ , \quad (G'+G')' = \big\| O \big\|$$

Now note the following algebra. Formally write aG+b where a and b are complex numbers and GG=0G.
Then $(aG+b)(cG+d) = (acGG + adG + bcG + bd) = (ad+bc)G + bd$.
Define $v(aG+b) = a/b$. Then we have that
$v((aG+b)(cG+d)) = (a/b) + (c/d)$ so that $v(XY) = v(X) + v(Y)$.
Since $v(G) = v(G + 0) = 1/0$, we see that the rule GG=0G corresponds directly to the formal addition $1/0 + 1/0 = 0/0$.
Multiplication of the forms aG+b corresponds to formal addition of fractions a/b and c/d. In this context it is natural to define the involution $(aG+b)' = b^*G + a^*$ where z^* denotes the complex conjugate of z.

This form of calculation corresponds to the bracket model for these tangles. In the bracket model, we can expand the tangle as a formal sum of tangles using the rules

$$ \diagup\!\diagdown = \sqrt{i} \;\asymp\; \oplus \tfrac{1}{\sqrt{i}}\;)($$

Letting G denote the 0 tangle and 1 denote the ∞ tangle, we then have that any tangle S expands to a formal sum $S = aG + b$ for some complex numbers a and b.

The algebra of forms aG+b can itself be regarded as a non-standard extension of the almost-Boolean logic of G and 1 with GG =0G. Here G'=1 and 1'=G. In order to have I=aG+a* and J=a*G+a such that IJ=1 we require that $1 = IJ = (aG + a^*)(a^*G + a) = (a^2 + a^{*2})G + aa^*$. Hence $a^* = a^{-1}$ and $0 = a^2 + a^{-2}$. Thus we take $a = \sqrt{i}$. This corresponds exactly to the choice of bracket expansion that gives the conductance. I and J correspond to the two types of crossing. Here they are motivated by the desire to construct elements I and J in a quasi-Boolean algebra such that I'=I, J'=J, J and I are two views of an alternation [a,b], and IJ=1. Compare this discussion with the description of the DeMorgan algebra at the end of section 2.

Remark. The requirement IJ=1 that makes all the difference. If we had asked that IJ = 0 we can achieve this end quite simply by taking {a,b} with a and b Boolean values of 0 and 1, with {a,b}' = {b',a'} and G={0,0}, 1={1,1}, {a,b}+{c,d} = {a+c,b+d}, I={0,1}, J={1,0}. This is the DeMorgan algebra mentioned in Remark 1 of section 2. Thematically, asking for IJ=1 is to ask that the "waveforms" I and J interfere destructively. That the solutions to this go beyond a Boolean context is not surprising. That they are intimately involved in topology is remarkable.

The Topological Deformation of Logic
Finally, we must remark that having allowed GG=0G for G an analogue of a dominant Boolean value, it is now a small step to try **GG=dG** for d an arbitrary (possibly non-zero) constant. We try our (waveform) philosophy again: I=aG+b, J=bG+a.

We want the equation **IJ=1**.
Then $1 = (aG+b)(bG + a) = abGG + a^2G + b^2G + ab = (abd+a^2+b^2)G + ab$. Thus $b=a^{-1}$ and $d = -a^2 - a^{-2}$. Thus we arrive at the topological conditions for the bracket polynomial transposed into a quasi-Boolean domain. (The specific properties of tangle fraction addition are special to the value d=0.) The resulting logic is deformed with deformation parameter a.

Its idempotencey law carries a remembrance of multiplicities in the the powers of $d = -a^2 - a^{-2}$.

XI. Logic and Circuit Design – Knot Automata

In using the interlock algebra, one regards the link diagram as a circuit whose parts (the arcs in the diagram) are both carriers of circuit values and operators that process these values. This duality is the core of the interrelationship with topology. In actual applications of digital circuitry, there is usually a sharp distinction between circuit elements as operators and circuit elements as carriers of signals. One exception to this is the phenomena of inductance and capacitance where the time dependent values in components of the circuit affect the way these components process the values. The close analogy of combinatorial knot theory with a combinatorial theory of digital circuits is worth pursuing even in the absence of inductance and capacitance. The purpose of this section is to outline such a theory of digital circuits for future reference and comparison with the knot theory.

In this section we consider a class of automata that are direct abstractions of digital circuitry.   A real digital circuit instantiates this structure into hardware.   The circuits that are described in this section are a well defined class of abstract automata. They are rich enough to build real computers, hence rich enough to construct universal Turing machines.

The basic digital element is an inverter, diagrammed as shown below.

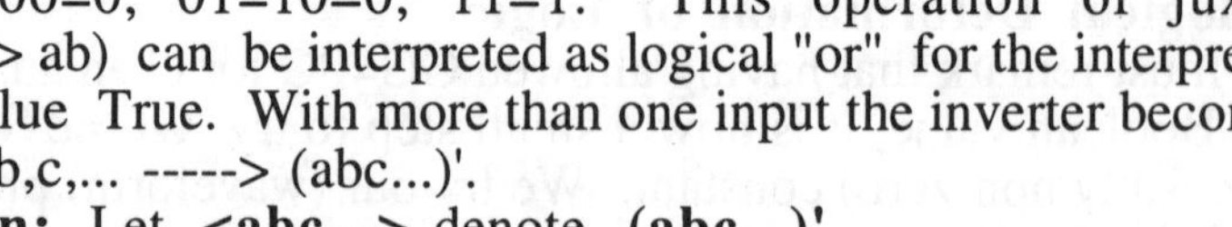

$$x \quad \rhd \quad x' = \langle x \rangle$$

Here we use two valued logic with values 0 and 1. We take $0'=1$ and $1'=0$, $00=0$, $01=10=0$, $11=1$.   This operation of juxtaposition $(a,b \text{-----> } ab)$ can be interpreted as logical "or" for the interpretation of 0 as the value True. With more than one input the inverter becomes a NOR gate: $a,b,c,... \text{-----> } (abc...)'$.

Notation: Let **<abc...>** denote **(abc...)'**.

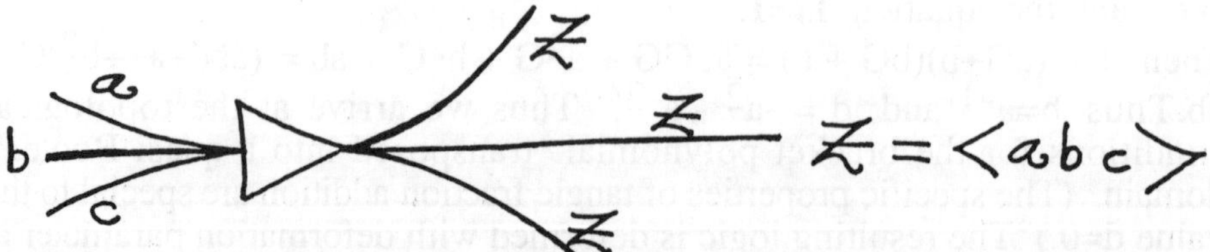

$$z = \langle a b c \rangle$$

In this convention, the value 0 is dominant among inputs to the NOR gate, since 01=0.

In a circuit diagram, a *state* is a coloring of the arcs that start from one inverter's output and terminate at another inverter's input. The colors are chosen from the set {0,1}. *All arcs emanating from a given inverter are colored identically in a given state.*   (In this model an inverter has only one output value in any given state.)
As a consequence of this stipulation we can write a single equation that describes the action of a given inverter in the circuit. Let z denote the label for the outgoing lines of the inverter. Let a,b,c,... denote the labels of its ingoing lines. Then z=(abc...)'=<abc...> (see the notational remark above) is the equation describing the action of the inverter. In a given state these equations may not be satisfied at some places in the circuit.

A state is said to be *balanced* if the equation z= <abc...> is satisfied   at every inverter in the diagram. Here z=<abc...> denotes the equation that defines the operation of the given inverter. Thus in the circuit below the balanced states are choices of values for a and b such that b=<a> and a=<b>.

This circuit has exactly two balanced states: a=0,b=1 and a=1,b=0.

If S is an unbalanced state of a circuit C, then there will be one or more equations of the form z = <abc...>   that are not satisfied by the coloring. A *transition*   consists in reassigning the value of z for the outgoing arcs z of one inverter at which there is an imbalance.   The new state achieved by the transition may or may not itself be balanced.

Example 1.   In the circuit below there are two possible transitions: a=1,b=1 -----> a=1,b=0   and a=1,b=1 -----> a=0,b=1.   The states that result from this transition are both balanced. Call this circuit a *memory*.   It has the equations **a=<b>, b=<a>**.

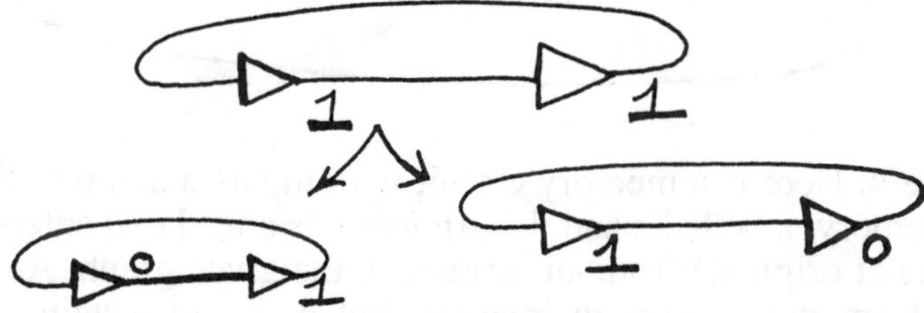

Example 2.   In the circuit below there is one possible transition a=1 -----> a=0, but the resulting state is not balanced, and its transition a=0 -----> a=1 returns the circuit to its original state.
This circuit has the equation   z=<z>, for which there are no Boolean solutions.

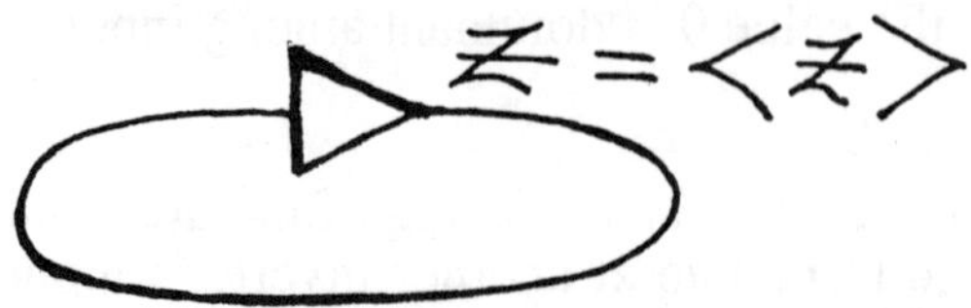

The circuit z=<z> embodies the Liar paradox. If z = 0 then z=1. If z=1, then z=0. Its behaviour is an oscillation between 0 and 1.

Circuit action   consists in a sequence of transitions from an (unbalanced) state of a given circuit. The action *terminates* when a balanced state is reached.

We are interested in designing circuits with given **behaviours**. The behaviour of a circuit consists in an appropriate summary of its circuit action — what balanced states it can achieve from a given set of unbalanced states that are relevant to the design problem. In this regard it is useful to say that a circuit action   is *determinate* if it has only one possible end state independent of the possible sequences of transitions that may lead to this end state. Thus we can ask of a given unbalanced state whether the resulting circuit action is determinate. In the first example above the action is not determinate. In the second example the action is determinate, but the set of possible balanced end-states is empty.

Example 3.   This example, a modified memory, has equations **a=<bc>,b=<a>, c=<b>**. Its only balanced state is a=1, b=0,c=1.If placed in any other state it transits to this balanced state. A sample transition is indicated below. This automaton is the abstract version of a machine that acts to turn itself off whenever it is turned on.

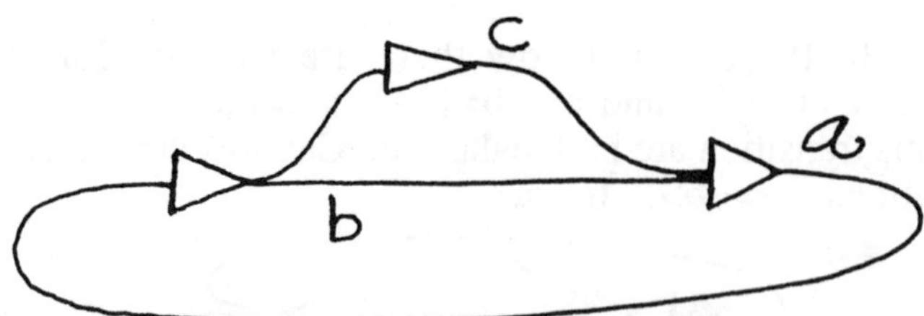

Example 4. Here is a memory circuit with inputs a and b to the two sides of the memory, labelled c and d. (An *input* is a lead that enters an inverter, but does not originate from an inverter in the given graph. An *output* is a lead that emanates from an inverter, but does not terminate at another inverter.) If we set a=0, b=1, c=1, d=1 then the circuit has a determinate transition to the end state a=0, b=1, c=0, c=1.

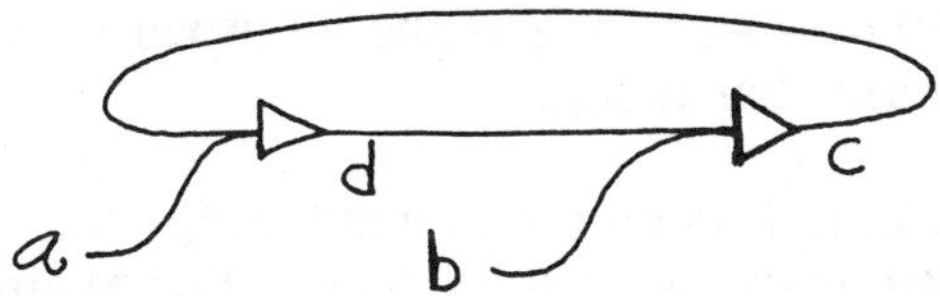

Note that input values do not change during a transition.

Example 5. The equations for this automaton, **M**, are

$$a = \langle biz \rangle$$
$$b = \langle ajz \rangle$$
$$c = \langle bd \rangle$$
$$d = \langle ac \rangle$$
$$i = \langle ad \rangle$$
$$j = \langle bc \rangle.$$

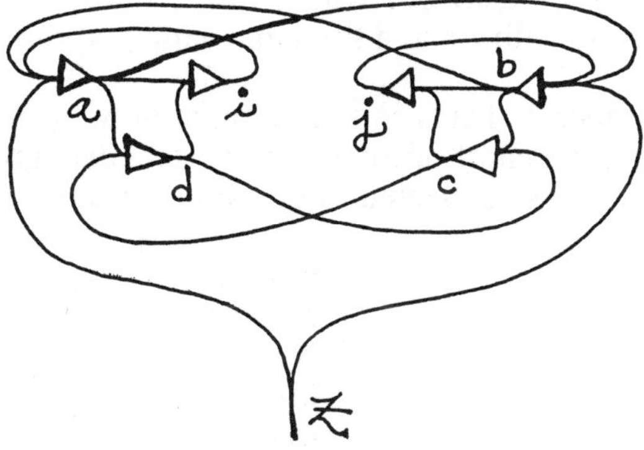

Here we regard z as an input to the system. For each value of z there are two balanced states of M. If z=0, then V = (a,b,c,d,i,j) =A or C where A= (1,1,0,1,0,1) and C= (1,1,1,0, 1,0). If z=1, thenV=B or D where and B= (1,0,1,0,1,1) and D=(0,1, 0,1, 1,1). One can then verify that for a given value of z and balanced state S, the transition that ensues upon changing z (from 0 to 1 or from 1 to zero) is determinate. The result is that the sequence of values z=0,1,0,1,0,1,... results in the the sequence of states   A,B,C,D,A,B,C,D, ... (Assuming that we start with z=0 in state A.)

As a model for action we assume that each change in z is held fixed long enough for the automaton to accomplish its transition to the next state. In terms of applications this means that the model assumes delays associated with each inverter. There are no delays associated with the connecting lines in the graph. This method of distributing the delays is a mathematical abstraction, but it is sufficiently realistic so that these circuits can actually work at the hardware level. In any given instantiation the delays are given up to the variation in the components. If the automaton is mathematically determinate (as in this example), then it will behave in the same way for any choice of actual delays — so long as the input varies more slowly than the time needed for internal balancing.

The circuit in this example converts an input oscillation   z: 010101... to internal oscillations of twice the period. For example we have in the above state sequence d:100110011001100....   By taking d as an output, we therefore obtain a black box B with input line z and output line d with this behaviour. This is exactly the behaviour needed to make circuits that count in binary. A series connection of n such black boxes produces an automaton that cycles through 2^{n+1} distinct states as the the input z oscillates between 0 and 1.

Discussion

Note the basic behaviour of our black box B. If z changes from 0 to 1 then the output d changes its value. If z changes from 1 to 0, then the output d does not change its value. Call a determinate automaton with this behaviour (or the corresponding behaviour with 0 and 1 interchanged, and also the possibility of starting with z and d the same value) a **reductor**.

Note that the number of leads in the automaton M can be read from its equations by making a chart of the inverters (labelled a,b,c,d,i,j) to which each inverter or input is connected. For our automaton **M** this chart takes the form

> **z:ab**
> **a:bdi**
> **b:acj**
> **c:dj**
> **d:ci**
> **i:a**
> **j:b**

Here each line in the chart is of the form
> **R: List of inverters to which R is connected.**

where R is either an inverter or an input (z). The number of leads (14) is the number of letters occurring after the colons in this chart.

Thus we have a notion of the complexity of a reductor in terms of the number of inverters and the number of leads. We shall say that M is of type (6,14), meaning that it has 6 inverters and 14 leads. Until recently I had thought that this design, which I discovered in 1978, was the reductor of minimal complexity. However, G. Spencer-Brown informed me in the Fall of 1992 that he has found a reductor of type (6,13) [SB-92]. It may be that (6,13) is the true minimum for this design. I conjecture this to be the case.

A more general conjecture is the following.

Conjecture: It is not possible to make a determinate (asynchronous) reductor with less than six inverters.

In this last conjecture, you are allowed to use as many leads as you please, but are requested to minimize the number of inverters.

The designs in common use such as the asynchronous JK flip flop [GI] tend to use more inverters (NOR gates or NAND gates) and more leads. The least number of inverters in a published flip flop design that I have encountered is nine. Nevertheless, it is the case that smaller working designs such as the reductor M are available, and could be used to save the number of transistors in the central processing units of digital computers by a factor of (2/3).

The most straightforward case for comparing the modes of thinking about circuit automata presented in this section with the knots discussed in the rest of the essay is to juxtapose the quandle description of a knot with the equational description of a circuit. Each structure is determined by a set of local equations that describes it interconnectedness and graphical structure. In the case of the topology of knots and links we have regarded the quandle equations as defining a possible coloring of the arcs in the knot diagram. This coloring is the analogue of a balanced state in a circuit automaton. In the topology we wanted to know that by perturbing the structure of the knot by a topological transformation (Reidemeister move) there was a natural balanced state for the new version of the knot corresponding to each balanced state of the old version. This led to an analysis of a very simple class of state transitions for the knot diagrams. In the circuit automata we do not change the structure of the network, but we do allow a great complexity of state transitions.

Knot Automata

Consider a class of circuit automata that are based on the theory of knots and links in three dimensional space. The basic circuit element for these automata has an equation of the form $z = xRy$ or $z = xLy$ with box depictions as shown below. Note the orientations on the lines.

$$x R y = z \qquad) \qquad x L y = z$$

Here R and L denote the two types of operations, depending upon left and right orientations in the plane. The circuit box for $z = xRy$ is a box with inputs y and x and outputs y and z. The box is regarded as passing without processing it, the value of y, while it transforms x to $z = xRy$ by some, as yet unspecified, rule. In this way, the action of the box is dependent upon the y value, but its action does not affect this value. It is part of the rules of the game, that the circuit diagram for such an automaton must be drawn in the plane, and that it must satisfy the following diagrammatic exchanges without affecting the balanced states of the automaton.

This means that if a given automaton has a balanced state, then all the automata obtained from it by transformations as shown will also have balanced states. By examining properties of the states of two given automata it is often possible to show that there is no sequence of transformations from one of them to the other due to differences in particularities of the states.

These structures have a topological interpretation because it is possible to associate a diagram for a knot or a link to each automaton, as shown below.

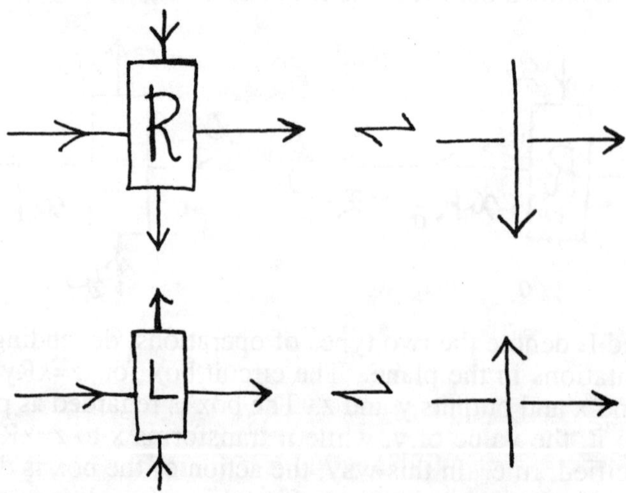

In this way the transformations that we have indicated become topological transformations of the diagrams, and these three types of transformation are known to generate all possible topological transformations of knots and links in three dimensional space. (See the discussion of the Reidemeister moves in section 3.)

Returning to the automata, the three moves translate into the demands

1. aRa = a, aLa = a
2. (aRb)Lb = a, (aLb)Rb = a
3. (aRb)Rc= (aRc)R(bRc), (aLb)Lc= (aLc)L(bLc)

The second and the third are the most significant demands, asking that the operations R and L are invertible and inverses of each other for any b, and that the operations R and L are self-distributive.The resulting algebraic structure is a quandle (See [J],[BR],[DH].)

For our purposes, the simplest example of a quandle is the structure aRb=aLb = 2b-a where a and b are elements of an additive abelian group. Thus the knottedness of the trefoil can be seen to be a consequence of using a three valued logic in the signals of an automaton associated with the diagram of the knot.

It remains to be seen how the transition behaviour of these automata is related to the topology.

XII. Pregeometry
John Wheeler coined the term pregeometry in relation to foundations of physics.

" Among all the principles that one can name out of the world of science, it is difficult to think of one more compelling than *simplicity*; and among all the simplicities of dynamics and life and movement none is starker than the binary choice yes-no or true-false. It in no way proves that this choice for a starting principle is correct, but at least it gives one some comfort in the choice that Pauli's "nonclassical two-valuedness " or "spin" so dominates the world of particle physics."

"It is one thing to have a start, a tentative construction of pregeometry: but how does one go on? ....   One suddenly realizes that a machinery for the combination of   yes-no or true-false elements does not have to be invented. It already exists. What else can pregeometry be, one asks oneself, than the calculus of propositions?"   ([MTW] pp. 1208-1209)

The diagrammatics of knots and links forms a natural domain for such a pre-geometric calculus of propositions . Links and their diagrams encode three dimensional manifolds. In this form a link is precisely a pregeometry. It is a distillation of the topological structure of a three dimensional manifold.

Knots and links form a calculus that is inherently self-referential and mutual. It is a pregeometry whose networks describe spaces and contain instructions for building these spaces. The knot and link diagrams are an intermediary domain between the realm of logical form and the geometry and topology of the perceived world.

In order to begin to understand how the diagrammatic languages for knots and links can be interpreted as pre-geometry, we must stand before these pictures with a new mind. These pictures, so redolent of images from familiar 3-dimensional space, are actually of another character entirely. They are traces of elementary action - the stroke of a hand, the movement of a brush. They are beginnings that fall back into void.   They cohere through rules we provided for them, and fall apart when we change these rules. They are a mirror of language. They are the basis of language. In the multiplicity of mathematical interpretations for these diagrams, we have traversed wide territory. Yet there are other realms prior to geometry, prior to logic, more akin to the emotions and the brush stroke of the artist. These too are in the diagrams, and the world is every bit as much constructed from such ground as the ground of reason.   It is necessary to start again and begin to draw a line ...

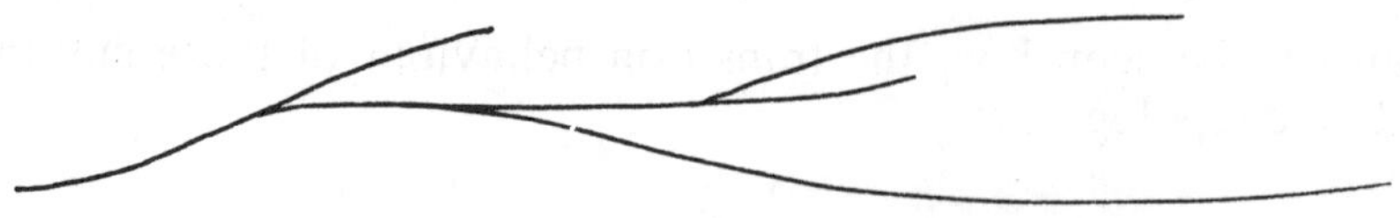

Pregeometry arose in the beginnings of things. In these beginnings, structures are unified because the distinctions that we use to tell them apart are not present. There seem to be hints of greater unifications at these points of beginning. It is here that one can start over again. In this sense, all the movements from nothing — from scientific descriptions of the creation of the universe to a writer's gropings before a blank sheet of paper — are all parts of the domain of pregeometry.

In this essay knots have been a touchstone in reconstructing logical ideas in a fusion with topology, recursion and quantum mechanics. Our attitude towards knots as pregeometry has been that of the mathematician standing before a clean blackboard and finding out what wants to be constructed. The idea of pregeometry arose in looking for a unification of gravity and relativity. Can knots be useful in that quest? Remarkably, there is a strong case for just that in the Ashtekar-Smolin-Rovelli theory of quantum gravity [Ash92],[ASH],[PUL],[SM],[SM88]. In that theory, knots take a fundamental role through the topology and geometry of the loop transform.

Quantum Gravity — The Loop Transform
We now discuss briefly the relationship of the Wilson loop <K|A> and quantum gravity as forged in the theory of Ashtekar, Rovelli and Smolin. In this theory the metric is expressed in terms of a spin connection A, and quantization involves considering wavefunctions $\Psi(A)$. Smolin and Rovelli analyze the loop transform $\Psi^\wedge(K) = \int dA\ \Psi(A)$ <K|A> where <K|A> denotes the Wilson loop for the knot or singular embedding K. Differential operators on the wavefunction can be referred, via integration by parts, to corresponding statements about the Wilson loop. It turns out that the condition that $\Psi^\wedge(K)$ be a knot invariant is equivalent to the so-called diffeomorphism constraint for these wave functions. In this way, knots and weaves and their topological invariants become a language for representing the states of quantum gravity. This effects a transformation between field theoretic and differential geometric formulations of gravity with formulations based upon functionals on loops in three dimensional space.

The key to this transition from classical gravity to quantum gravity is the movement to functions on arbitrary loops in space. In the classical mode, the Wilson loop around a very tiny loop about a point measures the curvature of the gauge field at that point. In this theory the Wilson loop around arbitrary loops contains extra information that is quantum mechanical. The constraints on the quantum theory demand that the loop functionals be topological invariants. This means that the question of size of a loop must disappear. This quantum theory does not discriminate between the macroscopic and the microscopic. In fact, it regards the entire three dimensional spatial universe as the analogue of a single particle. Size returns in the form of a mesh of measurements by loop or weave that fills the space. For a given classical metric there is an optimal weave ([ASH92], [SM88]) whose loops best approximate this metric. This means that the metrics on the space can be replaced (up to approximation) by weaves that fill the space. In this sense this theory takes to heart the old metaphors associated with the "fabric of spacetime".

Insertion, Lambda Calculus and Pregeometry

The Wilson line is the limit, over partitions of the loop K, of products of the matrices $(1 + A(x))$ where x runs over the partition. Thus one can write symbolically,

$$\langle K|A\rangle = \pi_{x\epsilon K}(1 + A(x)) = \pi_{x\epsilon K}(1 + A^a{}_k(x)T_a dx^k).$$

It is understood that a product of matrices around a closed loop connotes the trace of the product. The ordering is forced by the one dimensional nature of the loop. Insertion of a given matrix into this product at a point on the loop is then a well-defined concept. If T is a given matrix then it is understood that *$\langle K|A\rangle T$ denotes the insertion of T into some point of the loop.*

From the point of view of the discussions in this paper of lambda calculus in relation to knots, it is apparent that *the Wilson line provides the knot with the structure of a lambda operator* . In fact, within the confines of the conventions we have indicated for insertion, the notation could be prolonged to write $\lambda xyz.\langle K|A\rangle$ to indicate that insertions were to be performed at the positions x, y and z successively along the knot.

Our remarks imply the following formula for the variation of the Wilson loop with respect to the gauge field:

$$\delta\langle K|A\rangle/\delta(A^a{}_k(x)) = \lambda x.\langle K|A\rangle T_a dx^k.$$

Varying the Wilson loop with respect to the gauge field results in the insertion of an infinitesimal Lie algebra element into the loop.

Proof.

$$\delta\langle K|A\rangle/\delta(A^a{}_k(x))$$

$$= \delta\pi_{y\epsilon K}(1 + A^a{}_k(y)T_a dy^k)/\delta(A^a{}_k(x))$$

$$= [\pi_{y<x}(1+A^a{}_k(y)T_a dy^k)]\,[T_a dx^k][\pi_{y>x}(1+A^a{}_k(y)T_a dy^k)]$$

$$= \lambda x.\langle K|A\rangle T_a dx^k. \quad \textbf{QED.}$$

In practice, one tends to use this operator structure informally, with points of insertion indicated by the context. Nevertheless the lambda structure is crucial to the use of the loop transform. For example we now work out the transform of the operator $\Delta = \delta/\delta(A^a{}_k(x))$:

$$(\Delta\Psi)^{\wedge}(K) \;=\; \int dA \; (\Delta\Psi(A)) \; <K|A>$$

$$=\; \int dA \; (\delta\Psi(A)/\delta(A^a{}_k(x))) \; <K|A>$$

$$=\; -\int dA \; \Psi(A) \; (\delta<K|A>/\delta(A^a{}_k(x)))$$
(integration by parts)

$$(\Delta\Psi)^{\wedge}(K) \;=\; -\int dA \; \Psi(A) \; \lambda x.<K|A>T_a dx^k.$$

This example shows clearly how the loop transform takes differential operators on the wave functions $\Psi(A)$ and translates them into operations on Wilson loops that can be expressed in terms of this version of lambda calculus.

The knots form an underlying calculus of propositions for the the states in the Ashtekar-Smolin-Rovelli theory of quantum gravity. They are indeed a calculus of propositions forming pregeometry.

Coda on Reentry
Having pointed to knots and links as a form of pregeometry, it is necessary to ask whether this is too restrictive a point of view. It may be more appropriate to say that the domain of recursive forms and self-reference is the actual resting place of pregeometry, and that knots are a special case of this phenomena that interrelate directly with the structure of three dimensional space. In order to leave the reader with an example to ponder for this question, here is a sequence of reentry forms that we shall label $S_0, S_1, S_2, S_3, \ldots$.

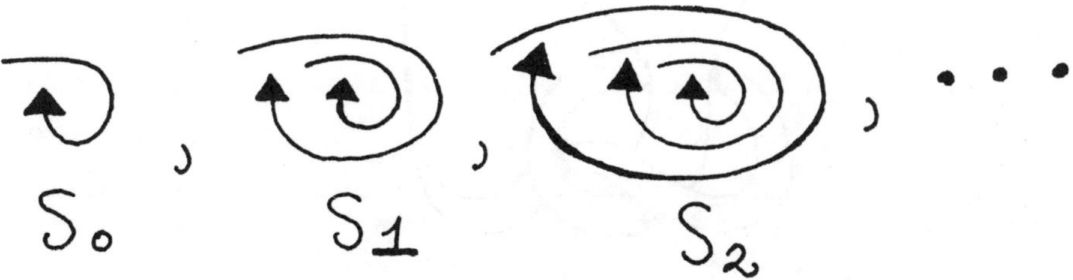

It is easy to see that the number of divisions of S_n at depth k , denoted $S_{n,k}$ is given by the formula $S_{n+1,k+1} = S_{n,k} + S_{n,k-1}$. From this it follows that $S_{n,k+1}$ is the number of divisions of Euclidean n-space by k hyperplanes in general position. The forms S_n are the representatives of the pregeometry of the Euclidean spaces. In particular, a point (dimension zero) is represented by an elementary self-reference.

By the same token, every knot is a labelling of the Fibonacci tree [K2, Chapter 6] represented by the reentry shown below.

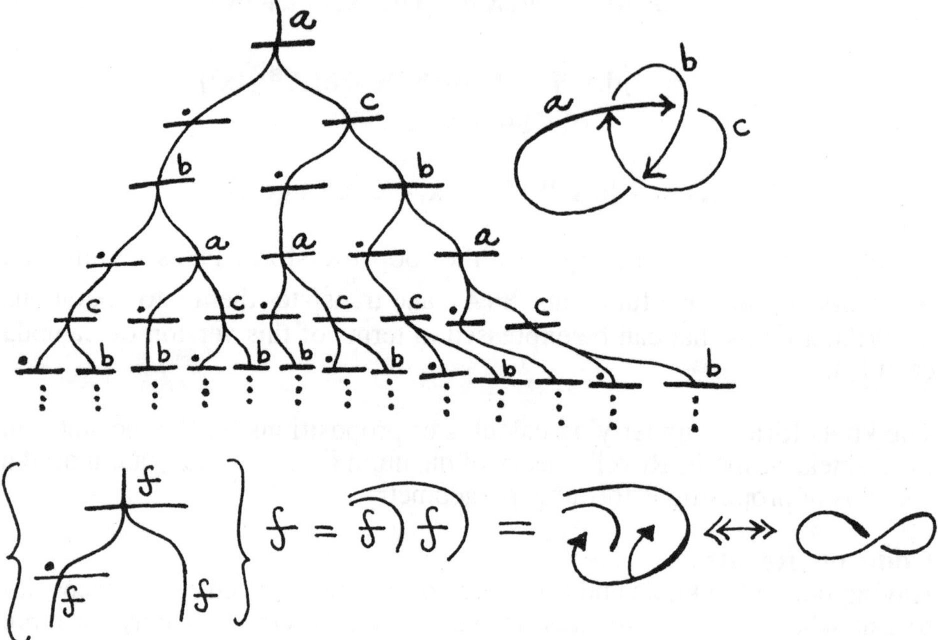

Deeper understanding of space, topology, geometry and physics is hidden in the properties of these recursive forms.

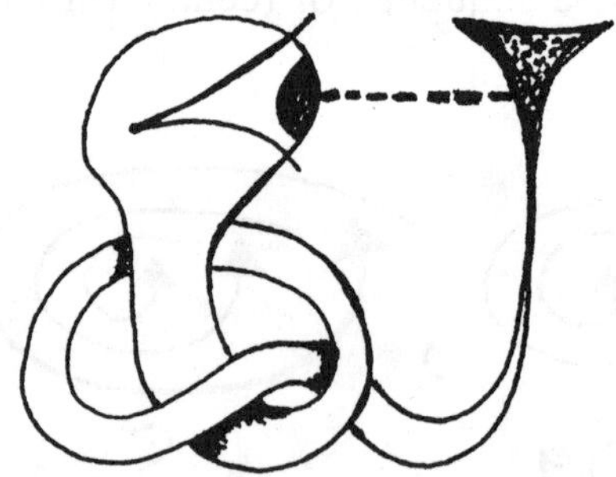

Appendix: The Bracket Polynomial

This appendix describes the construction and basic properties of the bracket polynomial [K3],[K4],[K6]. We have referred to the bracket in a number of places in this essay. The bracket polynomial is probably the simplest entry point into the study of invariants of knots and links. This invariant provides a model for the original Jones polynomial. The bracket formulation is related to models in statistical mechanics and to the Temperley Lieb algebra.

The idea is to first set up a well-defined recursive polynomial calculation on unoriented link diagrams. The calculation will depend upon three polynomial variables and we then investigate what specialization of these variables can yield invariance under the Reidemeister moves.

The recursion is based on the following idea: Given a link diagram and a crossing in that diagram, there are two ways to smooth the crossing to eliminate the crossing. See the diagram below.

We take the defining equation for the bracket calculation to be

$$< \asymp > \; = \; A< \asymp > \; + \; B< \, \supset\subset \, >$$

where the small diagrams indicate parts of larger diagrams that are identical except at the site of the crossing and its two smoothings.

Here A and B are commuting algebraic variables. Repeated performance of this calculation eventually eliminates all the crossings and demands an evaluation of collections of disjoint simple closed curves in the plane. We take the evaluation of such a collection S to be $d^{\|S\|}$ where d is a new algebraic variable commuting with A and B and $\|S\|$ is one less than the number of simple closed curves in the collection S. We can summarize this rule by the equations $< O\, K> = d<K>$, and $<O> = 1$. The first equation states that an extra curve in the link diagram multiplies the bracket by d. The second states that a single curve receives the evaluation of 1.

Thus we have the bracket defined by the axioms

$$< \asymp > \; = \; A< \asymp > \; + \; B< \, \supset\subset \, >$$

$$< O\, K> = d<K>$$

$$<O> = 1.$$

It is easy to see that this gives a well-defined 3-variable polynomial associated to any unoriented link diagram. One way to think about this calculation is to view it as a sum over "states" where a state S is a configuration of simple closed curves in the plane that is obtained from a given diagram K by replacing each of its crossings by one of the two smoothings decorated either by A or by B as shown below:

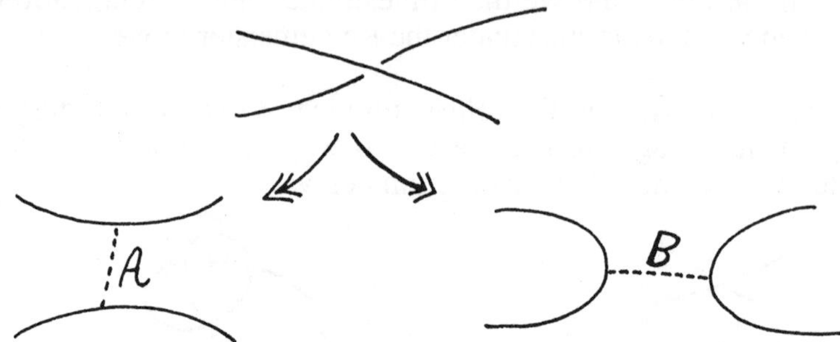

We define the A's and B's that decorate a state to be the *vertex weights* of that state, and take <K|S> to be the product of the vertex weights for the state S. We take ||S|| as defined above.

Then the bracket evaluation of K is given by the formula

$$\langle K \rangle = \sum_S \langle K|S\rangle d^{||S||}.$$

It is the summation of the product of the vertex weights times the loop evaluation over all the states of the diagram S.

This form of bracket evaluation is in direct analogy to formulas for partition functions in statistical mechanics. In fact for planar graphs, the bracket can be used to directly evaluate the partition function for the Potts model [K4] and also to evaluate chromatic polynomials for planar graphs.See [JO] and [K6] for more about the relationship of knot theory and statistical mechanics.

Returning to the topology, we see at once that

Lemma. $\langle$ ⤬ $\rangle$ = AB$\langle$ ⊃⊂ $\rangle$ + (ABd +A^2 +B^2)$\langle$ ⤢ $\rangle$

Proof.

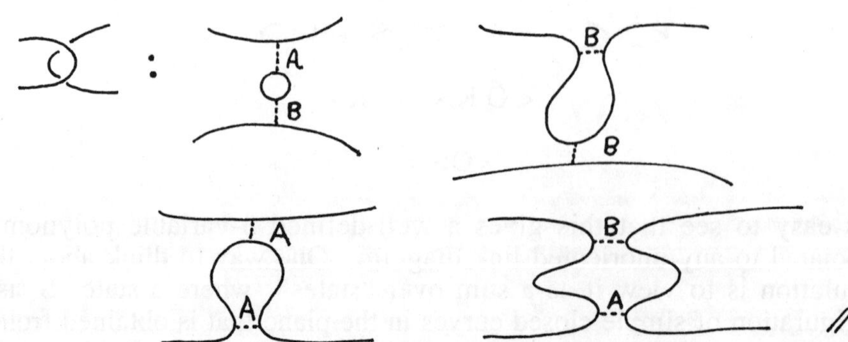

Hence we can achieve the invariance

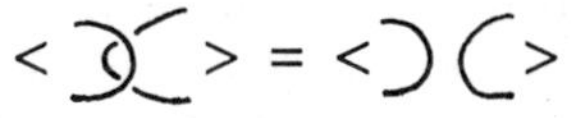

$$\langle \ \rangle = \langle \ \rangle \langle \ \rangle$$

by taking $B = A^{-1}$ and $d = -A^2 - A^{-2}$. A miracle happens, and we are granted invariance under the triangle move with no extra restrictions:

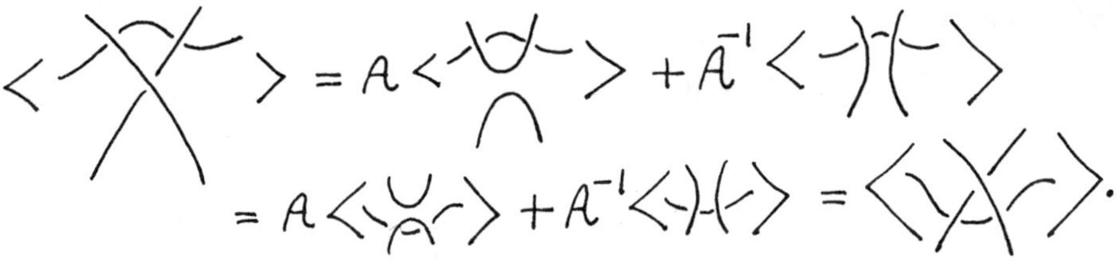

$$\langle \ \rangle = A \langle \ \rangle + A^{-1} \langle \ \rangle$$
$$= A \langle \ \rangle + A^{-1} \langle \ \rangle = \langle \ \rangle .$$

Call this invariant the bracket polynomial [K3].

Note that the bracket polynomial is not invariant under the first Reidemeister move. It should be regarded as an invariant of framed links, whose framing is expressed in the plane. We have the formulas

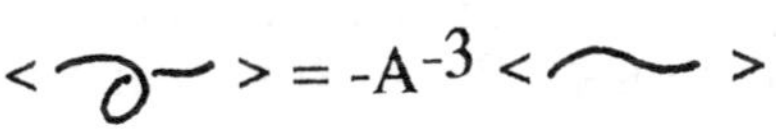

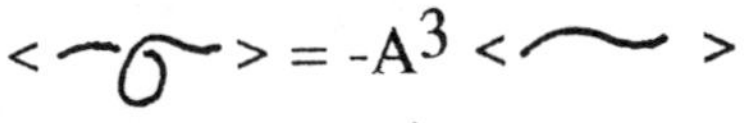

$$\langle \ \rangle = -A^{-3} \langle \ \rangle$$

$$\langle \ \rangle = -A^3 \langle \ \rangle$$

This allows normalization of the bracket by multiplication by a power of $(-A^3)$. Up to this normalization, the bracket gives a model for the original Jones polynomial [JO]. The Jones polynomial is denoted $V_K(t)$. The precise relationship with the bracket is [K3] that $V_K(t) = f_K(t^{-1/4})$ where $f_K(A) = (-A^3)^{-w(K)} \langle K \rangle (A)$ where $w(K)$ is the sum of the crossing signs of the oriented link K, and $\langle K \rangle$ is the bracket polynomial obtained by ignoring the orientation of K.

For braids, the bracket polynomial provides a representation, ρ, of the Artin braid group into the Temperley Lieb algebra with loop value $d = -A^2 - A^{-2}$.

$$\rho(\sigma_i) = A U_i + A^{-1}$$

$$\rho(\sigma_{i-1}) = A^{-1} U_i + A$$

The reader unfamiliar with this bracketology should compute the bracket polynomial for the trefoil knot and use this result to prove that the trefoil is topologically distinct from its mirror image. The mentions of the bracket polynomial in the body of the essay occur in sections 9 and 10. The Temperley Lieb algebra is introduced in section 2, remark 3. The representation of the Temperley Lieb algebra to the Artin Braid group is discussed at the end of section 7.

References

[A] Alexander, J.W. Topological invariants of knots and links. Trans. Amer. Math. Soc. 20 (1923), pp. 275-306.

[ASH92] Abhay Ashtekar, Carlo Rovelli, Lee Smolin. Weaving a classical geometry with quantum threads. (preprint 1992).

[AB] Aharonov,Y. and Bohm, D.. Significance of electromagnetic potentials in quantum theory. Phys. Rev. 115. (1959). pp. 485-491.

[AC] Aczel, P. **The Theory of Non-Well Founded Sets.** CSLI Lect. Notes #14.

[AH] Aharonov, Y. and Susskind, L. Observability of the sign change of spinors under 2π rotations. Phys. Rev. Vol. 158. No.5. June 1967, p. 1237.

[ASH] Ashtekar, A. **New Perspectives In Canonical Gravity**. Bibliopolis (1988).

[At] Atiyah, M.F. **The Geometry and Physics of Knots.** Cambridge Univ. Press. (1990).

[BaS] Barbour, J. and Smolin, L. Extremal variety as the foundation of a cosmological quantum theory. (preprint 1992).

[B] Barendregt, H. **The Lambda Calculus.** North Holland Pub. Co. (1978).

[BX] Baxter, R.J. **Exactly Solved Models in Statistical Mechanics.** Acad. Press. (1982)

[BCW] Bauer, W.R., Crick, F.H.C. and White, J.H. Supercoiled DNA. Scientific American. Vol.243. July 1980. pp. 118-133.

[BN1] Bar-Natan, D. Perturbative Chern-Simons Theory. (preprint 1990).

[BN2] Bar-Natan, D. On the Vassiliev knot invariants. (preprint 1992).

[BL] Birman, J. and Lin, X.S. Knot polynomials and Vassiliev's invariants. Invent. Math. (to appear).

[BO] Bondi, H. **Relativity and Common Sense.** Dover Pub. (1977).

[BR] Brieskorn, E. Automorphic sets and braids and singularities. Proceedings of Santa Cruz Conf. on Artin Braid Group. Contemp. Math. - AMS 78 (1988), pp. 45-115.

[BRI] Bricken, W. Boundary Logic and the Losp Parallel Deduction Engine. (manuscript 1984).

[C] Conway, J.H. An enumeration of knots and links. **Computational Problems in Abstract Algebra** (ed. J. Leech). Pergammon Press (1969). pp. 329 -358.

[DH] DeHornoy, P. Free distributive groupoids. J. Pure and Applied Algebra. 16 (1989), pp. 123-146.

[FI1] Finkelstein, D. Coherent quantum logic. Int. J. Theo. Phys., Vol 26, No. 2 (1987), pp. 109-129.

[FI2] Finkelstein, D. Algebras and manifolds: differential, difference, simplicial and quantum. Physica 18D (1986) pp 197-208.

[FI3] Finkelstein, D. First flash. (preprint).

[F] Fox, R. A quick trip through knot theory. In **Topology of 3-Manifolds.** edited by M.K.Fort, Prentice-Hall Inc. (1962). pp. 120-167.

[FR] Fenn,R.A. and Rourke,C.P.. On Kirby's Calculus of Links. Topology, 18 (1979), pp. 1-15.

[G] Gardner, M. **Knotted Doughnuts.** W.H.Freeman & Co. (1986). Chapter 5, pp. 55-67.

[GI] Gibson, J.R. Electronic Logic Circuits. Edward Arnold Pub. (1992).

[GK] Goldman,J.R. and Kauffman,L.H. Knots, tangles and electrical networks. Advances in applied mathematics **14**, 267-306 (1993).

[H1] Hellerstein. N. Diamond - A Logic of Paradox. Cybernetic - Vol.1 No.1 (1985).

[H2] Hellerstein. N. **N-Fold Logic or Paradox Island.** (manuscript 1993)

[HY] Hocking, J.G. and Young, G.S. **Topology**. Addison Wesley (1961) and reprinted by Dover Pub. (1988).

[JO] Jones, V.F.R. A Polynomial Invariant of Links via von Neumann Algebras. Bull. Amer. Math. Soc. 129 (1985), 103-112.

[JS] Joyal, A. and R. Street, R. Braided monoidal categories. Macquarie reports 86008 (1986).

[J] Joyce, D. A Classifying Invariant of Knots, the Knot Quandle. J. Pure and Appl. Alg. 23 (1983), 37-65.

[K1] Kauffman, L.H. **Formal Knot Theory**. Princeton University Press Mathematical Notes #30 (1983).

[K2] Kauffman, L.H. **On Knots**. Annals Study 115, Princeton University Press (1987).

[K3] Kauffman, L.H. State Models and the Jones Polynomial. Topology. 26 (1987), 395-407.

[K4] Kauffman, L.H. Statistical Mechanics and the Jones Polynomial. AMS Contemp. Math. Series (Proceedings of 1986 Conference on the Artin Braid Group - Santa Cruz, CA.) 78 (1989), 263-297.

[K5] Kauffman, L.H. An Invariant of Regular Isotopy. Trans. Amer. Math. Soc. 318, No.2 (1990), 417-471.

[K6] Kauffman, L.H. **Knots and Physics**. World Sci. Pub. (1991). Second Edition (1993).

[K7] Kauffman, L.H. Spin networks and knot polynomials. Intl. J. Mod. Phys. A. Vol. 5. No. 1. (1990). pp. 93-115.

[K8] Kauffman, L.H. Transformations in Special Relativity. Int. J. Theo. Phys. Vol. 24. No. 3. pp. 223-236. March 1985.

[K9] Kauffman,L.H. Knots, abstract tensors, and the Yang-Baxter equation. In **Knots, Topology and Quantum Field theories** - Proceedings of the Johns Hopkins Workshop on Current Problems in Particle Theory 13. Florence (1989). ed. by L. Lussana. World Scientific Pub. (1989). pp. 179-334.

[K10] Kauffman,L.H. Map coloring and the vector cross product. J. Comb. Theo. Ser. B. Vol. 48. No. 2. April 1990. pp. 145- 154.

[K11] Kauffman,L.H. From knots to quantum groups (and back). In **Proceedings of the CRM Workshop on Hamiltonian Systems, Transformation Groups and Spectral Transform Methods**. ed. by J.Harnad and J.E. Marsden. Les Publications CRM (1990). pp. 161-176.

[K12] Kauffman,L.H. **Map Reformulation**. Princelet Editions (1986).

[K13] Kauffman,L.H. An integral heuristic. Intl. J. Mod. Phys. A. Vol. 5. No. 7. (1990). pp. 1363-1367.

[K14] Kauffman,L.H. Special relativity and a calculus of distinctions. In **Proceedings of the 9th Annual International Meeting of the Alternative Natural Philosophy Association - Cambridge University, Cambridge, England (September 23, 1987)**. Published by ANPA West, Palo Alto, Calif. pp. 290-311.

[K15] Kauffman, L.H. Imaginary Values in Mathematical Logic. 17th International Symposium on Multiple-Valued Logic. IEEE Pub. (1987).

[K16] Kauffman,L.H. Self-reference and recursive forms. J. Soc. Bio. Strs. (1987), #10, pp. 53-72.

[K17] Kauffman, L.H. Reflections on reflexivity. (1990) (to appear).

[K18] Kauffman, L.H. Gauss Codes, Quantum Groups and Ribbon Hopf Algebras. Reviews in Mathematical Physics, Vol. 5, No. 4 (1993).

[K19] Kauffman, L.H. and Lins S. **Temperley Lieb Recoupling Theory and Invariants of 3-Manifolds**. Princeton University Press, Annals Series (1994).

[K20] Kauffman, L.H. and Magarshak, Y.B. Vassiliev knot invariants and the structure of RNA folding. (in **Knots and Applications**, ed. by L. Kauffman, World Scientific (1994)).

[K21] Kauffman, L.H. Map coloring, q-deformed spin networks, and Turaev-Viro invariants for 3-manifolds. Intl. J. Mod. Phys. B, Vol.6, Nos. 11&12 (1992), 1765-1794.

[K22] Kauffman, L.H. Spin networks, topology and discrete physics. In **Braid Group, Knot Theory and Statistical Mechanics II** (edited by Ge and Yang), World Sci. Pub. Co. (1994).

[KV] Kauffman, L.H. and Varela, F.J. Form dynamics. J. Social and Bio, Struct. (1980). Vol.3. pp. 171-206.

[K] Kelley, J. **General Topology**. Van Nostrand (1955).

108

[KH] Khovanov, M. New generalizations of braids and links. (preprint 1992).

[KI] Kirby, R. A calculus for framed links in S^3. Invent.Math.45 (1978),35-56.

[KM] Kervaire, M. and Milnor, J. Groups of homotopy spheres I, Ann. of Math. $\underline{77}$ (1963), pp. 504-537.

[KO] M. Kontsevich, M.. Graphs, homotopical algebra and low dimensional topology. (preprint 1992).

[L] Laver, R. The left distributive law and the freeness of an algebra of elementary embeddings. Advances in Math. 91 (1992), pp. 209-231.

[LI1] Lickorish, W.B.R. A representation of orientable combinatorial 3-manifolds. Ann. of Math. 76 (1962), pp. 531-540.

[LI2] Lickorish, W.B.R. The skein method for 3-manifold invariants. Journal of Knot Theory and Its Ramifications. Vol.2 No.2 (June 1993).

[MTW] Misner, C.W., Thorne, K.S. and Wheeler, J.A. **Gravitation**. W.H. Freeman and Co. (1971).

[MO] Moise, E. **Geometric Topology in Dimensions 2 and 3**. Graduate Texts in Math. $\underline{47}$. Springer-Verlag (1977).

[O] Oshins, E. **Quantum Psychology Notes. Vol. 1: A Personal Construct Notebook** (1987). (privately distributed).

[P1] Pedretti, A. (editor) **Self-Reference on the Isle of Wight** - Transcripts of the first International conference on Self-Reference - August 24 -27 (1979), Princelet Editions.

[P2] Pedretti, A. **Cybernetics of Language.** Princelet Editions (1981). **[In the I of Language.** (2nd edition of Cybernetics of Language) (to appear)].

[PI] Charles S. Peirce. **The New elements of Mathematics.** Edited by Carolyn Eisle. Houghton Publishers, The Hague - Paris (1976) [part of a published collection of the works of Charles Sanders Pierce].

[PUL] Pullin, J.. Knot theory and quantum gravity - a primer. (preprint 1993)

[R] Reidemeister, K. **Knotentheorie.** Julius Springer, Berlin (1932).

[RE]   Reshetikhin, N.Y. Quantized universal enveloping algebras, the Yang-Baxter equation and invariants of links, I and II. LOMI reprints E-4-87 and E-17-87, Steklov Institute, Leningrad, USSR.

[S]   Sumners, D.W.   Untangling DNA.   Math. Intell. Vol. 12. No. 3. (1990) pp. 71-80.

[RF] Fenn, R.A. and Rourke, C.P. Racks and Links in codimension two. Journal of Knot Theory and its Ramif. Vol. 1, No. 4 (1992). pp. 343-406.

[RZvL] Rosenfeld, I., Ziff, E. and van Loon, B. **DNA for Beginners.** Writers and Readers Pub. Inc. (1983).

[SH] Shannon, C.E. A symbolic analysis of relay and switching circuits. Trans. Am. Inst. Elec. Eng. **57** (1938), 713-723.

[S-B] Spencer-Brown, G. **Laws of Form**. George Allen and Unwin Ltd., London (1969).

[S-B(92)] Spencer-Brown, G. Private Communication

[S] Scott, D.S.   Continuous lattices.   In **Toposes, Algebraic Geometry and Logic,** Lecture Notes in Mathematics 274. Springer-Verlag, Berlin (1972). pp. 97-136.

[SM88] Smolin, L. Quantum gravity in the self-dual representation. Contemp. Math. Vol. 71 (1988), pp. 55-97.

[STA] Staley, F.J. **Void.** and **Three Neomathematical Essays.** (to appear)

[ST]   Stanford, T. Finite-type invariants of knots, links and graphs. (preprint 1992).

[SU]   Sumners, D.W. (editor). **New Scientific Applications of Geometry and Topology.** Proceedings of Symposia in Applied Mathematics. Vol. 45. American Math. Soc. (1992).
    Evolution of DNA topology: implications for its biological roles. (article by N. Cozzarelli). pp. 1-16.
    Geometry and topology of DNA and DNA -protein interactions (article by James H. White). pp. 17-38.
    Knot theory and DNA.(article by De Witt Sumners). pp. 39-72.

[V]   V. Vassiliev, V. Cohomology of knot spaces. In Theory of Singularities and Its Applications. (V.I.Arnold, ed.), Amer. Math. Soc. (1990), pp. 23-69.

[VF]   von Foerster, H. **Observing Systems**. Intersystems Publications. (1981). Notes for an epistemology for living things., p 268.

[Wa1] Walba, D. M.   Experimental studies on the hook and ladder approach to molecular knots: synthesis of a topologically chiral cyclized hook and ladder.

[Wa2] Walba, D. M. Topological stereochemistry. 9.^{1} synthesis and cutting "in half" of a molecular mobius strip   (to appear in the New J. Chem.)

[Wald] Waldhausen, F. Gruppen mit zentrum und 3-dimensionale mannigfaltigkeiten. Topology 6 (1967), 505-517.

[WhR] Whitehead, A.N. and Russell, B. **Principia Mathematica**.
Vol. 1, 2nd edition, Cambridge (1927).

[WIT] E.Witten. Quantum field theory and the Jones polynomial. Commun.Math.Phys. <u>121</u> , 351-399 (1989).

[WITT] Wittgenstein, L. **Tractatus Logico-Philosophicus.** Routledge & Kegan Paul - London and New York. (1921), (1961).

[W] Winker, S.W.   Quandles, Knot Invariants and the n-fold Branched Cover. Doctoral Thesis, University of Illinois at Chicago, Chicago, Illinois (1984).

Reprinted from Mathematical and Physical Papers, Vol. IV. Hydrodynamics and general dynamics, Cambridge Univ. Press, 1910.

Sir William Thomson, Baron Kelvin

HYDRODYNAMICS

1. ON VORTEX ATOMS.

[*Proceedings of the Royal Society of Edinburgh*, Vol. VI, pp. 94—105; reprinted in *Phil. Mag.* Vol. XXXIV, 1867, pp. 15—24.]

AFTER noticing Helmholtz's admirable discovery of the law of vortex motion in a perfect liquid—that is, in a fluid perfectly destitute of viscosity (or fluid friction)—the author said that this discovery inevitably suggests the idea that Helmholtz's rings are the only true atoms. For the only pretext seeming to justify the monstrous assumption of infinitely strong and infinitely rigid pieces of matter, the existence of which is asserted as a probable hypothesis by some of the greatest modern chemists in their rashly-worded introductory statements, is that urged by Lucretius and adopted by Newton—that it seems necessary to account for the unalterable distinguishing qualities of different kinds of matter. But Helmholtz has proved an absolutely unalterable quality in the motion of any portion of a perfect liquid in which the peculiar motion which he calls " Wirbelbewegung " has been once created. Thus any portion of a perfect liquid which has " Wirbelbewegung " has one recommendation of Lucretius's atoms —infinitely perennial specific quality. To generate or to destroy " Wirbelbewegung " in a perfect fluid can only be an act of creative power. Lucretius's atom does not explain any of the properties of matter without attributing them to the atom itself. Thus the "clash of atoms," as it has been well called, has been invoked by his modern followers to account for the elasticity of gases. Every other property of matter has similarly required an assumption of specific forces pertaining to the atom. It is as easy (and as improbable—not more so) to assume whatever specific forces may be required in any portion of matter which possesses the "Wirbelbewegung," as in a solid indivisible piece of matter; and hence the Lucretius atom has no *prima facie* advantage over the Helmholtz atom. A magnificent display of smoke-rings, which he recently had the pleasure of witnessing in Professor Tait's lecture-room, diminished by one the number of assumptions required to explain the properties of matter on the hypothesis that all bodies are composed of vortex atoms in a perfect homogeneous liquid. Two smoke-rings were frequently seen to bound obliquely from one another, shaking violently from the effects of the shock. The result was very similar to that observable in two large india-rubber rings striking one another in the air. The elasticity of each smoke-ring seemed no further from perfection than might be expected in a solid india-rubber ring of the same shape, from what we know of the viscosity of india-rubber. Of course this kinetic elasticity of form is perfect elasticity for vortex rings in a perfect liquid. It is at least as good a beginning as the "clash of atoms" to account for the elasticity of gases. Probably the beautiful investigations of D. Bernoulli, Herapath, Joule, Krönig, Clausius, and Maxwell, on the various thermodynamic properties of gases, may have all the positive assumptions they have been obliged to make, as to mutual forces between two atoms and kinetic energy acquired by individual atoms or molecules, satisfied by vortex rings, without requiring any other property in the matter whose motion composes them than inertia and incompressible occupation of space. A full mathematical investigation of the mutual action between two vortex rings of any given magnitudes and velocities passing one another in any two lines, so directed that they never come nearer one another than a large multiple of the diameter of either, is a perfectly solvable mathematical problem; and the novelty of the circumstances contemplated presents difficulties of an exciting character. Its solution will become the foundation of the proposed new kinetic theory of gases. The possibility of founding a theory of elastic solids and liquids on the dynamics of more closely-packed vortex atoms may be reasonably anticipated. It may be remarked in connexion with this anticipation, that the mere title of Rankine's paper on " Molecular Vortices," communicated to the Royal Society of Edinburgh in 1849 and 1850, was a most suggestive step in physical theory.

Diagrams and wire models were shown to the Society to illustrate knotted or knitted vortex atoms, the endless variety of which is infinitely more than sufficient to explain the varieties and allotropies of known simple bodies and their mutual affinities. It is to be remarked that two ring atoms linked together or one knotted in any manner with its ends meeting, constitute a system which, however it may be altered in shape, can never deviate from its own peculiarity of multiple continuity, it being impossible for the matter in any line of vortex motion to go through the line of any other matter in such motion or any other part of its own line. In fact, a closed line of vortex core is literally indivisible by any action resulting from vortex motion.

The author called attention to a very important property of the vortex atom, with reference to the now celebrated spectrum-analysis practically established by the discoveries and labours of Kirchhoff and Bunsen. The dynamical theory of this subject, which Professor Stokes had taught to the author of the present paper before September 1852, and which he has taught in his lectures in the University of Glasgow from that time forward, required that the ultimate constitution of simple bodies should have one or more fundamental periods of vibration, as has a stringed instrument of one or more strings, or an elastic solid consisting of one or more tuning-forks rigidly connected. To assume such a property in the Lucretius atom, is at once to give it that very flexibility and elasticity for the explanation of which, as exhibited in aggregate bodies, the atomic constitution was originally assumed. If, then, the hypothesis of atoms and vacuum imagined by Lucretius and his followers to be necessary to account for the flexibility and compressibility of tangible solids and fluids were really necessary, it would be necessary that the molecule of sodium, for instance, should be not an atom, but a group of atoms with void space between them. Such a molecule could not be strong and durable, and thus it loses the one recommendation which has given it the degree of acceptance it has had among philosophers; but, as the experiments shown to the Society illustrate, the vortex atom has perfectly definite fundamental modes of vibration, depending solely on that motion the existence of which constitutes it. The discovery of these fundamental modes forms an intensely interesting problem of pure mathematics. Even for a simple Helmholtz ring, the analytical difficulties which it presents are of a very formidable character, but certainly far from insuperable in the present state of mathematical science. The author of the present communication had not attempted, hitherto, to work it out except for an infinitely long, straight, cylindrical vortex. For this case he was working out solutions corresponding to every possible description of infinitesimal vibration, and intended to include them in a mathematical paper which he hoped soon to be able to communicate to the Royal Society. One very simple result which he could now state is the following. Let such a vortex be given with its section differing from exact circular figure by an infinitesimal harmonic deviation of order i. This *form* will travel as waves round the axis of the cylinder in the same direction as the vortex rotation, with an angular velocity equal to $(i-1)/i$ of the angular velocity of this rotation. Hence, as the number of crests in a whole circumference is equal to i, for an harmonic deviation of order i there are $i-1$ periods of vibration in the period of revolution of the vortex. For the case $i=1$ there is no vibration, and the solution expresses merely an infinitesimally displaced vortex with its circular form unchanged. The case $i=2$ corresponds to elliptic deformation of the circular section; and for it the period of vibration is, therefore, simply the period of revolution. These results are, of course, applicable to the Helmholtz ring when the diameter of the approximately circular section is small in comparison with the diameter of the ring, as it is in the smoke-rings exhibited to the Society. The lowest fundamental modes of the two kinds of transverse vibrations of a ring, such as the vibrations that were seen in the experiments, must be much graver than the elliptic vibration of section. It is probable that the vibrations which constitute the incandescence of sodium-vapour are analogous to those which the smoke-rings had exhibited; and it is therefore probable that the period of each vortex rotation of the atoms of sodium-vapour is much less than $\frac{1}{525}$ of the millionth of the millionth of a second, this being approximately the period of vibration of the yellow sodium light. Further, inasmuch as this light consists of two sets of vibrations coexistent in slightly different periods, equal approximately to the time just stated, and of as nearly as can be perceived equal intensities, the sodium atom must have two fundamental modes of vibration, having those for their respective periods, and being about equally excitable by such forces as the atom experiences in the incandescent vapour. This last condition renders it probable that the two fundamental modes concerned are approximately similar (and not merely different orders of different series chancing to concur very nearly in their periods of vibration). In an approximately circular and uniform disk of elastic solid the fundamental modes of transverse vibration, with nodal division into quadrants, fulfil both the conditions. In an approximately circular and uniform ring of elastic solid these conditions are fulfilled for the flexural vibrations in its plane, and also in its transverse vibrations perpendicular to its own plane. But the circular vortex ring, if created with one part somewhat thicker than another, would not remain so, but would experience longitudinal vibrations round its own circumference, and could not possibly have two fundamental modes of vibration similar in character and approximately equal in period. The same assertion may, it is probable*, be practically extended to any atom consisting of a single vortex ring, however involved, as illustrated by those of the models shown to the Society which consisted of only a single wire knotted in various ways. It seems, therefore, probable that the sodium atom may not consist of a single vortex line; but it may very probably consist of two approximately equal vortex rings passing through one another like two links of a chain. It is, however, quite certain that a vapour consisting of such atoms, with proper volumes and angular velocities in the two rings of each atom, would act precisely as incandescent sodium-vapour acts—that is to say, would fulfil the "spectrum test" for sodium.

The possible effect of change of temperature on the fundamental modes cannot be pronounced upon without mathematical investigation not hitherto executed; and therefore we cannot say

---

* *Note*, April 26, 1867.—The author has seen reason for believing that the sodium characteristic might be realized by a certain configuration of a single line of vortex core, to be described in the mathematical paper which he intends to communicate to the Society.

that the dynamical explanation now suggested is mathematically
demonstrated so far as to include the very approximate identity of
the periods of the vibrating particles of the incandescent vapour
with those of their corresponding fundamental modes at the lower
temperature at which the vapour exhibits its remarkable absorbing-
power for the sodium light.

A very remarkable discovery made by Helmholtz regarding
the simple vortex ring is that it always moves, relatively to the
distant parts of the fluid, in a direction perpendicular to its plane,
towards the side towards which the rotatory motion carries the
inner parts of the ring.   The determination of the velocity of this
motion, even approximately, for rings of which the sectional radius

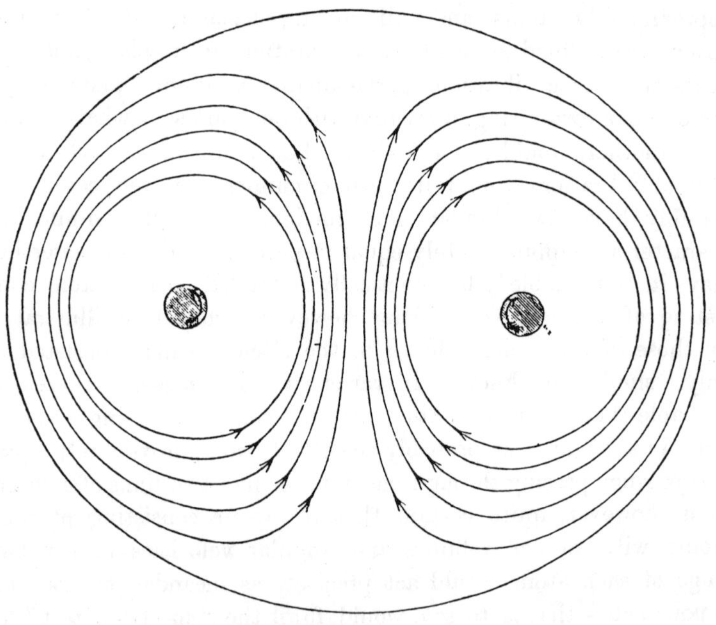

is small in comparison with the radius of the circular axis, has
presented mathematical difficulties which have not yet been over-
come*.   In the smoke-rings which have been actually observed, it
seems to be always something smaller than the velocity of the

* See, however, note added to Professor Tait's translation of Helmholtz's paper
(*Phil. Mag.* 1867, vol. xxxiii. Suppl.), where the result [see *infra*, p. 67] of a mathematical investigation which the author of the present communication has recently
succeeded in executing is given.

fluid along the straight axis through the centre of the ring; for the observer standing beside the line of motion of the ring sees, as its plane passes through the position of his eye, a convex* outline of an atmosphere of smoke in front of the ring. This convex outline indicates the bounding surface between the quantity of smoke which is carried forward with the ring in its motion and the surrounding air which yields to let it pass. It is not so easy to distinguish the corresponding convex outline behind the ring, because a confused trail of smoke is generally left in the rear. In a perfect fluid the bounding surface of the portion carried forward would necessarily be quite symmetrical on the anterior and posterior sides of the middle plane of the ring. The motion of the surrounding fluid must be precisely the same as it would be if the space within this surface were occupied by a smooth solid; but in reality the air within it is in a state of rapid motion, circulating round the circular axis of the ring with increasing velocity on the circuits nearer and nearer to the ring itself. The circumstances of the actual motion may be imagined thus:—Let a solid column of india-rubber, of circular section, with a diameter small in proportion to its length, be bent into a circle, and its two ends properly spliced together so that it may keep the circular shape when left to itself; let the aperture of the ring be closed by an infinitely thin film; let an impulsive pressure be applied all over this film, of intensity so distributed as to produce the definite motion of the fluid, specified as follows, and instantly thereafter let the film be all liquified. This motion is, in accordance with one of Helmholtz's laws, to be along those curves which would be the lines of force, if, in place of the india-rubber circle, were substituted a ring electromagnet†; and the velocities at different points

---

* The diagram represents precisely the convex outline referred to, and the lines of motion of the interior fluid carried along by the vortex, for the case of a double vortex consisting of two infinitely long, parallel, straight vortices of equal rotations in opposite directions. The curves have been drawn by Mr D. M'Farlane, from calculations which he has performed by means of the equation of the system of curves, which is

$$\frac{y^2}{a} = \frac{2x}{a} \cdot \frac{N+1}{N-1} - \left(1 + \frac{x^2}{a^2}\right), \text{ where } \log_\epsilon N = \frac{x+b}{a}.$$

The proof will be given in the mathematical paper which the author intends to communicate in a short time to the Royal Society of Edinburgh.

† That is to say, a circular conductor with a current of electricity maintained circulating through it.

8   HYDRODYNAMICS   [1

are to be in proportion to the intensities of the magnetic forces in the corresponding points of the magnetic field. The motion, as has long been known, will fulfil this definition, and will continue fulfilling it, if the initiating velocities at every point of the film perpendicular to its own plane be in proportion to the intensities of the magnetic force in the corresponding points of the magnetic field. Let now the ring be moved perpendicular to its own plane in the direction *with* the motion of the fluid through the middle of the ring, with a velocity very small in comparison with that of the fluid at the centre of the ring. A large approximately globular portion of the fluid will be carried forward with the ring. Let the velocity of the ring be increased; the volume of fluid carried forward will be diminished in every diameter, but most in the axial or fore-and-aft diameter, and its shape will thus become sensibly oblate. By increasing the velocity of the ring forward more and more, this oblateness will increase, until, instead of being wholly convex, it will be concave before and behind, round the two ends of the axis. If the forward velocity of the ring be increased until it is just equal to the velocity of the fluid through the centre of the ring, the axial section of the outline of the portion of fluid carried forward will become a lemniscate. If the ring be carried still faster forward, the portion of it carried with the india-rubber ring will be itself annular; and, relatively to the ring, the motion of the fluid will be backwards through the centre. In all cases the figure of the portion of fluid carried forward and the lines of motion will be symmetrical, both relatively to the axis and relatively to the two sides of the equatorial plane. Any one of the states of motion thus described might of course be produced either in the order described, or by first giving a velocity to the ring and then setting the fluid in motion by aid of an instantaneous film, or by applying the two initiative actions simultaneously. The whole amount of the impulse required, or, as we may call it, the effective momentum of the motion, or simply the momentum of the motion, is the sum of the integral values of the impulses on the ring and on the film required to produce one or other of the two components of the whole motion. Now it is obvious that as the diameter of the ring is very small in comparison with the diameter of the circular axis, the impulse on the ring must be very small in comparison with the impulse on the film, unless the velocity given to the ring is much greater

than that given to the central parts of the film. Hence, unless
the velocity given to the ring is so very great as to reduce the
volume of the fluid carried forward with it to something not in-
comparably greater than the volume of the solid ring itself, the
momenta of the several configurations of motions we have been
considering will exceed by but insensible quantities the momentum
when the ring is fixed. The value of this momentum is easily
found by a proper application of Green's formulæ. Thus the
actual momentum of the portion of fluid carried forward (being
the same as that of a solid of the same density moving with the
same velocity), together with an equivalent for the inertia of the
fluid yielding to let it pass, is approximately the same in all these
cases, and is equal to a Green's integral expressing the whole
initial impulse on the film. The equality of the effective momentum for different velocities of the ring is easily verified without
analysis for velocities not so great as to cause sensible deviations
from spherical figure in the portion of fluid carried forward. Thus
in every case the length of the axis of the portion of the fluid
carried forward is determined by finding the point in the axis of
the ring at which the velocity is equal to the velocity of the ring.
At great distances from the plane of the ring that velocity varies,
as does the magnetic force of an infinitesimal magnet on a point
in its axis, inversely as the cube of the distance from the centre.
Hence the cube of the radius of the approximately globular
portion carried forward is in simple inverse proportion to the
velocity of the ring, and therefore its momentum is constant for
different velocities of the ring. To this must be added, as was
proved by Poisson, a quantity equal to half its own amount, as an
equivalent for the inertia of the external fluid; and the sum is
the whole effective momentum of the motion. Hence we see not
only that the whole effective momentum is independent of the
velocity of the ring, but that its amount is the same as the
magnetic moment in the corresponding ring electromagnet. The
same result is of course obtained by the Green's integral referred
to above.

The synthetical method just explained is not confined to the
case of a single circular ring specially referred to, but is equally
applicable to a number of rings of any form, detached from one
another, or linked through one another in any way, or to a single
line knotted to any degree and quality of "multiple continuity,"

and joined continuously so as to have no end. In every possible such case the motion of the fluid at every point, whether of the vortex core or of the fluid filling all space round it, is perfectly determined by Helmholtz's formulæ when the shape of the core is given. And the synthetic investigation now explained proves that the effective momentum of the whole fluid motion agrees in magnitude and direction with the magnetic moment of the corresponding electromagnet. Hence, still considering for simplicity only an infinitely thin line of core, let this line be projected on each of three planes at right angles to one another. The areas of the plane circuit thus obtained (to be reckoned according to De Morgan's rule when autotomic, as they will generally be) are the components of momentum perpendicular to these three planes. The verification of this result will be a good exercise on "multiple continuity." The author is not yet sufficiently acquainted with Riemann's remarkable researches on this branch of analytical geometry to know whether or not all the kinds of "multiple continuity" now suggested are included in his classification and nomenclature.

That part of the synthetical investigation in which a thin solid wire ring is supposed to be moving in any direction through a fluid with the free vortex motion previously excited in it, requires the diameter of the wire at every point to be infinitely small in comparison with the radius of curvature of its axis and with the distance of the nearest of any other part of the circuit from that point of the wire. But when the effective moment of the whole fluid motion has been found for a vortex with infinitely thin core, we may suppose any number of such vortices, however near one another, to be excited simultaneously; and the whole effective momentum in magnitude and direction will be the resultant of the momenta of the different component vortices each estimated separately. Hence we have the remarkable proposition that the effective momentum of any possible motion in an infinite incompressible fluid agrees in direction and magnitude with the magnetic moment of the corresponding electromagnet in Helmholtz's theory. The author hopes to give the mathematical formulæ expressing and proving this statement in the more detailed paper, which he expects soon to be able to lay before the Royal Society.

The question early occurs to any one either observing the phenomena of smoke-rings or investigating the theory,—What conditions determine the size of the ring in any case? Helmholtz's investigation proves that the angular vortex velocity of the core varies directly as its length, or inversely as its sectional area. Hence the strength of the electric current in the electromagnet, corresponding to an infinitely thin vortex core, remains constant, however much its length may be altered in the course of the transformations which it experiences by the motion of the fluid. Hence it is obvious that the larger the diameter of the ring for the same volume and strength of vortex motions in an ordinary Helmholtz ring, the greater is the whole kinetic energy of the fluid, and the greater is the momentum; and we therefore see that the dimensions of a Helmholtz ring are determinate when the volume and strength of the vortex motion are given, and, besides, either the kinetic energy or the momentum of the whole fluid motion due to it. Hence if, after any number of collisions or influences, a Helmholtz ring escapes to a great distance from others and is then free, or nearly free, from vibrations, its diameter will have been increased or diminished according as it has taken energy from, or given energy to, the others. A full theory of the swelling of vortex atoms by elevation of temperature is to be worked out from this principle.

Professor Tait's plan of exhibiting smoke-rings is as follows:— A large rectangular box, open at one side, has a circular hole of 6 or 8 inches diameter cut in the opposite side. A common rough packing-box of 2 feet cube, or thereabout, will answer the purpose very well. The open side of the box is closed by a stout towel or piece of cloth, or by a sheet of india-rubber stretched across it. A blow on this flexible side causes a circular vortex ring to shoot out from the hole on the other side. The vortex rings thus generated are visible if the box is filled with smoke. One of the most convenient ways of doing this is to use two retorts with their necks thrust into holes made for the purpose in one of the sides of the box. A small quantity of muriatic acid is put into one of these retorts, and of strong liquid ammonia into the other. By a spirit-lamp applied from time to time to one or other of these retorts, a thick cloud of sal-ammoniac is readily maintained in the inside of the box. A curious and interesting experiment may be made with two boxes thus arranged, and placed either side by side close to one another or facing one another so as to project smoke-rings meeting from opposite directions—or in various relative positions, so as to give smoke-rings proceeding in paths inclined to one another at any angle, and passing one another at various distances. An interesting variation of the experiment may be made by using clear air without smoke in one of the boxes. The invisible vortex rings projected from it render their existence startlingly sensible when they come near any of the smoke-rings proceeding from the other box.

1868] (13)

2. On Vortex Motion.

[*Transactions of the Royal Society of Edinburgh*, Vol. xxv. 1869,
pp. 217—260. Read 29th April, 1867.]

(§§ 1—59 recast and augmented 28th August to 12th November, 1868.)

1. THE mathematical work of the present paper has been performed to illustrate the hypothesis, that space is continuously occupied by an incompressible frictionless liquid acted on by no force, and that material phenomena of every kind depend solely on motions created in this liquid. But I take, in the first place, as subject of investigation, a finite mass of incompressible friction-less* fluid completely enclosed in a rigid fixed boundary.

2. The containing vessel may be either *simply* or *multiply continuous*†. And I shall frequently consider solids surrounded by the liquid, which also may be either simply or multiply continuous. It will not be necessary to exclude the supposition that any such solid may touch the outer boundary over some finite area, in which case it is *not* surrounded by the liquid; but each such solid, whether surrounded by the liquid or not, and whether moveable or fixed, must be considered as a part of the whole boundary of the liquid.

3. Let the whole fluid be given at rest, and let no force, except pressure from the containing vessel, or from the surfaces of solids immersed in it, ever act on any part of it. Let there be any number of solids, perfectly incompressible, and of the same density as the fluid; but either perfectly rigid, or more or less

* A frictionless fluid is defined as a mass continuously occupying space, whose contiguous portions press on one another everywhere exactly in the direction perpendicular to the surface separating them.

† Helmholtz—*Ueber Integrale der hydrodynamischen Gleichungen, welche den Wirbelbewegungen entsprechen*, Crelle (1858); translated by Tait in *Phil. Mag.* 1867, I. Riemann—*Lehrsätze aus der Analysis situs, &c.*, Crelle (1857). See also § 58, below.

flexible, with perfect or imperfect elasticity. Some of these may at times be supposed to lose rigidity, and become perfectly liquid; and portions of the liquid may be supposed to acquire rigidity, and thus to constitute solids. Let the solids act on one another with any forces, pressures, frictions, or mutual distant actions, subject only to the law of "action and reaction." Let motions originate among them, and in the liquid, either by the natural mutual actions of the solids or by the arbitrary application of forces to them during some limited time. It is of no consequence to us whether these forces have reactions on matter outside the containing vessel, so that they might be called "natural forces" in the present state of science (which admits action and reaction at a distance); or are applied arbitrarily by supernatural action without reaction. To avoid circumlocution, and, at the same time, to conform to a common usage, we shall call them *impressed forces*.

4. From the homogeneousness as to density of the contents of the fixed bounding vessel, it follows that the centre of inertia of the whole system of liquid and solids immersed in it remains at rest; in other words, the integral momentum of the motion is zero. Hence (Thomson and Tait's *Natural Philosophy*, § 297) the time integral of the sum of the components of *pressure on the containing vessel*, parallel to any fixed line, is equal to the time-integral of the sum of the components of *impressed forces* parallel to the same line. This equality exists, of course, at each instant during the action of the impressed forces, and continues to exist for the constant values of their time integrals, after they have ceased. Thus, in the subsequent motion of the solids, and of the fluids compelled to yield to them, whatever pressure may come to act on the containing vessel, whether from the fluid or from some of the solids coming in contact with it, the components of this pressure, parallel to any fixed line, summed for every element of the inner surface of the vessel, must vanish for every interval of time during which no impressed forces act. If, for example, one of the solids strikes the containing vessel, there will be an impulsive pressure of the fluid over all the rest of the fixed containing surface, having the sum of its components parallel to any line, equal and contrary* to the corresponding component of the

---

* I shall use the word *contrary* to designate merely directional opposition; and reserve the unqualified word *opposite*, to signify *contrary and in one line*.

impulsive pressure of the solid on the part of this surface which it strikes [see § 8, and consider oblique impulse of an inner moving solid, on the fixed solid spherical boundary]. *But, after the impressed forces cease to act, and as long as the containing vessel is not touched by any of the solids, the integral amount of the component of fluid pressure on it, parallel to any line, vanishes.*

5. If now forces be applied to stop the whole motion of fluid and solids [as (§ 62) is done, if the solids are brought to rest by forces applied to themselves only], the time integrals of the sums of the components of these forces, parallel to any stated lines, *may or may not in general be equal and contrary* to the time integrals of the corresponding sums of components of the initiating impressed forces (§ 3). But we shall see (§§ 19, 21) that *if the containing vessel be infinitely large, and all of the moving solids be infinitely distant from it during the whole motion,* there must be not merely the equality in question between the time integrals of the components in contrary directions of the initiating and stopping impressed forces, but there must be (§ 21) *completely equilibrating opposition between the two systems.*

6. To avoid circumlocution, henceforth I shall use the unqualified term *impulse* to signify a system of impulsive forces, to be dealt with as if acting on a rigid body. Thus the most general impulse may be reduced to an impulsive force, and couple in plane perpendicular to it, according to Poinsot; or to two impulsive forces in lines not meeting, according to his predecessors. Further, I shall designate by *the impulse of the motion at any instant,* in our present subject, the system of impulsive forces on the moveable solids which would generate it from rest; or any other system which would be equivalent to that one if the solids were all rigid and rigidly connected with one another, as, for instance, the Poinsot resultant impulsive force and minimum couple. The line of this resultant impulsive force will be called the *resultant axis of the motion,* and the moment of the minimum couple (whose plane is perpendicular to this line) will be called the *rotational moment* of the motion.

7. But, having thus defined the terms I intend to use, I must, to warn against errors that might be fallen into, remark that the momentum of the whole motions of solids and liquid is *not equal* to what I have defined as *the impulse,* but (§ 4) is equal

to zero; being the force-resultant of "the impulse" and the impulsive pressure exerted on the liquid by the containing vessel during the generation of the motion: and that the moment of momentum of the whole motion round the centre of inertia of the contents of the vessel is *not* equal to the *rotational moment*, as I have defined it, but is equal to the moment of the couple constituted by "the impulse" and the impulsive pressure of the containing vessel on the liquid. It must be borne in mind that however large, and however distant all round from the moveable solids, the containing vessel may be, it exercises a finite influence on the momentum and moment of momentum of the whole motion within it. But if it is infinitely large, and infinitely distant all round from the solids, it does so by infinitely slow motion through an infinitely large mass of fluid, and exercises no finite influence on the finite motion of the solids or of the neighbouring fluid. This will be readily understood, if for an instant we suppose the rigid containing vessel to be not fixed, but quite free to move as a rigid body without mass. The momentum of the whole motion will then be not zero, but exactly equal to the force-resultant of the impulse on the solids; and the moment of momentum of the whole motion round the centre of inertia will be precisely equal to the resultant impulsive couple found by transposing the constituent impulsive forces to this point after the manner of Poinsot. But the finite motion of the immersed solids, and of the fluid in their neighbourhood which we shall call the *field of motion*, will not be altered by any finite difference, whether the containing vessel be held fixed or left free, provided it be infinitely distant from them all round. It is, therefore, essentially indifferent whether we keep it fixed or let it be free. The former supposition is more convenient in some respects, the latter in others; but it would be inconvenient to leave any ambiguity, and I shall adhere (§ 1) to the former in all that follows.

8. To further illustrate the impulse of the motion, and its resultant impulsive force and couple, according to the previous definitions, as distinguished from the momentum, and the moment of momentum, of the whole contents of the vessel, let the vessel be spherical. Its impulsive pressure on the liquid will always be reducible to a single resultant in a line through its centre, which (§ 4) will be equal and contrary to the force-resultant of "the

impulse"; and, therefore, with it will constitute in general a couple. The resultant, of this couple and the couple-resultant of the impulse, will be equal to the moment of momentum of the whole motion round the centre of the sphere (which is the centre of inertia). But if the vessel be infinitely large, and infinitely distant all round from the moveable solids, the moment of momentum of the whole motion is irrelevant; and what is essentially important, is the impulse and its force and couple-resultants, as defined above.

9. The following way of stating (§§ 10, 12), and proving (§§ 11—15), a fundamental proposition in fluid motion will be useful to us for the theory of the impulse, whether of the moveable solids we have hitherto considered or of vortices.

10. The moment of momentum of every spherical portion of a liquid mass in motion, relatively to the centre of the sphere, is always zero, if it is so at any one instant for every spherical portion of the same mass.

11. To prove this, it is first to be remarked, that the moment of momentum of that part of the liquid which at any instant occupies a certain fixed spherical space can experience no change, at that instant (or its *rate* of change vanishes at that instant), because the fluid pressure on it (§ 1), being perpendicular to its surface, is everywhere precisely towards its centre. Hence, if the moment of momentum of the matter in the fixed spherical space varies, it must be by the moment of momentum of the matter which enters it not balancing exactly that of the matter which leaves it. We shall see later (§§ 20, 17, 18) that this balancing is vitiated by the entry of either a moving solid, or of some of the liquid, if any there is, of which spherical portions possess moment of momentum, into the fixed spherical space; but it is perfect under the condition of § 10, as will be proved in § 15.

12. First, I shall prove the following purely mathematical lemmas; using the ordinary notation u, v, w for the components of fluid velocity at any point (x, y, z).

Lemma (1). The condition (last clause) of § 10 requires that $u\,dx + v\,dy + w\,dz$ be a complete differential*, at whatever instant and through whatever part of the fluid the condition holds.

---

* This proposition was, I believe, first proved by Stokes in his paper "On the Friction of Fluids in Motion, and the Equilibrium and Motion of Elastic Solids," *Cambridge Philosophical Transactions*, 14th April, 1845.

Lemma (2). If $u\,dx + v\,dy + w\,dz$ be a complete differential of a single valued function of x, y, z, through any finite space of the fluid, at any instant, the condition of § 10 holds through that space at that instant.

13. The following is Stokes' proof of Lemma (1):—First, for any motion whatever, whether subject to the condition of § 10 or not, let L be the component moment of momentum round OX of an infinitesimal sphere with its centre at O. Denoting by $\iiint$ integration through this space we have

$$L = \iiint (wy - vz)\,dx\,dy\,dz \quad\ldots\ldots\ldots\ldots\ldots(1).$$

Now let $(dw/dx)_0$, $(dw/dy)_0$, &c. denote the values at O of the differential coefficients. We have, by Maclaurin's theorem,

$$w = x \left(\frac{dw}{dx}\right)_0 + y \left(\frac{dw}{dy}\right)_0 + z \left(\frac{dw}{dz}\right)_0,$$

and so for v. Hence, remembering that $(dw/dx)_0$, &c. are constants for the space through which the integration is performed, we have

$$\iiint dx\,dy\,dz\,wy$$

$$= \left(\frac{dw}{dx}\right)_0 \iiint xy\,dx\,dy\,dz + \left(\frac{dw}{dy}\right)_0 \iiint y^2\,dx\,dy\,dz + \left(\frac{dw}{dz}\right)_0 \iiint zy\,dx\,dy\,dz.$$

The first and third of the triple integrals vanish, because every diameter of a homogeneous sphere is a principal axis; and if A denote moment of momentum of the spherical volume round its centre, we have for the second

$$\iiint y^2\,dx\,dy\,dz = \tfrac{1}{2}A.$$

Dealing similarly with vz in the expression for L, we find

$$L = \tfrac{1}{2}A \left[\left(\frac{dw}{dy}\right)_0 - \left(\frac{dv}{dz}\right)_0\right] \quad\ldots\ldots\ldots\ldots\ldots(2).$$

But L must be zero according to the condition of § 10; and, therefore, as the centre of the infinitesimal sphere now considered may be taken at any point of space through which this condition holds at any instant, we must have, throughout that space,

and similarly

$$\left.\begin{array}{l} \dfrac{dw}{dy} - \dfrac{dv}{dz} = 0 \\[2mm] \dfrac{du}{dz} - \dfrac{dw}{dx} = 0 \\[2mm] \dfrac{dv}{dx} - \dfrac{du}{dy} = 0 \end{array}\right\} \quad\ldots\ldots\ldots\ldots\ldots(3);$$

which proves Lemma (1).

14.   To prove Lemma (2), let

$$u = \frac{d\phi}{dx}, \quad v = \frac{d\phi}{dy}, \quad w = \frac{d\phi}{dz} \quad \ldots\ldots\ldots\ldots\ldots(4);$$

and let L denote the component moment of momentum round OX, through any spherical space with O in centre. We have [(1) of § 13],

$$L = \iiint dx\,dy\,dz\,(wy - vz) \quad \ldots\ldots\ldots\ldots\ldots(5),$$

$\iiint$ denoting integration through this space (not now infinitesimal). But by (4)

$$yw - vz = \left(y\,\frac{d}{dz} - z\,\frac{d}{dy}\right)\phi = \frac{d\phi}{d\psi} \ldots\ldots\ldots\ldots\ldots(6);$$

if $d/d\psi$ denote differentiation with reference to ψ, in the system of co-ordinates x, ρ, ψ, such that

$$y = \rho\cos\psi, \quad z = \rho\sin\psi \quad \ldots\ldots\ldots\ldots\ldots(7).$$

Hence, transforming (5) to this system of co-ordinates, we have

$$L = \iiint dx\,d\rho\,\rho\,d\psi\,\frac{d\phi}{d\psi} \quad \ldots\ldots\ldots\ldots\ldots(8).$$

Now, as the whole space is spherical, with the origin of co-ordinates in its centre, we may divide it into infinitesimal circular rings with OX for axis, having each for normal section an infinitesimal rectangle with dx and $d\rho$ for sides. Integrating first through one of these rings, we have

$$dx\,d\rho\,\rho \int_0^{2\pi} \frac{d\phi}{d\psi}\,d\psi,$$

which vanishes, because ϕ is a single-valued function of the co-ordinates. Hence $L = 0$, which proves Lemma (2).

15.   Returning now to the dynamical proposition, stated at the conclusion of § 11; for the promised proof, let R denote the radial component velocity of the fluid across any element, $d\sigma$, of the spherical surface, situated at (x, y, z); and let u, v, w be the three components of the resultant velocity at this point; so that

$$R = u\,\frac{x}{r} + v\,\frac{y}{r} + w\,\frac{z}{r} \quad \ldots\ldots\ldots\ldots\ldots(9).$$

The volume of fluid leaving the hollow spherical space across $d\sigma$ in an infinitesimal time dt is $Rd\sigma\,.\,dt$, and the moment of momentum of this moving mass round the centre has, for component round OX,

$$(wy - vz)\, Rd\sigma\, dt.$$

Hence, if L denote the component of the moment of momentum of the whole mass within the spherical surface at any instant, t, we have ($\S$ 11),

$$\frac{dL}{dt} = \iint (wy - vz)\, R\, d\sigma \dotfill (10).$$

Now, using Lemma (1) of $\S$ 12, and the notation of $\S$ 14, we have

$$wy - vz = \frac{d\phi}{d\psi},$$

and, by (9),

$$R = \frac{d\phi}{dr},$$

where d/dr denotes rate of variation per unit length perpendicular to the spherical surface, that is differentiation with reference to r, the other two co-ordinates being directional relatively to the centre. Hence, using ordinary polar co-ordinates, r, θ, ψ, we have

$$\frac{dL}{dt} = r^2 \iint \frac{d\phi}{dr}\frac{d\phi}{d\psi} \sin\theta\, d\theta\, d\psi \dotfill (11).$$

But the "equation of continuity" for an incompressible liquid (being

$$\frac{du}{dx} + \frac{dv}{dy} + \frac{dw}{dz} = 0),$$

gives* $\nabla^2 \phi = 0$, for every point within the spherical space; and therefore [Thomson and Tait, App. B]

$$\phi = S_0 + S_1 r + S_2 r^2 + \&c. \dotfill (12),$$

a converging series, where S_0 denotes a constant, and S_1, S_2, &c., surface harmonics of the orders indicated.

Hence $$R = \frac{d\phi}{dr} = S_1 + 2rS_2 + 3r^2 S_3 + \&c. \dotfill (13).$$

And it is clear from the synthesis of the most general surface harmonic, by zonal, sectioral, and tesseral harmonics [Thomson and Tait, $\S$ 781], that $dS_i/d\psi$ is a surface harmonic of the same

---

* By ∇^2 we shall always understand $d^2/dx^2 + d^2/dy^2 + d^2/dz^2$.

order as $S_i{}^*$: from which [Thomson and Tait, App. B (16)], it
follows that,

$$\iint S_i \frac{dS_{i'}}{d\psi} \sin\theta\, d\theta\, d\psi = 0,$$

except when $i' = i$. But this is true also when $i' = i$ because

$$S_i \frac{dS_i}{d\psi} = \tfrac{1}{2}\frac{d\,(S_i{}^2)}{d\psi},$$

and therefore, as in § 14, the integration for ψ, from $\psi = 0$ to
$\psi = 2\pi$ gives zero.   Hence (11) gives

$$\frac{dL}{dt} = 0.$$

This and § 11 establish § 10.

16.   Lemma (1) of § 11, and § 10 now proved, show that in
any motion whatever of an incompressible liquid, whether with
solids immersed in it or not, $udx + vdy + wdz$ is always a complete
differential through any portion of the fluid, for which it is a
complete differential at any instant, to whatever shape and position of space this portion may be brought in the course of the
motion.   This is the ordinary statement of the fundamental
proposition of fluid motion referred to in § 9, which was first
discovered by Lagrange.   (For another proof see § 60.)   I have
given the preceding demonstration, not so much because it is
useful to look at mathematical structures from many different
points of view, but (§ 19) because the dynamical considerations
and the formulæ I have used are immediately available for
establishing the theory of the impulse (§§ 3—8), of which a

* This follows, of course, from the known analytical theorem that the operations
∇^2 and $(y\,d/dz - z\,d/dy)$ are commutative, which is proved thus :—

By differentiation we have

$$\frac{d^2\left(y\dfrac{d\phi}{dz}\right)}{dy^2} = y\frac{d^2}{dy^2}\frac{d\phi}{dz} + 2\frac{d}{dy}\frac{d\phi}{dz};$$

and therefore, since $d/dy\,d\phi/dz = d/dz\,d\phi/dy$,

$$\nabla^2\left(y\frac{d\phi}{dz} - z\frac{d\phi}{dy}\right) = y\nabla^2\left(\frac{d\phi}{dz}\right) - z\nabla^2\left(\frac{d\phi}{dy}\right) = \left(y\frac{d}{dz} - z\frac{d}{dy}\right)\nabla^2\phi,$$

or

$$\nabla^2\left(y\frac{d}{dz} - z\frac{d}{dy}\right)\phi = \left(y\frac{d}{dz} - z\frac{d}{dy}\right)\nabla^2\phi,$$

ϕ being any function whatever.   Hence, if $\nabla^2\phi = 0$, we have

$$\nabla^2\left(y\frac{d\phi}{dz} - z\frac{d\phi}{dy}\right) = 0.$$

fundamental proposition was stated above (§ 5). To prove this proposition (in § 19) I now proceed.

17. Imagine any spherical surfaces to be described round a moveable solid or solids immersed in a liquid. The surrounding fluid can only press (§ 1) perpendicularly; and therefore when any motion is (§ 3) generated by impulsive forces applied to the solids, the moment round any diameter of the momentum of the matter within the spherical surface at the first instant, must be exactly equal to the moment of those impulsive forces round this line. And the moment round this line, of the momentum of the matter in the space between any two concentric spherical surfaces is zero, provided neither cuts any solid, and provided that, if there are any solids in this space, no impulse acts on them.

18. Hence, considering what we have defined as "the impulse of the motion," (§ 6), we see that its moment round any line is equal to the moment of momentum round the same line, of all the motion within any spherical surface having its centre in this line, and enclosing all the matter to which any constituent of the impulse is applied. This will still hold, though there are other solids not in the neighbourhood, and impulses are applied to them: provided the moments of momentum of those only which are within S are taken into account, and provided none of them is cut by S.

19. The statements of § 11, regarding fluid occupying at any instant a fixed spherical surface, are applicable without change to the fluids and solids occupying the space bounded by S, because of our present condition, that no solid is cut by S. Hence every statement and formula of § 15, as far as equation (11), may be now applied to the matter within S; but instead of (12) we now have [Thomson and Tait, § 736], if we denote by T_1, T_2, &c., another set of surface spherical harmonics,

$$\left.\begin{aligned}\phi = S_0 + S_1 r \; + S_2 r^2 \; + \&c.\\ + T_1 r^{-2} + T_2 r^{-3} + \&c.\end{aligned}\right\} \;\;\ldots\ldots\ldots\ldots(14)^*,$$

for all space between the greatest and smallest spherical surface concentric with S, and having no solids in it, because through all

---

* There is no term T_0/r, because this would give, in the integral of flow across the whole spherical surface, a finite amount of flow out of or into the space within, implying a generation or destruction of matter.

this space, § 16, and the equation of continuity prove that $\nabla^2\phi = 0$. Hence, instead of (13), we now have

$$R = \frac{d\phi}{dr} = S_1 + 2rS_2 + 3r^2S_3, \text{ \&c.}$$
$$\left. -\frac{2}{r^3}T_1 - \frac{3}{r^4}T_2 - \frac{4}{r^5}T_3 + \text{\&c.}\right\} \quad \dots\dots\dots(15).$$

Hence finally

$$\frac{dL}{dt} = \sum_{i=0}^{i=\infty}\iint\left[iS_i\frac{dT_i}{d\psi} - (i+1)T_i\frac{dS_i}{d\psi}\right]\sin\theta\,d\theta\,\psi \quad \dots(16).$$

Now if, as assumed in § 5, neither any moveable solids, nor any part of the boundary exist within any finite distance of S all round; S_1, S_2, &c., must each be infinitely small: and therefore (16) gives $dL/dt = 0$. This proves the proposition asserted in § 5: because a system of forces cannot have zero moment round every line drawn through any finite portion of space, without having force-resultant and couple-resultant each equal to zero.

20. As the rigidity of the solids has not been taken into account, all or any of them may be liquefied (§ 3) without violating the demonstration of § 19. To save circumlocutions, I now define *a vortex* as a portion of fluid having any motion that it could not acquire by fluid pressure transmitted through itself from its boundary. Often, merely for brevity, I shall use the expression *a body* to denote either a solid or a vortex, or a group of solids or vortices.

21. The proposition thus proved may be now stated in terms of the definitions of § 6, which were not used in § 5, and so becomes simply this:—*The impulse of the motion of a solid or group of solids or vortices and the surrounding liquid remains constant as long as no disturbance is suffered from the influence of other solids or vortices, or of the containing vessel.*

This implies, of course (§ 6), that the magnitudes of the force-resultant and the rotational moment of the impulse remain constant, and the position of its axis invariable.

22. In Poinsot's system of the statics of a rigid body we may pass from the resultant force and couple along and round the central axis to an equal resultant force along the parallel line through any point, and a greater couple the resultant of the former (or minimum) couple, and a couple in the plane of the two parallels, having its moment equal to the product of their distance

into the resultant force. So we may pass from the force-resultant and rotational moment of the impulse along and round its axis, to an equal force-resultant and greater moment of impulse, by transferring the former to any point, Q, not in the axis (§ 6) of the motion. This greater moment is (§ 18) equal to the moment of momentum round the point Q, of the motion within any spherical surface described from Q as centre, which encloses all the vortices or moving solids.

23. Hence a group of solids or vortices which always keep within a spherical surface of finite radius, or a single body, moving in an infinite liquid, can have no permanent average motion of translation in any direction oblique to the direction of the force-resultant of the impulse, if there is a finite force-resultant. For the matter within a finite spherical surface enclosing the moving bodies or body, cannot have moment of momentum round the centre increasing to infinity.

24. But there may be motion of translation when the force-resultant of the impulse vanishes; and there will be, for example, in the case of a solid, shaped like the screw-propeller of a steamer, immersed in an infinite homogeneous liquid, and set in motion by a couple in a plane perpendicular to the axis of the screw.

25. And when the force-resultant of the impulse does *not* vanish, there may be no motion of translation, or there may even be translation in the direction opposite to it. Thus, for example, a rigid ring, with cyclic motion, established (§ 63) through it, will, if left at rest, remain at rest. And if at any time urged by an impulse in either direction in the line of the force-resultant of the impulse of the cyclic motion, it will commence and continue moving with an average motion of translation in that direction; a motion which will be uniform, and the same as if there were no cyclic motion, when the ring is symmetrical. If the translatory impulse is contrary to the cyclic impulse, but less in magnitude, the translation will be contrary to the whole force-resultant impulse.

If the translatory impulse is equal and opposite to the cyclic impulse, there will be translation with zero force-resultant impulse—another example of what is asserted in § 24. In this case, if the ring is plane and symmetrical, or of any other shape such that the cyclic motion (which, to fix ideas, we have supposed given first, with the ring at rest) must have had only a force-resultant, and no rotational moment, we have a solid moving with a uniform motion of translation through a fluid, and both force and couple resultant of the whole impulse zero.

26.   From §§ 21 and 4, we see that, however long the time of application of the impressed forces may be—provided only that, during the whole of it, the solid or group of solids has been at an infinite distance from all other solids and from the containing vessel—the time integrals of the impressed forces parallel to three fixed axes, and of their moments round these lines, are equal to the six corresponding components of " the impulse " (§ 6).

27.   If two groups, at first so far asunder as to exercise no sensible influence on one another, come together, the " impulse " of the whole system remains unchanged by any disturbance each may experience from the other, whether by impacts of the solids, or through motion and pressure of the surrounding fluid; and (§ 6) it is always reducible to the force-resultant along the central axis, and the minimum couple-resultant, of the two impulses reckoned as if applied to one rigid body. The same holds, of course, if one group separates into two so distant as to no longer exert any sensible influence on one another.

28.   Hence whatever is lost of impulse perpendicular to a fixed plane, or of component rotational movement round a fixed line, by one group through collision with another, is gained by the other.

29.   Two of the moveable solids, or two groups, will be said to be *in collision* when, having been so far asunder as not to disturb one another's motions sensibly, they are so near as to do so. This disturbance will generally be supposed to be through fluid pressure only, but impacts of solids on solids may take place during a collision.

30.   We are now prepared to investigate (§§ 30, 31, 32) the influence of a fixed solid on the impulse of a moveable solid, or of a vortex, or of a group of solids or vortices, passing near it, thus— If during such collisions or separations as are considered in §§ 27, 28, forces are impressed on any one or more of the solids, their alteration of the whole impulse is (§ 26) to be reckoned by adding to each of its rectangular components the time integral of the

corresponding component of these impressed forces. Now, let us suppose such forces to be impressed on any one of the moveable solids as shall keep it at rest. These forces are zero as long as no moving solid is within a finite distance. But if a moving solid or vortex, or group of solids or vortices, passes near the fixed solid, the change of pressure due to the motion of the fluid will tend to move it, and the impression of force on it becomes necessary to keep it fixed. Let $d\sigma$ be an element of its surface; (x, y, z) the co-ordinates of the centre of this element; α, β, γ the inclinations of the normal at (x, y, z) to the three rectangular axes; and p the fluid pressure at time t, and point (x, y, z). The six components of force and couple required to hold the body fixed at time t, are

$$\left. \begin{array}{l} \iint d\sigma.\cos\alpha.p,\ \iint d\sigma.\cos\beta.p,\ \iint d\sigma.\cos\gamma.p; \\ \iint d\sigma(y\cos\gamma-z\cos\beta)p,\ \iint d\sigma(z\cos\alpha-x\cos\gamma)p,\ \iint d\sigma(x\cos\beta-y\cos x)p \end{array} \right\}$$
$$\dots\dots\dots(1).$$

If in these expressions we substitute

$$\int p\,dt \dots\dots\dots\dots\dots\dots\dots\dots\dots(2),$$

in place of p ($\int dt$ denoting a time integral from any era of reckoning before the disturbance became sensible, up to time t, which may be any instant during the collision, or after it is finished), we have the changes in the corresponding components of the impulse up to time t, provided there has been no impact of moveable solid on the fixed solid.

31. Let now the "velocity potential" (as we shall call it, in conformity with a German usage which has been adopted by Helmholtz), be denoted by ϕ; that is (§ 16), let ϕ be such a function of (x, y, z, t) that

$$u = \frac{d\phi}{dx}, \quad v = \frac{d\phi}{dy}, \quad w = \frac{d\phi}{dz}\dots\dots\dots\dots(3),$$

and let $\dot\phi$ (or $d\phi/dt$) denote its rate of variation per unit of time at any instant t, for the point (x, y, z) regarded as fixed.

Also, let q denote the resultant fluid velocity, so that

$$q^2 = u^2 + v^2 + w^2 = \frac{d\phi^2}{dx^2} + \frac{d\phi^2}{dy^2} + \frac{d\phi^2}{dz^2} \dots\dots\dots(4).$$

The ordinary hydrodynamical formula gives

$$p = \Pi - \dot\phi - \tfrac{1}{2}q^2\dots\dots\dots\dots\dots\dots(5);$$

where Π denotes the constant pressure in all sensibly quiescent parts of the fluid.

32. The constant term Π disappears from p in each of the integrals (1) of § 30, because a solid is equilibrated by equal pressure around. And in the time integral (2), we have

$$\int \dot{\phi}\, dt = \phi \quad \dots\dots\dots\dots\dots\dots\dots(6);$$

and therefore if $(XYZ)\,(LMN)$ denote the changes in the force- and couple-components of the impulse produced by the collision up to time t, we have

$$X = -\iint d\sigma \cos\alpha\,(\phi + \tfrac{1}{2}\int q^2 dt), \quad Y = \&\text{c.}, \quad Z = \&\text{c.} \quad \left.\right\}$$
$$L = -\iint d\sigma\,(y\cos\gamma - z\cos\beta)\,(\phi + \tfrac{1}{2}\int q^2 dt), \quad M = \&\text{c.}, \quad N = \&\text{c.} \left.\right\}$$
$$\dots\dots\dots(7).$$

But because the fluid is quiescent in the neighbourhood of the fixed body when the moving body or group of bodies is infinitely distant from it; it follows that before the commencement and after the end of the collision we have $\phi = 0$ at every point of the surface of the fixed body. Hence, for every value of t representing a time after the completion of the collision, the preceding expressions become

$$X = -\tfrac{1}{2}\iint d\sigma \cos\alpha \int q^2 dt, \quad Y = \&\text{c.}, \quad Z = \&\text{c.} \quad \left.\right\}\dots(8);$$
$$L = -\tfrac{1}{2}\iint d\sigma\,(y\cos\gamma - z\cos\beta)\int q^2 dt, \quad M = \&\text{c.}, \quad N = \&\text{c.}\left.\right\}$$

which express that *the integral change of impulse experienced by a body or group of bodies, in passing beside a fixed body without striking it, may be regarded as a system of impulsive attractions towards the latter, everywhere in the direction of the normal, and amounting to $\tfrac{1}{2}\int q^2 dt$ per unit of area.* But it must not be forgotten that the term ϕ in the expression [§ 31 (5)] for p produces, as shown in § 30 (1), an influence *during the collision*, the integral effect of which only disappears from the expression [§ 32 (7)] for the impulse *after the collision is completed;* that is (§ 29) after the moving system has passed away so far as to leave no sensible fluid motion in the neighbourhood of the fixed body.

33. Hence, and from § 23, we see that when there is no impact of moving solid against the fixed body, and when the moving solid or group of solids passes altogether on one side of the fixed body, the direction of the translation will be deflected, as if there were, on the whole, an *attraction towards* the fixed body,

or a *repulsion from it*, according as (§ 25) the translation is in
the direction of the impulse or opposite to it. For, in each
case, the impulse is altered by the introduction of an impulse
towards the fixed body upon the moving body or bodies as they
pass it; and (§ 23) the translation before and after the collision is
always along the line of the impulse, and is altered in direction
accordingly. This will be easily understood from the diagrams,
where in each case B represents the fixed body, the dotted line
$ITT'I'$ and arrow-heads II', the directions of the force-resultant
of the impulse at successive times, and the full arrow-heads TT',
the directions of the translation.

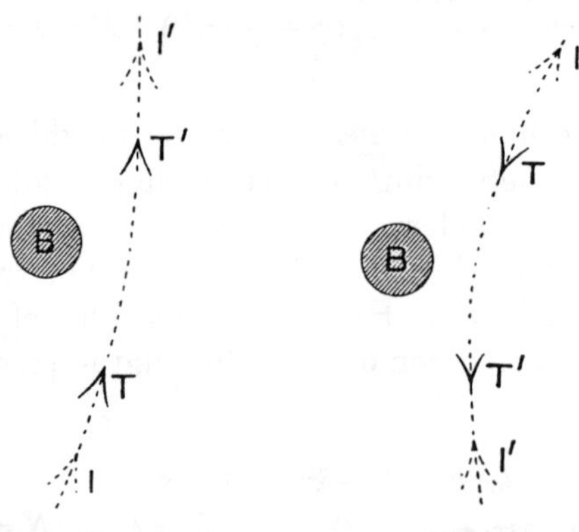

Fig. 1. Fig. 2.

All ordinary cases belong to the class illustrated by fig. 1.
The case of a rigid ring, with cyclic motion (§ 25) established
round it as core, belongs to the class illustrated by fig. 2, if the
ring be projected through the fluid in the direction perpendicular
to its own plane, and contrary to the cyclic motion through its
centre.

34. When (§ 66) we substitute vortices for the moving solids,
we shall see (§ 67) that the translation is probably always in the
direction *with* the impulse. Hence, as illustrated by fig. 1, there is
always the deflection, as if by *attraction*, when a group of vortices
pass all on one side of a fixed body. This is easily observed, for a
simple Helmholtz ring, by sending smoke-rings on a large scale,
according to Professor Tait's plan, in such directions as to pass
very near a convex fixed surface. An ordinary 12-inch globe,
taken off its bearings and hung by a thin cord, answers very well
for the fixed body.

35.   The investigation of §§ 30, 31, 32 is clearly applicable to a vortex or a moving body, or to a group of vortices or moving bodies, which keep always near one another (§ 23), passing near a projecting part of the fixed boundary, and being, before and after this collision (§ 29), at a very great distance from every part of the fixed boundary.   Thus a Helmholtz ring, projected so as to pass near a projecting angle of two walls, shows a deflection of its course, as if caused by attraction towards the corner.

36.   In every case the force-resultant of the impulse is, as we shall presently see (§ 37), determinate when the flow of the liquid across every element of any surface completely enclosing the solids or vortices is given; but not so, from such data, either the axis (§ 6) or the rotational moment, as we see at once by considering the case of a solid sphere (which may afterwards be supposed liquefied) set in motion by a force in any line not through the centre, and a couple in a plane perpendicular to it.   For this line will be the " axis," and the impulsive couple will be the rotational moment of the whole motion of the solid and liquid.   But the liquid, on all sides, will move exactly as it would if the impulse were merely an impulsive force of equal amount in a parallel line through the centre of the sphere, with therefore this second line for "axis" and zero for rotational moment.   For illustration of rotational moment remaining latent in a liquid (with or without solids) until made manifest by actions, tending to alter its axis, or showing effects of centrifugal force due to it, see § 66 and others later.

37.   The component impulse in any direction is equal to the corresponding component momentum of the mass enclosed within the surface S, containing all the places of application of the impulse, together with that of the impulsive pressure outwards on this surface.   But as the matter enclosed by S (whether all liquid or partly liquid and partly solid) is of uniform density, its momentum will be equal to its mass multiplied into the velocity of the centre of gravity of the space within the surface S supposed to vary so as to enclose always the same matter, and will therefore depend solely on the normal motion of S; that is to say, on the component of the fluid velocity in the direction of the normal at every point of S.   And the impulsive fluid pressure, corresponding to the generation of the actual motion from rest, being the time

integral of the pressure during the instantaneous generation of the motion, is (§§ 31, 32) equal to $-\phi$, the velocity potential; which (§ 61) is determinate for every point of S, and of the exterior space when the normal component of the fluid motion is given for every point of S.   Hence the proposition asserted in § 36.   Denoting by $d\sigma$ any element of S; N the normal component of the fluid velocity; α the inclination to OX, of the normal drawn *outwards* through $d\sigma$; and X the x-component of the impulse; we have for the two parts of this quantity considered above, and its whole value, the following expressions; of which the first is taken in anticipation from § 42—

x-momentum of matter, within S, $= \iint Nx\, d\sigma$ (8) of § 42

x-component of impulsive pressure
  on S, outwards, $= -\iint \phi \cos \alpha\, d\sigma$

$$\left. \vphantom{\begin{array}{c}a\\a\\a\\a\end{array}} \right\} \dots(1),$$

$$X = \iint (Nx - \phi \cos \alpha)\, d\sigma \quad \dots\dots\dots\dots\dots\dots\dots(2).$$

It is worthy of remark that this expression holds for the impulse of all the solids or vortices within S, even if there be others in the immediate neighbourhood outside: and that therefore its value must be zero if there be no solids or vortices within S, and N and ϕ are due solely to those outside.

38.   If ϕ be the potential of a magnet or group of magnets, some within S and others outside it, and N the normal component magnetic force, at any point of S, the preceding expression (2) is equal to the x-component of the magnetic moment of all the magnets within S, multiplied by 4π.   For let ρ be the density of any continuous distribution of positive and negative matter, having for potential, and normal component force, ϕ and N respectively, at every point of S.   We have [Thomson and Tait, § 491 (c)] $\rho = -1/4\pi \, \nabla^2 \phi$, and therefore

$$\iiint \rho x\, dx\, dy\, dz = -\frac{1}{4\pi} \iiint x \left(\frac{d^2\phi}{dx^2} + \frac{d^2\phi}{dy^2} + \frac{d^2\phi}{dz^2} \right) dx\, dy\, dz \dots(3).$$

Now, integrating by parts*, as usual with such expressions, we have

---

* The process here described leads merely to the equation obtained by taking the last two equal members of App. A (1) (Thomson and Tait) for the case $a = 1$, $U = \phi$, $U' = x$.

$$\iiint x \frac{d^2\phi}{dx^2}\, dx\,dy\,dz = \iint x \frac{d\phi}{dx}\, dy\,dz - \iiint \frac{d\phi}{dx}\, dx\,dy\,dz$$

$$= \iint \left(x \frac{d\phi}{dx} - \phi \right) dy\,dz.$$

Hence integrating each of the other two terms of (3) once simply, and reducing as usual [Thomson and Tait, App. A (a)] to a surface integral we have

$$\iiint \rho x\, dx\, dy\, dz = -\frac{1}{4\pi} \iint (Nx - \phi \cos \alpha)\, d\sigma \ldots\ldots\ldots(4);$$

which proves the proposition, and also, of course, that if there be no matter within S, the value of the second member is zero.

39. Hence, considering the magnetic and hydrokinetic analogous systems with the sole condition that at every point of some particular closed surface the magnetic potential is equal to the velocity potential, we conclude that 4π times the magnetic moment of all the magnetism within any surface, in the magnetic system, is equal to the force-resultant of the impulse of the solids or vortices within the corresponding surface in the hydrokinetic system; and that the directions of the magnetic axis and of the force-resultant of the impulse are the same. For the theory of magnetism, it is interesting to remark that indeterminate distributions of magnetism within the solids, or portions of fluid to which initiating forces (§ 3) were applied, or determinate distributions in infinitely thin layers at their surfaces, may be found, which through all the space external to them shall produce the same potential as the velocity-potential, and therefore the same distribution of force as the distribution of velocity through the whole fluid. But inasmuch as when the magnetic force in the interior of a magnet is defined in the manner explained in § 48 (2) of my *Mathematical Theory of Magnetism* *, it is expressible through all space by the differential coefficients of a potential; and, on the contrary, for the kinetic system $u\,dx + v\,dy + w\,dz$ is not a complete differential generally through the spaces occupied by the solids, the agreement between resultant force and resultant flow holds only through the space exterior to the magnets and solids in the magnetic and kinetic systems respectively. But if the other definition of resultant force within a magnet [*Math. Theory of*

* *Trans. R. S. Lond.* 1851 ; or *Thomson's Electrical Papers*, Macmillan, 1869.

Magnetism, § 77, foot-note, and § 78], published in preparation for a sixth chapter "On Electro-magnets" (still in my hands in manuscript, not quite completed), and which alone can be adopted for spaces occupied by non-magnetic matter traversed by electric currents, the magnetic force has not a potential within such spaces; and we shall see (§ 68) that determinate distributions of closed electric currents through spaces corresponding to the solids of the hydrokinetic system can be found which shall give for every point of space, whether traversed by electric currents or not, a resultant magnetic force, agreeing in magnitude and direction with the velocity, whether of solid or fluid, at the corresponding point of the hydrokinetic system. This thorough agreement for all space renders the electro-magnetic analogue preferable to the magnetic; and, having begun with the magnetic analogous system only because of its convenience for the demonstration of § 38, we shall henceforth chiefly use the purely eloctro-magnetic analogue.

40. To prove the formula used in anticipation, in § 37 (1) we must now (§§ 41, 42, 43) find the momentum of the whole matter —fluid, fluid and solid, or even solid alone—at any instant within a closed surface S, in terms of the normal component velocity of the matter at any point of this surface, or, which is the same, the normal velocity of this surface itself, if we suppose it to vary so as to enclose always the same matter.

41. Let V be the volume of the space bounded by any varying closed surface S. As yet we need not suppose V constant. Let $\bar{x}, \bar{y}, \bar{z}$ be the co-ordinates of the centre of gravity. We have

$$V\bar{x} = \tfrac{1}{2} \iint [x^2 dy\,dz] \dotfill (5),$$

where [ ] indicates that the expression within it is to be taken between proper limits for S. Now as S varies with the time, the area through which $\iint dy\,dz$ is taken will in general vary; but the increments or decrements which it experiences at different parts of the boundary of this area, in the infinitely small time dt, contribute no increments or decrements to $\iint [x^2 dy\,dz]$, as we see most easily by first supposing S to be a surface everywhere convex outwards. Hence

$$\frac{d}{dt}\iint [x^2 dy\,dz] = \iint \left[\frac{d\,(x^2)}{dt}dy\,dz\right] = 2\iint \left[x\frac{dx}{dt}dy\,dz\right]\dotfill (6).$$

But if N denote the velocity with which the surface moves in the direction of its outward normal at (x, y, z), we have, in the preceding expression

$$\frac{dx}{dt} = N \sec \alpha$$

if α be the inclination of the outward normal to OX. Hence

$$\frac{d(V\bar{x})}{dt} = \int\!\!\int [xN \sec \alpha \, dy \, dz].$$

But the conditions as to limits indicated by [ ] are clearly satisfied, if, $d\sigma$ denoting an element of the surface, such that

$$dy dz = \cos \alpha \, d\sigma,$$

we simply take $\int\!\!\int d\sigma$ over the whole surface. Thus we have

$$\frac{d(V\bar{x})}{dt} = \int\!\!\int xN d\sigma \quad \dots\dots\dots\dots\dots(7).$$

42. In any case in which V is constant, this becomes

$$V\frac{d\bar{x}}{dt} = \int\!\!\int xN d\sigma \quad \dots\dots\dots\dots\dots(8).$$

If now the varying surface, S, is the boundary of a portion of the matter—fluid or solid—of uniform density unity, with whose motions we are occupied, the x-component momentum of this portion is $V dx/dt$; and, therefore, equation (8) is the required (§ 40) expression.

43.   The same formulæ (7) and (8) are proved more shortly of course by the regular analytical process given by Poisson* and Green† in dealing with such subjects; thus, in short.   Let u, v, w be the components of velocity, of any matter, compressible or incompressible, at any point (x, y, z) within S; and let c denote the value at this point of $du/dx + dv/dy + dw/dz$, so that

$$\frac{du}{dx} = c - \left(\frac{dv}{dy} + \frac{dw}{dz}\right) \quad \dots\dots\dots\dots\dots(9).$$

We have, for the component momentum of the whole matter within S, if of unit density at the instant considered,

$$\int\!\!\int\!\!\int u\, dx\, dy\, dz = \int\!\!\int ux\, dy\, dz - \int\!\!\int\!\!\int x\frac{du}{dx}\, dx\, dy\, dz \dots\dots(10).$$

---

* *Théorie de la Chaleur*, § 60.
† *Essay on Electricity and Magnetism.*

But by (9)

$$\iiint x \frac{du}{dx}\, dx\,dy\,dz = \iiint cx\,dx\,dy\,dz - \iiint x \left(\frac{dv}{dy} + \frac{dw}{dz}\right) dx\,dy\,dz,$$

and by simple integrations,

$$\iiint x \left(\frac{dv}{dy} + \frac{dw}{dz}\right) dx\,dy\,dz = \iint x\,(v\,dx\,dz + w\,dx\,dy).$$

Using these in (10), and altering the expression to a surface integral, as in Thomson and Tait, App. A (a), we have

$$\iiint u\,dx\,dy\,dz = \iint x\,(u\,dy\,dz + v\,dz\,dx + w\,dx\,dy) - \iiint cx\,dx\,dy\,dz$$

$$= \iint x N\,d\sigma - \iiint cx\,dx\,dy\,dz \quad\ldots\ldots\ldots\ldots\ldots\ldots\ldots(11),$$

which clearly agrees with (7).

When this mass is incompressible, we have $c = 0$ by the formula so ill named the equation "of continuity" (Thomson and Tait, § 191), and we fall upon (8).

The proper analytical interpretation of the differential coefficients du/dx, &c., and of the equation of continuity, when, as at the surfaces of separation of fluid and solids, u, v, w are discontinuous functions having abruptly varying values, presents no difficulty.

44. In the theory of the impulse applied to the collision (§ 29) of solids or vortices moving through a liquid, the force-resultant of the impulse corresponds, as we have seen, precisely to the resultant momentum of a solid in the ordinary theory of impact. Some difficulty may be felt in understanding how the zero-momentum (§ 4) of the whole mass is composed; there being clearly positive momentum of solids and fluids in the direction of the impulse in some localities near the place of its application, and negative in others. [Consider, for example, the simple case of a solid of revolution struck by a single impulse in the line of its axis. The fluid moves in the direction of the impulse, before and behind the body, but in the contrary direction in the space round its middle.] Three modes of dividing the whole moving mass present themselves as illustrative of the distribution of momentum through it; and the following propositions (§ 45) with reference to them are readily proved (§§ 46, 47, 48).

45. I. Imagine any cylinder of finite periphery, not necessarily circular, completely surrounding the vortices (or moving solids), and any other surrounding none, and consider the in-

finitely long prisms of variously moving matter at any instant surrounded by these two cylinders. The component momentum parallel to the length of the first is equal to the component of the impulse parallel to the same direction; and that of the second is zero.

II.　Imagine any two finite spherical surfaces, one enclosing all the vortices or moving solids, and the other none. The resultant-momentum of the whole matter enclosed by the first is in the direction of the impulse, and is equal to $\frac{2}{3}$ of its value. The resultant-momentum of the whole fluid enclosed by the second is the same as if it all moved with the same velocity, and in the same direction, as at its centre.

III.　Imagine any two infinite planes at a finite distance from one another and from the field of motion, but neither cutting any solid or vortex. The component perpendicular to them of the momentum of the matter occupying at any instant the space between them (whether this includes none, some, or all of the vortices or moving solids) is zero.

46.　To prove these propositions :—

Consider in either case a finite length of the prism extending to a very great distance in each direction from the field of motion, and terminated by plane or curved ends. Then, the motion being, as we may suppose (§ 61) started from rest by impulsive pressures on the solids [or (§ 66) on the portions of fluid constituting the vortices]; the impulsive fluid pressure on the cylindrical surface can generate no momentum parallel to the length; and to generate momentum in this direction there will be, in case 1, the impressed impulsive forces on the solids, and the impulsive fluid pressures on the ends; but in case 2 there will be only the impulsive fluid pressure on the ends. Now, the impulsive fluid pressures on the ends diminish [§ 50 (15)] according to the inverse square of the distance from the field of motion, when the prism is prolonged in each direction, and are therefore infinitely small when the prisms are infinitely long each way. Whence the proposition I.

47.　By using the harmonic expansions § 19 (14), (15), in the several expressions § 37 (1), (2); and the fundamental theorem

$$\iint \mathfrak{H}_i \mathfrak{H}_{i'} \, d\sigma = 0,$$

of the harmonic analysis [Thomson and Tait, App. B (16)]; and putting $S_i = 0$ for one case, and $T_i = 0$ for the other; we prove the two parts of Prop. II., § 45, immediately.

48.   To prove Prop. III., § 45, the well-known theory of electric images in a plane conductor* may be conveniently referred to. It shows that if N_1 denotes the normal component force at any point of an infinite plane due to any distribution, μ, of matter in the space lying on one side of the plane, a distribution of matter over the plane having $N_1/2\pi$ for surface density at each point exerts the same force as μ through all the space on the other side of the plane, and therefore that the whole quantity of matter in that surface distribution is equal to the whole quantity of matter in μ†.   Hence, $\iint d\sigma$ denoting integration over the infinite *plane*,

$$\iint N_1 d\sigma = 0 \quad\quad\quad\quad\quad\quad(12),$$

if the whole quantity of matter in μ be zero.   Hence, if N be the normal force due to matter through space on both sides of the plane, provided the whole quantity of matter on each side separately is zero,

$$\iint N d\sigma = 0 \quad\quad\quad\quad\quad\quad(13);$$

since N is the sum of two parts, for each of which separately (12) holds.   This, translated into hydrokinetics, shows that the whole flow of matter across any infinite plane is zero at every instant when it cuts no solids or vortices.   Hence, and from the uniformity of density which (§ 3) we assume, the centre of gravity of the matter between any two infinite fixed parallel planes has no motion in the direction perpendicular to them at any time when no vortex or moving solid is cut by either: which is Prop. III. of § 45 in other words.

---

* Thomson, *Camb. and Dub. Math. Journal*, 1849; Liouville's *Journal*, 1845 and 1847; or Reprints of *Electrical Papers* (Macmillan, 1869).

† This is verified synthetically with ease, by direct integrations showing (whether by Cartesian or polar plane co-ordinates), that

$$\int_{-\infty}^{\infty} \int_{-\infty}^{\infty} \frac{a\,dy\,dz}{(a^2+y^2+z^2)^{\frac{3}{2}}} = 2\pi.$$

And taking d/da of this, we have

$$\int_{-\infty}^{\infty} \int_{-\infty}^{\infty} \frac{(y^2+z^2-2a^2)\,dy\,dz}{(a^2+y^2+z^2)^{\frac{5}{2}}} = 0,$$

the synthesis of (12).

49. The integral flow of matter across any surface whatever, imagined to divide the whole volume of the finite fixed containing vessel of § 1 into two parts is necessarily zero, because of the uniformity of density; and therefore the momentum of all the matter bounded by two parallel planes, extending to the inner surface of the containing vessel, and the portion of this surface intercepted between them has always zero for its component perpendicular to these planes, whether or not moving solids or vortices are cut by either or both these planes. But it is remarkable that when any moving solid or vortex is cut by a plane, the integral flow of matter across this plane (if the containing vessel is infinitely distant on all sides from the field of motion), converges to a generally *finite* value, as the plane is extended to very great distances all round from the field of motion, which are still infinitely small in comparison with the distances to the containing vessel; and diminishes from that finite value to zero by another convergence, when the distances to which the plane is extended all round begin to be comparable with, and ultimately become equal to, the distances of the curve in which it cuts the containing vessel. Hence we see how it is that the condition of neither plane cutting any moving solid or vortex is necessary to allow § 45, III. to be stated without reference to the containing vessel, and are reminded that the equality to zero asserted in this proposition is proved in § 48 to be approximated to when the planes are extended to distances all round, which, though infinitely short of the distances to the containing vessel, are very great in comparison with their perpendicular distances from the most distant parts of the field of motion.

50. The convergencies concerned in § 45, I., III., may be analysed thus. Perpendicular to the resultant impulse draw any two planes on the two sides of the field of motion, with all the moving solids and vortices between them, and divide a portion of the space between them into finite prismatic portions by cylindrical (or plane) surfaces perpendicular to them. Suppose now one of these prismatic portions to include all the moving solids and vortices, and without altering the prismatic boundary, let the parallel planes be removed in opposite directions to distances each infinite (or very great) in comparison with the distance of the most distant of the moving solids or vortices. By § 45, I., the momen-

tum of the motion within this prismatic space is (approximately) equal to the force-resultant, I, of the impulse, and that of the motion within any one of the others is (approximately) zero.

But the sum of these (approximately) zero values must, on account of § 45, III., be equal to $-I$, if the portions of the planes containing the ends of the prismatic spaces be extended to distances very great in comparison with the distance between the planes. To understand this, we have only to remark that if ϕ denotes the velocity potential at a point distant D from the middle of the field, and x from a plane through the middle perpendicular to the impulse, we have (§ 53) approximately,

$$\phi = -\frac{Ix}{4\pi D^3}$$

provided D be great in comparison with the radius of the smallest sphere enclosing all the moving solids or vortices. Hence, putting $x = \pm a$ for the two planes under consideration, denoting by A the area of either end of one of the prismatic portions, and calling D *the proper mean distance* for this area, we have (§ 45) for the momentum of the fluid motion within this prismatic space, provided it contains no moving solids or vortices,

$$-2\frac{Ia}{4\pi D^3}$$

This vanishes when A/D^2 is an infinitely small fraction (as a/D is at most unity); but it is finite if A/D^2 is finite, provided a/D be not infinitely small. And its integral value (compare § 48, footnote) converges to $-I$, when the portion of area included in the integration is extended till a/D is infinitely small for all points of its boundary.

51. Both as regards the mathematical theory of the convergence of definite integrals, and as illustrating the distribution of momentum in a fluid, it is interesting to remark that, u denoting component velocity parallel to x, at any point (x, y, z), the integral $\iiint u\,dx\,dy\,dz$, expressing momentum, may, as is readily proved, have any value from $-\infty$ to $+\infty$ according to the portions of space through which it is taken.

52. As a last illustration of the distribution of momentum, let the containing vessel be spherical of finite radius a.

We have, as in § 19,

$$\phi = S_0 + S_1 r + S_2 r^2 + \&c. \\ \qquad + T_1 r^{-2} + T_2 r^{-3} + \&c. \Bigg\} \dots\dots\dots\dots(14),$$

each series converging, provided r is less than a, and greater than the radius of the smallest concentric spherical surface enclosing all the solids or vortices. Now, by the condition that there be no flow across the fixed containing surface, we must have

$$\frac{d\phi}{dr} = 0, \text{ when } r = a \dots\dots\dots\dots(15),$$

which gives

$$S_i = \frac{i+1}{i} \frac{T_i}{a^{2i+1}} \dots\dots\dots\dots(16);$$

and (14) becomes

$$\phi = \frac{T_1}{r^2}\left(1 + 2\frac{r^3}{a^3}\right) + \frac{T_2}{r^3}\left(1 + \frac{3}{2}\frac{r^5}{a^5}\right) + \&c.\dots\dots(17).$$

But [§ 37 (1)] if the whole amount of the x-component of impulsive pressure exerted by the fluid within the spherical surface of radius r, upon the fluid round it be denoted by F, we have

$$F = -\iint \phi \cos\theta\, d\sigma \dots\dots\dots\dots(18),$$

θ being the inclination to OX of the radius through $d\sigma$. Now $\cos\theta$ is a surface harmonic of the first order, and therefore all the terms of the harmonic expansion, except the first, disappear in the integral, which consequently becomes

$$F = -\left(1 + 2\frac{r^3}{a^3}\right)\iint T_1 \cos\theta \frac{d\sigma}{r^2} \dots\dots\dots(19).$$

Now let

$$T_1 = -\frac{Ax + By + Cz}{r} \dots\dots\dots\dots(20),$$

this being [Thomson and Tait, App. B, §§ i, j] the most general expression for a surface harmonic of the first order. We have $\cos\theta = x/r$; and therefore (by spherical harmonics, or by the elementary analysis of moments of inertia of a uniform spherical surface),

$$-\iint T_1 \cos\theta \frac{d\sigma}{r^2} = \frac{A}{r^4}\iint x^2 d\sigma = \frac{4\pi A}{3} \dots\dots\dots(21);$$

and (19) becomes

$$F = \left(1 + 2\frac{r^3}{a^3}\right)\frac{4\pi A}{3} \dots\dots\dots\dots(22).$$

Whence, if X denote the x-momentum of the fluid at any instant in the space between the concentric spherical surfaces of radius r and r',

$$X = -\frac{2}{3}\frac{r^3 - r'^3}{a^3}\, 4\pi A \dots\dots\dots\dots\dots(23).$$

If r and r' be each infinitely small in comparison with a, this expression vanishes, as it ought to do, in accordance with § 45, II. But if

$$\left.\begin{array}{l}\dfrac{r'}{a} = 0, \text{ and } r = a,\\[2mm] \text{it becomes} \qquad X = -\tfrac{2}{3}.4\pi A\end{array}\right\}\dots\dots\dots\dots\dots(24),$$

fulfilling § 4, by showing in the fluid outside the spherical surface of radius r a momentum equal and opposite to that (§ 45, II.) of the whole matter, whether fluid or solid, within that surface.

53. Comparing § 47 and § 52, we see that if X, Y, Z be rectangular components of the force-resultant of the impulse, the term $T_1 r^{-2}$ of the harmonic expansion (14) is as follows:—

$$T_1 r^{-2} = \frac{Xx + Yy + Zz}{4\pi r^3}\ \dots\dots\dots\dots(25),$$

provided all the solids and vortices taken into account are within a spherical surface whose radius is very small in comparison with the distances of all other vortices or moving solids, and with the shortest distance to the fixed bounding surface.

54. Helmholtz, in his splendid paper on Vortex Motion, has made the very important remark, that a certain fundamental theorem of Green's, which has been used to demonstrate the determinateness of solutions in hydrokinetics, is subject to exception when the functions involved have multiple values. This calls for a serious correction and extension of elementary hydrokinetic theory, to which I now proceed.

55. In the general theorem (1) of Thomson and Tait, App. A, let $a = 1$. It becomes

$$\iiint\left(\frac{d\phi}{dx}\frac{d\phi'}{dx} + \frac{d\phi}{dy}\frac{d\phi'}{dy} + \frac{d\phi}{dz}\frac{d\phi'}{dz}\right)dx\,dy\,dz$$

$$= \iint d\sigma\,\phi\,\eth\phi' - \iiint dx\,dy\,dz\,\phi\nabla^2\phi' = \iint d\sigma\,\phi'\,\eth\phi - \iiint dx\,dy\,dz\,\phi'\nabla^2\phi$$

$$\dots\dots\dots(1),$$

which is true without exception if ϕ and ϕ' denote any two *single-valued* functions of x, y, z; $\iiint dx\,dy\,dz$ integration through the space enclosed by any finite closed surface, S; $\iint d\sigma$ integration over the area of this surface; and $\mathfrak{d}$ rate of variation per unit of length in the normal direction at any point of it. This is Green's original theorem, with Helmholtz's limitation added (in italics). The reader may verify it for himself.

56. But if either ϕ or ϕ' is a many-valued function, and the differential coefficients $d\phi/dx$, ..., $d\phi'/dx$, ..., each single-valued, the double equation (1) cannot be generally true. Its first member is essentially unambiguous; but the process of integration by which the second member or the third member is found, would introduce ambiguity if ϕ or ϕ' is many-valued. In one case the first member, though not equal to the ambiguous second, would be equal to the third, provided ϕ' is not also many-valued; and in the other, the first member, though not equal to the third, would be equal to the second, provided ϕ is not many-valued.

For example, let
$$\phi' = \tan^{-1}\frac{y}{x} \quad\dotfill(2),$$

and let S consist of the portions of two planes perpendicular to OZ, intercepted between two circular cylinders having OZ for axis, and the portions of these cylinders intercepted between the two planes. The inner cylindrical boundary excludes from the space bounded by S, the line OZ where ϕ' has an infinite number of values, and $d\phi'/dx$, and $d\phi'/dz$ have infinite values. We have

$$\frac{d\phi'}{dx} = \frac{-y}{x^2+y^2},\quad \frac{d\phi'}{dy} = \frac{x}{x^2+y^2}\quad\dotfill(3),$$

and at every point of S, $\mathfrak{d}\phi' = 0$. Then, if ϕ be single-valued, there is no failure in the process proving the equality between the first and second members of (1), which becomes

$$\iiint \frac{x\dfrac{d\phi}{dy} - y\dfrac{d\phi}{dx}}{x^2+y^2}\,dx\,dy\,dz = 0 \quad\dotfill(4).$$

Compare § 14 (6) to end.

The third member of (1) becomes

$$\iint d\sigma \tan^{-1}\frac{y}{x}\,\mathfrak{d}\phi - \iiint \tan^{-1}\frac{y}{x}\,\nabla^2\phi\,dx\,dy\,dz \dotfill(5),$$

which is no result of unambiguous integration of the first member through the space enclosed by S, as we see by examining, in this case, the particular meaning of each step of the ordinary process in rectangular co-ordinates for proving Green's theorem. It is thus seen that we must add to (5) a term

$$2\pi \iint dx\,dz \left(\frac{d\phi}{dy}\right)_{y=0},$$

if in its other terms the value of $\tan^{-1} y/x$ is reckoned continuously round from one side of the plane ZOX to the other: or

$$-2\pi \iint dy\,dz \left(\frac{d\phi}{dx}\right)_{x=0},$$

if the continuity be from one side of ZOY to the other; to render it really equal to the first member of (1). Thus, taking for example, the first form of the added term, we now have for the corrected double equation (1) for the case of $\phi' = \tan^{-1} y/x$, ϕ any single-valued function, and S the surface, composed of the two co-axal cylinders and two parallel planes specified above:

$$\iiint \frac{x\dfrac{d\phi}{dy} - y\dfrac{d\phi}{dx}}{x^2 + y^2}\, dx\,dy\,dz = 0 = 2\pi \iint dx\,dz \left(\frac{d\phi}{dy}\right)_{y=0} + \iint d\sigma \tan^{-1}\frac{y}{x}\, \mathfrak{d}\phi$$

$$- \iiint dx\,dy\,dz\, \tan^{-1}\frac{y}{x}\, \nabla^2\phi \quad \dotsfill (6).$$

But if we annex to S any barrier stopping circulation round the inner cylindrical core, all ambiguity becomes impossible, and the double equation (1) holds. For instance, if the barrier be the portion of the plane ZOX, intercepted between the co-axal cylinders and parallel planes constituting the S of § 55, so that $\iint d\sigma$ must now include integration over each side of this rectangular area; (6) becomes simply the strict application of (1) to the case in question.

57.   The difficulty of the exceptional interpretation of Green's theorem for the class of cases exemplified in §§ 55 and 56, depends on the fact that $\int F ds$ may have different values when reckoned along the lengths of different curves, drawn within the space bounded by S, from a point P to a point Q; ds being an infinitesimal element of the curve, and F the rate of variation of ϕ per unit of length along it. Let PCQ, $PC'Q$ be two curves for which the $\int F ds$ has different values; and let both lie wholly within S.

If we draw any curve from P to Q; make it first coincide with PCQ, and then vary it gradually until it coincides with $PC'Q$; it must in some of its intermediate forms cut the bounding surface S: for we have

$$F\,ds = \frac{d\phi}{dx}\,dx + \frac{d\phi}{dy}\,dy + \frac{d\phi}{dz}\,dz$$

throughout the space contained within S, and $d\phi/dx, d\phi/dy, d\phi/dz$ are each of them unambiguous by hypothesis; which implies that $\int F\,ds$ has equal values for all gradual variations of one curve between P and Q, each lying wholly within S. Now, in a simply continuous space, a curve joining the points P and Q may be gradually varied from any curve PCQ to any other $PC'Q$, and therefore if the space contained within S be simply continuous, the difficulty depending on the multiplicity of value of ϕ or ϕ' cannot exist. And however multiply continuous (§ 58) the space may be, the difficulty may be evaded if we annex to S a surface or surfaces stopping every aperture or passage on the openness of which its multiple continuity depends; for these annexed surfaces, as each of them occupies no space, do not disturb the triple integrations (1), and will, therefore, not alter the values of its first member; but by removing the multiplicity of continuity, they free each of the integrations by parts, by which its second or third members are obtained, from all ambiguity. To avoid circumlocution, we shall call β the addition thus made to S; and further, when the space within S is (§ 58) not merely doubly but triply, or quadruply, or more multiply, continuous, we shall designate by β_1, β_2; or $\beta_1, \beta_2, \beta_3$; and so on; the several parts of β required in any case to stop all multiple continuity of the space. These parts of β may be quite detached from one another, as when the multiple continuity is that due to detached rings, or separate single tunnels in a solid. But one part β_1 may cut through part of another, β_2, as when two rings (§ 58, diagram) linked into one another without touching constitute part of the boundary of the space considered. And we shall denote by $\iint ds$, integration over the surface β, or over any one of its parts, β_1, β_2, &c. Let now P and Q be each infinitely near a point B, of β, but on the two sides of this surface. Let κ denote the value of $\int F\,ds$ along any curve lying wholly in the space bounded by S, and joining PQ without cutting the barrier; this value being the same for all such curves, and for all positions of B to which it may be brought without leaving β, and without making either P or Q pass through any part of β. That is to say, κ is a single constant when the space is not more than doubly continuous; but it denotes one or other of n constants κ_1, κ_2, ... κ_n, which may be all different from one another, when the space is n-ply continuous. Lastly, let κ' denote the same element, relatively to ϕ', as κ relatively to ϕ. We find that the first steps of the integrations by parts now introduce, without ambiguity, the additions

$$\Sigma\kappa\iint ds\, \eth\phi' \quad \text{and} \quad \Sigma\kappa'\iint ds\, \eth\phi \quad \dots\dots\dots\dots(7),$$

to the second and third members of (1): Σ denoting summation of the integrations for the different constituents β_1, β_2, ...: of β; but only a single term when the space is (§ 58) not more than doubly continuous.   Green's theorem thus corrected becomes

$$\iint\left(\frac{d\phi}{dx}\frac{d\phi'}{dx} + \frac{d\phi}{dy}\frac{d\phi'}{dy} + \frac{d\phi}{dz}\frac{d\phi'}{dz}\right)dx\,dy\,dz$$

$$= \iint d\sigma\, \phi\eth\phi' + \Sigma\kappa\iint ds\, \eth\phi' - \iiint \phi\nabla^2\phi'\, dx\,dy\,dz$$

$$= \iint d\sigma\, \phi'\eth\phi + \Sigma\kappa'\iint ds\, \eth\phi - \iiint \phi'\nabla^2\phi\, dx\,dy\,dz\dots\dots(8).$$

58.   Adopting the terminology of Riemann, as known to me through Helmholtz, I shall call a finite position of space n-ply continuous when its bounding surface is such that there are n irreconcilable paths between any two points in it. To prevent any misunderstanding, I add (1), that by *a* portion of space I mean such a portion that any point of it may be travelled to from any other point of it, without cutting the bounding surface; (2), that the "paths" spoken of all lie within the portion of space referred to; and (3), that by irreconcilable paths between two points P and Q; I mean paths such, that a line drawn first along one of them cannot be gradually changed till it coincides with the other, being always kept passing through P and Q, and always wholly within the portion of space considered. Thus, when all the paths between any two points are reconcilable, the space is simply continuous. When there are just two sets of paths, so that each of one set is irreconcilable with any one of the other set, the space is doubly continuous; when there are three such sets it is triply continuous, and so on. To avoid circumlocutions, we shall suppose S to be the boundary of a hollow space in the interior of a solid mass, so thick that no operations which we shall consider

shall ever make an opening to the space outside it. A tunnel through this solid opening at each end into the interior space constitutes the whole space doubly continuous; and if more tunnels be made, every new one adds one to the degree of multiple continuity. When one such tunnel has been made, the surface of the tunnel is continuous with the whole bounding surface of the space considered; and in reckoning degrees of continuity, it is of no consequence whether the ends of any fresh tunnel be in one part or another of this whole surface. Thus, if two tunnels be made side by side, a hole anywhere opening from one of them into the other adds one to the degree of multiple continuity. Any solid detached from the outer bounding solid, and left, whether fixed or movable in the interior space, adds to the bounding surface an isolated portion, but does not interfere with the reckoning of multiple continuity. Thus, if we begin with a simply continuous space bounded outside by the inner surface of the supposed external solid, and internally by the boundary of the detached solid in its interior, and if we drill a hole in this solid we produce double continuity. Two holes, or two solids in the interior each with one hole (such as two ordinary solid rings), constitute triple continuity, and so on. A sponge-like solid whose pores communicate with one another, illustrates a high degree of multiple continuity, and it is of no consequence whether it is attached to the external bounding solid or is an isolated solid in the interior. Another type of multiple continuity, that presented by two rings linked in one another, was referred to in § 57.

When many rings are linked into one another in various combinations, there are complicated mutual intersections of the several partial barriers β_1, β_2, ... required to stop all multiple continuity. But without having any portion of the bounding solid detached, as in that case in which one at least of the two rings is loose, we have varieties of multiple continuity curiously different from that illustrated by a single ordinary straight or bent tunnel, illustrated sufficiently by the simplest types, which are obtained by boring a tunnel along a line agreeing in form with the axis of a cord or wire on which a simple knot is tied; and by fixing the two ends of wire with a knot on it to the bounding solid, so that the surface of the wire shall become part of the bounding surface of the space considered, the knot not being pulled tight, and the wire being arranged not to touch itself in any point; or by placing a knotted wire, with its ends united, in the interior of the space. No amount of knotting or knitting, however complex, in the cord whose axis indicates the line of tunnel, complicates in any way the continuity of the space considered, or alters the simplicity of the barrier surface required to stop the circulation. But it is otherwise when a knotted or knitted wire forms part of the bounding solid. A single simple knot, though giving only double continuity, requires a curiously self-cutting surface for stopping barrier: which, in its form of minimum area, is beautifully shown by the liquid film adhering to an endless wire, like the first figure, dipped in a soap solution and removed. But no complication of these types, or of combinations of them with one another, eludes the statements and formulæ of § 57.

Instalment, received Nov.—Dec. 1869 [§ 59—§ 64 (*f*)].

59. I shall now give a dynamical lemma, for the immediate object of preparing to apply Green's corrected theorem (§ 57) to the motion of a liquid through a multiply continuous space. But later we shall be led by it to very simple demonstrations of Helmholtz's fundamental theorems of vortex motion; and shall see that it may be used as a substitute for the common equations of hydrokinetics.

(Lemma.) An endless finite tube* of infinitesimal normal section, being given full of liquid (whether circulating round

* A finite length of tube with its ends done away by uniting them together.

through it, or at rest) is altered in shape, length, and normal section, in any way, and with any speed. The average value of the component velocity of the fluid along the tube, reckoned all round the circuit (irrespectively of the normal section), varies inversely as the length of the circuit.

59 (a)   To prove this, consider first a single particle of unit mass, acted on by any force, and moving along a smooth guiding curve, which is moved and bent about quite arbitrarily. Let ρ be the radius of curvature, and ξ, η the component velocities of the guiding curve, towards the centre of curvature, and perpendicular to the plane of curvature, at the point P, through which the moving particle is passing at any instant. Let ζ be the component velocity of the particle itself, along the instantaneous direction of the tangent through P. Thus ξ, η, ζ are three rectangular components of the velocity of the particle itself. Let $\mathbf{Z}$ be the component in the direction of ζ, of the whole force on P. We have, by elementary kinetics,

$$\frac{d\zeta}{dt} = \mathbf{Z} + \frac{\zeta\xi}{\rho} + \xi\frac{d\xi}{ds} + \eta\frac{d\eta}{ds} \quad \dots\dots\dots\dots(1)^*,$$

* This theorem (not hitherto published ?) will be given in the second volume of Thomson and Tait's *Natural Philosophy*. It may be proved analytically from the general equations of the motion of a particle along a varying guide-curve (Walton, *Cambridge Mathematical Journal*, 1842, February); or more synthetically, thus— Let l, m, n be the direction cosines of PT, the tangent to the guide at the point through which the particle is passing at any instant; (x, y, z) the co-ordinates of this point, and $(\dot{x}, \dot{y}, \dot{z})$ its component velocities parallel to fixed rectangular axes. We have

$$\zeta = l\dot{x} + m\dot{y} + n\dot{z}; \quad \text{and} \quad \mathbf{Z} = l\ddot{x} + m\ddot{y} + n\ddot{z},$$

and from this

$$\frac{d\zeta}{dt} = l\ddot{x} + m\ddot{y} + n\ddot{z} + \dot{l}\dot{x} + \dot{m}\dot{y} + \dot{n}\dot{z} = \mathbf{Z} + \dot{l}\dot{x} + \dot{m}\dot{y} + \dot{n}\dot{z}.$$

But it is readily proved (Thomson and Tait's *Natural Philosophy*, § 9, to be made more explicit on this point in a second edition) that the angular velocity with which PT changes direction is equal to $\sqrt{(\dot{l}^2 + \dot{m}^2 + \dot{n}^2)}$, and, if this be denoted by ω, that

$$\frac{\dot{l}}{\omega}, \ \frac{\dot{m}}{\omega}, \ \frac{\dot{n}}{\omega}$$

are the direction cosines of the line PK, perpendicular to PT in the plane in which PT changes direction, and on the side towards which it turns. Hence,

$$\frac{d\zeta}{dt} = \mathbf{Z} + \kappa\omega$$

if κ denote the component velocity of P along PK. Now, if the curve were fixed we should have $\omega = \zeta/\rho$, by the kinematic definition of curvature (Thomson and Tait, § 5); and the plane in which PT changes direction would be the plane of curvature. But in the case actually supposed, there is also in this plane an additional angular

48 HYDRODYNAMICS [2

where ρ denotes the radius of curvature, and $d\xi/ds$, $d\eta/ds$ rates of variation of ξ and η from point to point along the curve at one time.

59 (b)   Now, instead of a single particle of unit mass, let an infinitesimal portion, μ, of a liquid, filling the supposed endless tube, be considered.   Let ϖ be the area of the normal section of the tube in the place where μ is, and δs the length along the tube of the space occupied by it, at any instant; so that (as the density of the fluid is called unity),

$$\mu = \varpi \delta s.$$

Further, let dp/ds denote the rate of variation of the fluid pressure along the tube, so that

$$\mathcal{Z} = -\varpi \frac{dp}{ds} \delta s.$$

Thus we have, by (1)

$$\frac{d\zeta}{dt} = \frac{\zeta\xi}{\rho} + \xi\frac{d\xi}{ds} + \eta\frac{d\eta}{ds} - \frac{dp}{ds} \quad\quad\dots\dots\dots\dots\dots(2).$$

(c)   Now, because the two ends of the arc δs move with the fluid, we have, by the kinematics of a varying curve,

$$\frac{d\delta s}{dt} = \frac{d\zeta}{ds}\delta s - \frac{\xi}{\rho}\delta s \quad\quad\dots\dots\dots\dots\dots(3);$$

and, therefore, $$\frac{d(\zeta\delta s)}{dt} = \frac{d\zeta}{dt}\delta s + \zeta\left(\frac{d\zeta}{ds}\delta s - \frac{\xi}{\rho}\delta s\right)\dots\dots\dots\dots(4).$$

Substituting in this for $d\zeta/dt$ its value by (2), we have

$$\frac{d(\zeta\delta s)}{dt} = \left(\xi\frac{d\xi}{ds} + \eta\frac{d\eta}{ds} - \frac{dp}{ds} + \zeta\frac{d\zeta}{ds}\right)\delta s,$$

or $$\frac{d(\zeta\delta s)}{dt} = \delta\left(\tfrac{1}{2}q^2 - p\right) \quad\quad\dots\dots\dots\dots(5),$$

velocity equal to $d\xi/ds$, and a component angular velocity in the plane of PT and η, equal to $d\eta/ds$; due to the normal motion of the varying curve.   Hence the whole angular velocity ω is the resultant of two components,

$$\frac{\zeta}{\rho} + \frac{d\xi}{ds} \text{ in the plane of } \xi,$$

and $$\frac{d\eta}{ds} \text{ in the plane of } \eta.$$

Hence $$\xi\left(\frac{\zeta}{\rho} + \frac{d\xi}{ds}\right) + \eta\frac{d\eta}{ds} = \kappa\omega,$$

and the formula (1) of the text is proved.

if q denote the resultant fluid velocity; and δ the differences for the two ends of the arc δs. Integrating this through the length of any finite arc $P_1 P_2$ of the fluid, its ends P_1, P_2 moving with the fluid, we have

$$\frac{d\Sigma_1^2(\zeta\delta s)}{dt} = (\tfrac{1}{2}q^2 - p)_2 - (\tfrac{1}{2}q^2 - p)_1 \quad \ldots\ldots\ldots\ldots(6),$$

the suffixes denoting the values of the bracketed function, at the points P_2 and P_1, respectively; and Σ_1^2 denoting integration along the arc from P_1 to P_2. Let now P_2 be moved forward, or P_1 backward, till these points coincide, and the arc P_1P_2 becomes the complete circuit; and let Σ denote integration round the whole closed circuit. (6) becomes

$$\frac{d\Sigma(\zeta\delta s)}{dt} = 0 \quad \ldots\ldots\ldots\ldots\ldots\ldots(7);$$

and we conclude that $\Sigma\zeta\delta s$ remains constant, however the tube be varied. This is the proposition to be proved, as the "average velocity" referred to is found by dividing $\Sigma(\zeta\delta s)$ by the length of the tube.

59 (d) The tube, imagined in the preceding, has had no other effect than exerting, by its inner surface, normal pressure on the contained ring of fluid. Hence the proposition* at the beginning

---

* Equation (6), from which, as we have seen, that proposition follows immediately, may be proved with greater ease, and not merely for an incompressible fluid, but for any fluid in which the density is a function of the pressure, by the method of rectilineal rectangular co-ordinates from the ordinary hydrokinetic equations. These equations are

$$\frac{Du}{Dt} = -\frac{d\varpi}{dx}, \quad \frac{Dv}{Dt} = -\frac{d\varpi}{dy}, \quad \frac{Dw}{Dt} = -\frac{d\varpi}{dz},$$

if D/Dt denote rate of variation per unit of time, of any function depending on a point or points *moving with the fluid;* and $\varpi = \int dp/\rho$, ρ denoting density. In terms of rectilineal co-ordinates we have

$$\zeta\delta s = u\delta x + v\delta y + w\delta z.$$

Hence
$$\frac{D(\zeta\delta s)}{Dt} = \frac{Du}{Dt}\,\delta x + u\,\frac{D\delta x}{Dt} + \&c.$$

Now
$$\frac{D\delta x}{Dt} = \delta u, \quad \frac{D\delta y}{Dt} = \delta v, \quad \text{and} \quad \frac{D\delta z}{Dt} = \delta w.$$

These and the kinetic equations reduce the preceding to

$$\frac{D(\zeta\delta s)}{Dt} = u\delta u + v\delta v + w\delta w - \frac{d\varpi}{dx}\,\delta x - \frac{d\varpi}{dy}\,dy - \frac{d\varpi}{dz}\,\delta z = \delta\left[\tfrac{1}{2}(u^2 + v^2 + w^2) - \varpi\right]\ldots(8);$$

whence, by Σ integration, equation (6) generalised to apply to compressible fluids.

of § 59 is applicable to any closed ring of fluid forming part of an incompressible fluid mass extending in all directions through any finite or infinite space, and moving in any possible way; and the formulæ (5) and (6) are applicable to any infinitesimal or infinite arc of it with two ends not met.   Thus in words—

PROP. (1).   *The line-integral of the tangential component velocity round any closed curve of a moving fluid remains constant through all time.*

And, PROP. (2), The rate of augmentation, per unit of time, of the space integral of the velocity along any terminated arc of the fluid is equal to the excess of the value of $\frac{1}{2}q^2 - p$, at the end towards which tangential velocity is reckoned as positive, above its value at the other end.

59 (*e*)   The condition that $u\,dx + v\,dy + w\,dz$ is a complete differential [proved above (§ 13) to be the criterion of irrotational motion] means simply

That the flow [defined § 60 (*a*)] *is the same in all different mutually reconcilable lines from one to another of any two points in the fluid;* or, which is the same thing,

That the circulation [§ 60 (*a*)] *is zero round every closed curve capable of being contracted to a point without passing out of a portion of the fluid through which the criterion holds.*

From Proposition (1), just proved, we see that this condition holds through all time for any portion of a moving fluid for which it holds at any instant; and thus we have another proof of Lagrange's celebrated theorem (§ 16), giving us a new view of its dynamical significance, which [see for example § 60 (*g*)] we shall find of much importance in the theory of vortex motion.

(*f*)   But it is only in a closed curve, *capable of being contracted to a point without passing out of space occupied by irrotationally moving fluid,* that the circulation is necessarily *zero,* in irrotational motion.   In § 57 we saw that a continuous fluid mass, occupying doubly or multiply continuous space, may move altogether irrotationally, yet so as to have finite circulation in a closed curve $PP'QQ'P$, provided $PP'Q$ and $PQ'Q$ are "irreconcilable paths" between P and Q.   *That the circulation must be the same in all mutually reconcilable closed curves* (compare § 57), is an immediate consequence from the now proved [§ 59 (Prop. 2)] equality of the flows [§ 60 (*a*)] in all mutually reconcilable conterminous arcs. For by leaving one part of a closed curve unchanged, and varying the remaining arc continuously, no change is produced in the flow, in this part; and, by repetitions of the process, a closed curve may be changed to any other reconcilable with it.

60. *Definitions and elementary propositions.* (*a*) The line-integral of the tangential component velocity along any finite line, straight or curved, in a moving fluid, is called the flow in that line. If the line is endless (that is, if it forms a closed curve or polygon), the *flow* is called *circulation*. The use of these terms abbreviates the statements of Propositions (2) and (1) of § 59 to the following :—

[§ 59, Prop. (2)]. The rate of augmentation, per unit of time, of the flow in any terminated line which moves with the fluid, is equal to the excess of the value of $\frac{1}{2}q^2 - p$ at the end from which, above its value at the end towards which, positive flow is reckoned.

[§ 59, Prop. (1)]. The circulation in any closed line moving with the fluid, remains constant through all time.

(*b*) If any open finite surface, lying altogether within a fluid, be cut into parts by lines drawn across it, the circulation in the boundary of the whole is equal to the sum of the circulations in the boundaries of the parts. This is obvious, as the latter sum consists of an equal positive and negative flow in each portion of the boundary common to two parts, added to the sum of the flows in all the parts into which the single boundary of the whole is divided.

(*c*) Hence the circulation round the boundaries of infinitesimal areas, infinitely near one another in one plane, are simply proportional to these areas.

(*d*) *Proposition.* Let any part of the fluid rotate as a solid (that is, without changing shape); or consider simply the rotation of a solid. The " circulation " in the boundary of any plane figure moving with it is equal to twice the area enclosed, multiplied by the component angular velocity in that plane (or round an axis perpendicular to that plane). For, taking r, θ to denote polar co-ordinates of any point in the boundary, A the enclosed area, and

ω the component angular velocity in the plane, and continuing the notation of § 59, we have

$$\zeta = r\omega \frac{rd\theta}{ds},$$

and therefore

$$\Sigma \zeta \delta s = \omega \Sigma r^2 \frac{d\theta}{ds} \delta s = \omega \Sigma r^2 \delta\theta = \omega \times 2A.$$

60 (*e*) *Definition.* For a fluid moving in any manner, the circulation round the boundary of an infinitesimal plane area, divided by double the area, is called the *component rotation* in that plane (or round an axis perpendicular to that plane) of the neighbouring fluid.

In this statement, the single word "rotation" is used for *angular velocity of rotation:* and the definition is justified by (*c*) and (*d*); also by § 13 (2) above, applied to (*p*) below. It agrees, in virtue of (*p*), with the definition of rotation in fluid motion given first of all, I believe, by Stokes, and used by Helmholtz in his memorable *Vortex Motion,* also in Thomson and Tait's *Natural Philosophy,* §§ 182 and 190 (*j*).

(*f*) *Proposition.* If ξ, η, ζ be the components of rotation at any point, P, of a fluid, round three axes at right angles to one another, and ω the component round an axis, making with them angles whose cosines are l, m, n,

$$\omega = \xi l + \eta m + \zeta n.$$

To prove this, let a plane perpendicular to the last-mentioned axis cut the other three in A, B, C. The circulation in the periphery of the triangle ABC is, by (*b*), equal to the sum of the circulations in the peripheries PBC, PCA, and PAB. Hence, calling Δ and α, β, γ the areas of these four triangles, we have, by (*e*),

$$\omega\Delta = \xi\alpha + \eta\beta + \zeta\gamma.$$

But α, β, γ are the projections of Δ on the planes of the pairs of the rectangular axes; and so the proposition is proved.

It follows, of course, that the composition of rotations in a fluid fulfils the law of the compositions of angular velocities of a solid, of linear velocities, of forces, &c.

(*g*) Hence, in any infinitesimal part of the fluid, the circulation is zero in the periphery of every plane area passing through a certain line;—the resultant axis of rotation of that part of the fluid. But (a) the circulation remains zero in every closed line moving with the fluid, for which it is zero at any time. Hence

60 (h)   The axial lines [defined (i)] move with the fluid.

(i) *Definition.* An axial line through a fluid moving rotationally, is a line (straight or curved) whose direction at every point coincides with the resultant axis of rotation through that point.

(j) *Proposition.* The resultant rotation of any part of the fluid varies in simple proportion to the length of an infinitesimal arc of the axial line through it, terminated by points moving with the fluid. To prove this, consider any infinitesimal plane area, A, moving with the fluid. Let ω be the resultant rotation, and θ the angle between its axis and the perpendicular to the plane of A. This makes $\omega \cos \theta$ the component rotation in the plane of A ; and therefore $A\omega \cos \theta$ remains constant. Now, draw axial lines through all points of the boundary of A, forming a tube whose area of normal section is $A \cos \theta$. The resultant rotation must vary inversely as this area, and therefore (in consequence of the incompressibility of the fluid) directly as the length of an infinitesimal line along the axis.

(k)   Form a surface by axial lines drawn through all points of any curve in the fluid. The circulation is zero round the boundary of any infinitesimal area of this surface; and therefore (b) it is zero round the boundary of any finite area of it.

(l)   Let the curve of (k) be closed, and therefore the surface tubular. On this surface let $ABCA$, $A'B'C'A'$ be any two curves closed round the tube, and ADA' any arc from A to A'. The circulation in the closed path, $ADA'B'C'A'DACBA$, is zero by (h). Hence the circulation in $ABCA$ is equal to the circulation in $A'B'C'A'$—that is to say,
The circulations are equal in all circuits of a vortex tube.

(m) *Definitions.* An *axial surface* is a surface made up of axial lines. A *vortex tube* is an axial surface through every point of which a finite endless path, cutting every axial line it meets, can be drawn. Any such path, passing just once round, is called *a circuit* or, *the circuit* of the tube. The *rotation of a vortex tube* is the circulation in its circuit.   A *vortex sheet* is (a portion as it were of a collapsed vortex tube) a surface on the two sides of which the fluid moves with different tangential component velocities.

60 (*n*)   Draw any surface cutting a vortex tube, and bounded by it.   The surface integral of the component rotation round the normal has the same value for all such surfaces; and this common value is what we now call the rotation of the tube.

(*o*)   In an unbounded infinite fluid, an axial tube must be either finite and endless or infinitely long in each direction*.   In an infinite fluid with a boundary (for instance, the surface of an enclosed solid), an axial tube may have two ends, each in the boundary surface; or it may have one end in the boundary surface, and no other; or it may be infinitely long in each direction, or it may be finite and endless.   In a finite fluid mass, an axial tube may be endless, or may have one end, but, if so, must have another, both in the boundary surface.

(*p*)   *Proposition.*   Applying the notation of (*f*), to axes parallel to those of co-ordinates *x, y, z,* and denoting, as formerly, by *u, v, w,* the components of the fluid velocity at (*x, y, z*), we have

$$\xi = \tfrac{1}{2}\left(\frac{dw}{dy} - \frac{dv}{dz}\right), \quad \eta = \tfrac{1}{2}\left(\frac{du}{dz} - \frac{dw}{dx}\right), \quad \zeta = \tfrac{1}{2}\left(\frac{dv}{dx} - \frac{du}{dy}\right).$$

The proof is obvious, according to the plan of notation, &c., followed in § 13 above.

(*q*)   Hence by (*f*), (*e*), and (*b*)

$$\iint dS \left\{ l\left(\frac{dw}{dy} - \frac{dv}{dz}\right) + m\left(\frac{du}{dz} - \frac{dw}{dx}\right) + n\left(\frac{dv}{dx} - \frac{du}{dy}\right) \right\}$$
$$= \int (u\,dx + v\,dy + w\,dz),$$

where $\iint dS$ denotes integration over any portion of surface bounded by a closed curve; $\int (u\,dx + \&c.)$ integration round the whole of this curve; and (*l, m, n*) the direction cosines of any point (*x, y, z*) in the surface.   It is worthy of remark that the equation of continuity for an incompressible fluid does not enter into the

---

* Vortex tubes apparently ending in the fluid, for instance, a portion of fluid bounded by a figure of revolution, revolving round its axis as a solid, constitute no exception.   Each infinitesimal vortex tube in this case is completed by a strip of vortex sheet and so is endless.

demonstration of this proposition, and therefore u, v, w may be any functions whatever of x, y, z. In a purely analytical light the result has an important bearing on the theory of the integration of complete or incomplete differentials. It was first given, with the indication of a more analytical proof than the preceding, in Thomson and Tait's *Natural Philosophy*, § 190 (*j*).

60 (*r*)　Propositions (*h*) (*j*) (*n*) (*o*) of the present section (§ 60) are due to Helmholtz; and with his integration for associated rotational and cyclic irrotational motion in an unbounded fluid, to be given below, constitute his general theory of vortex motion. (*n*) and (*o*) are purely kinematical; (*h*) and (*j*) are dynamical.

(*s*)　Henceforth I shall call *a circuit* any closed curve not continuously reducible to a point, in a multiply continuous space. I shall call *different circuits*, any two such closed curves if mutually irreconcilable (§ 58), but different mutually reconcilable closed curves will not be called different circuits.

(*t*)　Thus, $(n+1)$ply continuous space, is a space for which there are n, and only n, different circuits. This is merely the definition of § 58, abbreviated by the definite use of the word circuit, which I now propose. The general terminology regarding simply and multiply continuous spaces is, as I have found since § 58 was written, altogether due to Helmholtz; Riemann's suggestion, to which he refers, having been confined to two-dimensional space. I have deviated somewhat from the form of definition originally given by Helmholtz, involving, as it does, the difficult conception of a stopping barrier*; and substituted for it the definition by reconcilable and irreconcilable paths. It is not easy to conceive the stopping barrier of any one of the first three diagrams of § 58, or to understand its singleness; but it is easy to see that in each of those three cases, any two closed curves drawn round the solid wire represented in the diagrams are reconcilable, according to the definition of this term given in § 58, and

---

* But without this conception we can make no use of the theory of multiple continuity in hydrokinetics (see §§ 61–63), and Helmholtz's definition is, therefore, perhaps preferable after all to that which I have substituted for it. Mr Clerk Maxwell tells me that J. B. Listing has more recently treated the subject of multiple continuity in a very complete manner in an article entitled " Der Census räumlicher Complexe."—*Königl. Ges. Göttingen,* 1861. See also Prof. Cayley "On the Partition of a Close."—*Phil. Mag.* 1861.

therefore, that the presence of any such solid adds only one to the degree of continuity of the space in which it is placed.

60 (*u*) If we call a *partition*, a surface which separates a closed space into two parts, and, as hitherto, a *barrier*, any surface edged by the boundary of the space, Helmholtz's definition of multiple continuity may be stated shortly thus:—

A space is (*n* + 1)*ply continuous if n barriers can be drawn across it, none of which is a partition.*

(*v*) Helmholtz has pointed out the importance in hydro-kinetics of many-valued functions, such as $\tan^{-1} y/x$, which have no place in the theories of gravitation, electricity, or magnetism, but are required to express electro-magnetic potentials, and the velocity potentials for the part of the fluid which moves irrotationally in vortex motion. It is, therefore, convenient, before going farther, that we should fix upon a terminology, with reference to functions of that kind, which may save us circumlocutions hereafter.

(*w*) A function $\phi(x, y, z)$ will be called *cyclic* if it experiences a constant augmentation every time a point P, of which x, y, z are rectangular rectilinear co-ordinates, is carried from any position round a certain circuit to the same position again, without passing through any position for which either $d\phi/dx$, $d\phi/dy$, or $d\phi/dz$ becomes infinite. The value of this augmentation will be called the cyclic constant for that particular circuit. The cyclic constant must clearly have the same value for all circuits mutually reconcilable (§ 58), in space throughout which the three differential coefficients remain all finite.

(*x*) When the function is cyclic with reference to several different mutually irreconcilable circuits, it is called *polycyclic*. When it is cyclic for only one set of circuits, it is called *monocyclic*.

Example.—The apparent area of a circle as seen from a point (x, y, z) anywhere in space, is a monocyclic function of x, y, z, of which the cyclic constant is 4π.

The apparent area of a plane curve of the (2*n*)th degree, consisting of *n* detached closed (that is finite endless) branches (some of which might be enclosed within others) is an *n*-cyclic function, of which the *n*-cyclic constants are essentially equal, being each 4π.

Algebraic equations among three variables (x, y, z) may easily be found to represent tortuous curves, constituting one or more finite, isolated, endless branches (which may be knotted, as shown in the first three diagrams of § 58, or linked into one another, as in the fourth and fifth). The integral expressing what, for brevity, we shall call the *apparent area* of such a curve, is a cyclic function, which, if polycyclic, has essentially equal values for all its cyclic constants. By the *apparent area of a finite endless curve* (tortuous or plane), I mean the *sum of the apparent areas of all barriers edged by it, which we can draw without making a partition*.

It is worthy of notice that every polycyclic function may be reduced to a sum of monocyclic functions.

60 (y)　Fluid motion is called *cyclic* unless the circulation is zero in every closed path through the fluid, when it is called *acyclic*. Rotational motion is (e) essentially cyclic.

(z)　Irrotational motion may [§ 59 (f)] be either acyclic or cyclic. If cyclic it is *monocyclic* if there is only one distinct circuit, or *polycyclic* if there are several distinct circuits, in which there is circulation. It is *purely cyclic* if the boundary of the space occupied by irrotationally moving fluid is at rest. If the boundary moves and the motion of the fluid is cyclic, it is *acyclic compounded with cyclic*.

61 (a)　We are now prepared to investigate the most general possible irrotational motion of a single continuous fluid mass, occupying either simply or multiply continuous space, with, for every point of the boundary, a normal component velocity given arbitrarily, subject only to the condition that the whole volume remains unaltered.

(b)　*Genesis of a cyclic motion.* Commencing, as in § 3, with a fluid mass at rest throughout, let all multiplicity of the continuity of the space occupied by it be done away with by temporary barrier surfaces, β_1, β_2... stopping the circuits, as described in § 57. The bounding surface of the fluid, which ordinarily consists of the inner surface of the containing vessel, will thus be temporarily extended to include each side of each of these barriers. Let now, as in § 3, any possible motion be arbitrarily given to the bounding surface. The liquid is consequently set in motion, purely through fluid pressure; and the motion is [§§ 10–15, or 60, 59] throughout irrotational. Hence irrotational motion fulfilling the prescribed surface conditions is possible, and the actual motion is, of course (as the solution of every real problem is), unambiguous. But from this bare physical principle we could not even suspect, what the following simple application of Green's equation proves, that the surface normal velocity at any instant determines the interior motion irrespectively of the previous history of the motion from rest.

61 (c) *Determinacy of irrotational motion in simply continuous space.* In § 57 (1), which is immediately applicable, as the volume is now simply continuous, make $\phi' = \phi$, and put $\nabla^2\phi = 0$, so that ϕ may be the velocity potential of an incompressible fluid. That double equation becomes the following single equation

$$\iiint\left(\frac{d\phi^2}{dx^2} + \frac{d\phi^2}{dy^2} + \frac{d\phi^2}{dz^2}\right) dx\,dy\,dz = \iint d\sigma\,\phi\,\eth\phi,$$

where the surface integration $\iint d\sigma$ must now include each side of each of the barrier surfaces $\beta_1, \beta_2 \ldots$. Hence, if $\eth\phi = 0$ for every point of the bounding surface, we must have

$$\iiint\left(\frac{d\phi^2}{dx^2} + \frac{d\phi^2}{dy^2} + \frac{d\phi^2}{dz^2}\right) dx\,dy\,dz = 0,$$

which requires that

$$\frac{d\phi}{dx} = 0, \quad \frac{d\phi}{dy} = 0, \quad \frac{d\phi}{dz} = 0 :$$

that is to say, if there is no motion of the boundary surface in the direction of the normal, there can be no motion of the irrotational species in the interior; whence it follows that there cannot be two different internal irrotational motions with the same surface normal component velocities. Thus, as a particular case, beginning with a fluid at rest, let its boundary be set in motion; and brought again to rest at any instant, after having been changed in shape to any extent, through any series of motions. The whole liquid comes to rest at that instant.

A demonstration of this important theorem, which differs essentially from the preceding, and includes what the preceding does not include, a purely analytical proof of the possibility of irrotational motion throughout the fluid, fulfilling the arbitrary surface-condition specified above, as was first published in Thomson and Tait's *Natural Philosophy*, § 317 (3), and is to be given below, with some variation and extension. In the meantime, however, we satisfy ourselves as to the *possibility* of irrotational motions fulfilling the various surface-conditions with which we are concerned, because the surface motions are possible and require the fluid to move, and [§§ 10–15, or § 59] because the fluid cannot acquire rotational motion through fluid pressure from the motion of its boundary; and we go on, by aid of Green's extended formula [§ 57 (7)], to prove the determinateness of the interior motion under conditions now to be specified for multiply continuous space, as we have done by his unaltered formula [§ 57 (1)] for simply continuous space.

62. *Genesis of cyclic irrotational motion.* In the case of motion considered in § 61, the value of the normal component velocity is not independently arbitrary over the whole boundary, but has equal arbitrary values, positive and negative, on the two sides of each of the barriers β_1, β_2, &c. We must now introduce a fresh restriction in order that, when the barriers are liquefied, the motion of the fluid may be irrotational throughout the space thus re-opened into multiple continuity. For although we have secured that the normal component velocity is equal everywhere on the two sides of each barrier, we have hitherto left the tangential velocity unheeded. If they are not equal on the two sides, and in the same direction, there will be a finite slipping of fluid on fluid across the surface left by the dissolution of the infinitely thin barrier membrane; constituting [§ 60 (m) above], as Helmholtz has shown, a " vortex sheet." The analytical expression of the condition of equality between the tangential velocities is that the variation of the velocity potential in tangential directions shall be equal on the two sides of each barrier. Hence, by integration, we see that the difference between the values of the velocity potential on the two sides must be the same over the whole of each barrier. This condition requires that the initiating pressure be equal over the whole membrane. For, at any time during the instituting of the motion, let p_1, p_2 be the pressures at two points P_1, P_2 of the fluid, and moving with the fluid, infinitely near one another on the two sides of one of the membranes, so that the pressure ϖ, which must be applied to the membrane to produce this difference of fluid pressure on the two sides, is equal to $p_1 - p_2$ in the direction opposed to p_1. And let

ϕ_1, ϕ_2 be the velocity potentials at P_1 and P_2, so that if $\int ds$ denote integration from P_1 to P_2, along any path $P_1 P P_2$ whatever from P_1 to P_2, altogether through the fluid (and therefore cutting none of the membranes), and ζ the component of fluid velocity along the tangent at any point of this curve, we have

$$\int \zeta ds = \phi_2 - \phi_1 \quad \ldots\ldots\ldots\ldots\ldots\ldots(1).$$

Hence, by (6) of § 59,

$$\frac{d\,(\phi_2 - \phi_1)}{dt} = \varpi - \tfrac{1}{2}\,(q_1{}^2 - q_2{}^2) \quad \ldots\ldots\ldots\ldots(2),$$

where q_1, q_2 denote the resultant fluid velocities at P_1 and P_2. Now, the normal component velocities at P_1 and P_2 are necessarily equal; and therefore, if the components parallel to the tangent plane of the intervening membrane are also equal, we have

$$q_1 = q_2$$

and the preceding becomes

$$\frac{d\,(\phi_2 - \phi_1)}{dt} = \varpi \quad \ldots\ldots\ldots\ldots\ldots\ldots(3).$$

But if the tangential component velocities at P_1 and P_2 are not only equal, but in the same direction, $\phi_2 - \phi_1$ must, as we have seen, be constant over the membrane, and therefore ϖ must also be constant.

Suppose now that after pressure has been applied for any time in the manner described, of uniform value all over the membrane at each instant, it is applied no longer, and the membrane (having no longer any influence) is done away with. The fluid mass is left for ever after in a state of motion, which is irrotational throughout, but cyclic. The "circulation" [§ 60 (a)], or the cyclic constant being equal to $\phi_2 - \phi_1$, for every circuit reconcilable with $P_1 P P_2 P_1$ is given by the equation

$$\phi_2 - \phi_1 = - \int \varpi dt \ldots\ldots\ldots\ldots\ldots\ldots(4),$$

$\int dt$ denoting a time-integral extended through the whole period during which ϖ had any finite value.

The same kind of operation may be performed, on each of the n barriers temporarily introduced in § 61 to reduce the $(n+1)$fold continuity of the space occupied by the fluid, to simple continuity.

The velocity potential at any point of the fluid will then be a polycyclic function [§ 60 (x)] equal to the sum of the separate values corresponding to the pressure separately applied to the several barriers. Thus we see how a state of irrotational motion, cyclic with reference to every one of the different circuits of a multiply continuous space, and having arbitrary values for the corresponding cyclic constants, or circulations, may be generated. But the proof of the possibility of fluid motion fulfilling such conditions, founded on this planning out of a genesis of it, leaves us to imagine that it might be different according to the infinitely varied choice we may make of surfaces for the initial forms of the barriers, or according to the order and the duration of the applications of pressure to them in virtue of which these figures may be changed more or less, and in various ways, before the initiating pressures all cease; and hitherto we have seen no reason even to suspect the following proposition to the contrary.

63. (PROP.) The motion of a liquid moving irrotationally within an $(n+1)$ply continuous space is determinate when the normal velocity at every point of the boundary, and the values of the circulations in the n circuits, are given.

This is proved by an application of Green's extended formula (7) of § 57, showing, as the simple formula (1) of the same section showed us in § 61 for simply continuous space, that the difference of the velocity potentials of two motions, each fulfilling this condition, is necessarily zero throughout the whole fluid. Let ϕ, ϕ' be the velocity potentials of two motions fulfilling the prescribed conditions, and let

$$\psi = \phi - \phi'.$$

At every point of the boundary (the barriers not included) the prescribed conditions require that $\eth\phi = \eth\phi'$, and therefore $\eth\psi = 0$. Again, the cyclic constants for ϕ' are equal to those for ϕ; those for ψ, being their differences, must therefore vanish. Hence, if the ϕ and ϕ' of § 57 (7) be made equal to one another and to avoid confusion with our present notation we substitute ψ for each, the second members of that double equation vanish, and it becomes simply

$$\iiint\left(\frac{d\psi^2}{dx^2} + \frac{d\psi^2}{dy^2} + \frac{d\psi^2}{dz^2}\right) dx\,dy\,dz = 0;$$

which, as before (§ 61), proves that $\psi = 0$, and therefore $\phi' = \phi$; and so establishes our present proposition.

EXAMPLE (1).   The solution $\phi = \tan^{-1} y/x$ considered in § 56, fulfils Laplace's equation, $\nabla^2\phi = 0$; and obviously satisfies the surface condition, not merely for the annular space with rectangular meridional section there considered, but for the hollow space bounded by the figure of revolution obtained by carrying a closed curve of any shape round any axis (OZ) not cutting the curve; which, for brevity, we shall in future call a *hollow circular ring*. Hence the irrotational motion possible within a fixed hollow circular ring is such that the velocity potential is proportional to the angle between the meridian plane through any point, and a fixed meridian.

EXAMPLE (2).   The solid angle, α, subtended at any point (x, y, z) by an infinitesimal plane area, A, in any fixed position, fulfils Laplace's equation $\nabla^2\alpha = 0$. This well-known proposition may be proved by taking A at the origin, and perpendicular to OX, when we have

$$\alpha = \frac{Ax}{(x^2 + y^2 + z^2)^{\frac{3}{2}}} = A\,\frac{d}{dz}\,\frac{-1}{(x^2 + y^2 + z^2)^{\frac{1}{2}}}\ \ \ \ldots\ldots\ldots(5),$$

for which $\nabla^2\alpha = 0$ is verified.

The solid angle subtended at (x, y, z) by any single closed circuit is the sum of those subtended at the same point by all parts into which we may divide any limited surface having this curve for its bounding edge. [Consider particularly curves such as those represented by the first three diagrams of § 58.] Hence if we call ϕ the solid angle subtended at (x, y, z) by this surface, Laplace's equation $\nabla^2\phi$ is fulfilled. Hence ϕ represents the velocity potential of the irrotational motion possible for a liquid contained in an infinite fixed closed vessel, within which is fixed, at an infinite distance from the outer bounding surface, an infinitely thin wire bent into the form of the closed curve in question.

The particular case of this example for which the curve is a circle, presents us with the simplest specimen of cyclic irrotational motion not confined [as that of Example (1) is] to a set of parallel planes. The velocity potential being the apparent area of a circular disc (or the area of a spherical ellipse) is readily found, and shown to be expressible readily in terms of a complete elliptic integral of the third class, and therefore in terms of incomplete elliptic functions of the first and second classes. The equi-potential surfaces are therefore traceable by aid of Legendre's tables.   But it is to Helmholtz that we owe the remarkable and useful discovery, that the equations of the *stream lines* (or lines perpendicular to the equi-potential surfaces) are expressible in terms of complete integrals of the first and second classes. They are therefore easily traceable by aid of Legendre's tables. The annexed diagram, of which we shall make much use later, shows these curves as calculated and drawn by Mr Macfarlane from Helmholtz's formula, expressed in terms of rectangular co-ordinates. An improved method of tracing them is described in a note by Mr Clerk Maxwell*, which he has kindly allowed me to append to this paper.

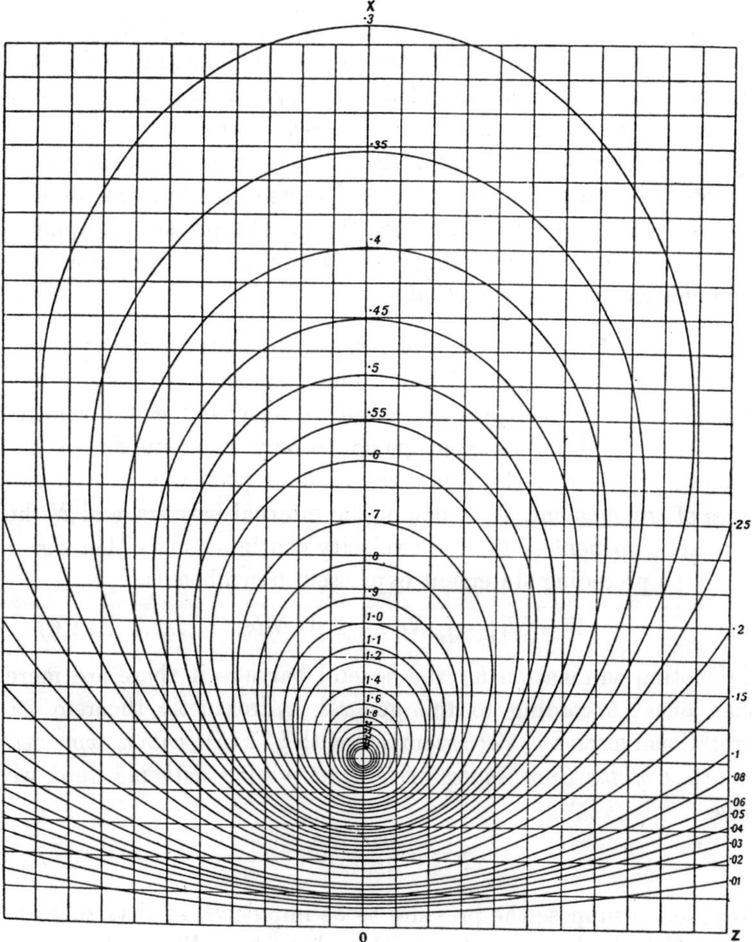

[* Cf. Maxwell's *Electricity and Magnetism*, vol. II.]

EXAMPLE (3). The motion described in Example (2) will remain unchanged outside any solid ring formed by solidifying and reducing to rest a portion of the fluid bounded by stream lines surrounding the infinitely thin wire. Thus we have a solid thick endless wire or bar forming a ring, or an endless knot as illustrated in the first three diagrams of § 59, of peculiar sectional figure depending on the stream lines round the arbitrary curve of Example (2); and the cyclic irrotational motion which, if placed in an infinite liquid it permits, is that whose velocity potential is proportional to the solid angle defined geometrically in the general solution given under Example (2).

64. *Kinetic energy of compounded acyclic and polycyclic irrotational motion—kinetico-statics.* The work done in the operation described in § 62 is calculated directly by summing the products of the pressure into an infinitesimal area of the surface, into the space through which the fluid contiguous with this area moves in the direction of the normal, for all parts of the surface, whether boundary or internal barrier, where the genetic pressure is applied, and for all infinitesimal divisions of the whole time from the commencement of the motion.

(*a*) Let w denote the work done, and $\int dt$ time-integration, from the beginning of motion up to any instant. At any previous instant let p be the pressure, q the velocity, and ϕ the velocity potential, of the fluid contiguous to any element $d\sigma$ of the bounding surface, k the difference of fluid pressures on the two sides of any element, $d\varsigma$, of one of the internal barriers, and N the normal component of the fluid velocity contiguous to either $d\sigma$ or $d\varsigma$. The preceding statement expressed in symbols is

$$W = \int dt \left[-\iint pN d\sigma + \Sigma \iint kN d\varsigma \right] \dots \dots \dots (6),$$

Σ denoting summation for the several barriers if there are more than one. According to the general hydrokinetic theorem for irrotational motion [§ 59 (6) compare with § 31 (5)], *with ϕ expressed in terms of the co-ordinates of a point moving with the fluid*, we have

$$p = -\frac{d\phi}{dt} + \tfrac{1}{2}q^2 \quad \dots \dots \dots \dots \dots \dots (7).$$

Now, let us suppose the pressure to be impulsive, so that there is infinitely little change of shape either of the bounding surface or of

the barriers during the time $\int dt$. This will also imply that $d\phi/dt$
is infinitely great in comparison with $\frac{1}{2}q^2$; so that

$$p = -\frac{d\phi}{dt} \quad \dots\dots\dots\dots\dots(8).$$

And according to the notation of § 57 we have

$$N = \eth\phi \quad \dots\dots\dots\dots\dots(9).$$

Also k is constant over each barrier surface.
Hence (6) becomes

$$W = \int dt \left[\iint \frac{d\phi}{dt}\eth\phi\, d\sigma + \Sigma k \iint \eth\phi\, d\varsigma \right] \quad \dots\dots\dots(10).$$

64 (b) The initiating motion of the bounding surface and
the pressures on the barriers may be varied quite arbitrarily from
the beginning to the end of the impulse; so that the history
within that period of the acquisition of the prescribed final
velocity may be altogether different, and not even simultaneous,
in different parts of the bounding surface. Thus k_1 and k_2 may
be quite different functions of t; provided only $\int k_1 dt$ and $\int k_2 dt$
have the prescribed values, which we shall denote by $\mathfrak{k}_1$ and $\mathfrak{k}_2$
respectively.

(c) But, for one example, we may suppose ϕ to have at each
instant of $\int dt$ everywhere one and the same proportion of its final
value; so that if the latter be denoted by Φ, and if we put

$$\frac{\phi}{\Phi} = m \quad \dots\dots\dots\dots\dots(11),$$

m is independent of co-ordinates of position, but may of course be
any arbitrary function of the time. Hence, observing that

$$\int dt\, m\, \frac{dm}{dt} = \tfrac{1}{2},$$

as the final value of m is 1, (10) becomes

$$W = \tfrac{1}{2}\left[\iint \Phi\eth\Phi\, d\sigma + \Sigma\mathfrak{k} \iint \eth\Phi\, d\varsigma \right] \quad \dots\dots\dots(12).$$

(d) The second member of this equation doubled agrees with
the two equal second members of (7) § 57 with ϕ and ϕ' each
made equal to Φ. And the first member of that equation becomes
twice the kinetic energy of the whole motion. Hence, when
$\phi' = \phi$, and $\nabla^2\phi = 0$, (7) of § 57 expresses the equation of energy

for the impulsive generation, of the fluid motion corresponding to velocity potential ϕ, by pressures varying throughout according to the same function of the time; the first member being twice the kinetic energy of the motion generated, and the second twice the work done in the process.

64 (e). As another example, let us suppose the initiating pressures to be so applied as first to generate a motion corresponding to velocity potential ϕ, and after that to change the velocity potential from ϕ to $\phi + \phi'$, denoting by ϕ and ϕ' any two functions, such that $\phi + \phi' = \Phi$, and each fulfilling Laplace's equation: and let the augmentation from zero to ϕ, and again from ϕ to $\phi + \phi'$ be uniform through the whole fluid. The work done in the first process, found as above (12),

$$\tfrac{1}{2}\left[\iint \phi\,\delta\phi\,d\sigma + \Sigma\kappa \iint \delta\phi\,ds\right] \quad \dots\dots\dots\dots(13),$$

if κ_1, κ_2, &c., denote the cyclic constants relative to ϕ, as $\Bbbk_1$, $\Bbbk_2$, &c., relatively to Φ, and the additional work done in the second process, similarly found, is

$$\tfrac{1}{2}\left[\iint \phi'\,(2\delta\phi + \delta\phi')\,d\sigma + \Sigma\kappa' \iint (2\delta\phi + \delta\phi')\,ds\right]\dots\dots(14).$$

(f) Now, as we have seen (§ 63) that the actual fluid motion depends at each instant wholly on the normal velocity at each point of the bounding surface and the values of the cyclic constants, it follows that the work done in generating it ought to be independent of the order and law of the acquisition of velocity at the bounding surface, and of the attainment of the values of the several cyclic constants. Hence, the sum of (13) and (14) ought to be equal to (12). But if for Φ in (12) we substitute $\phi + \phi'$, the difference between its value and that of the sum of (13) and (14) is found to be

$$\tfrac{1}{2}\left[\iint (\phi\,\delta\phi' - \phi'\,\delta\phi)\,d\sigma + \Sigma\,(\kappa \iint \delta\phi'\,ds - \kappa' \iint \delta\phi\,ds)\right]\dots(15);$$

which, being half the difference between the two equal second members of (7) § 57 for the case of

$$\nabla^2\phi = 0 \quad \text{and} \quad \nabla^2\phi' = 0,$$

is equal to zero. Hence, the equality of the second members of (7) § 57, constitutes the analytical reconciliation of the equations of energy for different modes of generation of the same fluid motion.

(67)

3. The Translatory Velocity of a Circular Vortex Ring.

[Appended to Prof. Tait's translation of Helmholtz's Memoir on Vortex Motion ; Phil. Mag. xxxiii. 1867, 511—512.]

FOLLOWING as nearly as may be Helmholtz's notation, let g be the radius of the circular axis of a uniform vortex-ring, and a the radius of the section of its core (which will be approximately circular when a is small in comparison with g), the vortex motion being so instituted that there is no molecular rotation in any part of the fluid exterior to this core, and that in the core the angular velocity of the molecular rotation is approximately ω, or rigorously

$$\frac{\omega \chi}{g}$$

for any fluid particle at distance χ from the straight axis.

I find that the velocity of translation is approximately equal to

$$\frac{\omega a^2}{2g}\left(\log \frac{8g}{a} - \frac{1}{4}\right)$$

(quantities of the same order as this multiplied by a/g being neglected).

The velocity of the liquid at the surface of the core is approximately constant and equal to ωa. At the centre of the ring it is $\pi \omega a^2/g$.

If these be denoted by Q and W respectively, and if T be the velocity of translation, we therefore have

$$T = \frac{a}{2g}\left(\log \frac{8g}{a} - \frac{1}{4}\right) Q$$

$$= \frac{\log \dfrac{8g}{a} - \dfrac{1}{4}}{2\pi} W.$$

Hence the velocity of translation is very large in comparison with the fluid velocity along the axis through the centre of the ring, when the section is so small that $\log 8g/a$ is large in comparison with 2π. But the velocity of translation is always small in comparison with the velocity of the fluid at the surface of the core, and the more so the smaller is the diameter of the section in comparison with the diameter of the ring.

These results remove completely the difficulty which has hitherto been felt with reference to the translation of infinitely thin vortex-filaments. I have only succeeded in obtaining them since the communication of my mathematical paper (April 29, 1867) to the Royal Society of Edinburgh, but hope to be allowed to add a proof of them to that paper should it be accepted for the Transactions.

                    (115)

10. VORTEX STATICS.

[From the *Proceedings of the Royal Society of Edinburgh*, Session 1875–76 ;
reprinted in *Phil. Mag.*, Aug. 1880.]

THE subject of this paper is *steady motion* of vortices.

1. Extended definition of "steady motion." The motion of
any system of solid, or fluid, or solid and fluid matter is said to be
steady when its configuration remains equal and similar, and the
velocities of homologous particles equal, however the configuration
may move in space, and however distant individual material
particles may at one time be from the points homologous to their
positions at another time.

2. Examples of steady and not steady motion :—

(1) A rigid body symmetrical round an axis, set to rotate
round any axis through its centre of gravity, and left free, performs
steady motion. Not so a body having three unequal principal
moments of inertia.

(2) A rigid body of any shape, in an infinite homogeneous
liquid, rotating uniformly round any, always the same, fixed line,
and moving uniformly parallel to this line, is a case of steady
motion.

(3) A perforated rigid body in an infinite liquid moving in
the manner of example (2), and having cyclic irrotational motion
of the liquid through its perforations, is a case of steady motion.
To this case belongs the irrotational motion of liquid in the
neighbourhood of any rotationally moving portion of fluid of the
same shape as the solid, provided the distribution of the
rotational motion is such that the shape of the portion endowed
with it remains unchanged. The object of the present paper
is to investigate general conditions for the fulfilment of this
proviso, and to investigate, further, the conditions of stability
of distributions of vortex motion satisfying the condition of
steadiness.

3. *General Synthetical Condition for Steadiness of Vortex Motion.* The change of the fluid's molecular rotation at any point fixed in space must be the same as if for the rotationally moving portion of the fluid were substituted a solid, with the amount and direction of axis of the fluid's actual molecular rotation inscribed or marked at every point of it, and the whole solid, carrying these inscriptions with it, were compelled to move in some manner answering to the description of example (2). If at any instant the distribution of any molecular rotation* through the fluid and corresponding distribution of fluid-velocity are such as to fulfil this condition, it will be fulfilled through all time.

4. *General Analytical Condition for Steadiness of Vortex Motion.* If, with (§ 24, below) vorticity and "impulse" given, the kinetic energy is a maximum or a minimum, it is obvious that the motion is not only steady, but stable. If, with same conditions, the energy is a maximum-minimum, the motion is clearly steady, but it may be either unstable or stable.

5. The simple circular Helmholtz ring is a case of stable steady motion, with energy maximum-minimum for given vorticity and given impulse. A circular vortex ring, with an inner irrotational annular core, surrounded by a rotationally moving annular shell (or endless tube), with irrotational circulation outside all, is a case of motion which is steady, if the outer and inner contours of the section of the rotational shell are properly shaped, but certainly unstable [if the shell be too thin]†. In this case also the energy is maximum-minimum for circular given vorticity and given impulse.

6. In these examples of steady motion, the "resultant-impulse" (V. M.‡ § 8) is a simple impulsive force, without couple: the corresponding rigid body of example (3) is a toroid; and its motion is purely translational and parallel to the axis of the toroid.

---

* One of Helmholtz's now well-known fundamental theorems shows that, *from the molecular rotation at every point of an infinite fluid*, the velocity at every point is determinate, being expressed synthetically by the same formulæ as those for finding the "magnetic resultant force" of a pure electromagnet. (Thomson's Reprint of *Papers on Electrostatics and Magnetism.*)

† [The phrase in [ ] is deleted in a copy annotated by Lord Kelvin.]

‡ My first series of papers on Vortex Motion in the *Transactions of the Royal Society of Edinburgh* will be thus referred to henceforth.

We have also exceedingly interesting cases of steady motion in which the impulse is such that, if applied to a rigid body, it would be reducible, according to Poinsot's method, to an impulsive force in a determinate line, *and a couple with this line for axis.* To this category belong certain distributions of vorticity giving longitudinal vibrations, with thickenings and thinnings of the core travelling as waves in one direction or the other round a vortex-ring, which will be investigated in a future communication to the Royal Society. In all such cases the corresponding rigid body of § 2 example (2) has both rotational and translational motion.

7. To find illustrations, suppose, first, the vorticity (defined below, § 24) and the force resultant of the impulse to be (according to the conditions explained below, § 29) such that the cross section is small in comparison with the aperture. Take a ring of flexible wire (a piece of block tin pipe* with its ends soldered together answers well), bend it into an oval form, and then give it a right-handed twist round the long axis of the oval, so that the curve comes to be not in one plane (fig. 1). A properly-shaped twisted ellipse of this kind [a shape perfectly determinate when the vorticity, the force resultant of the impulse, and the rotational moment of the impulse (V. M. § 6), are all given] is the figure of the core in what we may call the first† steady mode of single and simple toroidal vortex motion with

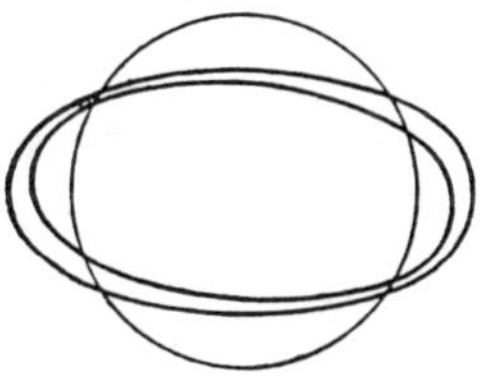

Fig. 1.

rotational moment. To illustrate the second steady mode, commence with a circular ring of flexible wire, and pull it out, at three points 120° from one another, so as to make it into as it were an equilateral triangle with rounded corners. Give now a right-handed twist, round the radius to each corner, to the plane of the curve at and near the corner; and, keeping the character of the twist thus given to the wire, bend it into a certain determinate shape proper for the data of the vortex‿motion. This is

* ["Block tin pipe" is substituted here for "very stout lead wire."]

† First or greatest, and second, and third, and higher modes of steady motion to be regarded as analogous to the first, second, third, and higher fundamental modes of an elastic vibrator, or of a stretched cord, or of steady undulatory motion in an endless uniform canal, or in an endless chain of mutually repulsive links.

the shape of the vortex-core in the second steady mode of single and simple toroidal vortex motion with rotational moment. The third is to be similarly arrived at, by twisting the corners of a square having rounded corners; the fourth, by twisting the corners of a regular pentagon having rounded corners; the fifth, by twisting the corners of a hexagon, and so on.

In each of the annexed diagrams of toroidal helices a circle is introduced to guide the judgment as to the relief above and depression below the plane of the diagram which the curve represented in each case must be imagined to have. The circle may be imagined in each case to be the circular axis of a toroidal core on which the helix may be supposed to be wound.

To avoid circumlocution, I have said "give a right-handed twist" in each case. The result in each case, as in fig. 1, illustrates a vortex motion for which the corresponding rigid body describes left-handed helices, by all its particles, round the central axis of the motion. If now, instead of right-handed twists to the plane of the oval, or the corners of the triangle, square, pentagon, &c., we give left-handed twists, as in figs. 2, 3, 4, the result in each case will be a vortex motion for which the corresponding

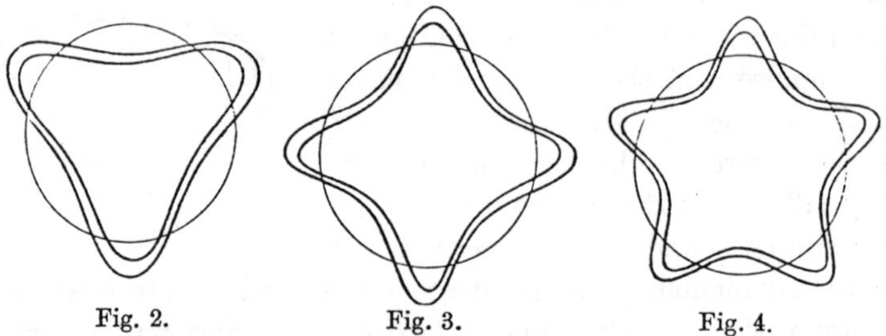

Fig. 2.   Fig. 3.   Fig. 4.

rigid body describes right-handed helices. It depends, of course, on the relation between the directions of the force resultant and couple resultant of the impulse, with no ambiguity in any case, whether the twists in the forms, and in the lines of motion of the corresponding rigid body, will be right-handed or left-handed.

8. In each of these modes of motion the energy is a maximum-minimum for given force resultant and given couple resultant of impulse. The modes successively described above are successive solutions of the maximum-minimum problem of § 4—a determinate problem with the multiple solutions indicated above, but no other

solution, when the vorticity is given in a single simple ring of the liquid.

9. The problem of steady motion, for the case of a vortex-line with infinitely thin core, bears a close analogy to the following purely geometrical problem :—

Find the curve whose length shall be a minimum with given resultant projectional area, and given resultant areal moment (§ 27 below). This would be identical with the vortex problem if the energy of an infinitely thin vortex-ring of given volume and given cyclic constant were a function simply of its apertural circumference. The geometrical problem clearly has multiple solutions answering precisely to the solutions of the vortex problem.

10. The very high modes of solution are clearly very nearly identical for the two problems (infinitely high modes identical), and are found thus :—

Take the solution derived in the manner explained above, from a regular polygon of N sides, when N is a very great number. It is obvious that either problem must lead to a form of curve like that of a long regular spiral spring of the ordinary kind bent round till its two ends meet, and then having its ends properly cut and joined so as to give a continuous endless helix with axis a circle (instead of the ordinary straight line-axis), and N turns of the spiral round its circular axis. This curve I call a toroidal helix, because it lies on a toroid*, just as the common

---

* I call a circular toroid a simple ring generated by the revolution of any singly-circumferential closed plane curve round any axis in its plane not cutting it. A "tore," following French usage, is a ring generated by the revolution of a circle round any line in its plane not cutting it. Any simple ring, or any solid with a single hole through it, may be called a toroid; but to deserve this appellation it had better be not very unlike a tore.

The endless closed axis of a toroid is a line through its substance passing somewhat approximately through the centres of gravity of all its cross sections. An apertural circumference of a toroid is any closed line in its surface once round its aperture. An apertural section of a toroid is any section by a plane or curved surface which would cut the toroid into two separate toroids. It must cut the surface of the toroid in just two simple closed curves, one of them completely surrounding the other on the sectional surface: of course it is the space between these curves which is the actual section of the toroidal substance; and the area of the inner one of the two is a section of the aperture.

A section by any surface cutting every apertural circumference, each once and only once, is called a cross section of the toroid. It consists essentially of a simple closed curve.

regular helix lies on a circular cylinder. Let a be the radius of the circle thus formed by the axis of the closed helix; let r denote the radius of the cross section of the ideal toroid on the surface of which the helix lies, supposed small in comparison with a; and let θ denote the inclination of the helix to the normal section of the toroid. We have

$$\tan \theta = \frac{2\pi a}{N \cdot 2\pi r} = \frac{a}{Nr},$$

because $2\pi a/N$ is, as it were, the step of the screw, and $2\pi r$ is the circumference of the cylindrical core on which any short part of it may be approximately supposed to be wound.

Let κ be the cyclic constant, I the given force resultant of the impulse, and μ the given rotational moment. We have (§ 28) approximately

$$I = \kappa \pi a^2, \qquad \mu = \kappa N \pi r^2 a.$$

Hence

$$a = \sqrt{\frac{I}{\kappa \pi}}, \qquad r = \sqrt{\frac{\mu}{N \kappa^{\frac{1}{2}} \pi^{\frac{3}{2}} I^{\frac{1}{2}}}},$$

$$\tan \theta = \sqrt{\frac{I^{\frac{3}{2}}}{N \mu \kappa^{\frac{1}{2}} \pi^{\frac{1}{2}}}}.$$

11. Suppose now, instead of a single thread wound spirally round a toroidal core, we have two separate threads forming, as it were, a "two-threaded screw," and let each thread make a whole number of turns round the toroidal core. The two threads, each endless, will be two helically tortuous rings linked together, and will constitute the core of what will now be a double vortex-ring. The formulæ just now obtained for a single thread would be applicable to each thread, if κ denoted the cyclic constant for the circuit round the two threads, or twice the cyclic constant for either, and N the number of turns of either alone round the toroidal core. But it is more convenient to take N for the number of turns of both threads (so that the number of turns of one thread alone is $\frac{1}{2}N$), and κ the cyclic constant for either thread alone, and thus for very high steady modes of the double vortex-ring,

$$I = 2\kappa \pi a^2, \qquad \mu = \kappa N \pi r^2 a,$$

$$\tan \theta = \sqrt{\frac{(\frac{1}{2}I)^{\frac{3}{2}}}{N \mu \kappa^{\frac{1}{2}} \pi^{\frac{1}{2}}}}.$$

Lower and lower steady modes will correspond to smaller and smaller values of N; but in this case, as in the case of the single vortex-core, the form will be a curve of some ultra-transcendent character, except for very great values of N, or for values of θ infinitely nearly equal to a right angle (this latter limitation leading to the case of infinitely small transverse vibrations).

12. The gravest steady mode of the double vortex-ring corresponds to $N = 2$. This with the single vortex-core gives the case of the twisted ellipse (§ 7). With the double core it gives a system which is most easily understood by taking two plane circular rings of stiff metal linked together. First, place them as nearly coincident as their being linked together permits (fig. 5). Then separate them a little, and incline their planes a little, as shown in the diagram. Then bend each into an unknown shape, determined by the strict solution of the transcendental problem of analysis to which the hydro-kinetic investigation leads for this case.

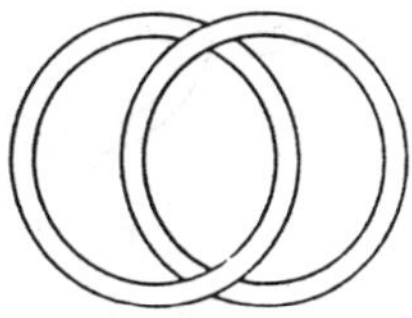

Fig. 5.

13. Go back now to the supposition of § 11, and alter it to this :—

Let each thread make one turn and a half, or any odd number of half turns, round the toroidal core : thus each thread will have an end coincident with an end of the other. Let these coincident ends be united. Thus there will be but one endless thread making an odd number N of turns round the toroidal core. The cases of $N = 3$ and $N = 9$ are represented in the annexed diagrams (figs. 8 and 9)*.

Imagine now a three-threaded toroidal helix, and let N denote the whole number of turns round the toroidal core; we have

$$I = 3\kappa\pi a^2, \qquad \mu = \kappa N \pi r^2 a,$$

$$\tan\theta = \sqrt{\frac{(\tfrac{1}{3}I)^{\frac{3}{2}}}{N\mu\kappa^{\frac{1}{2}}\pi^{\frac{1}{2}}}}.$$

Suppose now N to be divisible by 3; then the three threads form three separate endless rings linked together. The case of $N = 3$ is illustrated by the annexed diagram (fig. 6), which is repeated from the diagram of V. M. § 58. If N be not divisible

---

* The first of these was given in § 58 of my paper on Vortex Motion. It has since become known far and wide by being seen on the back of the "Unseen Universe."

by 3, the three threads run together into one, as illustrated for the case of $N = 14$ in the annexed diagram (fig. 7).

Fig. 6.　　　　　Fig. 7.　　　　　Fig. 8. "Trefoil Knot."

14.　The irrotational motion of the liquid round the rotational cores in all these cases is such that the fluid-velocity at any point is equal to, and in the same direction as, the resultant magnetic force at the corresponding point in the neighbourhood of a closed galvanic circuit, or galvanic circuits, of the same shape as the core or cores.　The setting-forth of this analogy to people familiar, as modern naturalists are, with the distribution of magnetic force in the neighbourhood of an electric circuit, does much to promote a clear understanding of the still somewhat strange fluid-motions with which we are at present occupied.

15.　To understand the motion of the liquid in the rotational core itself, take a piece of Indian-rubber gas-pipe stiffened internally with wire in the usual manner, and with it construct any of the forms with which we have been occupied, for instance the symmetrical trefoil knot (fig. 8, § 13), uniting the two ends of the tube carefully, by tying them firmly, by an inch or two of straight cylindrical plug; then turn the tube round and round, round its sinuous axis. The rotational motion of the fluid vortex-core is thus represented.　But it must be remembered that the outer form of the

Fig. 9. "Nine-leaved Knot."

core has a motion perpendicular to the plane of the diagram, and a rotation round an axis through the centre of the diagram and perpendicular to the plane, in each of the cases represented by the preceding diagrams. The whole motion of the fluid, rotational and irrotational, is so related in its different parts to one another, and to the translational and rotational motion of the shape of the core, as to be everywhere slipless.

16. Look to the preceding diagrams, and, thinking of what they represent, it is easy to see that there must be a determinate particular shape for each of them which will give steady motion; and I think we may confidently judge that the motion is stable in each, provided only the core is sufficiently thin. It is more easy to judge of the cases in which there are multiple sinuosities by a synthetic view of them (§ 3) than by consideration of the maximum-minimum problem of § 8.

17. It seems probable that the two- or three- or multiple-threaded toroidal helix motions cannot be stable, or even steady, unless I, μ, and N are such as to make the shortest distances between different positions of the core or cores considerable in comparison with the core's diameter. Consider, for example, the simplest case (§ 12, fig. 5) of two simple rings linked together.

18. Go back now to the simple circular Helmholtz ring. It is clear that there must be a shape of absolute maximum energy for given vorticity and given impulse, if we introduce the restriction that the figure is to be a figure of revolution—that is to say, symmetrical round a straight axis. If the given vorticity be given in this determinate shape, the motion will be steady; and there is no other figure of revolution for which it would be steady (it being understood that the impulse has a single force resultant without couple). If the given impulse, divided by the cyclic constant, be very great in comparison with the two-thirds power of the volume of liquid in which the vorticity is given, the figure of steadiness is an exceedingly thin circular ring of large aperture and of approximately circular cross section. This is the case to which chiefly attention is directed by Helmholtz. If, on the other hand, the impulse divided by the cyclic constant be very small compared with the two-thirds power of the volume, the figure becomes like a long oval bored through along its axis of revolution and with the ends of the bore rounded off (or trumpeted) symmetrically, so as to give a figure something like the handle of a child's skipping-rope, but symmetrical on the two sides of the plane through its middle perpendicular to its length. It is certain that, however small the impulse, with given vorticity the figure of steadiness thus indicated is possible, however long in

the direction of the axis and small in diameter perpendicular to the axis and in aperture it may be. I cannot, however, say at present that it is certain that this possible steady motion is stable; for there are figures not of revolution, deviating infinitely little from it, in which, with the same vorticity, there is the same impulse and the same energy, and consideration of the general character of the motion is not reassuring on the point of stability when rigorous demonstration is wanting*.

19. Hitherto I have not indeed succeeded in rigorously demonstrating the stability of the Helmholtz ring in any case. With given vorticity, imagine the ring to be thicker in one place than in another. Imagine the given vorticity, instead of being distributed in a symmetrical circular ring, to be distributed in a ring still with a circular axis, but thinner in one part than in the rest. It is clear that, with the same vorticity and the same impulse, the energy with such a distribution is greater than when the ring is symmetrical. But now let the figure of the cross section of the ring, instead of being approximately circular, be made considerably oval. This will diminish the energy with the same vorticity and the same impulse. Thus from the figure of steadiness we may pass continuously to others with same vorticity, same impulse, and same energy. Thus, we see that the figure of steadiness is, as stated above, a figure of maximum-minimum, and not of absolute maximum, nor of absolute minimum energy. Hence, from the maximum-minimum problem we cannot derive proof of stability.

20. The known phenomena of steam-rings and smoke-rings show us enough of, as it were, the natural history of the subject to convince us beforehand that the steady configuration, with ordinary proportions of diameters of core to diameter of aperture, is stable; and considerations connected with what is rigorously demonstrable in respect to stability of vortex columns (to be given in a later communication to the Royal Society) may lead to a rigorous demonstration of stability for a simple Helmholtz ring, if of thin-enough core in proportion to diameter of aperture. But at present neither natural history nor mathematics gives us perfect assurance of stability when the cross section is considerable in proportion to the area of aperture.

* [Prove steady, W. T., May 10, 1887.]

21. I conclude with a brief statement of general propositions, definitions, and principles used in the preceding abstract, of which some appeared in my series of papers on vortex motion communicated to the Royal Society of Edinburgh in 1867, –68 and –69, and published in the *Transactions* for 1869. The rest will form part of the subject of a continuation of that paper, which I hope to communicate to the Royal Society before the end of the present session.

Any portion of a liquid having vortex motion is called *vortex-core*, or, for brevity, simply " core." Any finite portion of liquid which is all vortex-core, and has contiguous with it over its whole boundary irrotationally moving liquid, is called *a vortex*. A vortex thus defined is essentially a ring of matter. That it must be so was first discovered and published by Helmholtz. Sometimes the word *vortex* is extended to include irrotationally moving liquid circulating round or moving in the neighbourhood of vortex-core ; but as different portions of liquid may successively come into the neighbourhood of the core, and pass away again, while the core always remains essentially of the same substance, it is more proper to limit the substantive term *a vortex* as in the definition I have given.

22. *Definition I*. The circulation of a vortex is the circulation [V. M. § 60 (*a*)] in any endless circuit once round its core. Whatever varied configurations a vortex may take, whether on account of its own unsteadiness (§ 1 above), or on account of disturbances by other vortices, or by solids immersed in the liquid, or by the solid boundary of the liquid (if the liquid is not infinite), its "circulation" remains unchanged [V. M. § 59, Prop. (1)]. The circulation of a vortex is sometimes called its *cyclic constant*.

Definition II. An axial line through a fluid moving rotationally, is a line (straight or curved) whose direction at every point coincides with the axis of molecular rotation through that point [V. M. § 59 (2)].

Every axial line in a vortex is essentially a closed curve, [being of course wholly without a vortex]*.

23. *Definition III*. A closed section of a vortex is any section of its core cutting normally the axial lines through every

---

* [Phrase in [ ] deleted by Lord Kelvin.]

point of it. Divide any closed section of a vortex into smaller areas; the axial lines through the borders of these areas form what are called vortex-tubes. I shall call (after Helmholtz) a vortex-filament any portion of a vortex bounded by a vortex-tube (not necessarily infinitesimal). Of course a complete vortex may be called therefore a vortex-filament; but it is generally convenient to apply this term only to a part of a vortex as just now defined. The boundary of a complete vortex satisfies the definition of a vortex-tube.

A complete vortex-tube is essentially endless. In a vortex-filament infinitely small in all diameters of cross sections "rotation" varies [V. M. § 60 (e)] from point to point of the length of the filament, and from time to time, inversely as the area of the cross section. The product of the area of the cross section into the rotation is equal to the circulation or cyclic constant of the filament.

24. Vorticity will be used to designate in a general way the distribution of molecular rotation in the matter of a vortex. Thus, if we imagine a vortex divided into a number of infinitely thin vortex-filaments, the vorticity will be completely given when the volume of each filament and its circulation, or cyclic constant, are given; but the shapes and positions of the filaments must also be given, in order that not only the vorticity, but its distribution, can be regarded as given.

25. The vortex-density at any point of a vortex is the circulation of an infinitesimal filament through this point, divided by the volume of the complete filament. The vortex-density remains always unchanged for the same portion of fluid. By definition it is the same all along any one vortex-filament.

26. Divide a vortex into infinitesimal filaments inversely as their densities, so that their circulations are equal; and let the circulation of each be $1/n$ of unity. Take the projection of all the filaments on one plane. $1/n$ of the sum of the areas of these projections is (V. M. §§ 6, 62) equal to the component impulse of the vortex perpendicular to that plane. Take the projections of the filaments on three planes at right angles to one another, and find the centre of gravity of the areas of these three sets of projections. Find, according to Poinsot's method, the resultant axis, force, and couple of the three forces equal respectively to

$1/n$ of the sums of the areas, and acting in lines through the three centres of gravity perpendicular to the three planes. This will be the resultant axis, the force resultant of the impulse, and the couple resultant of the vortex.

The last of these (that is to say, the couple) is also called the rotational moment of the vortex (V. M. § 6).

27. *Definition IV.* The moment of a plane area round any axis is the product of the area multiplied into the distance from that axis of the perpendicular to its plane through its centre of gravity.

Definition V. The area of the projection of a closed curve on the plane for which the area of projection is a maximum will be called the area of projection of the curve, or simply the area of the curve. The area of the projection on any plane perpendicular to the plane of the resultant area is of course zero.

Definition VI. The resultant axis of a closed curve is a line through the centre of gravity, and perpendicular to the plane of its resultant area. The resultant areal moment of a closed curve is the moment round the resultant axis of the areas of its projections on two planes at right angles to one another, and parallel to this axis. It is understood, of course, that the areas of the projections on these two planes are not evanescent generally, except for the case of a plane curve, and that their zero-values are generally the sums of equal positive and negative portions. Thus their moments are not in general zero.

Thus, according to these definitions, the resultant impulse of a vortex-filament of infinitely small cross section and of unit circulation is equal to the resultant area of its curve. The resultant axis of a vortex is the same as the resultant axis of the curve; and the rotational moment is equal to the resultant areal moment of the curve.

28. Consider for a moment a vortex-filament in an infinite liquid with no disturbing influence of other vortices, or of solids immersed in the liquid. We now see, from the constancy of the impulse (proved generally in V. M. § 19), that the resultant area, and the resultant areal moment of the curve formed by the filament, remain constant however its curve may become contorted; and its resultant axis remains the same line in space.

Hence, whatever motions and contortions the vortex-filament may experience, if it has any motion of translation through space this motion must be on the average along the resultant axis.

29. Consider now the actual vortex made up of an infinite number of infinitely small vortex-filaments. If these be of volumes inversely proportional to their vortex-densities (§ 25), so that their circulations are equal, we now see from the constancy of the impulse that the sum of the resultant areas of all the vortex-filaments remains constant; and so does the sum of their rotational moments: and the resultant areal axis of them all regarded as one system is a fixed line in space. Hence, as in the case of a vortex-filament, the translation, if any, through space is on the average along its resultant axis. All this, of course, is on the supposition that there is no other vortex, and no solid immersed in the liquid, and no bounding surface of the liquid, near enough to produce any sensible influence on the given vortex.

JOURNAL OF MATHEMATICAL PHYSICS   VOLUME 9, NUMBER 11   NOVEMBER 1968

Connection between Spin, Statistics, and Kinks

DAVID FINKELSTEIN*†
Belfer Graduate School of Science, Yeshiva University, New York

AND

JULIO RUBINSTEIN‡
Department of Physics, Columbia University, New York

(Received 7 August 1967)

Sufficiently nonlinear classical fields admit modes called kinks, whose number is strictly conserved in virtue of boundary conditions and continuity of the field as a function of space and time. In a quantum theory of such fields, with canonical commutation (not anticommutation) relations, kinks and their conservation still persist, and even if the intrinsic angular momentum is an integer, a rotating kink can have half-odd angular momentum, if double-valued state functionals are admitted. We formulate a natural concept of exchange appropriate for kinks. The principal result is that for fields with integer-valued intrinsic angular momentum, the observed relation between spin and (exchange) statistics follows from continuity alone, parastatistics being excluded. It is likely that in the theories with even (odd) exchange statistics, suitable creation operators will commute (anticommute). We show that, while the rotational spectrum of a kink will in general possess both integer and half-odd spin states, in fields with integer-valued intrinsic angular momentum only one of these two possibilities will ever be observed for each kind of kink, and that there is a nonzero "particle number" (strictly conserved, additive, scalar quantum number) attached to half-odd-spin kinks of each kind. It then follows that a boson and a fermion kink will always differ in at least one particle number, as well as in spin, and that, in particular, every fermion kink will have some nonzero particle number. These results are consistent with the hypothesis that the spinor fields usually employed to describe half-odd-spin quanta are not fundamental, but are useful "point-limit" approximations to operators creating or annihilating excitations in a nonlinear field of particular kinds of kinks in particular internal states.

INTRODUCTION

Quanta of integer spin are described by fields obeying canonical commutation relations, ramifications of the basic relation

$$pq - qp = \hbar/i,$$

while since the work of Jordan and Wigner,[1] quanta of half-odd spin have been described by fields obeying anticommutation relations like

$$pq + qp = \hbar/i.$$

For brevity, let us call these *canonical* and *anticanonical* quantizations, respectively. We have been exploring[2,3] a single kind of field quantization which appears to embrace both the canonical and anticanonical, and to reduce to them in appropriate circumstances. In this kind of quantization, for which we employ the term *multivalued quantization*, all the fundamental fields are supposed to obey commutation relations, as in canonical quantization, but the state-functionals on which they act are permitted to be multivalued. More succinctly, a multivalued quantization of a classical system S is a canonical quantization of the "covering system" $\bar{S}$, and the definition of the covering system $\bar{S}$ is formulated in close analogy to that of the (universal) covering space in topology.

It is well known that half-odd spin is related to a "quantum-mechanical double valuedness" under rotation. Here we show that Fermi–Dirac statistics can be related to a "quantum-mechanical double valuedness" under a different process which we call field exchange.

The most significant results are the following two:

1. In the framework of the usual two quantizations, it is well known that the spin and statistics of quanta are independent *per se* and further considerations (analyticity, Lorentz invariance, etc.) are necessary to account for the observed correlation. In multivalued quantized field theories, continuity arguments suffice to show that among theories with integer-valued intrinsic spin, the cases that admit half-odd spin states at all are the same as the cases that admit odd statistics. The correlation found in nature appears to follow purely from continuity considerations. Moreover, since there are only two "kinds" of spin, integer and half-odd integer, it follows that in this framework there are only two kinds of statistics and "parastatistics" is impossible.

2. In multivalued quantization, there is always a conserved additive integer attached to structures possessing half-odd spin. This coupling between the values of rotational and nonrotational quantum numbers corresponds well with the empirical fact that there exists a conserved "particle number" (strictly conserved, additive, scalar quantum number) that is nonzero for every fermion, e.g., $N_B + N_e + N_\mu$, where N_B is the baryon number, N_e the electron

* Supported by the National Science Foundation (Grant No. GP 6137).

† Young Men's Philanthropic League Professor of Physics.
‡ Supported by the U.S. Atomic Energy Commission. Present address: Dept. of Physics, University of Saskatchewan.

[1] P. Jordan and E. P. Wigner, Z. Physik **47**, 631 (1928).
[2] T. H. R. Skyrme, Nucl. Phys. **31**, 556 (1962).
[3] D. Finkelstein, J. Math. Phys. **7**, 1218 (1966).

number, and N_μ the muon number ($N_\mu + N_\mu$ is a "lepton number"). In brief, there are no truly neutral fermions. No boson and fermion can have exactly the same set of values for their conserved "particle numbers."

We may summarize our methods as follows:

Nonlinearity is crucial for our development. Loosely speaking, when a field is nonlinear enough, there are shapes it can assume which cannot be wiped out by any continuous process whatever. We have called these classically indestructible objects "kinks" and proven their possibility and conservation in quantum theory as well, by primitive considerations of continuity which ultimately reduce to homotopy calculations, recourse to detailed dynamical calculations being unnecessary.[2,3] Equally primitive is the distinction between multivalued quantized systems which can possess state functions double-valued under 2π rotation, and those which cannot: briefly, theories with half-odd spin vs integer spin theories. For example, the rigid rotator with 3 degrees of freedom is a theory with half-odd spin, but the dipole rotator with 2 degrees of freedom is not. This distinction also reduces to a homotopy calculation. Again, suitably nonlinear field theories are found to belong to the half-odd-spin type.[3] Indeed, that part of the total angular momentum of a field which is generally called the orbital angular momentum is itself found to possess half-odd-integer eigenvalues, an attribute usually supposed to be reserved for the other part of the angular momentum, the intrinsic. It is for this reason that we have refined the usual terminology, slightly reserving the words *half-odd spin* merely for the phenomenon of a sign change under 2π rotation, and calling the two parts of the total angular momentum **J** of a field *intrinsic* and *extrinsic*: $\mathbf{J} = \mathbf{J}^i + \mathbf{J}^e$.

It is possible to study the exchange of two field structures by the same methods that were used to study the 2π rotation of one structure. We ask whether a field theory admits states which are odd under exchange just as under 2π rotation, setting up an *odd/even* classification of field theories that parallels the previous *half-odd-spin/integer-spin* classification. We show that the existence of kinks is necessary (but not sufficient) for a multivalued quantized field with integer-valued intrinsic angular momentum to be of the half-odd-spin type or of the odd (Fermi–Dirac)-statistics type.

FIG. 1. The first rubber-band lemma. This relates a rotation and an exchange. See text.

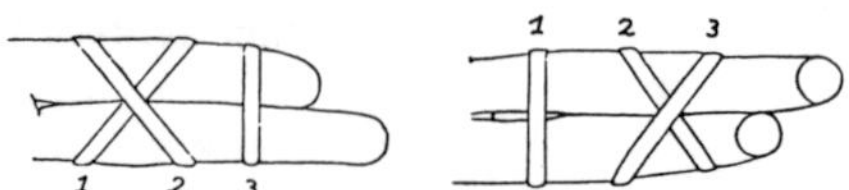

FIG. 2. The second rubber-band lemma. This relates two exchanges. See text.

Sometimes we call our work "rubber-band physics." In the first place, the simplest model of a kink is a 2π twist in a rubber band. In the second place, the topological calculations leading to the spin–statistics result are not completely transparent, and a heurism that we found indispensable is the following rubber-band lemma:

If a rubber band is wrapped twice about a rod, it exhibits at least two deformities: a self-crossing and a 2π twist.

We suggest the reader perform the experiment. If one uses a finger or two for the rod, with a little work he can get the self-crossing on the palm side, and the 2π twist concentrated in one of the two parallel strands on the back side (see Fig. 1). The crossing serves as a graph of an exchange process in which the contents of two regions of space are interchanged by a continuous deformation. (The angle around the rod is the deformation parameter.) The twist is a graph of a kind of 2π rotation of one of the two regions. The triviality of the rubber band convinced us that these two processes are homotopic, and a complete formalization and proof of this result makes up much of this paper.

The general relation proven is not between the total J and exchange, but between J^e and exchange. Just in the case of integer j^i, this coincides with the observed spin–statistics relation.

The nonexistence of parastatistics in multivalued quantized theories became clear from a second rubber-band lemma:

A rubber band can be wrapped three times around a rod with no twists and two crossings.

As graphs of exchanges among three regions 1, 2, 3 (see Fig. 2), this demonstrates that the exchange 1–2 and the exchange 2–3 are homotopic. Since parastatistics is characterized by the existence of nonequivalent exchanges, we inferred that parastatistics would be excluded. The rigorous proof given of this fact is not based on the second rubber-band lemma, however, for a simpler proof was found. But we found the second rubber-band lemma heuristically useful in another connection (see Fig. 11).

The paper is organized as follows. In Sec. I, the theory of kinks is presented. Since this is not the first exposition, the presentation is compressed and formal In Sec. II, a similar presentation is given of the idea of

D. FINKELSTEIN AND J. RUBINSTEIN

multivalued quantization of field theories in general, and field theories with kinks in particular. In Sec. III, the spin concept is defined and the topological criterion for half-odd spin is formulated. Section IV deals with exchange, and the main feature is that there is a close step-for-step correspondence between the spin treatment of Sec. III and the exchange treatment of Sec. IV. Then the connection between spin, exchange, and kinks is formulated and proven in Sec. V and in the Appendices, to which some useful but noxious lemmas are relegated.

The results described here still permit us to entertain the hypothesis that the stable elementary particles are kinks in a nonlinear field, in the sense that their "particle numbers" are the numbers of kinks present in the quantum. According to this hypothesis, the linear or almost linear spinor fields usually employed to describe half-odd-spin quanta are not fundamental, but are useful "point-limit" approximations to operators creating or annihilating excitations in a nonlinear field of particular internal states. It is regrettable that the dynamical calculations required to try this hypothesis more severely are so difficult and so likely to diverge, but we are not abandoning the question yet. We suspect in particular that the present methods can be used to construct, from the underlying commutative-field creation, operators that anticommute or commute appropriately, linking our exchange statistics to the usual field-theoretic concept of statistics. When quantum mechanics was discovered, the obstinate dualism of matter and field, particle and wave, at last dissolved into the unity of the quantum, only to be reincarnated subsequently in the distinction between Bose–Einstein and Fermi–Dirac quanta with their two different quantization recipes. The present work suggests that it is finally possible to unify these two into one nonlinear field, possessing two kinds of statistics, as a direct consequence of the topology of the rotation group in three dimensions.

I. CLASSICAL FIELDS

1. A *classical field* $\varphi(x)$ on a flat (Minkowskian) space–time $\{x\}$ is a continuous mapping of $\{x\}$ into some topological space Φ. We shall assume Φ to be a connected[4] manifold, so that the value φ can be locally represented by n real numbers φ_α, and any two points φ_0, φ_1 in Φ can be joined by a path in Φ, i.e., a continuous mapping $f: I \to \Phi$ of the unit interval into Φ with $f(0) = \varphi_0$, $f(1) = \varphi_1$.

2. Let $X = \{\mathbf{x}\} = \{x \mid t = \text{const}\}$ be ordinary 3-dimensional space and let Φ^X be the set of all contin-

uous mappings $\varphi: X \to \Phi$, and $\Phi^X(\varphi_0)$ the subset of Φ^X with $\varphi(\mathbf{x}) \to \varphi_0$ as $|\mathbf{x}| \to \infty$, φ_0 an arbitrary fixed point of Φ. The field $\varphi(x)\big|_{t=\text{const}} \equiv \varphi(\mathbf{x}, t)\big|$ is supposed to belong to $\Phi^X(\varphi_0)$ at any time its continuous evolution being determined by classical field equations and appropriate initial conditions.

3. We suppose that $\varphi(\mathbf{x}, t)$ has a single-valued law of transformation under the Poincaré group. This characteristic will be retained when we quantize the field, making our approach basically different from the usual ones, in which spin-$\frac{1}{2}$ particles are represented by double-valued fields.

4. Two fields $\varphi_1(\mathbf{x})$, $\varphi_2(\mathbf{x})$ in $\Phi^X(\varphi_0)$ are said to be homotopic to each other, $\varphi_1(\mathbf{x}) \sim \varphi_2(\mathbf{x})$, if there is a continuous mapping $\varphi(\mathbf{x}, u): X \otimes I \to \Phi$ such that $\lambda(\mathbf{x}, u)$ is in $\Phi^X(\varphi_0)$ for any given u in I, and

$$\varphi(\mathbf{x}, 0) = \varphi_1(\mathbf{x}),$$
$$\varphi(\mathbf{x}, 1) = \varphi_2(\mathbf{x});$$

$\varphi(\mathbf{x}, u)$ is called a homotopy between $\varphi_1(\mathbf{x})$ and $\varphi_2(\mathbf{x})$; "$\sim$" is an equivalence relation, and its equivalence classes are called the homotopy classes of $\Phi^X(\varphi_0)$.

5. Let us denote $\Phi^X(\varphi_0)$ by Q, and a general element $\varphi(\mathbf{x})$ in Q by q. We shall make of Q a topological space by introducing in it the compact-open topology (Ref. 5, p. 73), which coincides with the metric topology on Q, $d(q, q') = \max_{x \in X} d[\varphi(x) - \varphi'(x)]$, when X is compact and Φ metric (Ref. 5, p. 102). Q may or may not be connected, and its connected components Q_n are just the homotopy classes defined in Sec. I, part 4; a homotopy is a path $q(u)$ within some Q_n, and so is, by Sec. I, part 4, a time evolution $\varphi(x) = \varphi(\mathbf{x}, t)$, showing that dynamical variables that depend only on the homotopy class of $\varphi(\mathbf{x})$ are constants of motion. If there is more than one component Q_n we say that the field *admits kinks*.

6. Kinks form a group, in the following sense. The components Q_n are the elements of the group. We shall denote by $\pi_3(\Phi, \varphi_0)$ the set of all the Q_n. Starting from any $\varphi_1(\mathbf{x})$ in Q_1 and $\varphi_2(\mathbf{x})$ in Q_2 we may construct, through homotopies if necessary, representatives $\varphi_1'(\mathbf{x})$ in Q_1 and $\varphi_2'(\mathbf{x})$ in Q_2 such that

$$\varphi_1'(\mathbf{x}) = \varphi_0, \quad x_3 \geq 0,$$
$$\varphi_2'(\mathbf{x}) = \varphi_0, \quad x_3 \leq 0;$$

we define

$$Q_3 = Q_1 + Q_2 \tag{1}$$

as the homotopy class of

$$\varphi_3(\mathbf{x}) = \varphi_1'(\mathbf{x}), \quad x_3 \leq 0$$
$$= \varphi_2'(\mathbf{x}), \quad x_3 \geq 0. \tag{2}$$

---

[4] By "connected" we shall mean what is usually known as "arcwise connected" or "path connected"; except for this, we follow the standard terminology of topology.

[5] S. T. Hu, *Homotopy Theory* (Academic Press Inc., New York, 1959).

Under the operation (1) $\pi_3(\Phi, \varphi_0)$ is an Abelian group (Ref. 5, p. 109), the third homotopy of the pair $(\Phi, \varphi \in \Phi)$. It can be shown (Ref. 5, p. 126) that for any φ_0, φ_0' in Φ (connected), $\pi_3(\Phi, \varphi_0) \approx \pi_3(\Phi, \varphi_0')$. Accordingly, when concerned with only the group structure of $\pi_3(\Phi, \varphi_0)$, we shall denote this group by $\pi_3(\Phi)$. $\varphi_0(\mathbf{x}) = \varphi_0$, for all $\mathbf{x}$ belongs to the identity of $\pi_3(\Phi, \varphi_0)$, and if $\varphi(\mathbf{x})$ is in Q_n, then $\varphi(-\mathbf{x})$ is in Q_n^{-1} (or $-Q_n$, in the additive notation).

7. Since $\pi_3(\Phi)$ is Abelian, if it is finitely generated (Ref. 5, X, Corollary 8.3) its elements can be labeled by a set of numbers

$$(n_1, \cdots, n_k, n_{k+1}, \cdots, n_l) \equiv n, \qquad (3)$$

such that $n_1, \cdots, n_k$ range over the integers and $n_{k+1}, \cdots, n_l$ range over the integers modulo

$$r_i \quad (i = k + 1, \cdots, l),$$

r_i being a factor of r_{i+1}. The l components of n are conserved (Sec. I.5) and additive, in the following sense: if a field $\varphi_1(\mathbf{x})$ in $Q_n(1)$ is juxtaposed to $\varphi_2(\mathbf{x})$ in $Q_n(2)$ [see (2)], the resulting field $\varphi_3(\mathbf{x})$ belongs to the class $Q_{n(3)}$, with

$$n^{(3)} = (n_1^{(1)} + n_1^{(2)}, \cdots, n_l^{(1)} + n_l^{(2)}) \equiv n^{(1)} + n^{(2)}; \quad (4)$$

the n_i ($i = 1, \cdots, l$) can then be considered as particle numbers or "charges." The possibility of having "modulo" conservations (the n_i's for $i > k$) has arisen also in different context,[6] but is not supported by experimental evidence so far. From now on we shall discuss only Φ's for which $l = k$ in (3), but in general our results are valid. We shall say that $\varphi(\mathbf{x})$ is one "kink of type i" if $\varphi(\mathbf{x})$ is in Q_{N_i}, where

$$N_i = (0, 0, \cdots, 1, \cdots, 0), \qquad (5)$$

1 being in the ith place; then $\varphi(-\mathbf{x})$ is in Q_{-N_i} and will be called an antikink of type i. Any field $\varphi(\mathbf{x})$ in Q_n with n given by (3) can be considered as composed of a certain number of kinks or antikinks of the various types N_i.

8. In the literature the nth homotopy group of Φ is variously defined as the set of homotopy classes of mappings $(S^n, p_0) \to (\Phi, \varphi_0)$ of an n sphere S^n into Φ, with $p_0(\epsilon S^n) \to \varphi_0$, of mappings $(I^n, \partial I^n) \to (\Phi, \varphi_0)$ of an n cube I^n into Φ, with its boundary ∂I^n going into φ_0, etc. All those definitions are equivalent among themselves (this is proven by showing that after proper identifications the domains of the mappings are homeomorphic) and to ours for $n = 3$: we can, for instance, deform X into a 3-ball: $\mathbf{x} \to \mathbf{x}/1 + |\mathbf{x}|$, which is homeomorphic to a 3-sphere upon identification of its boundary $|\mathbf{x}| = 1$ into one point p_0. Notice that this is allowed by the boundary conditions introduced in Sec. I.2.

---

[6] See, for instance, M. Gell-Man, Phys. Letters **8**, 214 (1964).

TABLE I. Homotopy groups for the underlying manifolds of classical Lie groups.

	SO_3	SO_4	SO_5	SO_6	$SO_{n \geq 7}$	SU_2	$SU_{n \geq 3}$	$Sp_{n \geq 1}$
π_1	Z_2	Z_2	Z_2	Z_2	Z_2	0	0	0
π_2	0	0	0	0	0	0	0	0
π_3	Z	$Z+Z$	Z	Z	Z	Z	Z	Z
π_4	Z_2	Z_2+Z_2	Z_2	0	0	Z_2	0	Z_2
π_5	Z_2	Z_2+Z_2	Z_2	Z	0	Z_2	Z	Z_2

9. Any Abelian group π_3 that may be dictated by phenomenological considerations can be realized as the third homotopy group of some space Φ. Moreover, it can be shown (Ref. 5, p. 169) that given a sequence of groups

$$\pi_1, \pi_2, \cdots, \pi_n, \cdots,$$

such that all except possibly the first are Abelian and π_1 is a group of automorphisms of all the others, there exists a connected space Φ such that

$$\pi_n(\Phi) \approx \pi_n \quad \text{for all } n.$$

In particular, there are $C = 2^{\aleph_0}$ nonhomeomorphic spaces with the same π_3!

10. Some homotopy groups of spheres are tabulated in Ref. 3, Table I, and many more are given by Toda.[7] Table I shows the first five homotopy groups for the underlying manifolds of classical Lie groups; it has been compiled from several sources.[7]

The relation $\pi_n(\Phi \otimes \Psi) = \pi_n(\Phi) \oplus \pi_n(\Psi)$ is a useful tool for the construction of *ad hoc* spaces (Sec. I.9).

II. QUANTIZATION

1. When a classical field theory is quantized, the statement "the measurement of the field at time t will give the values $\varphi(\mathbf{x}, t)$" is replaced by "the probability amplitude for getting the result $\varphi(\mathbf{x})$ when measuring the field at time t is $\Psi[\varphi(\mathbf{x})](t)$," Ψ being a complex-valued time-dependent functional of $\varphi(\mathbf{x})$, continuous in t and φ. Notice that in this picture (Schrödinger) and representation [diagonal in $\varphi(\mathbf{x})$], the argument of the functional is a c-number field, an element of the function-set Φ^X of Sec. I.2.

2. We shall assume that, at any time,

$$\Psi[\varphi(\mathbf{x})](t) = 0 \quad \text{if} \quad \varphi(\mathbf{x}) \notin \Phi^X(\varphi_0) \equiv Q, \quad (6)$$

so we can use $Q = \Phi^X(\varphi_0)$ instead of Φ^X as our configuration space; this implies restrictions on the dynamics of the system, but we shall not elaborate

---

[7] H. Toda, *Composition Methods in Homotopy Groups of Spheres* (Princeton University Press, Princeton, N.J., 1962); A. Borel, in *Seminaire H. Cartan* (2) (Ecole Normale Superieure, Paris, 1956). L. Pontrjagin, *Topological Groups* (Princeton University Press, Princeton, N.J., 1946); V. G. Boltyanskii, Transl. Am. Math. Soc. 7, 135 (1957).

this point in the present work. Since for φ in Q we can write $\Psi[\varphi](t) = \Psi(q, t)$, a continuous complex-valued function on $Q \otimes R$ (R = real number space spanned by t), another way of putting (6) is

$$\int_Q |\Psi(q, t)|^2 \, dq = 1, \qquad (7)$$

provided that the concept of integral on Q can be defined; even if not, we shall use expressions of the type in (7) as symbols for expressions of the type in (6). We call Ψ the state function(al).

3. If we accept the Feynman scheme for quantization, which uses integrals over continuous histories $\varphi(\mathbf{x}, t)$ (in our case paths in Q), it can be shown (Ref. 3, Sec. III.2) that

$$\int_{Q_n} |\Psi(q, 0)|^2 \, dq = 1 \quad \text{implies} \quad \int_{Q_n} |\Psi(q, t)|^2 \, dq = 1; \qquad (8)$$

then, if the "charges" defined in (Sec. I.7) are interpreted as nonlocal observables in the quantum theory, (8) tells us that they are conserved in time. In the following, whenever we talk about a homotopic conservation law it shall be understood in the quantum sense given by (8). In short, (6) and (8) allow us to extend the results of Sec. I to a quantum theory of φ in which $\varphi(\mathbf{x})$ can still be treated as having a c-number field of eigenvalues. We assume for simplicity that the particle numbers n_i obey a superselection rule,[8] i.e., that the support of any realizable physical state $\Psi(q)$ is one of the Q_n.

4. In what follows we shall be concerned with two kinds of discrete operations that leave physical systems unchanged but may multiply the state vectors by -1: a 2π rotation and the exchange of two identical subsystems. The 2π rotation can be realized continuously by a succession of infinitesimal transformations in an obvious way, and we shall give in Sec. IV an analogous realization for an operation related to the exchange.

Let $q \to F_s q$ represent a flow in Q, i.e., a one-parameter family of 1–1 mappings F_s satisfying the conditions

$$F_0 q = q, \qquad (9)$$

so that $F_s q$ ($0 \leq s \leq 1$) is a path at q. Any such flow in Q induces a continuous one-parameter family of linear transformations of the state function $\Psi(q)$:

$$\Psi(q) \to \Psi(q, s) = \Psi(F_s^{-1} q) = \Psi(q_s), \quad 0 \leq s \leq 1. \qquad (10)$$

If

$$F_1 q = q, \qquad (11)$$

the flow is called *closed*. Its paths are then closed paths or *loops*. (11) is the mathematical expression of the "leaving the system unchanged" used above. Since until now we have considered Ψ to be single-valued, (9) and (11) imply

$$\Psi(F_0 q) = \Psi(F_1 q). \qquad (12)$$

Now we wish to consider that the state function $\Psi(q)$ might be multiple-valued; this point is elaborated upon in Sec. II.5, and (12) no longer holds.

5. Our first problem is to find whether the domain Q_n of the state function $\Psi(q)$ admits multiple-valued continuous functions. Intuitively speaking, these have more than one value at each q in Q_n (the value set changing continuously with q), such that any two values Ψ_0 and Ψ_1 at q can be connected by a continuous succession of values $\Psi[q(s)]$ by traveling some closed path $q(s)[q(0) = q(1) = q]$. It is evident that no simply connected Q_n admits such functions.[9] Moreover, since the value of a multivalued function depends on the point q and the way q is reached from some standard point, such a function on Q_n can be defined as a (single-valued) function on CQ_n, the universal covering space of Q.

We define CQ_n as follows:

We choose a base point q_{n_0} in Q_n and consider the paths $q(s)$ in Q_n ($0 \leq s \leq 1$) such that $q(0) = q_{n_0}$; two paths $q(s)$ and $q'(s)$ are equivalent if $q(1) = q'(1) = q$ (say) and there exists a homotopy relative to $\{0, 1\}$ between $q(s)$ and $q'(s)$, i.e., a continuous mapping $q(s, u): I^2 \to Q_n$ ($0 \leq s \leq 1$, $0 \leq u \leq 1$) such that

$$q(s, 0) = q(s), q(s, 1) = q'(s),$$
$$q(0, u) = q_{n_0}, q(1, u) = q.$$

The equivalence classes of paths $q(s)$ are the points of CQ_n, if we introduce the following topology: given $q(s)$ and an arcwise-connected open set U that contains $q(1)$, the union over U of the equivalence classes of the $q'(s)$ such that

$$q'(s) = q(2s) \quad (0 \leq s \leq \tfrac{1}{2}),$$
$$q'(s) \subset U \quad (\tfrac{1}{2} \leq s \leq 1),$$

is a set belonging to CQ_n; the collection of all such sets is taken to be a basis for the topology of CQ_n.

We now define a multivalued function on a space Q to be a single-valued function on the covering space CQ. In what follows, such functions will further be required to be continuous unless otherwise mentioned.

---

[8] G. C. Wick, Phys. Rev. **88**, 101 (1952).

[9] We remark that we are talking about state functions, i.e., complex-valued functions on Q. The statement is of course not true for the spinor wave function of ordinary quantum mechanics; indeed, its domain is simply connected, but its components do not satisfy the transformation law (10) when subject to rotations.

Q_n admits multivalued functions if and only if CQ_n is not homeomorphic to Q_n, which is equivalent to $\pi_1(Q_n) \neq 0$.

6. For any connected space Y, we shall write $\tilde{y}, \tilde{y}', \cdots$ for *different* points of CY associated with the end points $y \in Y$ and $\tilde{y}(s)$ for a representative path of the class $\tilde{y}$. We also say that $\tilde{y}, \tilde{y}', \cdots$ "cover" y, and denote by Cy the set of all the covering points of y; it is easy to see that the number of elements of Cy is constant over Y. In the following a multivalued function Ψ on Q_n shall be written as a function on CQ_n:

$$\Psi = \Psi(\tilde{q}). \tag{13}$$

7. A group that acts on Q cannot in general be defined to act on CQ continuously. For example, the action of the rotation group on spinors is not uniquely definable. However, the action of a flow on Q uniquely defines its action on CQ. If $F_s q$ is a flow on Q, then $F_s \tilde{q}$ can be defined in a natural way for each $s = s_0$ as the equivalence class of a path that consists of a path representing $\tilde{q}$ followed by the path $F_s q$, $0 \leq s \leq s_0$. Accordingly a flow on Q acts also on $\tilde{Q}$, and therefore on state-functions $\Psi(\tilde{q})$ in such a way that the value of Ψ is invariant:

$$F_s[\Psi(\tilde{q})] = \Psi(F_s^{-1}\tilde{q}) \equiv \Psi_s(\tilde{q}), \tag{14}$$

or

$$\Psi_s(\tilde{q}_s) = \Psi(\tilde{q})$$

where $\tilde{q}_s = F_s \tilde{q}$.

8. Computation of $\pi_1(Q_n)$. For Q_0, the component of $Q = \Phi^X(\varphi_0)$ that contains the field $\varphi_0(\mathbf{x}) \equiv \varphi_0$, we have (Ref. 3, Sec. V.2)

$$\pi_1(Q_0) \approx \pi_4(\Phi), \tag{15}$$

which formally solves the problem for Q_0, or at least reduces it to a standard one. Once $\pi_4(\Phi)$ is known, the problem is completely solved because it can be shown[10] that for any two connected components

$$Q_j, Q_k \text{ of } Q = \Phi^X(\varphi_0), \ \pi_1(Q_j) \approx \pi_1(Q_k),$$

so

$$\pi_1(Q_n) \approx \pi_4(\Phi) \quad \text{for all } n. \tag{16}$$

Our main interest in this result is that if $\pi_4(\Phi) \neq 0$, all Q_n admit multivalued functions; otherwise, none does.

9. Assuming that (13) is properly multivalued,

$$\Psi(\tilde{q}) \neq \Psi(\tilde{q}'), \tag{17}$$

in order to have a theory of the type described in Sec. II.4 we still have to show that the multivaluedness of $\Psi(q)$ is realized by two special flows q_s, namely, the 2π rotation of a field $\varphi(\mathbf{x})$ (Sec. III) and the "exchange" of two fields identical to $\varphi(\mathbf{x})$, a concept to be defined

---

[10] G. W. Whitehead, Ann. Math. **47**, 460, statement 2.6 (1946). We are indebted to Professor S. T. Hu for this reference.

in Sec. IV. These are in principle two independent problems, and it is one of the main purposes of this paper to prove that their solutions are interdependent, and that, in a sense specified in Sec. V, an affirmative (negative) answer for any one of them implies an affirmative (negative) answer to the other. In this sense, the "spin" of a system determines its "statistics" and vice versa. As we shall see (Sec. V), our proofs require only continuity of the fields $\varphi(\mathbf{x})$ and of the state functionals (13) with $q = [\varphi(\mathbf{x})]$.

III. SPIN

1. The angular momentum of a field theory is defined as the generator of infinitesimal rotations

$$\delta\mathbf{x} = \delta\boldsymbol{\theta} \times \mathbf{x},$$

where $\delta\boldsymbol{\theta}$ gives the axis and angle of rotation. The change in the field resulting from this rotation may be written as

$$\delta\varphi = \frac{\partial\varphi}{\partial\mathbf{x}} \cdot \delta\mathbf{x} + \frac{\partial\varphi}{\partial\boldsymbol{\theta}}\bigg|_{\theta=0} \cdot \delta\boldsymbol{\theta} \equiv \delta^e\varnothing + \delta^i\varnothing,$$

where the first term gives the "extrinsic" change due to the spatial variation of the φ field, and the second term is due to the "intrinsic" law of transformation

$$\boldsymbol{\theta}: \varphi \to \varphi' = \varphi'(\varphi, \boldsymbol{\theta})$$

of the φ field under a rotation parametrized by a vector $\boldsymbol{\theta}$. The corresponding decomposition of the generator of this rotation is written

$$\mathbf{J} = \mathbf{J}^e + \mathbf{J}^i.$$

The field theory exhibits half-odd-integer angular momentum if and only if the operator of a 2π rotation,

$$W = e^{2\pi i J}z,$$

possesses eigenvalues -1 as well as $+1$.

For brevity, half-odd-integer angular momentum will be referred to as *half-odd spin*. If a component Q_n of Q supports an eigenfunction Ψ of $W = -1$, we say Q_n (and the theory) admits half-odd spin.

2. In familiar field theories, J^e has integer eigenvalues and does not contribute to W,

$$W^e = \exp(2\pi i J_z^e) = 1.$$

Therefore half-odd angular momentum is usually attributed entirely to the transformation law of the field

$$W \approx \exp(2\pi i J_z^i).$$

In quantum-field theory, the generators $\mathbf{J}$ may be defined in terms of the transformation of the state functional of the field, rather than the field itself. If multivalued state functionals are admitted, then we have shown there exist quantum-field theories for which $\mathbf{J}^e$ contributes to W: While the field itself does not change sign under a 2π rotation, the quantum state of the field does. Heuristically, such a spin can

D. FINKELSTEIN AND J. RUBINSTEIN

be thought to come from the actual rotation of very asymmetric field structures having some of the properties of a quasirigid rotator. In this kind of theory the familiar decomposition of the angular momentum of the electron,

$$\mathbf{J} = \mathbf{L} + \mathbf{S},$$

is a feature of a phenomenological model and does not correspond to the decomposition

$$\mathbf{J} = \mathbf{J}^e + \mathbf{J}^i;$$

S itself may result from "orbital" angular momentum $\mathbf{J}^e$ of the kink representing the electron.

3. We now specify the flow $q \to (W^e)_s q$ of (10) for a continuous 2π extrinsic rotation of a field $\varphi(\mathbf{x})$. Let $W^e(s)$ be an extrinsic rotation through an angle $2\pi s$ around some axis through $\mathbf{x} = 0$, acting in the usual way on the argument $\mathbf{x}$ and leaving invariant the value φ,

$$W^e(s)\varphi = \varphi \quad \text{(for all } s \text{ and any axis).} \tag{18}$$

We then define the flow in Q,

$$q \to (W^e)_s q \equiv W^e(s) \cdot [\varphi(\mathbf{x})] \equiv \varphi_W(\mathbf{x}, s)$$
$$= \varphi[(W^e)^{-1}(s) \cdot \mathbf{x}] \quad (\theta \leq s \leq 1), \tag{19}$$

and (1) becomes, for rotations,

$$\Psi(q) \to W^e(s) \cdot \Psi(q) \equiv \Psi[(W^e)^{-1}(s) \cdot q]$$
$$= \Psi[(W^e)_{1-s}q]. \tag{20}$$

The effect of a 2π rotation on a single-valued function $\Psi(\tilde{q})$ is defined by Sec. II.7.

4. Since the $W^e(s)$ (for all s and any axis) constitute the elements of SO_3, the rotation group in 3 dimensions, W^e can be represented as a loop in the SO_3 manifold. SO_3 has a well-known graphical representation as a 3-dimensional ball of radius π with opposite points of the surface identified. A maximum cross section of this construction, including a representative path $W^e(s)$ of W^e, is shown in Fig. 3.

Any deformation of a path described by the $W^e(s)$ induces a deformation in the corresponding path described by the q's in Q, via (19); in particular, if a loop in SO_3 is homotopic relative to its end point[11] to the trivial loop, so is its induced loop $q(s)$ (but the reciprocal is not always true). A well-known example of such a loop is $(W^e)^2$, a 4π rotation, i.e.,

$$(W^e)^2 = 1. \tag{21}$$

The fact that we are dealing with a 3-dimensional theory is crucial for this. The result does not hold for the rotations in 2 dimensions, a 1-dimensional subgroup of SO_3 with the topology of S^1.

---

[11] By "relative to its end point" we mean that the two loops can be connected by a continuous family of *loops* with a common end point. Unless otherwise stated, whenever two loops (on X, Q, etc.) are said to be homotopic, it will mean homotopic relative to their end point.

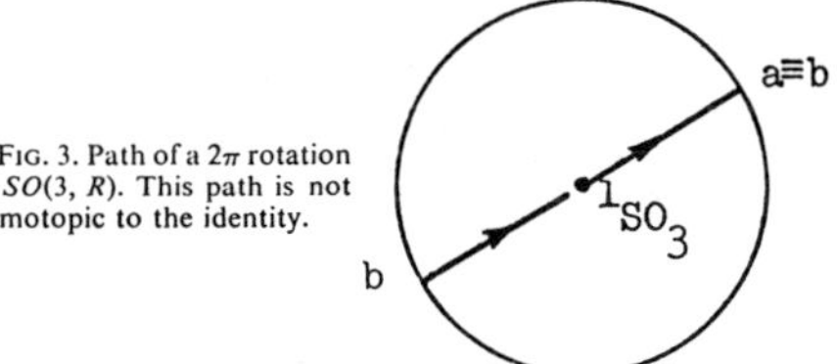

Fig. 3. Path of a 2π rotation in $SO(3, R)$. This path is not homotopic to the identity.

5. Notice that $W = \pm 1$ holds only for W as operator on state functions, whereas for the fields (or rather for their coverings), Eq. (21) only means that

$$W\tilde{q} = \tilde{q}' \Leftrightarrow W\tilde{q}' = \tilde{q}. \tag{22}$$

For Q_n to admit half-odd spin there must be a neighborhood in CQ whose points $\tilde{q}$ obey

$$W \cdot \tilde{q} \neq \tilde{q}; \tag{23}$$

this is a consequence of the following.

6. *Lemma:* Eq. (23) holds or fails to hold, simultaneously for all $\tilde{q} \in CQ_n$.

The proof of this lemma is given in Appendix A. The lemma implies that in order to see whether a given Q_n admits half-odd spin, it is enough to check if (23) holds for *any one* $\tilde{q} \in CQ_n$.

7. In what follows we give a few results helpful in finding whether or not a given Q_n admits half-odd extrinsic spin J^e.

A necessary condition for Q_n to admit half-odd J^e is that $\pi_1(Q_n)$ contain at least one element of order 2. This is an immediate consequence of (21) and (23).

A necessary condition for Q_n to admit half-odd J^e is that no $\varphi(x)$ in Q_n be axisymmetric, since for such a field, $W \cdot \tilde{q} = \tilde{q}$. In particular, Q_0 does not admit half-odd J^e, because $\varphi(\mathbf{x}) = \varphi_0$ is axisymmetric.

In other words, kinks $[\pi_3(\Phi) \neq 0]$ are necessary for half-odd extrinsic spin (Ref. 3, Sec. V.13). More strongly: kinks must be present in a state with half-odd extrinsic spin.

IV. FIELD EXCHANGE

1. In order to define an exchange operation on a field $\varphi(\mathbf{x})$, we have to identify the objects to be exchanged. Since we have neither a procedure for locating particles in a general field $\varphi(\mathbf{x})$ in Q nor the creation operators of the usual quantum-field theories, we shall at first restrict ourselves to "union" fields defined as follows. By the support sup φ of a field $\varphi(x)$, we mean the set of points x at which $\varphi(x) \neq \varphi_0$. $\varphi_1(x)$ and $\varphi_2(x)$ are fields of disjoint supports sup φ_1, sup φ_2, then their "union,"

$$\varphi = \varphi_1 \cup \varphi_2, \tag{24}$$

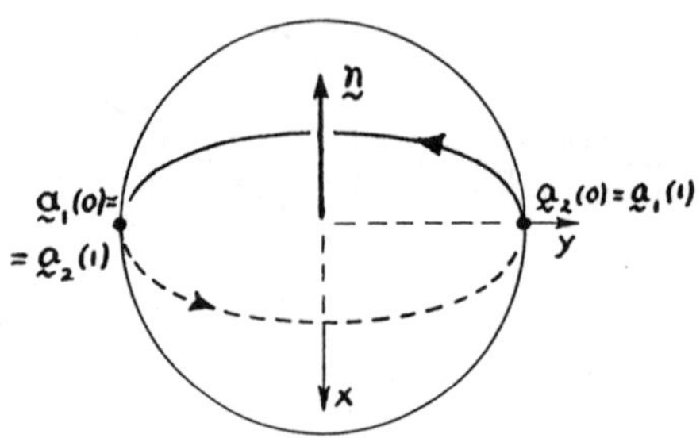

FIG. 4. Path in X of the exchange carried out on two kinks at $\mathbf{a}_1$ and $\mathbf{a}_2$.

is defined by

$$\varphi(x) = \varphi_1(\mathbf{x}), \quad \mathbf{x} \in \sup \varphi_1,$$
$$= \varphi_2(\mathbf{x}), \quad \mathbf{x} \in \sup \varphi_2, \tag{25}$$

otherwise,

$$= \varphi_0.$$

Now let $\varphi_n(x)$ be a kink of type n with $\sup \varphi_n \subset \{|\mathbf{x}| < \epsilon\}$. We take

$$\varphi(\mathbf{x}) = \varphi_n(\mathbf{x} - \mathbf{a}_1) \cup \varphi_n(\mathbf{x} - \mathbf{a}_2), \tag{26}$$

with

$$|\mathbf{a}_1 - \mathbf{a}_2| \geq 2\epsilon. \tag{27}$$

The $\mathbf{a}_i$ are finite, and (25) and (27) guarantee that (26) is well defined, i.e., that the supports of the kinks (of type n) $\varphi_n(\mathbf{x} - \mathbf{a}_i)$ do not overlap.

We can now specify the $q(s)$ of Sec. II.4 for an exchange of the structures at $\mathbf{a}_1$ and $\mathbf{a}_2$. Let $\mathbf{n}$ be unit vector perpendicular to $\mathbf{a}_1 - \mathbf{a}_2$, and (see Fig. 4)

$$\mathbf{a}_i(s) = \tfrac{1}{2}(\mathbf{a}_1 + \mathbf{a}_2) - (-1)^i \exp s\pi\mathbf{n} \times \tfrac{1}{2}(\mathbf{a}_1 - \mathbf{a}_2); \tag{28}$$

here $\mathbf{n} \times$ is the linear operator on vectors $\mathbf{v}$ defined by $(\mathbf{n} \times)\mathbf{v} = \mathbf{n} \times \mathbf{v}$. Then

$$q_n(s) = \varphi_n[\mathbf{x} - \mathbf{a}_1(s)] \cup \varphi_n(\mathbf{x} - \mathbf{a}_2(s)), \quad 0 \leq s \leq 1 \tag{29}$$

has the desired property of being a loop starting and ending at $\varphi(\mathbf{x})$, with the structure originally at $\mathbf{a}_1$ now at $\mathbf{a}_2$, and vice versa (Fig. 4).

2. We now define a closed flow (cf. Sec. II.4) X^m in the function space Q. Physically speaking, X^m makes a clearing at infinity, creates there two kink–antikink pairs of type m, exchanges the two kinks, annihilates the kinks with their antikinks, and restores the clearing to its original form.

Symbolically, we compose X^m out of the following five flows Y_i, in the order

$$X^m = Y_5 Y_4 Y_3 Y_2 Y_1, \tag{30}$$

where Y_1 is a shrinkage of the field to a support of radius ϵ, Y_2 is a creation of two kink–antikink pairs outside the support, Y_3 is the exchange of the two

created kinks (see Fig. 5), Y_4 is the annihilation of the two pairs, the inverse of Y_2, and Y_5 is an expansion of the field, the inverse of Y_1. Let

$$\varphi_{-n}(\mathbf{x}) = \varphi_n(-x, y, z), \quad \mathbf{b} = (b, 0, 0),$$
$$\epsilon = (\epsilon, 0, 0), \quad b > 0.$$

$\mathbf{b}$ and $\boldsymbol{\epsilon}$ are vectors along the x axis of length b and ϵ, and let $\mathbf{n}$ now be a unit vector along the z axis. Let $r = |\mathbf{x}|$, then the Y_i are defined as follows:

$$Y_1 : \varphi(\mathbf{x}, s) = \varphi\left(\frac{\mathbf{x}}{1 - \dfrac{s}{s_1\epsilon}r}\right), \quad r \leq \frac{s_1}{s}\epsilon$$
$$= \varphi_0, \quad r \geq \frac{s_1}{s}\epsilon \qquad 0 \leq s \leq s_1,$$

$$Y_2 : \varphi(\mathbf{x}, s)$$
$$= \varphi_m\left[\mathbf{x} - \left(\mathbf{b} + \frac{s - s_1}{s_2 - s_1}\boldsymbol{\epsilon}\right)\right], \quad y \geq b$$
$$= \varphi_{-m}\left[\mathbf{x} - \left(\mathbf{b} - \frac{s - s_1}{s_2 - s_1}\boldsymbol{\epsilon}\right)\right], \quad \epsilon < y \leq b$$
$$= \varphi_{-m}\left[\mathbf{x} - \left(-\mathbf{b} + \frac{s - s_1}{s_2 - s_1}\boldsymbol{\epsilon}\right)\right], \quad -b \leq y < -\epsilon$$
$$= \varphi_m\left[\mathbf{x} - \left(-\mathbf{b} - \frac{s - s_1}{s_2 - s_1}\boldsymbol{\epsilon}\right)\right], \quad y \leq -b$$
$$= \varphi(\mathbf{x}, s_1), \quad -\epsilon \leq y \leq \epsilon \qquad s_1 \leq s \leq s_2,$$

$$Y_3 : \varphi(x, s)$$
$$= \varphi_m\left[\mathbf{x} - \exp\left(2\pi\frac{s - s_2}{s_3 - s_2}\mathbf{n} \times\right) \times (\mathbf{b} + \boldsymbol{\epsilon})\right]$$
$$\cup \varphi_m\left[\mathbf{x} + \exp\left(2\pi\frac{s - s_2}{s_3 - s_2}\mathbf{n} \times\right) \times (\mathbf{b} + \boldsymbol{\epsilon})\right], \quad r \geq b$$
$$= \varphi(\mathbf{x}, s_2), \quad r \leq b \qquad s_2 \leq s \leq s_3,$$

and

$$Y_5 Y_4 : \varphi(\mathbf{x}, s) = \varphi\left(\mathbf{x}, s_2\frac{1 - s}{1 - s_3}\right) \qquad s_3 \leq s \leq 1.$$

Notice that $\varphi(\mathbf{x}, s)$ is continuous at $s = s_3$ because $\varphi(\mathbf{x}, s_2) = \varphi(\mathbf{x}, s_3)$; also, during $Y_3 \varphi(\mathbf{x}, s)$ is precisely the loop $q_m(s)$ of Eq. (29) performed upon the two

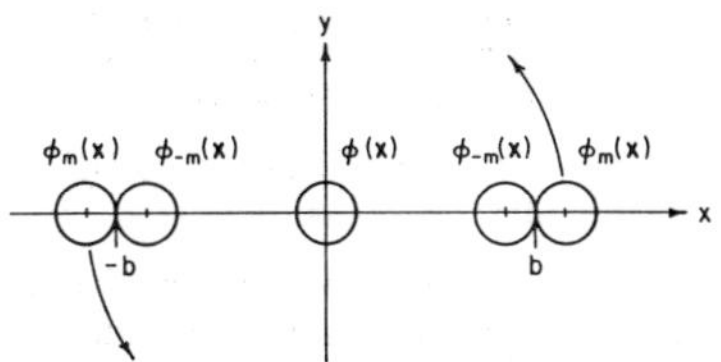

FIG. 5. Exchange operator. The field $\varphi(x)$ is shrunk into a finite sphere to leave "working space." Two kink–antikink pairs of the same kind are created in the working space. The two kinks are exchanged. The new pairs are allowed to annihilate, and the field $\varphi(x)$ expands to refill the now empty working space.

kinks φ_m created in the step Y_2, with $\mathbf{a}_2 = \mathbf{b} + \boldsymbol{\epsilon} = -\mathbf{a}_1$.

3. If the state $\varphi(\mathbf{x})$ already contains kinks of type m, as say a union field (Sec. IV.1), the exchange of two of these kinks is homotopic to $X^m\varphi(\mathbf{x})$. This is proven in Sec. V.3.

4. A double exchange of two m kinks, i.e., $(X^m)^2$, is trivial:

$$(X^m)^2 = 1. \tag{31}$$

Proof: Any deformation of a trajectory described by the $\mathbf{a}$'s induces a deformation (homotopy) in the corresponding path described by q via (29); in particular, if a set of trajectories of the $\mathbf{a}$'s is homotopic relative to its end points to the trivial trajectory [$\mathbf{a}_i(s) = \mathbf{a}_i$ for all s], so is its induced loop $q(s)$.

In Fig. 6(a) we show, slightly displaced in order to distinguish them from one another, the trajectories $\mathbf{a}_1(s)$ corresponding to the loop $q_m(s)$ given by (28) and (29) with s replaced by $2s$ in the exponent, i.e., a representative loop of $(X^m)^2$.

Figure 6(b) is an intermediate stage of a deformation of the trajectories of Fig. 6(a) into the trivial trajectories of Fig. 6(c). If at every stage of the deformation $\mathbf{a}_1(s)$ and $\mathbf{a}_2(s)$ move on their respective circumferences with uniform speed, the distance between them is always $|\mathbf{a}_1 - \mathbf{a}_2|$. Since the supports of $\varphi_1[\mathbf{x} - \mathbf{a}_1(s)]$ and $\varphi_2[\mathbf{x} - \mathbf{a}_2(s)]$ do not overlap initially, they do not overlap during the process. Figure 6 constitutes a proof of (31). In symbols, let

$$\mathbf{n}(t) = [(1 - t)\mathbf{n} + t(\mathbf{a}_1 - \mathbf{a}_2)]/|[\ ]| \tag{32}$$

(the denominator is the magnitude of the numerator). Then

$$\mathbf{a}_i(s, t) = \tfrac{1}{2}(\mathbf{a}_1 + \mathbf{a}_2) - (-1)^i$$
$$\times \exp[2\pi s n(t) \times] \tfrac{1}{2}(\mathbf{a}_1 - \mathbf{a}_2) \tag{33}$$

satisfies

$$\mathbf{a}_i(s, 0) = \mathbf{a}_i(2s),\ \mathbf{a}_i(s, 1) = \mathbf{a}_i.\quad \text{Q.E.D.} \tag{34}$$

Again, our proof of (31) is valid only in 3-dimensional theories! In Appendix B we show that, in general, $(X^m)^2 \neq 1$ in two dimensions. This, together with the

other similarities between X and W, suggests the possibility of a relationship between W and X; the existence and nature of such a relation is the subject matter of Sec. V.

5. It is simple to prove that the homotopic triviality of the flow $X^m\varphi(\mathbf{x})$ depends only on m and not on $\varphi(\mathbf{x})$.

6. When

$$X^m\tilde{q} \neq \tilde{q} \tag{35}$$

is obtained for X^m acting as defined in Secs. II.7 and IV.2, we say that the theory, and in particular Q_m (the component of Q that contains the kink being exchanged), admit negative field exchange, or "odd statistics" for short. The reason for the term "negative" or "odd" is that (31) and (35) imply that Q admits state functions $\Psi(\tilde{q})$ with the property

$$X^m\Psi(\tilde{q}) = -\Psi(\tilde{q}). \tag{36}$$

Notice that $X^m = \pm 1$ holds only as an eigenvalue equation for X^m as an operator on state functions, whereas for fields (or rather for their coverings), Eq. (31) means only that

$$X^m\tilde{q} = \tilde{q}' \Leftrightarrow X^m\tilde{q}' = \tilde{q}. \tag{37}$$

7. As in Sec. III, we can test whether Q_m admits odd statistics by checking if

$$X^m\tilde{q} \neq \tilde{q} \tag{38}$$

is obtained for any one $\tilde{q} \in Q$, as shown by the following.

Lemma: Equation (38) holds or does not hold simultaneously for all $\tilde{q} \in Q$.

Proof: Exact analog of the proof of Lemma 3.

In other words, this lemma says the parity of a structure under field exchange depends only on the component of Q_m to which it belongs, and not on the surrounding structures.

8. A single kink of type m will admit both odd and even exchange states if it admits odd exchange states.

The conserved baryon and lepton numbers do not behave this way: for example, a single baryon can be a fermion but not a boson, in that there is a fermion but no boson with

$$N_{\text{B}} = 1,\ N_e = N_\mu = 0. \tag{39}$$

Why? Or does the kink model lead us to expect an

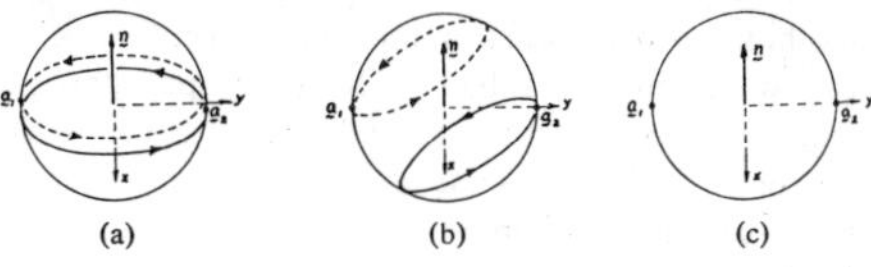

FIG. 6. (a) Path in X of an iterated exchange and (b) its deformation to (c) the identity.

excited state of the nucleon with the quantum numbers (39) but with integer spin and Bose statistics?

Our answer consists in two topological laws that combine to prevent kinks from ever exhibiting both of the statistics they may admit in principle: the invariance of statistics with respect to (a) the partition of the system into subsystems and (b) the passage of time.

In a universe of kinks of type m, the measurement of the exchange operator for any pair may give the value $+1$ or -1 in principle. However, if the result is -1, then an immediate measurement of the exchange operator for any other pair of kinks of type m will give the same result -1 by (a), and any subsequent measurement of this exchange will continue to give the result -1 by (b). Thus the statistics will appear to be a universal and permanent attribute of the type of kink. Let us now state more carefully and prove the topological laws mentioned here:

(a) Consider an eigenstate Ψ such that φ_{op} is a union field,

$$\varphi_{\mathrm{op}}\Psi = \cup \, \varphi_\alpha(x)\Psi,$$

with $\varphi_1, \cdots, \varphi_N$ kinks of the same type m. Then the $N(N-1)/2$ exchanges $\varphi_i \Leftrightarrow \varphi_j$, as continuous processes, are all homotopic to each other, and to the effect of the exchange operator X^m (defined in Sec. IV.2). This is a consequence of Sec. IV.3.

(b) The exchange operator X^m is a constant of the motion. Indeed, the state functional $\Psi[\varphi](t)$ is continuous in (φ, t) (Sec. II.1), so when φ undergoes a continuous closed deformation $\varphi \to \varphi(s)$ (in our case an exchange of two type m kinks) we get a function $\Psi[\varphi(s)](t)$ of s and t, continuous in (s, t), with the restriction [for $\varphi \to X^m(s)\varphi$] that

$$\Psi[\varphi(0)](t) = \pm\Psi[\varphi(1)](t), \quad \text{all } t. \quad (40)$$

But $\Psi[\varphi(1)](t)$ is continuous in t, ergo the sign is $+$ for all t or $-$ for all t.

9. Parastatistics. As our terminology suggests, we intend to interpret the operator X^m as the "performer" of the exchange of certain real particles. Accepting provisionally this meaning of X^m, the relation

$$(X^m)^2 = 1 \quad (31)$$

implies that *parastatistics are forbidden within the present theory.*

Ordinary quantum field theory does not exclude the possibility of parastatistics, and some recent work has been done on the subject. However, paraparticles seem to be absent from nature,[12] and it is suggestive that they do not even arise as a possibility in theories of the kind defined here.

---

[12] H. S. Green, Phys. Rev. **90**, 270 (1953); A. M. L. Messiah and O. W. Greenberg, Phys. Rev. **136**, B248 (1964).

V. CONNECTIONS BETWEEN "SPIN," "STATISTICS," AND NUMBER OF KINKS

1. *Theorem:* Q_m admits half-odd extrinsic spin if and only if it admits odd statistics.

The proof involves some rather specialized topological calculations and comprises Sec. V.2 and Appendix C. The reader willing to accept the theorem can pass on to Sec. V.3.

2. *Proof:* (a) Let $O \cdot \varphi(\mathbf{x})$ stand for the trivial loop in Q starting and ending at $\varphi(\mathbf{x})$; our theorem then reads [see (19) and Sec. IV.2]

$$X^m(s) \cdot [\varphi(\mathbf{x})] \equiv \varphi_X(\mathbf{x}, s) \quad (\text{say}) \sim O \cdot \varphi(\mathbf{x}), \quad (41)$$

if and only if

$$W^e(s) \cdot [\varphi_1(\mathbf{x})] \equiv \varphi_{1W}(\mathbf{x}, s) \quad (\text{say}) \sim O \cdot \varphi_1(\mathbf{x}),$$
$$\varphi_1(\mathbf{x}) \in Q_m. \quad (42)$$

(b) We may express (42) more explicitly. Let X stand for the Euclidean space spanned by $\mathbf{x}$, and I for the unit interval. Then (42) means that there exists a continuous function $\varphi(\mathbf{x}, s, t)$ defined on $X \otimes I \otimes I$, such that

$$\begin{aligned}
\varphi(\infty, s, t) &= \varphi_0, \\
\varphi(\mathbf{x}, s, 0) &= \varphi_X(\mathbf{x}, s), \\
\varphi(\mathbf{x}, s, 1) &= 0 \cdot \varphi(\mathbf{x}), \\
\varphi(\mathbf{x}, 0, t) &= \varphi(\mathbf{x}, 1, t) = \varphi(\mathbf{x}).
\end{aligned} \quad (43)$$

We can assume without loss of generality that the two structures exchanged by $X^m(s)$ are identical to $\varphi_1(\mathbf{x})$ and, for any s, lie completely within the 3-dimensional unit cube or 3-cube

$$I^3 = \{x, y, z \mid 0 \le x, y, z \le 1\}, \quad (44)$$

and that $\varphi(\mathbf{x})$ contains no other structures, since by Secs. IV.3 and IV.5 they are irrelevant for the statistics criterion; then (41) is equivalent to the existence of a function $\varphi(\mathbf{x}, s, t)$ on $I^3 \otimes I \otimes I = I^5$, such that ($\partial\Omega$ stands for the boundary of the region Ω):

$$\begin{aligned}
\varphi(\partial I^3, s, t) &= \varphi_0, \\
\varphi(\mathbf{x}, s, 0) &= \varphi_X(\mathbf{x}, s), \\
\varphi(\mathbf{x}, s, 1) &= 0 \cdot \varphi(\mathbf{x}), \\
\varphi(\mathbf{x}, 0, t) &= \varphi(\mathbf{x}, 1, t) = \varphi(\mathbf{x}).
\end{aligned} \quad (45)$$

Notice that (45) completely specifies a continuous function on the boundary ∂I^5 of $I^5 = I^3 \otimes I \otimes I$, the unit 5-cube spanned by $\mathbf{x}, s, t$; we shall denote that function by $\varphi_0[\partial I^5]$.

(c) We can also apply the above considerations to the rotation loop (42), and state that (42) means that there exists a function $\varphi_1(\mathbf{x}, s, t)$ on I^5 with boundary

D. FINKELSTEIN AND J. RUBINSTEIN

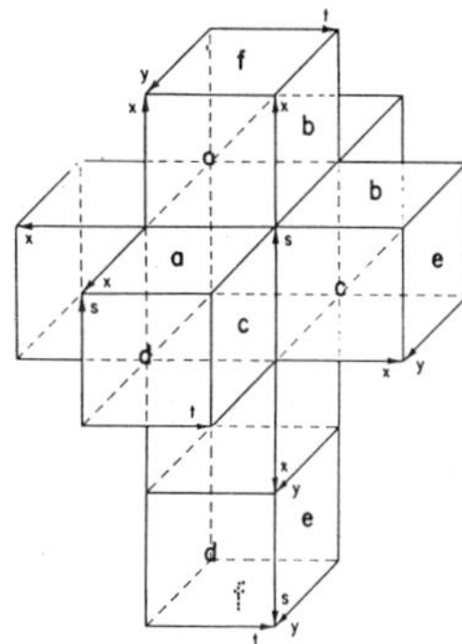

Fig. 7. Boundary of the 4-cube, or tesseract, $0 \leq (x, y, s, t) \leq 1$, where x, y are space coordinates, s is a loop parameter associated with the path of an exchange or a 2π rotation, and t is a deformation parameter associated with the homotopy of one loop into another.

values

$$\varphi_1(\partial I^3, s, t) = \varphi_0,$$
$$\varphi_1(\mathbf{x}, s, 0) = \varphi_{1IV}(\mathbf{x}, s),$$
$$\varphi_1(\mathbf{x}, s, 1) = 0 \cdot \varphi_1(\mathbf{x}), \tag{46}$$
$$\varphi_1(\mathbf{x}, 0, t) = \varphi_1(\mathbf{x}, 1, t) = \varphi_1(\mathbf{x}).$$

We shall denote the function on ∂I^5 defined by (46) as $\varphi_1[\partial I^5]$.

The existence of a continuous function $\varphi(\mathbf{x}, s, t) \equiv \varphi$ defined on I^5 and with prescribed boundary values $\varphi[\partial I^5]$, is by no means trivial, and is in fact an example of a "fundamental problem in topology, the extension problem" (Ref. 5, Sec. I.1). φ is called an extension of $\varphi[\partial I^5]$ over I^5 and (Ref. 5, Sec. I.4) there are spaces Φ such that not every $\varphi[\partial I^5]$ admits an extension over I^5.

(d) The following result is the basis for our proof[13]:

The homotopy-extension theorem: Let K be a finite simplicial complex, and L a closed subcomplex. Let $f_0 : K \to Y$ and $g_u : L \to Y$ be such that $g_0 = f_0|_L$. Then f_0 admits a homotopy $f_u : K \to Y$ such that $f_u|_L = g_u$ $(0 \leq u \leq 1)$.

Since I^5 is a finite simplicial complex and ∂I^5 a closed subcomplex, the theorem applies for $K = I^5$ and $L = \partial I^5$ (note that $Y = \Phi$ is not restricted at all). Suppose that we can find a homotopy $g_u[\partial I^5]$, i.e., a family of functions on ∂I^5, such that $g_0[\partial I^5] = \varphi_0[\partial I^5]$ and $g_1[\partial I^5] = \varphi_1[\partial I^5]$; then, by the extension theorem, if $\varphi_0[\partial I^5]$ can be extended over I^5, there exists a homotopy $f_u[I^5]$ such that $f_u[I^5]|_{\partial I^5} = g_u[\partial I^5]$ so, in particular, there exists a function $f_1(I^5)$ such that

$$f_1[I^5]|_{\partial I^5} = \varphi_1[\partial I^5]. \tag{47}$$

This would prove the "only if" of the theorem in Sec. V.1, the "if" being proved by exchanging the subindices 0 and 1 in the above process.

(e) It will be convenient to represent $\varphi_0[\partial I^5]$ graphically, but this is, of course, not possible since

[13] P. J. Hilton, *An Introduction to Homotopy Theory* (Cambridge University Press, Cambridge, 1964).

∂I^5 is a 4-dimensional manifold. However, for our proof we shall only need 2 dimensions of X, or I^3, so we can represent each $z = $ const (say) cross section of I^5 and ∂I^5 independently: $I^5|_{z=\text{const}}$ is a 4-cube, I^4, and $\partial I^5|_{z=\text{const}}$ is its boundary ∂I^4, a 3-dimensional manifold homomorphic to S^3, the 3 sphere. Thus, we cannot imbed ∂I^4 in X (so we cannot give a visualization of it) but we can "open" it along some "edges" (here squares) and "flatten" it, in much the same way as the surface (boundary) of a 3-cube can be brought into a flat 2-dimensional figure; at this point our ∂I^4 can be imbedded in X, so we can give a perspective visualization of it: Fig. 7. Some of the pairs of "edges" (squares) to be identified with each other are labeled with a common letter; each of the "faces" (3-cubes) is spanned by three of the variables (x, y, s, t), each going from 0 to 1 in such a way as to match, when we make the indicated identifications between the squares they span.

(f) In Fig. 8 we show only the four "faces" in which $\varphi_0[\partial I^5]|_{z=\text{const}} \equiv \varphi_0[\partial I^4] \neq \varphi_0$, two spanned by (x, y, s) and two by (x, y, t), and the "face" $(0, y, s, t)$, which has common "edges" with the former four. The support (Sec. IV.1) of $\varphi_1(\mathbf{x})|_{z=\text{const}}$ is taken to be (see Sec. V.2b) the square $|x| < \epsilon/\sqrt{2}$, $|y| < \epsilon/\sqrt{2}$. In the $(x, y, s, 0)$ 3-cube of Fig. 8 we show the exchange of two φ_1 structures, for a given z and for $\mathbf{n}$ of (28) parallel to the $-z$ axis, the exchange of two φ_1 structures, not exactly as prescribed by (29) but in a way clearly homotopic to it; the "pipes" of Fig. 8 are generated by the supports of $\varphi_1[\mathbf{x} - \mathbf{a}_1(s)]|_{z=\text{const}}$ and

$$\varphi_1[\mathbf{x} - \mathbf{a}_2(s)]|_{z=\text{const}}.$$

Analogously the $(x, y, s, 1)$ 3 cube shows the trivial

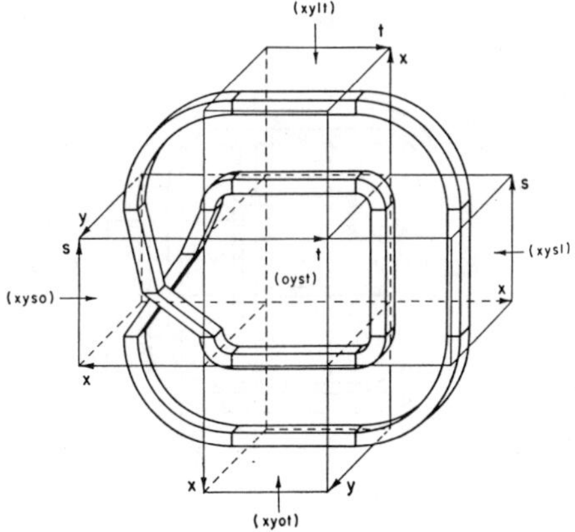

Fig. 8. Boundary values on the 4-cube of Fig. 7 whose extendibility into the interior of the 4-cube expresses the triviality of an exchange.

loop 0. $(\varphi_1(\mathbf{x} - \mathbf{a}_1) \cup \varphi_1(\mathbf{x} - \mathbf{a}_2))$. The two remaining 3-cubes with $\varphi \not\equiv \varphi_0$ represent the last line of (45). The "curved pipes" only indicate how to identify the structures contained in identical squares; they can also be thought of as generated by the supports of the structures contained in one of the squares in which they end, if that square is "flopped over" to make it coincide with the identical one.

From now on we may speak of I^4, ∂I^4, $f[I^4]$, $g[\partial I^4]$ instead of I^5, ∂I^5, $f[I^5]$, $g[\partial I^5]$, in the understanding that we always refer to the $z = $ const cross sections of the latter.

(g) We now construct a homotopy $g_u[\partial I^4]$ with the required properties (Sec. V.2.d). In all of ∂I^4, except the five 3-cubes shown in Fig. 8, $g_u[\partial I^4] = \varphi_0$ for all u, and $g_0[\partial I^5] = \varphi_0[\partial I^5]$ is given in Fig. 8.

In going from Fig. 8 to Fig. 9 we just "pull" $g_u[\partial I^4]$ towards the $(0yst)$ cube, in which g_u is defined in such a way as to match the boundary conditions generated when the supports of g_u in the surrounding cubes are cut by the common boundaries.

Next the "inner handle" of Fig. 9 undergoes a rigid π rotation around the α axis, in the sense shown there, with the values of the g_u rigidly carried along; the rest of g_u in the $(xy0t)$, $(xy1t)$, and $(xys1)$ cubes is unchanged. The homotopy is more complicated (not rigid) in the $(xys0)$ cube, where g_u is defined in a way that matches the boundary values for all u; then, since the shaded regions undergo a π rotation each, the two "pipes" in the $(xys0)$ cube are subject to a π twist each. By translating the content of the $(0yst)$ cube into the $(xys0)$ cube, straightening the resulting $(xys0)$ "pipe," and distributing uniformly along the s axis the two π-twists concentrated at its lower and upper ends, we

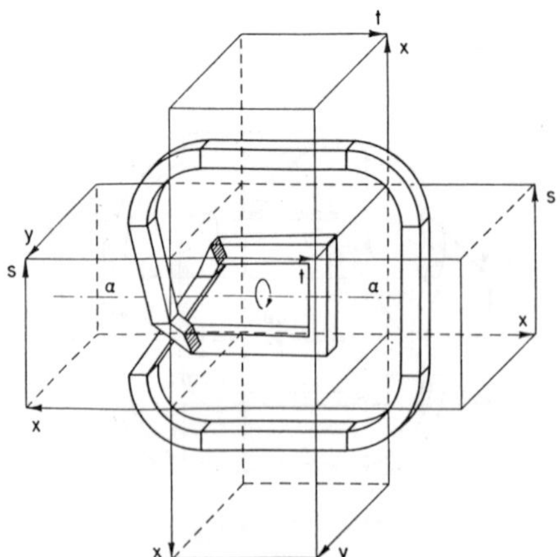

FIG. 9. Intermediate step in the deformation of the boundary values of Fig. 8 into those of Fig. 10.

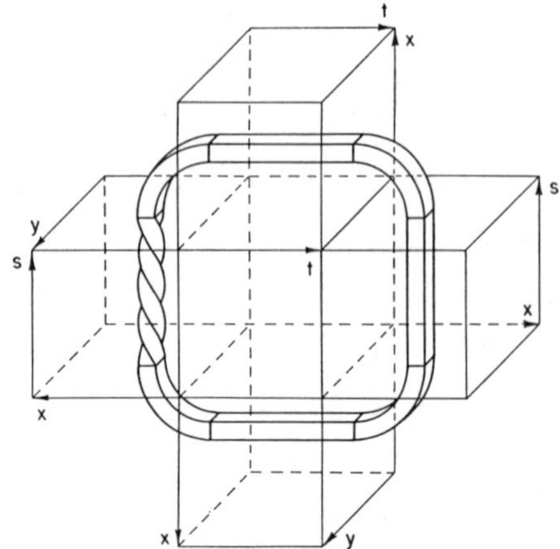

FIG. 10. Boundary values on the 4-cube of Fig. 7 whose extendability into the interior of the 4-cube expresses the triviality of a 2π rotation.

finally obtain, in Fig. 10, a graphical representation of $\varphi_1[\partial I^4] = (46)$, and thus, according to Sec. 1.d, we have proven the theorem. The detailed analytical construction of $g_u(\partial I^5)$ is given in Appendix C.

3. Consistency of the definition of X^m. We shall now prove Sec. IV.3, i.e., that if φ is the union (Sec. IV.1) of more than one pair of identical fields $\varphi_m(\mathbf{x}) \in Q_m$, the exchange of the two kinks in one pair is homotopic relative to its end point[11] to the exchange of the two kinks in any of the other pairs. The analytic proof is rather tedious, and similar to the one given in proving the theorem in Sec. V.1 (Appendix C); we shall omit it, and give only the corresponding graphic constructions (see Sec. V.2 for meaning of figures).

(a) We will first prove our statement for a particular case, i.e., when the two pairs have one kink in common. In Fig. 11 we show a field containing 3 identical kinks A, B, C. In $(x, y, s, 0)$ we exchange A, B, and in $(x, y, s, 1)$ we exchange B, C; in $(x, y, 0, t)$ and $(x, y, 1, t)$ we leave A, B, C fixed. The $g_0[\partial I^5]$ of Fig. 11 can be extended into I^5. This can be seen by treating the "inner handle" as in Sec. V.2 (Figs. 8–10): it is clear that we will obtain the $g[\partial I^5]$ of Fig. 12; again, treating the inner handle of Fig. 12 as in Sec. V.2 we are finally left with the $g[\partial I^5]$ of Fig. 13; since it can obviously be extended into I^5, our $g_0[\partial I^5]$ can too, so the exchange of A, B is homotopic to the exchange of B, C.

In case we had performed the B, C exchange in the opposite direction, we would have ended up with a (W^e) loop in the $(xys1)$ cube, instead of the trivial loop of Fig. 13. This would not invalidate our result, since $(W^e)^2 \sim 1$ (Sec. III.4).

D. FINKELSTEIN AND J. RUBINSTEIN

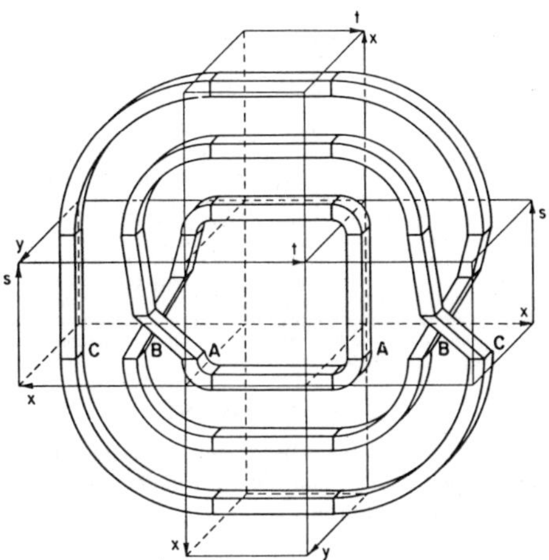

FIG. 11. Boundary values for the homotopy of one exchange (AB) into another (BC). A, B, C are identical kinks. This is a formalization of the second rubber-band lemma.

(b) In Fig. 14 we show only one space dimension (x) and two pairs of kinks, AB and CD, with

$$\varphi_B(\mathbf{x} + \mathbf{b}) = \varphi_A(\mathbf{x} + \mathbf{a})$$

and

$$\varphi_D(\mathbf{x} + \mathbf{d}) = \varphi_C(\mathbf{x} + \mathbf{c}),$$

and all of them belonging to Q_m; to distinguish them we assume the support of $\varphi_A(\mathbf{x})$ is very small and the support of $\varphi_C(\mathbf{x})$ is small, but bigger than the support of $\varphi_A(\mathbf{x})$. Our purpose will be accomplished by showing that the boundary conditions on the "cube"

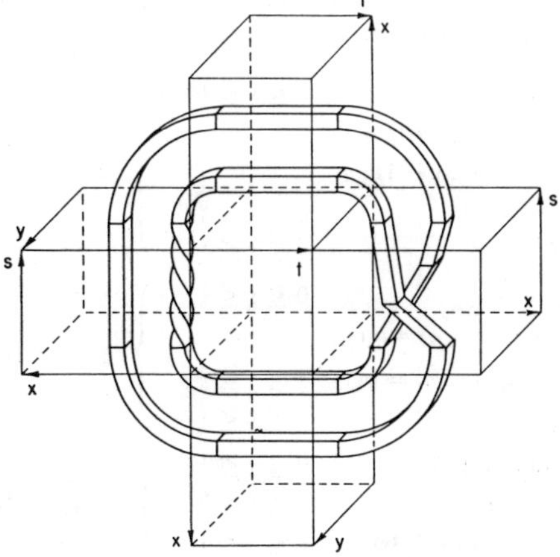

FIG. 12. Intermediate step in the deformation of Fig. 11 into Fig. 13. Note the resemblance to the first rubber-band lemma.

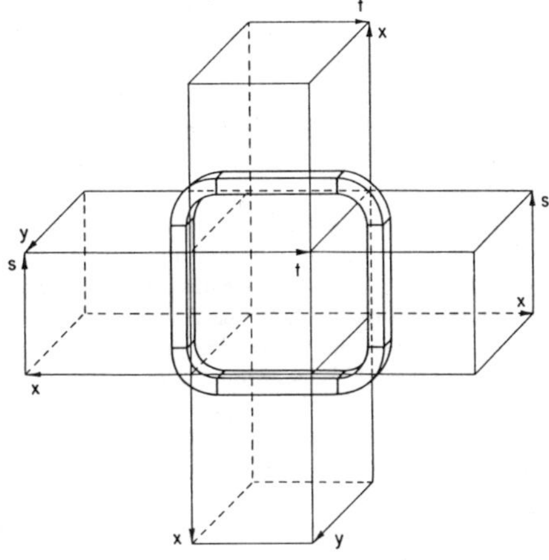

FIG. 13. Boundary values of the identity deformation on the identity loop. These values are obviously extendable into the interior.

of Fig. 14 can be extended into its interior (the t scale has been taken larger than the s, x scales to facilitate the drawing) by some function $\varphi[\partial I^5]$.

In going from $t = 0$ to $t = t_1$ we gradually transform the identity loop on C, D into the loop consisting in deforming C, D, into C', D' [with $\varphi_{D'}(\mathbf{x} + \mathbf{d}) = \varphi_{C'}(\mathbf{x} + \mathbf{c}) = \varphi_B(\mathbf{x} + \mathbf{b}) = \varphi_A(\mathbf{x} + \mathbf{a})$] for $0 \leq s \leq s_1$, C', D' remaining fixed for $s_1 \leq s \leq s_2$, and C', D' becoming C, D for $s_2 \leq s \leq 1$; also, the exchange of A, B is concentrated in the region $s_1 \leq s \leq s_2$.

In $t_1 \leq t \leq t_2$ we take $\varphi[I^5]$ identical to its value at $t = t_1$, for $0 \leq s \leq s_1$ and $s_2 \leq s \leq 1$. For $s_1 \leq s \leq s_2$ we have a situation to which we can apply the result just obtained: the loop at t_1, in which A, B are exchanged and C', D' remain fixed, is homotopic to a loop in which B, C' are exchanged and A, D' remain fixed, which in turn is homotopic to a loop in which C', D' are exchanged, whereas A, B remain fixed, at $t = t_2$.

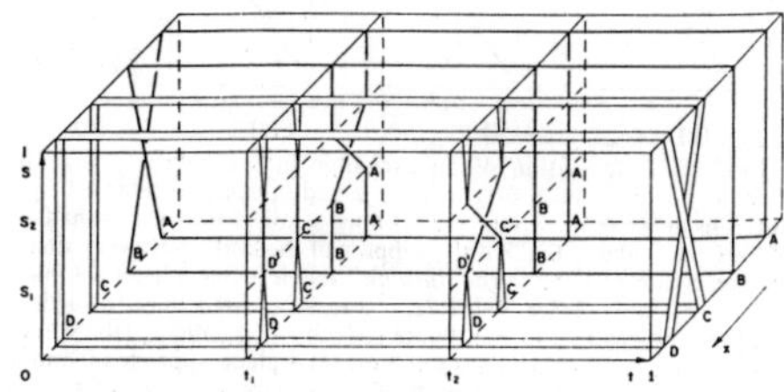

FIG. 14. Proof that two exchanges are equivalent (uniformity of statistics). Here A, B, are identical kinks of type m, and C, D are identical kinks of type m, but A and C need not be identical. At $t = 0$, the exchange AB is represented. At $t = 1$, the exchange CD is represented. The intermediate steps demonstrate their equivalence.

For $t_2 \leq t \leq 1$, $\varphi[I^5]$ consists simply in distributing uniformly along s the exchange performed at $t = t_2$, and at the same time "unmaking" the transformation made on C, D in the step $0 \leq t \leq t_1$. This completes the construction of an homotopy $\varphi[\partial I^5]$ between the exchange of A, B (at $t = 0$) and the exchange of C, D, at $t = 1$.   Q.E.D.

4. In Sec. V.3a we have obtained the boundary function of Fig. 12 as an intermediate step and then proved that it can always be extended into the interior of I^5. This means that for a two-kink field, exchanging the kinks is homotopic[11] to leaving one of them fixed and rotating the other one through 2π.

Without using the homotopy extension theorem the existence of an homotopy between the above mentioned loops is not quite evident, so it is interesting to know that we can actually construct the homotopy; its detailed exposition would be too lengthy (since the interest of the problem is rather academic) so we only give, in Fig. 15, a rough sketch of the method.

5. A $2n$-kink state does not admit half-odd spin: if in $n = (n_1, \cdots, n_k)$ all the n_i are even, by Sec. III.6 we can check the statement using as a test field a union field

$$\varphi(\mathbf{x}) = \varphi_m(\mathbf{x}) \cup \varphi_m(\mathbf{x} - \mathbf{a}),$$

$$m = \tfrac{1}{2}n;$$

taking $\mathbf{a}$ and $\mathbf{n}$ along the x axis, after a trivial homotopy the above process can be symbolized by Fig. 16(a), a simplified diagram of the type used in Figs. 11–15. The homotopy (a) $\to$ (b) is obvious, and in (b) $\to$ (c) we use the result analyzed in Sec. V.4. Since (c) is simply $(X^m)^2$, (c) $\to$ (d) is proven by $(X^m)^2 = 1$, Eq. (31).

In conventional quantum-field theories, the corresponding statement would follow from the laws of composition of angular momentum: an even number of half-odd-spin particles can only have integer spin.

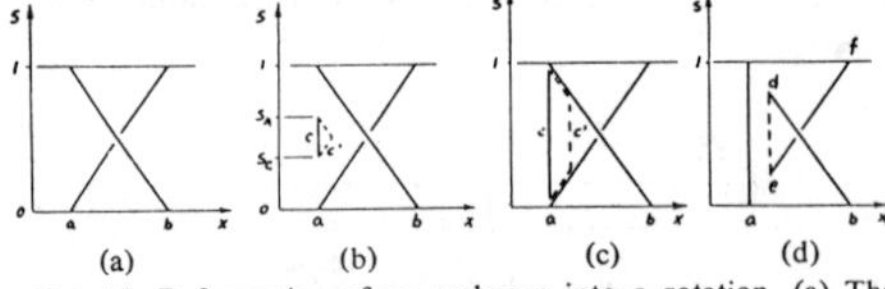

FIG. 15. Deformation of an exchange into a rotation. (a) The exchange of two kinks, originally at a and b, each represented by a dot; this is what we would see looking straight into the front face of the $(xysl)$ cube of Fig. 8 if the support of the kinks were very small. (b) We create at a (after the "a kink" has left that place) a kink c and its antikink c' (see Sec. I.7); c' moves (dotted line) towards b and then returns to a to annihilate c. In b $\to$ c s_C and s_A, the values of s at which creation and annihilation take place approach 0 and 1 respectively. (c) Part of the path of c' overlaps the path of the "a kink" and part of the path of the "b kink"; in the corresponding regions the field is equal to φ_0: (d) The rest of the homotopy (and the most involved part of it) consists of deforming the loop $bdef$ of d into the rotation of a kink sitting at b, but at this point we would need more elaborate figures.

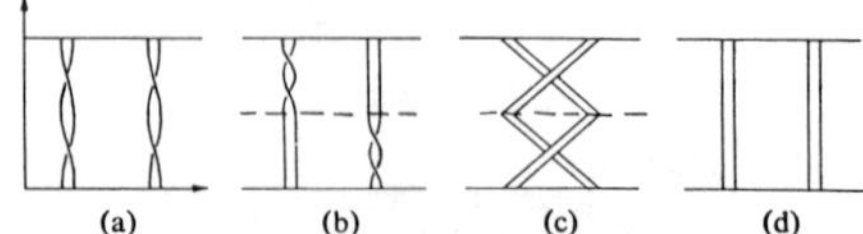

FIG. 16. Schematic of a deformation. The starting point (not shown) is a rigid rotation of two identical kinks about the line of centers. Then the axis of rotation of each kink is twined perpendicular to the line of centers, giving the deformation (a), where each kink is represented by a segment, whose endpoints traverse the intertwining helices shown. The intermediate step (b) is obtained in an obvious way from (a). In going from (b) to (c), the extendability of Fig. 12 into the 4-cube is applied to the upper and lower halves of (b) independently. Since (c) represents X^2, the deformation to the identity (d) has already been shown.

ACKNOWLEDGMENT

We thank Charles Misner for criticisms of the manuscript which helped our own understanding, as well as the presentation.

APPENDIX A: PROOF OF LEMMA 3

Let q, $q_1 \in\in Q_m$, $q(s)$ be a path between $q = q(0)$ and $q_1 = q(1)$, $W^e(s)$ a $2\pi s$ extrinsic rotation operator around some axis, and $q_1(s) = W(s) \cdot q$ a homotopically trivial loop (so $W \cdot \tilde{q}_1 = \tilde{q}_1$). Then $q(s, t) = W^e(s) \cdot q$,

$$q(s, t) = W^e(s) \cdot q, \qquad 0 \leq s < 1, \quad t = 0,$$

$$\left.\begin{array}{ll} q, & 0 \leq s \leq t \\[4pt] W^e\left(\dfrac{s - t}{1 - 2t}\right) \cdot q, & t \leq s \leq 1 - t \\[6pt] q, & 1 - t \leq s \leq 1 \end{array}\right\} \quad 0 < t < \tfrac{1}{4},$$

$$\left.\begin{array}{ll} q[(4t - 1)3s], & 0 \leq s \leq \tfrac{1}{3} \\[4pt] W^e(3s - 1) \cdot q(4t - 1), & \tfrac{1}{3} \leq s \leq \tfrac{2}{3} \\[4pt] q[(4t - 1)(3 - 3s)], & \tfrac{2}{3} \leq s \leq 1 \end{array}\right\} \quad \tfrac{1}{4} \leq t < \tfrac{2}{4},$$

$$\left.\begin{array}{ll} q(3s), & 0 \leq s \leq \tfrac{1}{3} \\[4pt] W^e(3s - 1) \cdot q(1) & \\ \quad = W(3s - 1) \cdot q_1 & \\ \quad = q_1(s), & \tfrac{1}{3} \leq s \leq \tfrac{2}{3} \\[4pt] q(3 - 3s), & \tfrac{2}{3} \leq s \leq 1 \end{array}\right\} \quad t = \tfrac{2}{4},$$

$$\left.\begin{array}{ll} q(3s), & 0 \leq s \leq \tfrac{1}{3} \\[4pt] q_1(s, t) : q_1(s, \tfrac{2}{4}) = q_1(s), & \\ \quad q_1(s, \tfrac{3}{4}) = q_1 = q(1), & \tfrac{1}{3} \leq s \leq \tfrac{2}{3} \\[4pt] q(3 - 3s), & \tfrac{2}{3} \leq s \leq 1 \end{array}\right\} \quad \tfrac{2}{4} \leq t \leq \tfrac{3}{4},$$

$$\left.\begin{array}{ll} q[(4 - 4t)3s], & 0 \leq s \leq \tfrac{1}{3} \\[4pt] q(4 - 4t), & \tfrac{1}{3} \leq s \leq \tfrac{2}{3} \\[4pt] q[(4 - 4t)(3 - 3s)], & \tfrac{2}{3} \leq s \leq 1 \end{array}\right\} \quad \tfrac{3}{4} \leq t \leq 1,$$

$$q(0) = q, \qquad 0 \leq s \leq 1, \qquad t = 1,$$

is a homotopy between $W^e(s) \cdot q$ and the trivial loop, for all $q \in Q_m$.

APPENDIX B

By composition law we mean a unique and continuous prescription to define a union field $\varphi = \cup \varphi_i$ when the supports of the φ_i overlap.[14] Then

Theorem: If Q does not admit a composition law, $(X^m)^2 = 1$ is not true in 2 dimensions.

Proof: Take $a_1 = (-a, 0)$, $a_2 = (a, 0)$ and rotate them around the origin of the (xy) plane. The trajectories $\mathbf{a}_1$, $\mathbf{a}_2$ corresponding to $(X^m)^2$ [Fig. 6(a)] are

$$a_{1x}(s) = -a \cos 2\pi s, \quad a_{2x}(s) = a \cos 2\pi s,$$
$$a_{1y}(s) = -a \sin 2\pi s, \quad a_{2y}(s) = a \sin 2\pi s, \quad \text{(B1)}$$

and admit the following homotopy in the xy plane [Fig. 17(a)]:

$$
\begin{aligned}
a_{1x}(s, t) &= -(1 - t)a \cos 2\pi s - ta, \\
a_{2x}(s, t) &= (1 + t)a \cos 2\pi s - ta, \\
a_{1y}(s, t) &= -(1 - t)a \sin 2\pi s, \\
a_{2y}(s, t) &= a \sin 2\pi s.
\end{aligned}
\quad \text{(B2)}
$$

At $t = 0$ (B2) gives (B1) and, at $t = 1$, [Fig. 17(b)]

$$a_{1x}(s, 1) = -a, \quad a_{2x}(s, 1) = -a(1 - 2\cos 2\pi s),$$
$$a_{1y}(s, 1) = 0, \quad a_{2y}(s, 1) = a \sin 2\pi s;$$

at no (s, t) do $\mathbf{a}_1(s, t)$ and $\mathbf{a}_2(s, t)$ coincide. For the trajectories at $t = 1$ to be deformed into the trivial trajectory, $\mathbf{a}_1$ has to be left still and $\mathbf{a}_2(s, 1)$ shrunk into $\mathbf{a}_2(0, 1)$. This is clearly impossible to do in the x, y plane without at some stage having $\mathbf{a}_2$ and $\mathbf{a}_1$ overlap [see Fig. 17(a)].

Should (B1) be deformable into the trivial loop (in the x, y plane and without overlapping), the inverse of the homotopy (B2) followed by such a deformation would be a deformation of the trajectory of Fig.

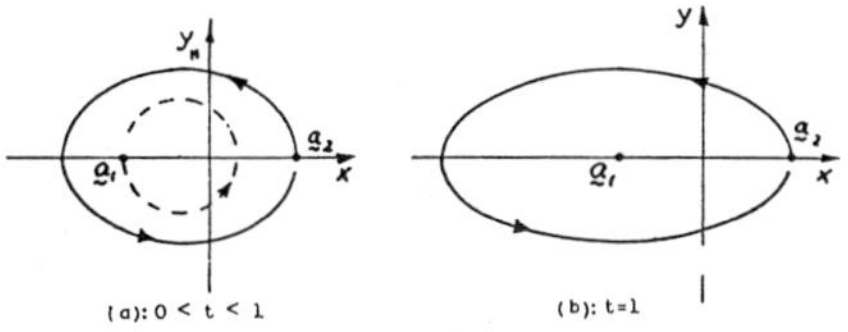

FIG. 17. A deformation of the $(X^m)^2$ loop in the plane, showing that in two dimensions, in general, $(X^m)^2 \neq 1$.

17(b) into the trivial one; since this is not possible, the theorem is proved.

As a by-product, by means of motions out of the (x, y) plane, Fig. 17 provides an easy alternative proof of the relation $(X^m)^2 = 1$.

APPENDIX C: ANALYTIC PROOF OF THE THEOREM IN SEC. 1

The following conventions are used in the proof:

(1) (x, y) stands for $\mathbf{x} = (x, y, z)$; z remains unaffected in all transformations.

(2) x, y, s, t, u (but not z) range all between 0 and 1, if not otherwise indicated.

(3) $s \leq s_1$ stands for $0 \leq s \leq s_1$ and $s \geq s_2$ for $s_2 \leq s \leq 1$; analogously for the other variables.

(4) Figures 8–10 represent part of cross sections $z = \text{const}$ of ∂I^5, i.e., ∂I^4, by showing five of its eight sectors: $(x, y, s, 0)$, $(x, y, s, 1)$, etc. For fixed u, $\varphi(x, y, s, t, u)$ is a function on ∂I^5; it is defined as equal to φ_0 at those points of ∂I^5 at which its formal expression given below is meaningless, and is equal to φ_0 on all the sectors not mentioned explicitly.

(5) $\varphi_1(\mathbf{x}) = \varphi_0$ for $|\mathbf{x}| > \epsilon$; $a < \frac{1}{2} - 2\epsilon$ is a fixed positive number, and

$$b = 1 - a; \quad 0 < u_1 < u_2 < \cdots < u_8 < 1$$

are eight real fixed numbers.

(6) $\varphi(x, y, s, t, u)$ is the $g_u[\partial I^5]$ of the text, Sec. V.2. Let

$$
\begin{aligned}
f_1(s) &= (1 - a) - (1 - 2a)s, & g_1(s) &= \tfrac{1}{2} - (1 - 2a)s, & & s \leq \tfrac{1}{2}, \\
& & &= a + (1 - 2a)(s - \tfrac{1}{2}), & & s \geq \tfrac{1}{2}, \\
f_2(s) &= a + (1 - 2a)s, & g_2(s) &= \tfrac{1}{2} + (1 - 2a)s, & & s \leq \tfrac{1}{2}, \\
& & &= (1 - a) - (1 - 2a)(s - \tfrac{1}{2}), & & s \geq \tfrac{1}{2}.
\end{aligned}
$$

Then (Fig. 8)

$$\varphi(x, y, s, 0, 0) = \varphi_1(x - f_1(s), y - g_1(s)) \cup \varphi_1(x - f_2(s), y - g_2(s)),$$
$$\varphi(x, y, s, 1, 0) = \varphi(x, y, 0, t, 0) = \varphi(x, y, 1, t, 0) = \varphi_1(x - a, y - \tfrac{1}{2}) \cup \varphi_1(x - b, y - \tfrac{1}{2}).$$

Let

$$
\left.
\begin{aligned}
f_1(s, u) &= f_1(s), & g_1(s, u) &= (1 - p)g_1(s) + pa, & g_3(s, u) &= (1 - p)\tfrac{1}{2} + pf_1(s) \\
f_2(s, u) &= f_2(s), & g_2(s, u) &= (1 - p)g_2(s) + pb, & g_4(s, u) &= (1 - p)\tfrac{1}{2} + pf_2(s)
\end{aligned}
\right\}
\quad p = u/u_1, \quad 0 \leq u \leq u_1.
$$

---

[14] For instance, when Φ is the manifold of a group, and φ_0 its identity element, a natural prescription is $\varphi_1(\mathbf{x}) \cup \varphi_2(\mathbf{x}) = \varphi_1(\mathbf{x}) \cdot \varphi_2(\mathbf{x})$, with "$\cdot$" indicating the product in Φ at each $\mathbf{x}$.

Then

$$\left.\begin{aligned}
\varphi(x, y, s, 0, u) &= \varphi_1(x - f_1(s, u), y - g_1(s, u)) \cup \varphi_1(x - f_2(s, u), y - g_2(s, u)) \\
\varphi(x, y, s, 1, u) &= \varphi_1(x - a, y - g_3(s, u)) \cup \varphi_1(x - b, y - g_4(s, u)) \\
\varphi(x, y, 0, t, u) &= \varphi_1(x - b, y - a) \cup \varphi_1(x - a, y - b) \\
\varphi(x, y, 1, t, u) &= \varphi_1(x - a, y - a) \cup \varphi_1(x - b, y - a)
\end{aligned}\right\} \quad 0 \le u \le u_1 .$$

Let

$$f_1(s, u) = (1 - g)f_1(s) + gf_1'(s),$$
$$f_2(s, u) = (1 - g)f_2(s) + gf_2'(s),$$
$$f_3(s, u) = (1 - g)a + gf_3'(s),$$
$$g_3'(s, u) = (1 - g)f_1(s) + gg_3'(s),$$

$$g_3'(s) = b,$$
$$= b - (s - (\tfrac{1}{2} - a + \epsilon))\frac{1 - 2a}{2(a - \epsilon)},$$
$$= a,$$

$$\left.\begin{aligned}
f_1'(s) &= b - \frac{\tfrac{1}{2} - a - \epsilon}{\tfrac{1}{2} + a - \epsilon}\, s, && s \le \tfrac{1}{2} + a - \epsilon \\
&= \tfrac{1}{2} + \epsilon - (s - (\tfrac{1}{2} + a - \epsilon)), && s \ge \tfrac{1}{2} + a - \epsilon \\
f_2'(s) &= a + s, && s \le \tfrac{1}{2} - a + \epsilon \\
&= \tfrac{1}{2} + \epsilon + \frac{\tfrac{1}{2} - a - \epsilon}{(\tfrac{1}{2} + a - \epsilon)(s - g)}, && s \ge \tfrac{1}{2} - a + \epsilon \\
f_3'(s) &= a + s, && s \le \tfrac{1}{2} - a + \epsilon \\
&= \tfrac{1}{2} + \epsilon, \quad \tfrac{1}{2} - a + \epsilon \le && s \le \tfrac{1}{2} + a - \epsilon \\
&= \tfrac{1}{2} - (s - (\tfrac{1}{2} + a - \epsilon)), && s \ge \tfrac{1}{2} + a - \epsilon
\end{aligned}\right\} \quad g = \frac{u - u_1}{u_2 - u_1}$$

$$u_1 \le u \le u_2 .$$

Then

$$\left.\begin{aligned}
\varphi(x, y, s, 0, u) &= \varphi_1(x - f_1(s, u), y - a) \cup \varphi_1(x - f_2(s, u), y - b) \\
\varphi(x, y, s, 1, u) &= \varphi_1(x - f_3(s, u), y - g_3(s, u)) \cup \varphi_1(x - b, y - f_2(s)) \\
\varphi(x, y, 0, t, u) &= \varphi(x, y, 0, t, u_1) \\
\varphi(x, y, 1, t, u) &= \varphi(x, y, 1, t, u_1)
\end{aligned}\right\} \quad u_1 \le u \le u_2 ,$$

$$\left.\begin{aligned}
\varphi(x, y, s, 0, u) &= \varphi(x + r, y, s, 0, u_2) \\
\varphi(x, y, s, 1, u) &= \varphi(x + r, y, s, 0, u_2) \\
\varphi(x, y, 0, t, u) &= \varphi(x + r, y, 0, t, u_2) \\
\varphi(x, y, 1, t, u) &= \varphi(x + r, y, 1, t, u_2) \\
\varphi(0, y, s, t, u) &= \varphi(r, y, s, 0, u_2)
\end{aligned}\right\} \quad r = \frac{1}{2}\frac{u - u_2}{u_3 - u_2} \quad u_2 \le u \le u_3 .$$

Let

$$\omega = y + is \quad \text{stand for } (y, s) \text{ in} \quad \varphi(x, y, s, t, u); \quad \alpha = \tfrac{1}{2} + \tfrac{1}{2}i \quad \text{and}$$

$$\left.\begin{aligned}
\omega_1(0) &= a + i(\tfrac{1}{2} + a) - i\epsilon = \beta_1 - i\epsilon, && \omega_1(u) = y_1(u) + is_1(u) = \alpha + e^{iv}(\beta_1 - \alpha) - i\epsilon \\
\omega_2(0) &= (1 - a) + i(\tfrac{1}{2} - a) + i\epsilon = \beta_2 + i\epsilon, && \omega_2(u) = y_2(u) + is_2(u) = \alpha + e^{iv}(\beta_2 - \alpha) + i\epsilon \\[4pt]
f_1(s, u) &= \tfrac{1}{2} - a - \frac{\tfrac{1}{2} - a - \epsilon}{s_1(u)}\, s, && g_1(s, u) = a - \frac{a - y_1(u)}{s_1(u)}\, s, && s \le s_1(u) \\
&= \epsilon - (s - s_1(u)), && = y_1(u), && s \ge s_1(u) \\
f_2(s, u) &= \epsilon + (s - s_2(u)), && g_2(s, u) = y_2(u), && s \le s_2(u) \\
&= \epsilon + \frac{\tfrac{1}{2} - a - \epsilon}{1 - s_2(u)}(s - s_2(u)), && = y_2(u) - \frac{y_2(u) - (1 - a)}{(1 - s_2(u))(s - g)}, && s \ge s_2(u) \\
\lambda_1(s, u) &= s/s_1(u), \quad s \le s_1(u), && \lambda_2(s, u) = 1, && s \le s_2(u) \\
&= 1, \qquad s \ge s_1(u), && = \frac{1 - s}{1 - s_2(u)}, && s \ge s_2(u)
\end{aligned}\right\} \quad v = \pi\frac{u - u_3}{u_4 - u_3}$$

$$\Omega_i = (x - f_i(s, u)) + i(y - g_i(s, u)) = (X_i + iY_i)(x, y, s, u), \qquad i = 1, 2$$

$$u_3 \le u \le u_4 .$$

D. FINKLESTEIN AND J. RUBINSTEIN

Then

$$\varphi(x, y, s, 0, u) = \varphi_1(e^{iv\lambda_1(s_1 u)}\Omega_1(x, y, s, u)) \cup \varphi_1(e^{-iv\lambda_2(s_1 u)}\Omega_2(x, y, s, u))$$

$$\varphi(x, y, s, 1, u) = \varphi(x, \alpha + e^{-iv}(\omega - \alpha), 1, u_3), \quad x \leq 2\epsilon$$

$$= \varphi(x, y, s, 1, u_3), \qquad x \geq 2\epsilon$$

$$\varphi(x, y, 0, t, u) = \varphi(x, y, 0, t, u_3)$$

$$\varphi(x, y, 1, t, u) = \varphi(x, y, 1, t, u_3) \qquad\qquad u_3 \leq u \leq u_4,$$

$$\varphi(0, y, s, t, u) = \varphi(0, \alpha + e^{-iv}(\omega - \alpha), t, u_3)$$

$$\varphi(x, y, s, 0, u) = \varphi(x, y, s, 0, u_4)$$

$$\varphi(x, y, s, 1, u) = \varphi(x + w, y, s, 1, u_4), \qquad x \leq 2\epsilon$$

$$= \varphi(x, y, s, 1, u_4), \qquad x \geq 2\epsilon$$

$$\varphi(x, y, 0, t, u) = \varphi(x, y, 0, t, u_4) \qquad\qquad w = 2\epsilon \frac{u - u_4}{u_5 - u_4} \quad u_4 \leq u \leq u_5.$$

$$\varphi(x, y, 1, t, u) = \varphi(x, y, 1, t, u_4)$$

$$\varphi(0, y, s, t, u) = \varphi(0, y, s, t, u_4), \qquad t \leq 1 - w$$

$$= \varphi(t + w - 1, y, s, 1, u_4), \quad t > 1 - w$$

$$\varphi(x, y, s, 0, u) = \varphi(x, y, s, 0, u_5)$$

$$\varphi(x, y, s, 1, u) = \varphi(x, y, s, 1, u_5)$$

$$\varphi(x, y, 0, t, u) = \varphi(x, y, 0, t, u_5) \qquad\qquad \mu = (1 - 2\epsilon) \frac{u - u_5}{u_6 - u_5} \quad u_5 \leq u \leq u_6.$$

$$\varphi(x, y, 1, t, u) = \varphi(x, y, 1, t, u_5)$$

$$\varphi(0, y, s, t, u) = \varphi(0, y, s, t + \mu, u_5)$$

$$\varphi(x, y, s, 0, u) = \varphi(0, y, s, v - x, u_6), \quad x < v$$

$$= \varphi(x - v, y, s, 0, u_6), \quad x > v$$

$$\varphi(x, y, s, 1, u) = \varphi(x - v, y, s, 1, u_6)$$

$$\varphi(x, y, 0, t, u) = \varphi(x - v, y, 0, t, u_6) \qquad v = 2\epsilon \frac{u - u_6}{u_7 - u_6} \quad u_6 \leq u \leq u_7.$$

$$\varphi(x, y, 1, t, u) = \varphi(x - v, y, 1, t, u_6)$$

$$\varphi(0, y, s, t, u) = \varphi(0, y, s, t + v, u_6)$$

Let

$$f(s) = s, \qquad\qquad 0 \leq s \leq \tfrac{1}{2} - a + \epsilon,$$

$$= \tfrac{1}{2} - a + \epsilon, \qquad\qquad \tfrac{1}{2} - a + \epsilon \leq s \leq \tfrac{1}{2} + a - \epsilon,$$

$$= (\tfrac{1}{2} - a + \epsilon) - (s - (\tfrac{1}{2} + a - \epsilon)), \qquad \tfrac{1}{2} + a - \epsilon \leq s1,$$

$$g(s) = \frac{1 - 2a}{\tfrac{1}{2} - a - \epsilon} s, \qquad\qquad 0 \leq s \leq \tfrac{1}{2} - a - \epsilon,$$

$$= 1 - 2a, \qquad\qquad \tfrac{1}{2} - a - \epsilon \leq s \leq \tfrac{1}{2} - a + \epsilon,$$

$$= 1 - 2a - \frac{1 - 2a}{2(a - \epsilon)} (s - (\tfrac{1}{2} - a - \epsilon)), \quad \tfrac{1}{2} - a + \epsilon \leq s \leq \tfrac{1}{2} + a - \epsilon,$$

$$= 0, \qquad\qquad \tfrac{1}{2} + a - \epsilon \leq s \leq \tfrac{1}{2} + a + \epsilon,$$

$$= \frac{1 - 2a}{\tfrac{1}{2} - a - \epsilon} (s - (\tfrac{1}{2} + a + \epsilon)), \qquad \tfrac{1}{2} + a + \epsilon \leq s \leq 1.$$

Then

$$\varphi(x, y, s, 0, u) = \varphi(x - \rho f(s), y + \rho g(s), s, u, u_7)$$

$$\varphi(x, y, s, 1, u) = \varphi(x, y + \rho f_2(s), s, 1, u_7)$$

$$\varphi(x, y, 0, t, u) = \varphi(x, y, 0, t, u_7) \qquad \rho = \frac{u - u_7}{u_8 - u_7}, \quad u_7 \leq u \leq u_8,$$

$$\varphi(x, y, 1, t, u) = \varphi(x, y + \rho(1 - 2a), 1, t, u_7)$$

$$\left.\begin{aligned}
\varphi(x, y, s, 0, u) &= \varphi(x, y, \sigma s, 0, u_8), & s &\leq \tfrac{1}{2}\\
&= \varphi(x, y, \sigma(s-1)+1, 0, u_8), & s &\geq \tfrac{1}{2}\\
\varphi(x, y, s, 1, u) &= \varphi(x, y, s, 1, u_8)\\
\varphi(x, y, 0, t, u) &= \varphi(x, y, 0, t, u_8)\\
\varphi(x, y, 1, t, u) &= \varphi(x, y, 1, t, u_8)
\end{aligned}\right\} \quad \sigma = 1 - \frac{u - u_8}{1 - u_8}\, 2(a - \epsilon), \quad u_8 \leq u \leq 1.$$

Let $(a + 2\epsilon, a, 0) = \mathbf{c}$ and $W(s)$ be a rotation operator around the z axis (Sec. III). Collecting the above expressions we get (Fig. 10)

$$\begin{aligned}
\varphi(x, y, s, 0, 1) &= \varphi_1(W(s) \cdot (\mathbf{x} - \mathbf{c})),\\
\varphi(x, y, s, 1, 1) &= \varphi_1(\mathbf{x} - \mathbf{c}),\\
\varphi(x, y, 0, t, 1) &= \varphi_1(\mathbf{x} - \mathbf{c}),\\
\varphi(x, y, 1, t, 1) &= \varphi_1(\mathbf{x} - \mathbf{c}), \qquad\qquad\qquad \text{Q.E.D.}
\end{aligned}$$

Reprinted from:

PHYSICAL REVIEW D   VOLUME 6, NUMBER 2   15 JULY 1972

Flux Quantization and Particle Physics

Herbert Jehle

*Physics Department, George Washington University, Washington, D. C. 20006**

(Received 27 September 1971; revised manuscript received 27 December 1971)

Quantized flux has provided an interesting model for muons and for electrons: One closed flux loop of the form of a magnetic dipole field line is assumed to adopt alternative forms which are superposed with complex probability amplitudes to define the magnetic field of a source lepton. The spinning of that loop with an angular velocity equal to the *Zitterbewegung* frequency $2mc^2/\hbar$ implies an electric Coulomb field, (negative) positive, depending on (anti) parallelism of magnetic moment and spin. The model implies *CP* invariance. A quark may be represented by a quantized flux loop if interlinked with another loop in the case of a meson, with two other loops in the case of a baryon. Because of the link, their spinning is very different from that of a single loop (lepton). The concept of a single quark does not exist accordingly, and it is seen that a baryon with a symmetric spin-isospin function in the SU(2) × SU(3) quark representation might not violate the Pauli principle because the wave function representing the relative position of linked loops may be chosen antisymmetric. Weak interactions may be understood to occur when the flux loops involved in the interaction have to cross over themselves or over each other. Strangeness is readily interpreted in terms of the trefoil character of a λ quark: Strangeness-violating interactions imply crossing of flux lines and are thus weak and parity-nonconserving. $\Delta S = \Delta Q$ is favored in such interactions. Intrinsic symmetries may be interpreted in terms of topology of linked loops. Sections I and II give a short résumé of the 1971 paper.

I. INTRODUCTION

In recent years, several attempts have been made to move from an abstract description of quarks [successfully achieved in terms of the SU(3) and SU(6) symmetries] to a more specific model which might relate the internal and the external symmetries. In this connection the question has arisen whether magnetic monopoles can be considered as the physical counterparts to the formal definition of quarks. To extend this type of approach to a more conservative viewpoint, the obvious suggestion has been made that a quark may be considered as a closed quantized flux loop if interlinked with other flux loops. To verify the details of such a model, the known classifications of particles have been discussed in terms of the topological structure of linked quantized flux loops.

In our previous work we succeeded in formulating a charged-lepton theory (i.e., a muon or an electron) in terms of quantized flux. It was proposed that the lepton's magnetic field may be represented by the superposition of alternative forms which a quantized flux loop may adopt. These alternative loopforms should be superimposed with complex probability amplitudes in a manner similar to the superposition of alternative path histories in Feynman's space-time approach to quantum mechanics. With the appropriate choice of the spatial distribution of these complex amplitudes, the magnetic field of a muon-magneton or a Bohr-magneton source may be reconstructed. Furthermore, the electric Coulomb field has been shown to result from the spinning of the loop. In our previous work we have also made a quantitative proposal to explain the relation between the charge e of the electron and the Planck constant h.

To move from this heuristic to a more familiar and concise formulation of the theory, the probability-amplitude distributions of manifolds of loopforms are expressed in terms of a wave equation for these amplitudes (cf. the Appendix of the present paper). It has, however, been found that the preceding heuristic formulation of the theory and the analysis of its consistency[1] were essential in anticipating a more sophisticated formulation.

In the present paper we have proposed a topological configuration of the linked quantized flux loops to represent mesons and baryons. Mesons and baryons are represented by two and three interlinked quantized flux loops, respectively. We should note that the manifold of infinitely many loopforms of this type (defining a fibration of space) is needed for a full description of the particle. The superposition of this manifold, weighted appropriately with complex probability amplitudes, defines the internal structure of the particle.

Topologically these loops may be characterized by their winding numbers about the circular and about the straight axis (Sec. III, Figs. 1–5); they are assumed to be of the type of torus knots to avoid unnecessary singularities, and to permit independent spinning of the different quarks which make up a particle. The loop-antiloop (quark-antiquark) dichotomy is assumed to correspond to left-handedness–right-handedness which in the case of a neutrino or in the case of a λ quark implies the forms of left-handed-right-handed trefoils.

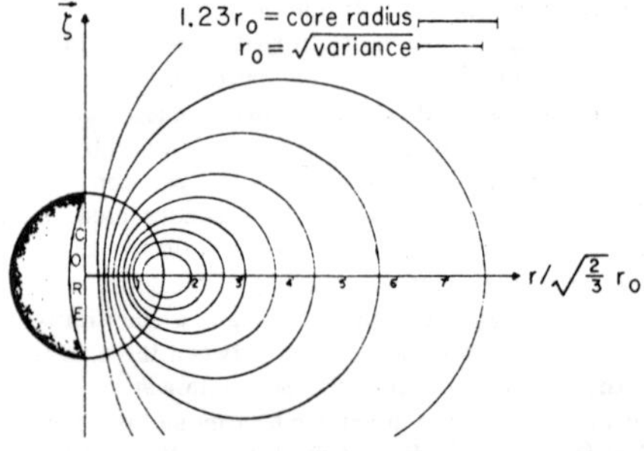

FIG. 1. Spinning-top model (spinning about a straight axis and a circular axis), referring to a muon or an electron. The figure shows alternative loopforms (magnetic flux lines) of an extended source corresponding to a spherically symmetric Gaussian distribution of polarization in $\bar\zeta$ direction. The variance of the distribution is denoted by r_0^2, the core radius is chosen so as to go through the circular axis of the magnetic field which is located at 1.23r_0 for the Gaussian distribution. The total amount of flux is subdivided into 10 equal sheaves by the nine toroidal surfaces (which in the figure look like flux lines). This extended source model is supposed to visualize the quasinonlocal nature of a "single-particle" source, implied by the Pryce-Tani-Foldy-Wouthuysen transformation.

We assume that reactions which imply a loop crossing over itself, or over the loop with which it interacts, is slow (a weak process). Strangeness may therefore be interpreted in terms of left- or right-handedness of the trefoil (λ) knot. The reason for this is that two mirror-related trefoils may readily annihilate without flux-loop crossings.

Due to the topological constraints, the spinning of interlinked loops differs very much from that of free loops. Consequently the unlinkage of a quark (conditional to conservation laws) is not considered to lead to a "free quark" but to a lepton.

It was shown earlier[1] that the spinning of loops implies an electric potential by the very same assumption which defined flux in the first place. The signature of equivalent charge depends on parallelism (+) or antiparallelism (−) of magnetic moment and spin.

The quark loops are assumed to spin about both the circular and about the straight axes with the same angular velocity $2m_q c^2/\hbar$, in a left-handed spin for left-handed loops, and right-handed spin for right-handed loops so as to minimize electric field energy production. Thus the electric field will be seen to be proportional to the difference of the winding numbers which characterize a loop.

The ratio of absolute value of equivalent electric charge for a $\mathfrak{N}$ loop of winding numbers (2, 1), for a $\mathcal{P}$ loop of winding numbers (3, 1), and a λ loop of winding numbers (3, 2) are $2 - 1 = 1$ to $3 - 1 = 2$ to $3 - 2 = 1$; the signatures of the charges depend on the magnetic moment versus spin orientation, and will be determined by the following consideration.

Charge conservation (in reactions involving a replacement of a quark by another quark with accompanying change $\pm e$ or 0 of charge) implies that the different quark charges be ± 1 or 0 units of e apart. Only the assignment $-\frac{1}{3}e$, $+\frac{2}{3}e$, $-\frac{1}{3}e$ satisfies these conditions and implies the integer-charge spectra 0, $\pm e$, $\pm 2e$ for $q\bar q$ and for qqq, but not for qq or $qq\bar q$ nor their antiparticles.

II. REVIEW OF A LEPTON'S ELECTROMAGNETIC FIELD IN TERMS OF QUANTIZED FLUX

Flux Quantization

Flux quantization arises from the possibility that the wave function of an electrically charged field particle, even though single-valued, may have a phase ϑ which is single-valued only modulo 2π. Lines in ordinary three-dimensional space around which the phase of a field particle changes by $\pm 2\pi$ define quantized flux. Indeed, if we set the gauge-invariant combinations of four-potential $A_k = (V, -\vec{A})$ with the derivative of the phase ϑ of any field particle's ψ function $|\psi|\exp i\vartheta$,

$$A_k - (\hbar c/e)\partial_k\vartheta = \mathbf{a}_k = 0, \tag{1}$$

equal to zero, we simply state that a ψ field with a singularity of $\partial_k\vartheta$, characterized by a ϑ change of 2π, implies a singular potential A_k of quantized flux:

$$-2\pi = \oint \nabla\vartheta \cdot d\vec{r}$$
$$= (e/\hbar c) \oint \sum_1^3 A_k dr^k$$
$$= -(e/\hbar c)\Phi_q. \tag{2}$$

We want to understand a source particle in terms of quantized flux. In particular we assumed that a source lepton is to be understood as a single, closed, quantized flux loop which takes on the form of a closed magnetic field line of a magnetic dipole source.

Superposition of Loopforms

In a manner somewhat analogous to Feynman's space-time approach to quantum mechanics which constructs a quantum-mechanical path from a superposition of "alternative path histories" of classical paths, a superposition with complex probability amplitudes, we construct the magnetic dipole field of a source lepton from a complex amplitude superposition of the alternative loopforms which the quantized flux loop may adopt. (Such a superposition is not a superposition of different quantum states, but a superposition of alternative classical or semiclassical loopforms to construct a quantum state.)

A particular closed loop with its implication concerning the phase ϑ of the field particle's wave functions is to be considered as only one of a manifold of "loopforms" characterizing the source lepton (in terms of a corresponding manifold of field particle wave functions). The superposition of these loopforms is assumed to be made with probability amplitudes, continuous functionals of the loopforms, and result in a continuous magnetic and (as we shall see) electric field.

The flux loop is considered to be "attached" to the source, i.e., to the point at which the Maxwell-Lorentz equations have inhomogeneous terms. The probability amplitudes are to be chosen such as to result in a field satisfying the Maxwell-Lorentz equations. This leaves their phases still largely undetermined, a fact which was seen[1] to be altogether important for the possibility of constructing a consistent theory of leptons. The phases of the probability amplitudes have nothing to do with the above-mentioned phase ϑ of a field

particle's wave function.

The Pertinent Fields

We have to distinguish three types of fields. First, the ψ functions of one of many field particles, its phase ϑ defining quantized flux (and as we shall see later, also an electric field). Second, the probability amplitudes of the loopform attached to one source lepton; they, *in toto*, represent the quantum-mechanical state of the source. Third, the electromagnetic field defined by the former fields; this field is a quantum-mechanical observable.

Heuristic Approach to Determine Probability Amplitudes

A central problem of the present and of the previous work[1] is how to make the appropriate choice for the probability amplitudes of the alternative loopforms. The usual way to do this is to find the appropriate (in this case "internal") wave equation for the system and use conventional techniques to carry through the analysis. In the Appendix to this paper a sketch of such an approach to this problem is given. This facilitates the transition from the ordinary techniques of the particle picture to the loop picture with all its topological refinements.

As it is inadvisable to speculate about possible wave equations and equations describing topological manifolds without previous study of a heuristic model, we took it as our task in the previous and present work to develop such a heuristic model which should determine the structure of the loop model and the physical meaning of the fields involved in it. This essentially qualitative model is made to fit the Maxwell-Lorentz equations, and the conservation laws and other data of particle physics. This heuristic step may help to find a precise quantitative formulation of the model, a formulation which might make use of fiber-space topology and differentiable manifolds. In the present paper we would like to check the consistency of that model and show how such a heuristic model makes it possible to understand some basic open issues in particle physics.

Magnetic Monopoles, Quantized Flux Loops, and Their Structure

Considerations similar to those relating to quantized flux had previously been made in order to introduce magnetic monopoles. With the introduction of magnetic charge, the physical basis for electromagnetic theory receives a drastic change, however: It may be an unnecessary complication of the well-defined discipline of quantum electrodynamics.

The introduction of closed quantized flux loops

as a basic entity implies new mathematical techniques and concepts. Whereas magnetic monopoles may use the formalisms developed for point charges, the flux loops need, in order to describe their probability amplitude distributions, techniques which imply interesting topological concepts. But a model of particles in terms of flux loops does not imply a change from Maxwell-Lorentz theory as monopoles do. It seems therefore obvious that this more conservative loop model should be developed.

The basically new features which enter in the present theory are structural. In the previous work an essential issue was to characterize the manifolds of flux loopforms in terms of statistically independent bundles of loopforms of a lepton; it was essentially by geometrical means that these bundles were defined. In the present paper it is the topological characterization of linked loops of mesons and baryons which brings the interesting results. One may, with a certain amount of oversimplification, say that the present program is one of reducing problems of particle physics to topological issues of quantized magnetic flux loops.

Electromagnetic Field

Let us, for the moment, assume that this magnetic dipole is constructed such as to correspond to a moment $e\hbar/2mc$, and discuss later how the probability amplitudes will have to be chosen to properly relate this dipole moment to the quantized flux Φ_q. The effective (averaged) magnetic vector potential $\vec{A}_{\text{eff}}$ will be time-independent for a fixed lepton.

By the same definition, $\alpha_k = 0$, which defined quantized flux through

$$\vec{A} = -(\hbar c/e)\nabla\vartheta, \tag{3a}$$

we find that moving quantized field lines define an electric potential

$$V = +(\hbar c/e)\partial\vartheta/\partial ct. \tag{3b}$$

We assume that every flux loopform performs a spinning motion with the *Zitterbewegung* angular velocity

$$\Omega = 2mc^2/\hbar \tag{4}$$

about its flux-orientation axis. The field $\vartheta(t, x, y, z)$ of the field lepton is carried along with that spinning motion (otherwise one would get an entanglement).

Any point which is "linked" with that loopform (Fig. 2 of Ref. 1), i.e., which is inside the perimeter of the loopform, experiences a $\partial\vartheta/\partial ct$ which amounts to a unidirectional rate of change of ϑ, a change $\pm 2\pi$ per spin period. An outside point Q, not linked with that loopform, experiences only

periodical fluctuations of ϑ. We may thus calculate the expectation value of $V = (\hbar c/e)\partial\vartheta/\partial ct$ if we assume that the probability amplitude of the flux loopform corresponds to the total field of dipole moment $e\hbar/2mc$.

It may be shown that the manifold of loopforms of such a source lepton then gives rise to a Coulomb field, and this is not due to any additional assumptions but holds by virtue of the very same definition of the four-potential $A_k = (\hbar c/e)\partial_k\vartheta$ which defined quantized flux in the first place. Positive and negative charge of the source lepton corresponds to parallelism and antiparallelism, respectively, of magnetic moment and spin, again in agreement with well-known facts.

To illustrate this result in a greatly simplified manner we may evaluate, for a point P on the equator of the source lepton, the fraction F (of quantized flux $\Phi_q = 2\pi\hbar c/e$) of loopforms which are linked with P, as

$$F = (e^2/2mc^2)/r \tag{5}$$

because the effective magnetic dipole field implies

$$\int_r^\infty B_{\text{eff}}\, 2\pi r\, dr = (e\hbar/2mc)2\pi/r. \tag{6}$$

As the phase ϑ changes by $\pm 2\pi$ for each passage of a flux of the amount Φ_q, we have

$$\begin{aligned}
V_{\text{eff}} &= (\hbar c/e)(\partial\vartheta/\partial ct)_{\text{eff}} \\
&= \pm(\hbar c/e)(1/c)(2\pi)F(\Omega/2\pi) \\
&= \pm(\hbar c/e)(1/c)(e^2/2mc^2 r)(2mc^2/\hbar) \\
&= \pm e/r. \tag{7}
\end{aligned}$$

The result of a potential V_{eff} corresponding to an equivalent charge e, after we assumed the dipole moment to be $e\hbar/2mc$, does not surprise us; it is simply the reverse of Dirac's result of intrinsic moment $e\hbar/2mc$ which arises when a relativistic electron of charge e is considered.

A more detailed discussion[1] shows the isotropy of the electric field.

It is of great importance to note that the mass m of the lepton cancels out, rigourously, indeed; the effective electric charge of muon and of electron is thus identical.

The motion of an electric potential implicitly arising from spinning flux loops is one of the essential points of the theory. We derived this potential from the basic equation (1) rather than from an inappropriate application of the induction law to a situation of loops in spinning motion, and that with linear velocities beyond the velocity of light.

In the later sections of the present paper we shall make ample use of this simple relationship between magnetic moment and effective electric charge, and of the proportionality of electric charge to the number of "wings" of a loop (expressed in terms of its effective winding numbers).

The issue of the preceding work[1] was to account for the relationship of quantized flux $\Phi_q = 2\pi\hbar c/e$ and magnetic moment $e\hbar/2mc$ and also, to make the model consistent, to show how the electromagnetic energy accounts for mc^2 and the electromagnetic angular momentum for $\frac{1}{2}\hbar$.

Quasi-Nonlocality

The question now arises: How can we, with the simple loopform concepts, at least qualitatively satisfy these requirements; i.e., can we find a reasonable interpretation of charged leptons in terms of superposition of quantized flux loopforms, given these requirements.

The first obvious step in such a picture of the lepton in terms of semiclassical loopforms is the construction of the source lepton. If the source lepton (when we attach a manifold of semiclassical loopforms to it) were considered as a point source, not only would there be a true singularity of the magnetic field $\vec{B}_{\text{eff}}$ (unlike other points of the field at which only infinitesimal probability amplitudes contribute to each flux loopform), but also, for a given finite magnetic flux, the magnetic moment would be zero because almost all flux loopforms are of infinitesimally small size.

Though this circumstance seems to present a formidable problem, it has an obvious solution. It should be remembered that a source lepton, if used as a source to which semiclassical loopforms are attached, will have to be considered as a "single particle." Therefore, when we use a description of a source lepton in terms of a space-time distribution of loopforms, we necessarily have to attach those loopforms to the "mean position" of the lepton, the position of a single particle. The Pryce-Tani-Foldy-Wouthuysen[2] transformation of the Dirac electron from representation in ordinary position into a single-particle representation makes the particle's mean position an operator which is nonlocal of the extent $\hbar/mc$ in ordinary position space. For a stationary single particle this implies a nonlocality $\hbar/mc$ for the position of the particle as well as for the field lines emanating from it. As the underlying theory is truly local we might term this effect "quasi-nonlocality." We might then perhaps make the terribly crude hypothesis that we may take account of this quasi-nonlocality of the source by considering the source as an extended source of size $\hbar/mc$.

This then seems to imply a magnetic dipole moment of the order of $(\Phi_q/4\pi)(\hbar/mc)$, a moment 2 orders of magnitude too large compared with the

Bohr or muon magneton, respectively. That expression for magnetic dipole moment would be valid if the total effective magnetic flux of the dipole field were equal to Φ_q. We assume, however, that such a statement were only correct if the complex probability amplitudes of the alternative loopforms were all in phase. Their actual phase difference is bringing about a superposition corresponding to a reduction factor in the calculation of effective flux from quantized flux, as well as in the calculation of electromagnetic energy and angular momentum.

We do not, in this paper, recapitulate the essential objective of the previous work, i.e., the answer to the aforementioned questions. This program also provided, as seen in Sec. X of Ref. 1, for an understanding of the electromagnetic interaction constant $e^2/\hbar c$, and was applied to the electron-muon problem.

Neutrino

The neutrino is proposed to be a loop of the form of a left-handed trefoil, the antineutrino a right-handed one. It is proposed to spin through space like a coasting three-blade propeller. (A surface $\vartheta = 0$, $\vartheta = 2\pi$ representing the "cut surface" of the multivalued pseudo gauge field ϑ may be chosen in the following way: Deform, i.e., dilate

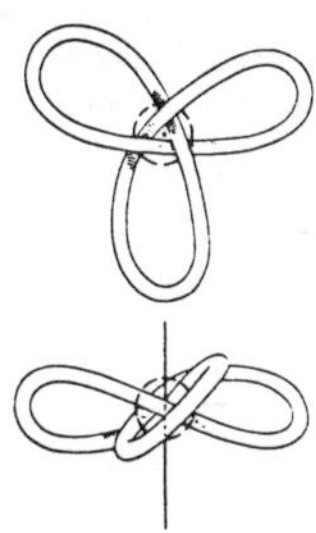

FIG. 2. A trefoil representing a neutrino loop which, like a coasting three-bladed propeller, moves in a helical spinning motion in the direction of the spin axis. In this and in subsequent figures, flux loops are drawn as double lines merely to better visualize the form of the loops. The loops are singular lines, the alternative forms of which define fibration of space. The question of orientation of the magnetic flux is still open; a neutrino might even be a superposition, not only of different loopforms, but also of both signatures of magnetic flux orientation. The difference between electron and muon neutrino is discussed in Sec. IV and in Appendix II of Ref. 1; the distinction is in regard to phase-related versus random-phased probability amplitudes superposition of the contributions of loopform bundles. A *single* loop of this form never represents anything else but a neutrino.

the central loop region which in the upper Fig. 2 projection shows a triangular form, into a form which has a circular projection; build this circle into a cylinder which is coaxial with the spin axis and which reaches up toward infinity. This cylindrical surface, together with three wing surfaces attached to the cylinder and extending out to the three loopwings, respectively, may represent this "cut surface." The helical coasting motion of that surface then implies no "sweeping," i.e., no $\partial \vartheta/\partial t$, and thus no electric field.)

This neutrino model is suggested by the possibility of explaining the helicity of the neutrino in terms of the seemingly general tendency of flux loops to spin in such a manner as to produce a minimal (in this case zero) electric field which is attained by the coasting rather than sweeping motion of the wings.

As the handedness of a neutrino does not change with any Lorentz transformation, whereas a neutrino of finite rest mass would change its helicity with a Lorentz transformation beyond its rest system, our model implies zero rest mass if helicity is interpreted in terms of loop handedness.

The knot character of the neutrino is also suggested because weak interactions involving neutrinos seem to imply the creation or destruction of a knotted loop: It is *assumed* that the crossing of a flux loop over itself or over a loop with which it is interacting, is a slow, i.e., weak process. The creation or destruction of a neutrino implies such processes. It is also to be remarked that

two trefoils of opposite handedness, e.g., neutrino and antineutrino, have a means to annihilate each other or be pair-created without the crossing of loops. The neutrino-antineutrino dichotomy implies, apart from left-handedness–right-handedness of the trefoil loops, also opposite signature of the frequencies of their probability amplitudes.

The smallness of interaction of a neutrino with matter may be understood in terms of its zero electric field, and in terms of the weakness of a process involving the change of a neutrino trefoil into an ordinary loop.

As to the question of the "size" of such a loop one might consider that its lab energy indicates spinning frequency (we should note that the spinning frequency $2mc^2/\hbar$ referred to particles in their rest frame and may be interpreted as 2 lab energy/$\hbar$). As the essentially important ("first shell") loopforms may spin with linear velocities of the order of magnitude c, the radius (size) of a neutrino loop might be of the order of $\hbar c/$lab energy.

III. MESONS AND BARYONS, SPINNING-TOP MODEL

We shall outline two models of linked flux loops to represent hadrons. The basic assumptions are quite similar for both models. The first, the "spinning-top" model (Figs. 3–5), is a development of the model which we sketched in Appendix II of Ref. 1, now formulated in closer relationship to topology. The second, the "symmetric-axes"

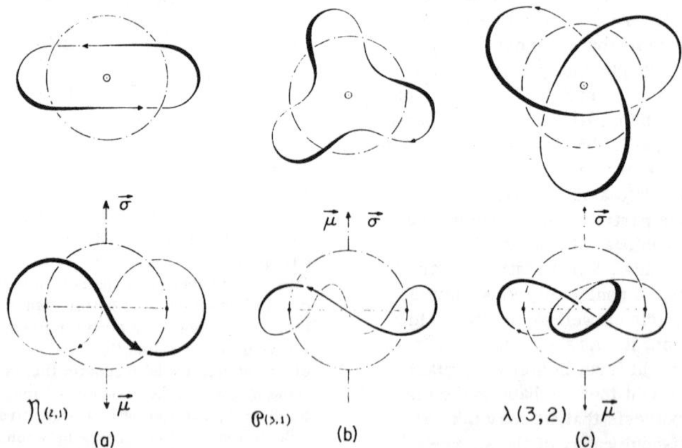

FIG. 3. Forms of quarks in the spinning-top model. These loops represent quarks only if interlinked with other loops as shown in Figs. 4 and 5. The difference of winding numbers about the two dash–dot–dash axes, i.e., $2-1=1(\mathfrak{N})$, $3-1=2(\mathfrak{P})$, $3-2=1(\lambda)$, multiplied with the signature of spin with respect to magnetic moment, is proportional to the equivalent electric charge of the respective quarks. Quarks are assumed to be left-handed, antiquarks to be right-handed. Winding numbers have obviously a simple group-theoretical interpretation.

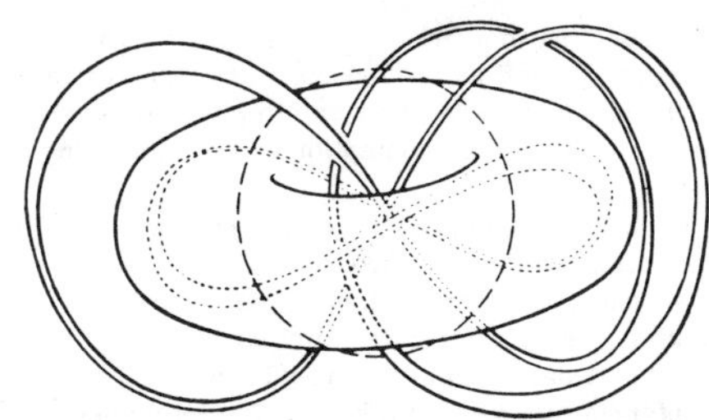

FIG. 4. Spinning-top model. λ and $\bar{\pi}$ quark interlinked, contributing to a meson. To illustrate the topological (knot-theoretical) relationships of the two loops, space is here subdivided by a toroidal surface [dashed lines in Fig. 4(a) which show a doughnut cut in half]. The λ is located entirely outside this doughnut shaped surface, the $\bar{\pi}$ entirely inside. This surface is dividing the fibrated space of λ loopforms from that of $\bar{\pi}$ loopforms; this toroidal interface may arbitrarily shrink or extend itself. Both loops pass through the spherical core region which is indicated by the dashed circle; the two loops may spin independently in a rolling-spinning motion about both the circular and the straight axes.

model (Figs. 6–11), is a generalization of the first model. We describe the basic ideas in terms of the spinning-top model which we consider the first choice to represent mesons and baryons. The symmetric-axes model is an alternative possibility; it also serves to illustrate the topological issues of flux loop linkage.

Linked Loops: Axes and Core

We assumed that a quark is a quantized flux loop if linked with another loop (to make a meson) or two other loops (to make a baryon). In terms of quantized flux loops, different quarks are defined by the form and orientation (direction of mag-

netic flux) of the linked loop, and by their spinning. A linked loop's mode of spinning is very different from that of a single loop. Accordingly the concept of an isolated single quark has no meaning in this theory. Loops, if able to dissociate themselves (in accord with conservation laws), behave as leptons.

We assumed that the linked loops are confined to regions between toroidal surfaces (cf. Figs. 4 and 5), which makes it possible for them to spin independently; this is an obvious requirement for loops representing quarks with spin.

For simplicity we assumed that such toroidal surfaces have the symmetry of a doughnut, i.e., that they have a circular axis and a straight central axis [which is perpendicular to the plane of the doughnut (Figs. 4 and 5)]. These toroidal surfaces are by no means fixed; they may shift altogether towards one or the other axis.

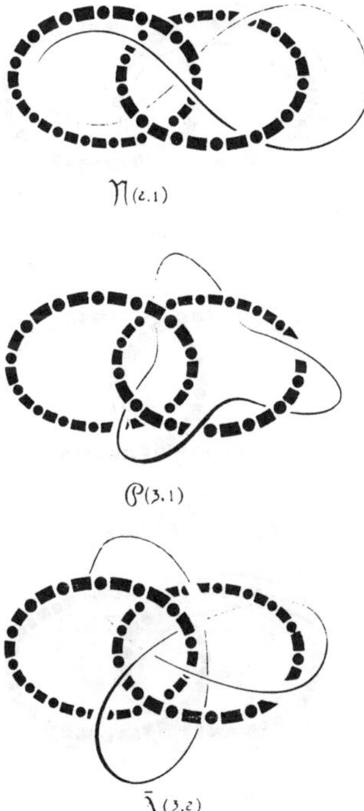

FIG. 6. Symmetric-axes model. Two interlinked axes are shown by thick dash-dot-dash lines. This is a generalization of the spinning-top model for which one of the axes is straightened out to reach $\pm$ infinity. The quarks $\pi(2,1)$, $\wp(3,1)$, and $\bar{\lambda}(3,2)$ are shown in relation to one of the two axes.

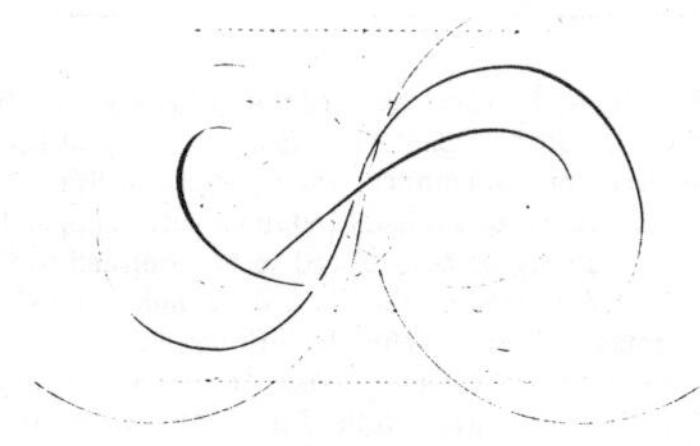

FIG. 5. Spinning-top model of a $\bar{\pi}\,\bar{\pi}\,\bar{\lambda}$ contribution to a baryon. The three loops define these fibrated space regions separated by the toroidal (dashed) interfaces. The subdivision of space permits independent spinning of the three quarks.

In the model of the lepton we made the obvious assumption that the *Zitterbewegung* amplitude $\hbar/mc$, which corresponds to the linear extent of the quasi-nonlocality (which defines the radius of the "core" of the source lepton) is of the size of the radius of the circular axis (Fig. 1); this in turn is the extent of the inhomogeneous Maxwell-Lorentz region. In other words, what we called the "core equatorial ring" coincides with the circular axis. In the case of mesons and baryons, loops are linked in the manner illustrated in Figs. 4 and 5. In view of the discussion of (quasi-) nonlocality in Sec. V and Appendix I of Ref. 1, the region extending to the circular axis is to be considered again as the core, i.e., the "position" of the particle.

As every loop has, of course, to pass through this core (source) region, one may consider that as amounting to a linkage with the core. This linkage, as well as the linkage with the other loops of the same particle, may be simply expressed in terms of the above assumption and the following one:

To make an orderly linkage and motion possible it is assumed that the linked loops all share the same axes, and while spinning, do so with instantaneous spinning axes which are coaxial.

Some remarks about superposition of alternative loopforms may be useful here. The interlinked loops of a hadron are, if we refer to one of their "alternative forms," just two or three loops in space. The superposition of a continuous manifold of alternative (similar) forms which such a linked loop doublet or triplet may adopt, is formulated in terms of a superposition of products of three probability amplitude functions. The superposition defines a magnetic field in three-space. This may be considered to be a fibration of space, but a special one: It is made up of doublets or triplets of closed loopforms by definition because the superposition refers to alternative closed loopforms.

The superposition of the continuous manifold of loopforms, resulting in a fibration of space and formulated in terms of probability amplitude waves, implies a corresponding interpretation of coaxiality.

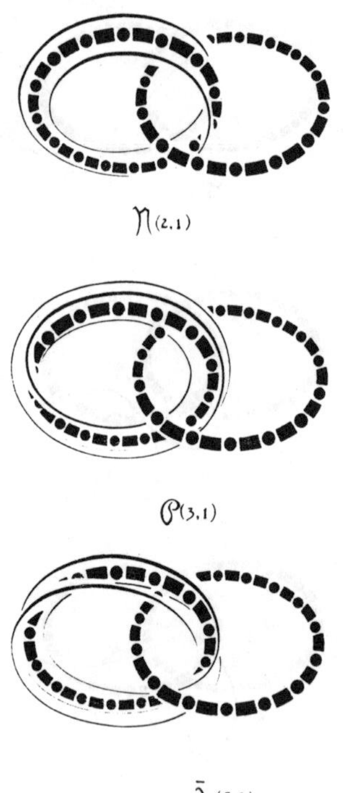

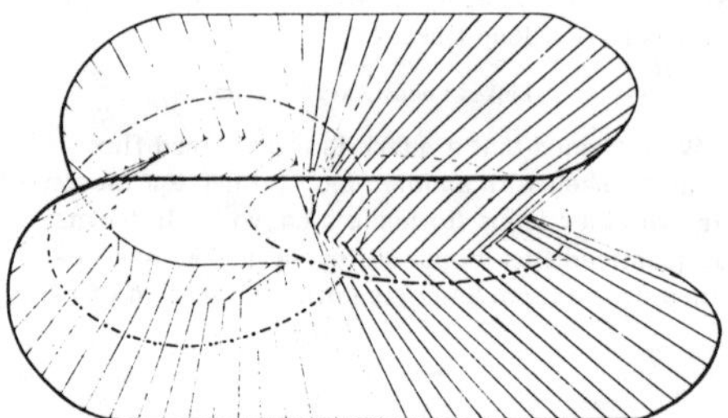

FIG. 8. The symmetric axes of the quarks shown in Fig. 6 are indicated as dash-dot-dash lines. The two flux loops are not shown here; they appear in Fig. 9. In order to permit the two meson loops to spin independently, these loops may be considered to be confined to the two domains separated by the shaded surface. To visualize this surface which reaches to infinity, it is in this figure, bounded by one long-winded line. The opening, connecting front to rear with the far right region, is free from shadings; we see an axis passing through that opening. The other opening, connecting upper to lower with the far left region, is hidden behind the shaded surface; the other axis passes through that opening. The surface is, as in the spinning-top model, not fixed at all. The spinning is again, as in the spinning-top model, a rolling, whirling motion about both axes, for each of the loops.

FIG. 7. The same quark loops as in Fig. 6 are shown here in relation to the other axes. This setting involves closer crowding of magnetic field lines and is thus expected to be less favored than the setting of Fig. 6, which shows the preferred setting of the loops.

Form of Quark Loops; Winding Numbers

We shall assume that every quark loop is a torus knot, i.e., a closed loop which, if projected onto one of the doughnuts, does not intersect itself (Fig. 3). As a "loop" stands for the corresponding type of fibration of space, such an assumption permits us to avoid unwanted singularities of that fibration. (In our previous work[1] we did not yet adhere to this requirement.)

Every torus knot has two "winding numbers" which indicate how often it winds about the torus. We may denote by $(1, 0)$ an electron or muon loop (Fig. 1) and by $(2, 1)$ a loop which winds once about the hole of the doughnut, twice about the doughnut itself [Fig. 3(a)].

These winding numbers characterize the fibration. A fibration with rotational symmetry about a straight axis has also an axis which is circular; it may represent a topological singularity.

Handedness of Quark Loops

A loop (i.e., a fibration of space) has a handedness (right or left) which is defined by attaching two arrows to the (straight and circular) "axes," i.e., giving an orientation to the magnetic field at or near these axes. Handedness is, in the topologist's language, the "orientation" of the fibrated three-dimensional space.

A special case of handedness arises if the loop is a knotted one, i.e., a left-handed or right-handed trefoil [clover leaf knot, Fig. 3(c)]. We then assume that strangeness of a quark is represented by the topological character of the λ, $\bar{\lambda}$ loops. They are assumed to be of trefoil shape (λ and $\bar{\lambda}$ of opposite handedness) because an annihilation or pair creation of strange quarks must be fast if there is strangeness conservation: Indeed, two oppositely handed trefoils of opposite magnet-ic flux orientation may annihilate each other rapidly without crossing of flux loops. This is not so for two trefoils of equal handedness or for a trefoil and a simple nonstrange loop.

We assumed that the helicity of a neutrino, i.e., the handedness represented by the ψ function of the neutrino, corresponds to the handedness of the flux loop.

As the particle-antiparticle character of a neutrino (lepton) loop and of a λ quark loop is characterized by left-handedness and right-handedness, respectively, we might assume that this holds for all quarks.

As the segments of a line which form a loop seem to tend to spread out, "repelling each other" as magnetic field lines do to minimize magnetic field energy, the loops of type $(2, 1)$, $(3, 1)$, $(3, 2)$ (which look like Figs. 6 if the left axes may be thought of as corresponding to the straight axes) seem to have preference over the loops $(1, 2)$, $(1, 3)$, $(2, 3)$ (which look like Figs. 7 if the right axes may be thought of as corresponding to the straight axes) which more often go around the straight axes (around the holes of the doughnuts).

It is assumed that a flux loop crosses over itself, or over a loop with which it is interacting, in a weak process. As the unknotting of a knotted flux loop implies such flux loop crossing, we may look at a strangeness-nonconserving weak process as implying, in the simplest case, a transition from a trefoil to a simple loop. We therefore assumed the λ quark $S = -1$ to have the form of a (left-handed) trefoil and the $\bar{\lambda}$ ($S = +1$) the mirror form. Those two may annihilate each other or be

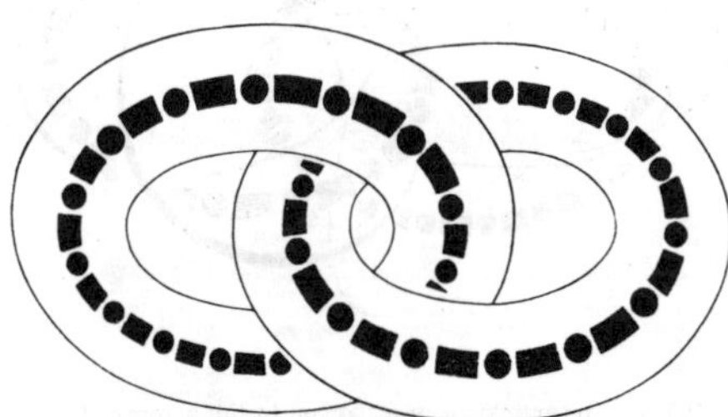

FIG. 10. Two toroidal surfaces separating space into three regions, for a baryon in the symmetric-axes model. The loops are not shown here, they are shown in the corresponding meson case of the spinning-top model Fig. 4(b). In the present case one loop is located altogether inside one doughnut (wound about its dash–dot–dash axis), another inside the other doughnut, and the third in between, i.e., the region which covers all outer space. Independent spinning is possible in this way.

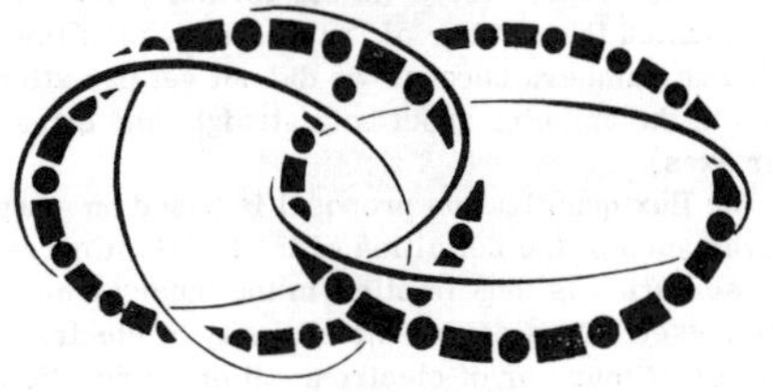

FIG. 9. A meson's loop-antiloop contribution in the symmetric-axes model (an $\mathfrak{N}\bar{\lambda}$ contribution). The interface between the fibrated space regions belonging to $\mathfrak{N}$ and $\bar{\lambda}$, respectively, is of the type shown in Fig. 8.

pair-created without crossing of loops.

A strangeness-violating weak process then implies crossing of loops. It is interesting to note that such a process is, in terms of the topology of the loops, parity-nonconserving.

Frequencies of Probability Amplitudes

We mentioned handedness (and thus also strangeness) characterizing quark with respect to antiquark. For the muon we assumed that the particle-antiparticle character corresponds also to positive-negative frequencies of their probability amplitudes. Not only is this what might be expected in analogy with relativistic quantum mechanics; this frequency assignment also makes it plausible that particle-antiparticle annihilation and pair creation may occur while there is a loop conseration law, i.e., number of loops minus number of antiloops being conserved.

Spinning of the Loops

The spinning of the loopforms is assumed to occur with angular velocity $2m_q c^2/\hbar$ about both the straight axis and about the circular axis, in the latter case causing a rolling motion of the loopform about that circular axis.

The relationship of spinning motion to fibration

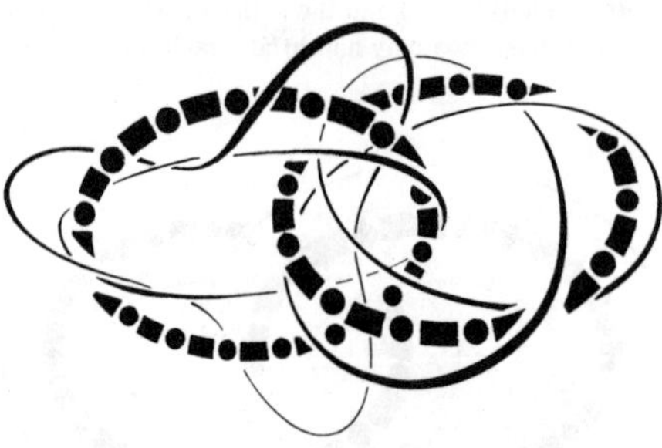

FIG. 11. Interlinkage of a baryon in the symmetric-axes model; an $\mathfrak{N}\,\mathfrak{N}\lambda$ contribution to a baryon (they are left-handed loops). The fibrated space corresponding to these quark loops has the interfaces represented by Fig. 10. In this symmetric-axes model of Figs. 9 and 11, orbital angular momentum might perhaps be accounted for by the rotational motion of the two axes about each other, spin being a matter of motion of the loop manifold about the two axes. The symmetric-axes model (Figs. 6–11) is shown to illustrate the topology of linkage; the spinning-top model (Figs. 3–5) has the advantage of simplicity.

of the flux loop field is assumed to be such that both have equal handedness. Such an assumption is equivalent to the assumption already made in regard to the neutrino, i.e., that the motion of the loopforms is such as to produce as little electric field energy as possible at the given spinning angular velocity. Such a kind of an assumption, similar to the assumption of forms of flux loops which correspond to a minimal magnetic field energy, is akin to a Maxwell-Lorentz field. One may shortly characterize this equalization of handedness of spinning and fibration as a tendency towards a coasting type of motion.

Spinning and Equivalent Electric Charge

There is still the alternative of the resulting spin being parallel or antiparallel to the orientation of the resulting magnetic moment which, as we shall see, corresponds to positively charged quarks or negatively charged quarks, respectively (cf. Fig. 3).

It is most interesting to note that given, e.g., parallel orientation, the signature of the electric potential produced by the spinning depends on the difference of the winding numbers, i.e., is proportional to $+2 -1 = +1$ for a loop with winding numbers $(2, 1)$. It is the difference because the coasting motion implies that the second winding number, i.e., 1 in case of $(2, 1)$ [Fig. 3(a)] (with spinning parallel to flux orientation) counteracts the electric effect of the first winding number, i.e., 2 in this case of an $\mathfrak{N}$ quark $(2, 1)$ in Fig. 3(a): The simultaneous spinning about both axes is exactly equivalent to a loop with "effective winding number" $2 - 1 = 1$ spinning only about the straight axis.

If we consider the absolute values of equivalent quark charges as proportional to the effective winding numbers, the equivalent electric charges of $(2, 1)$, $(3, 1)$, $(3, 2)$ loops are proportional to 1 to 2 to 1, respectively. (In our former proposal[1] we counted the number of "wings" instead of the winding numbers because we did not yet pay attention to the spinning about both straight and circular axes).

The flux quantization proposal is based on gauge covariance of the definition of the fields. Charge conservation is thus implicit in the theory from the outset. Considering the integrity of electric charge of muon or of electron, all other reactions, directly or indirectly involving a muon or an electron, may only occur with integer changes of charge. As these reactions imply a quark-antiquark annihilation or production, [equivalent to a replacement of an (anti) quark by another (anti) quark] the difference of the equivalent electric charge of any two quarks should be integer, ±1, or 0, indeed.

Magnetic Moment

This charge assignment means that for $\mathfrak{N}$, $\mathcal{P}$, and λ quarks the spin and magnetic moments are parallel, antiparallel, and parallel, respectively. The opposite holds for the antiparticles.

A little careful consideration of the signatures of the field and loop quantities shows that with the loops of Figs. 3 and 6 the signature of the equivalent electric charge is indeed given by the parallelism or antiparallelism of spin and magnetic moment; in the case of the electron and muon that relationship was trivial.

It may also be noted that on the basis of the present assumptions, the proportionality between electric charge and magnetic moment is the general basic relation derived at the beginning of the previous paper.[1]

It is therefore no longer necessary to introduce an assumption (Appendix II of Ref. 1) relating the effective magnetic moment to the number of core traverses. This circumstance permits us now to define the forms of quark loops in terms of the topologically straightforward winding number assumptions which we formulated above and which we illustrated in Fig. 3, in perfect accord with the properties of the conventional quark model.

We may refer to further detailed discussions about baryons, mesons, neutrinos, electrons, and muons, given in the main body and Appendixes of Ref. 1.

IV. SOME PUZZLES OF THE QUARK MODEL

We noted above the difference of quark charges to be ±1 or 0 (due to charge conservation in interactions with leptons). We also noted that the ratio of the absolute values of the equivalent quark charges of $\mathfrak{N}$, $\mathcal{P}$, λ are 1 to 2 to 1, while their signatures depend on magnetic moment being parallel or antiparallel to spin, an open choice so far. One may ask what quark charge assignments and what quark combinations may satisfy these conditions and result in particle charges 0, ±e, ±2e. Among the simple combinations it is only the $q\bar{q}$ and qqq, with charge assignments $-\frac{1}{3}e$, $+\frac{2}{3}e$, $-\frac{1}{3}e$ for q or for $\bar{q}$, which are compatible with those conditions.

We noted that the hypothesis of loop crossing being a slow process (when it comes to a loop trying to cross over itself or over the loop with which it may interact) leads to a topological interpretation of strangeness, to some understanding of weak interactions, and to an understanding of why strangeness-nonconserving weak interactions violate parity. We assumed that handedness of a quark loop itself, not only of the (probability amplitude) wave function referring to the loop, is related to parity.

Another interesting point is the following:

From the consideration of models of flux loops, their intrinsic handedness and their link with the axes, it becomes obvious that between a loop (3, 2) and a loop (3, 1), both of equal handedness, a transition is simple compared with a transition between (3, 2) and a (2, 1) loop. This fact may provide for an understanding of the ΔS-versus-ΔQ rule.

We discussed the muon-electron decay and noted that the presence of two types of neutrinos may be understood in terms of the muon-electron dichotomy. As a muon's probability amplitudes are assumed to be random phased, while the electron's are phase related, and because the transition muon to electron requires that the internal (and translational) k, ω distributions on both sides of the equation should match, a random-phased as well as a phase-related neutrino probability distribution must necessarily enter the picture.

In connection with loop crossing, the following issue is to be discussed. When a baryon's quark interacts with a meson's antiquark leading to an actual or virtual process involving annihilation and/or pair creation, the question arises whether the other (nonparticipating) quark loops are in the way, blocking any such $q\bar{q}$ interactions. It seems appropriate to assume that for an annihilation process the frequencies of the quark and of the antiquark with which it interacts (e.g., λ and $\bar{\lambda}$) are equal and of opposite signatures. In that case the wave function for the combination $q\bar{q}$ does not show beats, the absolute value of the wave function for $q\bar{q}$, in the course of time, does not frequently pass through zero (as other quark product wave functions do). Under such circumstances there is no cancellation of that $q\bar{q}$'s contribution to the magnetic field. Consequently for a possibility of fast annihilation the topological conditions will have to be met, i.e., the condition of opposite handedness of λ and of $\bar{\lambda}$, i.e., of strangeness conservation, so that annihilation may proceed without actual crossing of flux loops. For other quark products, the interference terms [written out in Eqs. (A1) and (A2) of Ref. 1] lead to many and frequent zero field values for the magnetic field. This circumstance should permit crossing of flux loops corresponding to such interfering pairs, and should permit the passage of the former $q\bar{q}$ quarks over regions in which other quarks

222

are located. Accordingly annihilation of $q\bar{q}$ pairs becomes possible if the frequencies of the interacting quarks are equal, but then only between quark and antiquark of the same type; annihilation occurs under disregard of the presence of the other loops linked with q or with $\bar{q}$.

The consideration is also of relevance when it comes to the formation of the spatial part of quark wave functions. As they accordingly may exchange places in their distribution over the toroidal regions between the two axes, we may set up antisymmetric "spatial" wave functions as regards this distribution, or symmetric ones. We shall come back to this when discussing the applicability of the Pauli principle to quarks.

The loop picture has raised the following question: Clearly, a λ and $\bar{\lambda}$, being of opposite handedness, may readily annihilate when they approach each other. How may they coexist when attached to the same core of the meson? A simple model shows that in the latter case their opposite handedness prevents ready annihilation without flux lines crossing each other.

The question also arises about the absence of spin-$\frac{1}{2}$ baryons of the $\mathfrak{N}\mathfrak{N}\mathfrak{N}$, of the $\mathcal{P}\mathcal{P}\mathcal{P}$, and of the $\lambda\lambda\lambda$ type. With total spin $=\frac{1}{2}$, there is in such a qqq always at least one pair of neighboring q's of opposite spin. As they are of equal charge, their magnetic field orientation is opposite. We may assume that those cannot coexist as nearest neighbors, attached to the same core, because they would repel each other.

The Question of Giant Quarks

We pointed out[1] that this spinning-top model brings up an interesting issue in regard to the higher-lying meson states (i.e., those above the pseudoscalar and the vector moments), as well as in regard to the higher baryon states. A spinning top has no orbital angular momentum. To introduce a $q\bar{q}$ orbiting about another $q\bar{q}$ introduces far too many unobservable states unless a plausible rule may be found to exclude them. It seems that the existence of the giant quarks might account for these higher lying states. A quark discussed as a spinning top indeed not only permits, but actually invites a spectrum of spins. A detailed discussion of this possibility has not been made.

Spin-Isospin Functions Without Violating the Pauli Principle

It should from the outset be remarked that it is an open question whether or not one may assign definite values of spin to quarks. As an individual isolated quark is a meaningless object in the present theory, it might also be questionable whether it may be given a definite spin.

The simple picture of quarks of spin $\frac{1}{2}$ and of magnetic moments proportional to their electric charges has, however, had such spectacular success with $SU(2)\times SU(3)$ that we shall stick to it for lower-lying meson and baryon states. The one drastic shortcoming was that the successful derivation of the ratio of magnetic moments of proton to neutron was in conflict with the Pauli principle. The successful symmetric spin-isospin functions for proton and neutron violate the Pauli principle because the quarks could not well be thought of as adopting an antisymmetrical orbital wave function as such a one is not expected to pertain to the lowest states as the nucleons are.

We like to consider now what new situation arises when the loop's distribution over the regions between the axes is considered. There is an innermost loop (i) (next to the doughnut's circular axis), a middle (m) and outermost loop (o) (closest to the straight axis). Were the loops confined to either (i) or (m) or (o), a simple product spin-wave function (as long as it corresponds to total spin $\frac{1}{2}$) would be appropriate for this loop triplet. If, however, there is a possibility of loops switching, i.e., permuting their locations (i), (m), and (o), we shall have to describe the situation again by the same spin-isospin function as in the case of nonlocalizable quark "particles." The same consideration holds in regard to quark loops spinning about linked axes, Fig. 11.

The proton's spin-isospin function, symmetric in the quarks, is

$$(18)^{-1/2}\{2\mathfrak{N}{\downarrow}\mathcal{P}{\uparrow}\mathcal{P}{\uparrow} - \mathfrak{N}{\uparrow}\mathcal{P}{\downarrow}\mathcal{P}{\uparrow} - \mathfrak{N}{\uparrow}\mathcal{P}{\uparrow}\mathcal{P}{\downarrow}$$
$$+ 2\mathcal{P}{\uparrow}\mathfrak{N}{\downarrow}\mathcal{P}{\uparrow} - \mathcal{P}{\uparrow}\mathfrak{N}{\uparrow}\mathcal{P}{\downarrow} - \mathcal{P}{\downarrow}\mathfrak{N}{\uparrow}\mathcal{P}{\uparrow}$$
$$+ 2\mathcal{P}{\uparrow}\mathcal{P}{\uparrow}\mathfrak{N}{\downarrow} - \mathcal{P}{\uparrow}\mathcal{P}{\downarrow}\mathfrak{N}{\uparrow} - \mathcal{P}{\downarrow}\mathcal{P}{\uparrow}\mathfrak{N}{\uparrow}\} .$$

This function[3] (cf. Bég, Thirring, and Weisskopf[3]) thus applies not only to nonlocalized particle quarks, but also to localizable loops if these are occasionally interchanging their locations.

The spatial part of the three-quark wave functions could not be expected to be antisymmetric in the conventional model of quark particles. In the present model of quark loops there may, however, be no objection against antisymmetric spatial distribution (over the toroidal regions spanning from one axis to the other, of Fig. 5 or 11). There is accordingly no conflict with the Pauli principle and no need for introducing parastatistics.

The same arguments as for the proton hold for the neutron spin-isospin function, also, and the beautiful $SU(3) \times SU(2)$ result,

magnetic moment of neutron to that of proton $= -\frac{2}{3}$,

holds again, but now we might avoid a conflict with the Pauli principle.

We may finally remark that it might not be surprising if there would be some connection between the present quark proposal and the ones which consider dipole pairs of positive and negative magnetic monopoles as discussed by Barut.[4] It might be suggested that in this respect a meson might be represented by two dipoles, and a baryon by three dipoles. Magnetic monopoles, if they should exist at all, would not be expected to represent quarks.

V. SYMMETRIC-AXES MODEL

We want to discuss some modifications and generalizations of the spinning-top model simply to clarify a number of topological issues.

We may generalize the assumptions about the axes by bending the straight central axis so that we now consider two interlinked axes as the axes about which spinning occurs. (The straight form of one of the axes represents a special case of linkage of axes). Unless we state the contrary, we shall discuss the case of two equal size linked axes (Figs. 6 and 7). We may then show a few simple loops with winding numbers $(2, 1)$, $(3, 1)$, and $(3, 2)$ in Fig. 6. These same loops also appear in the form of Fig. 7 (when their relation to the other axis is considered); they are topologically identical with the corresponding ones of Fig. 6. We shall later discuss why these loops may represent $\mathfrak{N}$, $\mathcal{P}$, and λ quarks, respectively.

The core maintains its significance as the region of inhomogeneous (i.e., source) terms of the Maxwell-Lorentz equations. It is now assumed to be bounded by the two axes.

These loops are either left-handed or right-handed, i.e., the fibrated space characterizing the magnetic field has a handedness. This handedness is again defined by attaching an arrow, i.e., an orientation to each of the two axes. Handedness characterizes a loop whether it is knotted (as a trefoil) or plain; for knotted loops it implies strangeness. One might suggest a quark to relate to a left-handed fibration, an antiquark to a right-handed fibration, in analogy to neutrino, antineutrino.

To generalize the assumption about spinning, we recall that one mode of spinning is a rolling, whirling motion about one axis; this motion, if smoothly continued throughout space, implies a

tangential translation (circumferential) motion along the other axis, causing it to be displaced congruently upon itself. A second mode of spinning is possible with the role of the two axes interchanged.

We again assume that spinning occurs about both axes simultaneously. The spinning motion might then be left-handed or right-handed. We might again assume handedness of motion to coincide with handedness of fibration so as to minimize electric field energy.

In order to discuss the alternative of the forms of Fig. 3 or the forms of Fig. 6 as regards the axes, we first consider this question in regard to the muon or electron. Their field with winding numbers $(1, 0)$ may be simply the dipole field of Fig. 1, only the spinning about that dipole axis matters.

We proceed to discuss the question of axes in the case of hadrons from now on. We first consider the relationship of a single quark loop to the two axes as illustrated in Figs. 6 and 7.

It is evident that *both* motions which a loop may perform with respect to the axes (the spinning-rolling motion about the two axes, be it Fig. 3 or be it Figs. 6 and 7) contribute to the generation of electric potential – both by the same argument which was made in sequel to Eq. (3b) and in Ref. 1. These contributions are expected to be proportional to the winding numbers.

The symmetric-axes model differs as follows from the spinning-top model: In the loop settings of the type of Fig. 7 the spinning about the left axis makes no electrical contribution in the limit of the loop converging toward the axis. On the other hand the loops of Fig. 3, loops setting close to the straight axis, contribute much to the electric field because these loops are big and sweep over large areas with each turn of the spinning motion. For those spinning top loops, wherever their setting, it is always the difference, $3 - 2$ in the present example, which determines the electric potential.

With the above comments taken into consideration, it is, in the symmetric-axes model, effectively again the difference of the winding numbers which, as in the spinning-top model, is responsible for the equivalent electric charge.

Thus, as we already remarked, the two spinning motions occur simultaneously; we arrive again at the most interesting relationship between winding numbers and their equivalent charge. With this topological definition of quark loops and their properties, we achieve a definition of the ratios of the quark's electric charges, and also of their magnetic moments.

The forms of the antiquarks as previously mentioned are the mirror forms of the quarks. We might take as the mirror plane a plane perpendicular to the paper plane of Figs. 6 and 7, a plane which bisects both doughnuts, e.g., goes through the major axes of the elliptic projections of the interlinked axes.

The spinning motions may occur without the axes moving in space (except for the sliding motions along the axes, referred to above); there are furthermore the motions of the two linked axes in space. Ignoring deformations of those two axes and considering them for simplicity as equivalent in size and mutual relationship, these axes might perform a rigid body motion in space, characterized, apart from translations, by the rates of change of three Euler angles.

It might be possible that this latter motion may represent orbital angular momentum whereas the first two motions might correspond to spin. If that is a correct interpretation, there would no longer be need (as was suggested in Appendix II of Ref. 1) to assume giant quarks with higher spin to understand the higher-lying mesons, as well as the baryons. The spinning-top model on the other hand is preferable because of its simplicity, and may be more amenable to quantitative discussion.

ACKNOWLEDGMENTS

I am deeply indebted to the Research Corporation for an individual research grant and I would like to thank very much the George Washington University Committee on Research for their support through an institutional NSF grant. I am grateful for the help I received from many colleagues through the years, in particular from Professor W. C. Parke, Professor A. E. Ruark, and from Dr. A. Ghovanlou, and for the drawings in this and the preceding paper by Dr. T. Furlow, Miss J. Furlow, and Professor Parke. I owe many thanks to my mathematics colleagues, Professor R. H. Fox, Professor Ch. H. Giffen, and Professor N. Smythe, for clarification of topological issues relating to linked loops. The present project was started at the Summer Institute for Theoretical Physics, University of Colorado, at Boulder.

APPENDIX: WAVE EQUATION FOR LOOP-FORMS OF QUANTIZED FLUX

The heuristic, geometrical picture, in terms of which we developed the present program, was designed to show the consistency of the idea of the quantized flux loop model. It permitted achieving approximate numerical results and it indicates to us how to develop a straightforward quantum-mechanical model.

It was evident from the beginning that a more axiomatic, rather than heuristic, formulation was eventually to be achieved. It would, however, have been too difficult to guess a correct analytical theory without first formulating the geometrical, heuristic model and showing its consistency. On this basis the appropriate choices and assumptions for an axiomatic analytic theory may then be made.

We start that effort by indicating what kinds of wave equations may appropriately chosen to describe the probability amplitudes of loopforms.

A general description of loopform probability amplitudes involves functionals, i.e., probability amplitudes for the various forms which a quantized flux loop may adopt. We assume that the probability amplitudes should reconstruct Maxwell-Lorentz electrodynamics, i.e., in the present case of muons or electrons, an ordinary point dipole field

$$\vec{B} = \mu[3r^{-5}(\hat{z} \cdot \vec{r})\vec{r} - r^{-3}\hat{z}]. \tag{A1}$$

Expression (A1) is the resultant magnetic field; it is a superposition of contributions from sheaves of magnetic moment in the direction $\vec{\zeta}$, $|\vec{\zeta}| = 1$,

$$\vec{B}_{\zeta} = \mu[1 + \cos(\vec{\zeta}, \hat{z})][3r^{-5}(\vec{\zeta} \cdot \vec{r})\vec{r} - r^{-3}\vec{\zeta}]. \tag{A2}$$

For a given $\vec{\zeta}$, expression (A2) represents a 2-parametric manifold (aximuth α and size parameter σ of loopforms). Therefore (A1) can be considered as the resultant of a 4-parametric manifold of loopforms (flux orientation $\vec{\zeta}$, aximuth α, and size parameter σ; ζ, α are Euler angles), cf. Figs. 1 and 4 of Ref. 1. The size of a loopform was characterized by the size parameter σ which measures the "aphelion" distance from the source point of the loopform in question.

We notice that the shape of a loopform is the same for all sizes of loopforms, and evidently also for all orientation parameters $\vec{\zeta}$ and azimuth parameters α. This permits us to replace a functional description of the manifold of loopforms by a description in terms of probability amplitudes, functions of three angle parameters $\vec{\zeta}$, α and a size parameter σ.

In our previous paper we have discussed the motion of a lepton's loopforms and we have seen that the lepton can be appropriately described in terms of just one loop. Mesons and baryons imply two- and three-quark loops. When the motion of loopforms of quark loops is under consideration, we have to remember that the quark loops are assumed to spin about the main straight axis *and* about the circular axis (Figs. 3–5). This latter, rolling motion is equivalent to a spinning about the straight axis with a commensurable spinning frequency, the commensurability is determined by the ratio of the winding numbers. Thus, with caution, we may apply the spinning-top model even in the case of quark loops.

Returning to the simplest case of a muon or electron loop, we may start with the question: What may be the wave equation for the loopforms generating the magnetic dipole field of a *point* source, and thereafter discuss the issue of quasi-nonlocality of that source.

Considering the invariance of the loopforms with respect to $\vec{\xi}$, α, i.e., with regard to the 3-parametric rotation group O_3, we follow Casimir's[5] spherical-top discussion. The homogeneous coordinates in terms of the unit vector $\vec{a}$ of the axis of rotation and of the angle ϕ of rotation are

$$\xi/\mathfrak{r} = a_x \sin(\phi/2),$$

$$\eta/\mathfrak{r} = a_y \sin(\phi/2),$$

$$\rho/\mathfrak{r} = a_z \sin(\phi/2), \tag{A3}$$

$$\chi/\mathfrak{r} = \cos(\phi/2),$$

$$\mathfrak{r}^2 = \xi^2 + \eta^2 + \rho^2 + \chi^2. \tag{A4}$$

The angular part of the wave-equation operator for the symmetric top, analogous to the three-dimensional case for a mass point

$$r^2\left(\frac{\partial^2}{\partial x^2} + \frac{\partial^2}{\partial y^2} + \frac{\partial^2}{\partial z^2}\right) - \left(\frac{\partial}{\partial r}\right) r^2 \left(\frac{\partial}{\partial r}\right)$$

$$= \left(\frac{1}{\sin\theta}\right)\left(\frac{\partial}{\partial\theta}\right)\sin\theta\left(\frac{\partial}{\partial\theta}\right) + \left(\frac{1}{\sin^2\theta}\right)\left(\frac{\partial^2}{\partial\varphi^2}\right), \tag{A5}$$

is

$$\mathcal{L}^2 = \mathfrak{r}^2\left(\frac{\partial^2}{\partial\xi^2} + \frac{\partial^2}{\partial\eta^2} + \frac{\partial^2}{\partial\rho^2} + \frac{\partial^2}{\partial\chi^2}\right) - \mathfrak{r}^{-1}\left(\frac{\partial}{\partial\mathfrak{r}}\right)\mathfrak{r}^3\left(\frac{\partial}{\partial\mathfrak{r}}\right) \tag{A6}$$

and has the eigenvalues

$$-4l(l+1). \tag{A7}$$

Whereas (A5) represents the Laplacian operating on the two-dimensional space of θ, φ, (A6) represents the Laplacian operating on the three-dimensional hypersurface spanned out by the

three Euler angles, i.e., $\vec{\xi}$, α; (A6) is the appropriate operator for the rigid spherical top.

Looking at the form (A6), (A4), and (A3) of the $\mathcal{L}^2$ operator, we find that it permits the assignment of physical interpretation to a fourth parameter, i.e., $\mathfrak{r}$, along with the three Euler angles $\vec{\xi}$, α. As the shape of the loopforms is not only the same for all values of $\vec{\xi}$, α but also for all values of size σ, we may extend the 3-parametric loopform characterization (by $\vec{\xi}$, α) to a 4-parametric characterization (by $\vec{\xi}$, α, σ) and assume the interpretation

$$\mathfrak{r} \propto \sigma. \tag{A8}$$

We may therefore associate with every loopform a point in the four-dimensional space of the variables ξ, η, ρ, χ.

And we may assume that instead of the Laplacian operating on the 3-parametric manifold [Eq. (A6)], the Laplacian now operates on the 4-parametric manifold. The operators

$$\frac{\partial^2}{\partial\xi^2} + \frac{\partial^2}{\partial\eta^2} + \frac{\partial^2}{\partial\rho^2} + \frac{\partial^2}{\partial\chi^2},$$

$$\mathfrak{r}^2\left(\frac{\partial^2}{\partial\xi^2} + \frac{\partial^2}{\partial\eta^2} + \frac{\partial^2}{\partial\rho^2} + \frac{\partial^2}{\partial\chi^2}\right) \tag{A9}$$

may now characterize this $O(4)$-invariant problem which characterizes the loopforms' invariance with respect to $\vec{\xi}$, α, and σ or ξ, η, ρ, and χ. We may thus assume a wave equation

$$\left\{\mathfrak{r}^2\left(\frac{\partial^2}{\partial\xi^2} + \frac{\partial^2}{\partial\eta^2} + \frac{\partial^2}{\partial\rho^2} + \frac{\partial^2}{\partial\chi^2}\right) + \tilde{\omega}^2 - C\right\}\psi = 0. \tag{A10}$$

The solution of such a wave equation corresponds to angular velocities Ω (of the loopforms) which are independent of $\mathfrak{r}$ because this equation is homogeneous of degree zero in $\mathfrak{r}$; ξ, η, ρ, and χ are proportional to $\mathfrak{r}$ [Eq. (A4)] This corresponds to our basic assumption of spinning of flux loops with *Zitterbewegung* angular velocity $2mc^2/\hbar$.

At this point it may be appropriate to comment on the proportionality (A8) between $\mathfrak{r}$ and σ. In the calculation of the electric field of the spinning flux loop, the mass of the electron or the muon cancels out rigorously. This implies the equality of the electric charge of these two leptons. The cancellation of the mass means that there is a scale invariance. Considering this we may write

$$\mathfrak{r} = \sigma/(\hbar/mc) \tag{A11}$$

as a parameter indicating the size of a loopform.

A Maxwell-Lorentz field $\vec{B}(\mathfrak{r}, \alpha)$ of the dipole form (A1) may be reconstructed by giving the probability amplitude $\psi(\xi, \eta, \rho, \chi)$ an $\mathfrak{r}$ dependence such that

$$B\mathfrak{r}d\mathfrak{r} \propto |\psi|^2 \mathfrak{r}^3 d\mathfrak{r}, \qquad (A12)$$

i.e.,

$$|\psi| \propto \mathfrak{r}^{-5/2}. \qquad (A13)$$

To discuss the solutions of a wave equation, we may be reminded of (A6) and (A7) which permit the solution of (A10) by separation of variables, leading to

$$\left[\mathfrak{r}^{-1}\left(\frac{d}{d\mathfrak{r}}\right)\mathfrak{r}^3\left(\frac{d}{d\mathfrak{r}}\right) - 4l(l+1) + \bar{\omega}^2 - C\right]R(\mathfrak{r}) = 0. \qquad (A14)$$

Since we want the ψ function to represent a dipole field, we have to take $R_{(\mathfrak{r})}$ to be proportional to $\mathfrak{r}^{-5/2}$,

$$R \propto \mathfrak{r}^\kappa = \mathfrak{r}^{-5/2}, \qquad (A15)$$

$$-4l(l+1) + \bar{\omega}^2 - C = -\kappa(\kappa + 2)$$

$$= -\tfrac{5}{4}. \qquad (A16)$$

This represents the point dipole solution.

The solution (A15) has a singularity at $\mathfrak{r} = 0$, which corresponds to a point dipole source. Considering the Pryce-Tani-Foldy-Wouthuysen representation of a stationary single particle, the latter appears in ordinary position as smeared out. Replacing, accordingly, the point dipole source (a crude substitute for a transformation from mean position to position), we get rid of that singularity. We may effect this by introducing into the wave equation (A10) or (A14) a "potential" $U(\mathfrak{r})$ which is positive in the "core" region $0 \leqslant \mathfrak{r} \leqslant 1$ and goes to zero at the core surface $\mathfrak{r} \approx 1$ and is zero outside, $1 \leqslant \mathfrak{r} < \infty$, which ensures the $\mathfrak{r}^{-5/2}$ behavior of $R(\mathfrak{r})$ for large $\mathfrak{r}$.

Considering these, Eq. (A14) may be written as

$$\left[\mathfrak{r}^{-1}\left(\frac{\partial}{\partial\mathfrak{r}}\right)\mathfrak{r}^3\left(\frac{\partial}{\partial\mathfrak{r}}\right) - U(\mathfrak{r}) - 4l(l+1) + \bar{\omega}^2 - C\right]R = 0. \qquad (A17)$$

This equation might represent the wave equation for the loopform of an electron or a muon. The choice $C = \frac{9}{4}$ gives, by (A16) to achieve commensurability,

$$\bar{\omega} = 2l + 1 = 1, \ 2, \ 3, \ \ldots . \qquad (A18)$$

This relates to the spherical top. The eigenvalues of the symmetric top show a 2-parametric spectrum. Commensurabilities of $\bar{\omega}$ permit phase-correlated motion of loopform amplitudes (angular group velocity) in the case of the electron, distin-guished from random-phased muon amplitudes (angular phase velocity).

The wave equation is presumably to be written in a linearized form. The group-theoretical analysis of Eq. (A10) is particularly promising; Barut's analysis is expected to contain many of the relevant results.

We found in Sec. VIII B of Ref. 1 the important result that lepton-antilepton pairs, represented by generalized spherical harmonics, have the correct transformation properties under CP conjugation. These harmonics form bases of the irreducible representations of the continuous group O(3). Equation (A10) admits the O(4) group. The question arises whether the bundling of the continuous manifold of flux loopforms, Secs. VII, X, XI, and XIV of Ref. 1, into a discrete number of statistically independent bundles may be formulated in terms of the *discrete* subgroups of the aforementioned continuous groups. The counting in terms of the pentagondodecahedron/icosahedron, Fig. 8 of Ref. 1, has already pointed in that direction.

And the question arises about the role of irreducible representations of these discrete subgroups in the description of the electron, muon, and other particles. The representation of the continuous groups, in particular the generalized spherical harmonics, should, however, first of all be considered for a description of bundling of loopforms.

This bundling is an important issue because it was shown in Ref. 1 that the concept of superposition of complex probability amplitudes, with different phases for different bundles, may permit us to derive effective magnetic moments (= Bohr or muon magneton) and electric charge (= e), electromagnetic energy (= mc^2), and electromagnetic angular momentum (= $\hbar/2$), all from quantized flux Φ_q (= hc/e).

Note added. In our paper[1] we characterized the statistical independence (of the probability amplitudes) of the loopforms by assuming that a difference (in size, in orientation, in azimuth) greater than 1 rad makes them to be independent, whereas closely neighboring loopforms are correlated as regards their amplitudes. We thereby were led to group the loopforms into 207 bundles. This was done by simplified geometrical means, using graphical illustrations like a pentagondodecahedron whose corners are about 1 rad apart (or the faces of an icosahedron), and a flux tube picture. A more formal treatment is sketched in this Appendix.

To the qualitative discussion of electron versus muon we might note that the (angular) group and phase velocities of the terms bilinear in probability amplitudes are as 1 to 207. As the linear velocities of the spinning loops are of the order c to c and as the sizes (radii of the cores) stand in the ratio of 207 to 1, their electromagnetic energies are of the order of 1 to 207.

---

*Reprint requests to H. Jehle, 1208 Sherwood Rd., Charlottesville, Va. 22901.

[1]H. Jehle, Phys. Rev. D **3**, 306 (1971).

[2]L. Foldy, in *Quantum Theory*, edited by D. R. Bates (Academic, New York, 1962); S. Tani, Progr. Theoret. Phys. (Kyoto) **6**, 267 (1951); K. M. Case, Phys. Rev. **95**, 1323 (1954); M. H. L. Pryce, Proc. Roy. Soc. (London) **A195**, 62 (1948); A. O. Barut and S. Malin, Rev. Mod. Phys. **40**, 632 (1968); L. Foldy and S. Wouthuysen, Phys. Rev. **78**, 29 (1950); H. Jehle and Wm. C. Parke, *ibid.* **137**, B760 (1965); R. L. Ingraham, Nuovo Cimento **34**, 182 (1964); in *Boulder Lectures in Theoretical Physics*, edited by W. E. Brittin *et al.* (Gordon and Breach, New York, 1964), Vol. VI, p. 112.

[3]M. A. B. Bég, Phys. Rev. Letters **13**, 514 (1964); W. Thirring, in *Internationale Universitätswochen für Kernphysik. Fourth Schladming Winter School in Physics, 1965*, edited by P. Urban (Springer, Berlin, 1965); Acta Phys. Austriaca, Suppl. **II**, 1965; in *Subnuclear Phenomena* (Erice Lectures, 1970), edited by A. Zichichi (Academic, New York, 1970), Part A; L. Pauling, Proc. Natl. Acad. Sci. U.S. **56**, 1676 (1966); V. F. Weisskopf, CERN Report No. CERN-TH-66-19, 1966 (unpublished); A. Dolgov, Phys. Letters **15**, 84 (1965); D. Horn, Nuovo Cimento **62A**, 581 (1969); J. Franklin, Phys. Rev. **181**, 1984 (1969).

[4]A. O. Barut, in *Boulder Lectures in Theoretical Physics*, edited by A. O. Barut and W. E. Brittin (Colorado Associated Univ. Press, Boulder, 1971), Vol. XIII; in *Boulder Lectures in Theoretical Physics*, edited by W. E. Brittin *et al.* (Gordon and Breach, New York, 1967), Vol. IX B, p. 273; Vol. IX A, p. 125; in *Boulder Lectures in Theoretical Physics*, edited by W. E. Brittin and A. O. Barut (Gordon and Breach, New York, 1968), Vol. X B, p. 377; in *Proceedings of the Second Coral Gables Conference on Fundamental Interactions at High Energy II*, edited by A. Perlmutter *et al.* (Gordon and Breach, New York, 1970), p. 199; Phys. Rev. D **3**, 1747 (1971); in *Topics in Modern Physics*, Tribute to E. U. Condon (Colorado Associated Univ. Press, Boulder, 1971), p. 15; in *Springer Tracts in Modern Physics*, edited by G. Höhler (Springer, New York, 1969), Vol. 50; A. O. Barut and A. Böhm, J. Math. Phys. **11**, 2938 (1970). A. O. Barut and H. Kleinert, in *Proceedings of the Fourth Coral Gables Conference on Symmetry Principles at High Energies*, edited by A. Perlmutter and B. Kurşunoğlu (Freeman, San Francisco, 1967), p. 76; in *Proceedings of the Fifth Coral Gables Conference on Symmetry Principles at High Energies, University of Miami, 1968*, edited by A. Perlmutter, C. Angas Hurst, and B. Kurşunoğlu (Benjamin, New York, 1968); in *Proceedings of the Coral Gables Conference on Fundamental Interactions at High Energy II*, edited by A. Perlmutter, G. J. Iverson, and R. M. Williams (Gordon and Breach, New York, 1970); Phys. Rev. **156**, 1546 (1967); **161**, 1464 (1967); A. O. Barut, D. Corrigan, and H. Kleinert, *ibid.* **167**, 1527 (1968); H. C. Corben, *Classical and Quantum Theories of Spinning Particles* (Holden Day, San Francisco, 1968); Phys. Rev. Letters **15**, 268 (1965); Nuovo Cimento **47**, 482 (1967); Phys. Rev. **131**, 2219 (1963); A. O. Barut, Phys. Rev. **156**, 1538 (1967).

[5]H. G. B. Casimir, *Rotation of a Rigid Body in Quantum Mechanics* (Wolters, Groningen-The Hague, 1931); F. Bopp and R. Haag, Z. Naturforsch. **5a**, 644 (1950); L. H. Thomas (private communication); Nature **117**, 514 (1926); Phil. Mag. **3**, 1 (1927); Ann. Math. **42**, 113 (1941); J. L. Lopez, *Lectures on Symmetries* (Gordon and Breach, New York, 1969); N.-P. Chang and Ch. A. Nelson (private communication); A. E. Ruark (private communication). Further references: E. P. Wigner, Am. J. Phys. **38**, 1005 (1970); P. Havas, Phys. Rev. **74**, 456 (1948); **87**, 309 (1952); **91**, 997 (1953); J. J. Loferski, thesis, University of Pennsylvania, 1949 (unpublished); E. R. Speer, Ann. Math. Studies **62**, 1 (1969); B. Deaver, in *Methods in Experimental Physics, Solid State*, edited by L. L. Marton and R. V. Coleman (Academic, New York, 1972); C. Kittel, in *Superconductivity*, edited by M. Tinkham (Gordon and Breach, New York, 1969); E. J. Post (private communication); L. H. Thomas, in *Quantum Theory of Atoms, Molecules, and the Solid State*, edited by P.-O. Löwdin (Academic, New York, 1966), p. 93; L. Tisza (private communication); L. Tisza and C. K. Whitney (private communication); F. Rohrlich, Phys. Rev. **150**, 1104 (1966); R. L. Ingraham (private communication).

General Relativity and Gravitation, Vol. 8, No. 3 (1977), pp. 175–196

Knot Wormholes in Geometrodynamics?[1]

ECKEHARD W. MIELKE

Joseph Henry Laboratories, Princeton University, Princeton, New Jersey 08540[2]

Received May 25, 1976

Abstract

The familiar wormhole model of geometrodynamics is extended to allow for knotted embeddings of the initial hypersurface. It is shown that topology change is not only a means to modify the connectivity of the space, but also the knot invariants of its embedding. In a probabilistic framework the process of "wormhole scattering" can be expressed by creation and annihilation operators acting on the wave function of quantum geometrodynamics. Implications concerning Wheeler's exciton model of elementary particles, the f-gravity approach to hadronic matter, and interrelations with Jehle's flux quantization program are discussed.

§(1): *Introduction and Summary*

This work is concerned with the long-term vision of Clifford and Einstein: *There is nothing in the world except curved empty space* [56]. The metaphysical reason for this belief can be traced back to, for example, Spinoza: "If the world can be conceptually comprehended and derived *more geometrico*, its order has to be geometrical itself." This and other fascinating philosophical aspects of the interrelation between "Geometry and Reality" have found extensive studies by Kanitscheider [27] and by Graves [21, 24].

Although there is a variety of interesting geometrical models of quantum

[1] Work supported by a grant of the Studienstiftung des deutschen Volkes and in part by National Science Foundation grant No. GP 30799 X to Princeton University.
[2] Present address: Mathematical Institute, University of Oxford, 24–29 St. Giles, Oxford OX1 3LB, England.

176   MIELKE

physics as for example the theory of kinks [9] or the twistor formalism [41] , to
mention only a few, only one line of approach will be followed here: Wheeler's
quantum geometrodynamics (QGMD) [56–59] . On the classical level the
Einstein–Maxwell equations are the basic set-up of geometrodynamics (GMD).
They are studied in otherwise *empty* space-time admitting multiconnected manifolds as spacelike hypersurfaces. Charge may thus be regarded as "flux lines
trapped in topology" and mass as manifestation of the energy of the electric and
magnetic field [38].

The main extension of this familiar model is to assume that these wormhole
initial hypersurfaces $W^3 = S^1 \times S^2$ are in general knotted embedded in space-
time, thus taking up Wheeler's proposal [55] to describe an elementary particle
as "a knotted-up region of high curvature." Accordingly, it is the program of
this and further work to study the bearings of knotting on the fine structure of
space-time and on the fundamental structure of elementary particles.

First, the concept of *knot wormholes* together with its main mathematical
properties has been analyzed in a 2 + 1 space-time model because in higher dimensions the mathematical difficulties increase considerably. A further handicap
for later physical interpretations is that in the 3 + 1 case no list of the knot invariants of the "simplest" knotted 2-spheres exists except for a subclass that can
be constructed from knots.

The interest in knotting results partly from the possibility of introducing
concisely the concept of "scattering" of embedded manifolds. As shown in an
example and made plausible on general grounds, "*wormhole scattering*" is a
means to change besides the topology also the knot invariants of the scattered
hypersurface; i.e., intuitively, the in- and outcoming manifolds are differently
knotted and linked. Although it is not at present possible to calculate the dynamics of these invariants, a characterization of links by the number of their
basic constituents with respect to topology change can be established as a first
outcome.

In the probabilistic framework of QGMD it is here suggested that we represent this "topological scattering process" by creation and annihilation operators
obeying a complex algebraic structure that is reduced in the case of trivially embedded handles to that of the canonical anticommutation relations (CAR) of
second quantization.

Only a few steps of this program of "pregeometry as the calculus of propositions" can be outlined here. There are, however, indications that quantum statistics of wormholes will be an effective tool to put further flesh on the "geometrodynamical exciton" model of elementary particles [56]. However, there could
exist alternative applications. If the gravitational constant is changed as in
f-gravity [46] , the knotting of f-wormholes could become responsible for the im-
prisonment of quarks in hadrons. In this context, Jehle's interpretation [26] of
knots should find analogs in f-QGMD. Whether or not the observed particle spec-

trum really has any bearing on the knotted fine structure of the internal manifolds of the hadrons is a difficult issue for the future. The same can be said of Sakharov's hypothesis [45] that the observed parity violation in elementary particle physics is a consequence of the amphicheirality of some knots.

The plan of the paper is as follows: In Section 2 the wormhole model of classical charge and its realization in GMD by the Reissner–Nordstrøm metric are reviewed. Section 3 deals with the difficult embedding problem of wormholes in space-time. "Topology change" and a "scattering theory" of manifolds are analyzed in Section 4. Some attempts to formulate a quantum geometrodynamical representation of these topological phenomena are presented in Section 5. Accordingly, the physical interpretation can only be given as an outlook in Section 6; too much work still has to be done before any conclusions concerning the models' explanatory success in elementary particle physics can be given.

Taking a very critical point of view one can easily imagine that the knot wormhole model will share the (well-known!) fate of Kelvin's theory of vortex atoms, which regarded atoms (roughly) as knots tied in the vortex lines of the ether (from Conway [5]). Finally, a word of warning may be in order here. As was stressed in a talk [32] previously given on this subject: Every statement in this work that does not follow from the cited literature is at least highly speculative!

$$\S(2): \quad \textit{Classical Wormholes: A Review}$$

In order to establish a concept of *classical* charge (compare v. Westenholz [54]) that is directly connected with the global topology of a hypersurface M^3 embedded in an orientable space-time manifold M^4 (which is not necessarily Riemannian) the two 2-forms on M^4

$$\omega \equiv \tfrac{1}{2} F_{\mu\nu} dx^\mu \wedge dx^\nu \tag{2.1}$$

$$^*\omega \equiv \tfrac{1}{2} {}^*F_{\mu\nu} dx^\mu \wedge dx^\nu, \qquad {}^*F_{\mu\nu} \equiv \tfrac{1}{2} \epsilon_{\mu\nu\rho\sigma} F^{\rho\sigma} \tag{2.2}$$

will be considered. They are closed,

$$d\omega = d{}^*\omega = 0 \tag{2.3}$$

owing to Maxwell's equations in a *source-free* space. According to de Rham's existence theorem [22] there exists for given periods (charges) $m^{(i)}$ a closed p-form $\omega^p \in H^p(M^n)$ (which is unique up to an exact form) with respect to a homology basis of β_p independent cycles $c_p^{(i)} \in H_p(M^n)$ such that the following bilinear mapping from the product of the pth homology group and the pth cohomology group to the real numbers can always be specified:

$$H_p(M^n) \times H^p(M^n) \longrightarrow \int_{c_p^{(i)}} \omega^p = 4\pi m^{(i)} \tag{2.4}$$

178   MIELKE

As an application, assume [38] that the magnetic charges $m^{(i)}$ corresponding to
the 2-form $\omega \in H^2(M^4)$ vanish, i.e., demand that the periods $m^{(i)}$ are zero for
every basis element of the homology group $H_2(M^4)$. Then, the second theorem
of de Rham guarantees the existence of a vector potential $A = (A^\mu)$ such that ω
is exact:

$$\omega = dA \tag{2.5}$$

If no "topology change" of M^4 occurs, the electric charges $q^{(i)}$ stay "constant
in time," i.e., the periods corresponding to $*\omega$ take the same values on $t = \text{const}$
surfaces c_2 that are homologous:

$$4\pi q' \equiv \alpha \int_{c_2'} *\omega = \alpha \int_{c_2 + \partial c_3} *\omega$$

$$= 4\pi q + \alpha \int_{c_3} d*\omega = 4\pi q \tag{2.6}$$

In the derivation above, Stoke's theorem and the closeness of $*\omega$ have been employed (If "monopoles" do exist [42], the same argument would carry through
for ω, but a unique vector potential does not exist).

The preceding analysis shows that space-times admitting nonzero electrical
charges represented by "flux lines trapped in topology" have to allow $\beta_2 \geqslant 1$
independent cycles c_2. In general the *Betti numbers* $\beta_p = \dim H^p(M^n)$ are related
to the Poincaré polynomial of M^n by

$$f_M(t) \equiv \sum_{p=0}^{n} \beta_p t^p \tag{2.7}$$

The Euler–Poincaré characteristic is

$$\chi \equiv f_M(-1) = 2(g - 1) \tag{2.8}$$

where g is the *genus* of the orientable manifold [48]. If M^n is connected, it is
known that $H^0(M^n) = R$, i.e., $\beta_0 = 1$ and if compactness and orientability is
assumed,

$$\beta_p = \beta_{n-p} \tag{2.9}$$

is valid, owing to the Poincaré duality.

Three dimensional manifolds M_c^3 that are closed, symmetric, and prime have
been classified [10]. They are either S^3, P^3, T^3, W^3, a polyhedral manifold or a
lens space $L(p, q)$. The space-times M^4 that will be of further interest in this
work are $R \times S^3$ and $R \times W^3$, where the spacelike hypersurface in the second
topological product represents a "*wormhole*":

$$W^n \equiv S^1 \times S^{n-1} \tag{2.10}$$

KNOT WORMHOLES IN GEOMETRODYNAMICS? 179

The two-dimensional torus $T^2 = W^2$ will serve, at the same time, as a wormhole model in a $2 + 1$-dimensional space-time $M^{2+1} = R \times W^2$.

The calculation of the Betti numbers will be facilitated by the observation that the graded cohomology algebras $H(R \times M_c^n)$ and

$$H(M_c^n) = \sum_{p=0}^{n} H^p(M_c^n)$$

are isomorphic, because R is contractible to a point, i.e., $\pi_1(R) = 0$.

Since the cohomology groups of spheres S^n, $n \geq 1$ are given by

$$H^0(S^n) = H^n(S^n) = R \tag{2.11}$$

and

$$H^p(S^n) = 0, \qquad 1 \leq p \leq n - 1 \tag{2.12}$$

the corresponding Poincaré polynomial is

$$f(t) = 1 + t^n \tag{2.13}$$

Employing the Künneth isomorphism for products of compact manifolds M_c and N_c which is reflected in the formula

$$f_{M_c \times N_c}(t) = f_{M_c}(t) f_{N_c}(t) \tag{2.14}$$

the polynomial can be easily calculated in the case of a wormhole:

$$f_{W^n}(t) = (1 + t)(1 + t^{n-1})$$

$$= 1 + t + t^{n-1} + t^n \tag{2.15}$$

Thus, the nonzero Betti numbers are

$$\beta_0 = \beta_1 = \beta_{n-1} = \beta_n = 1 \tag{2.16}$$

Since $\chi(R \times W^n) = \chi(R \times S^{2m+1}) = 0$, such space-times are in addition time-orientable [60] with respect to pseudo-Riemannian metrics with Lorentzian signature $(-, +, \ldots, +)$. The multiconnected character of the manifolds considered can be precisely read off from the homotopy groups of M^4, which for $k > 1$ are Abelian on pathwise connected manifolds [25]:

$$\pi_0(R \times W^n) = 0$$

$$\pi_1(R \times W^n) = \begin{cases} Z \oplus Z & \text{if } n = 2 \\ \pi_1(S^1) = Z & \text{if } n > 2 \end{cases}$$

$$\pi_2(R \times W^n) = \pi_2(S^{n-1}) = \begin{cases} Z & \text{if } n = 3 \\ 0 & \text{otherwise} \end{cases}$$

$$\pi_k(R \times W^n) = \pi_k(S^{n-1}) = \delta_{k,n-1} Z \qquad \text{if } 2 < k \leq n - 1$$

$$\pi_k(R \times W^3) = \pi_k(S^2) = \pi_k(S^3), \qquad k > 2$$

180   MIELKE

The next step in this analysis is to ensure the existence of 3-metrics on hypersurfaces M^3 endowed with the wormhole topology W^3 within the framework of the Rainich–Misner–Wheeler "Already Unified Field Theory" or *geometrodynamics* [38]. Because of the troubles in the null field case [15], as a more familiar starting point the Einstein-Maxwell equations (in geometrical units) will be considered:

$$R_{\mu\nu} - \tfrac{1}{2} g_{\mu\nu} R = F_\mu^\alpha F_{\nu\alpha} + {}^*F_\mu^\alpha {}^*F_{\nu\alpha} \tag{2.17}$$

With respect to the line element

$$ds^2 = -dt^2 + g_{ab} dx^a dx^b \tag{2.18}$$

of the pseudo-Riemannian manifold M^4 and the second fundamental form

$$S_{ab} \equiv \frac{1}{2} \frac{\partial}{\partial t} g_{ab} \tag{2.19}$$

on the hypersurface M^3, (2.17) splits into the *initial value equations*

$$\nabla^a S_{ab} - \nabla_b S = -2\epsilon_{bce} E^c B^e \tag{2.20}$$

$$S^{ab} S_{ab} - S^2 - {}^{(3)}R = -2(E^a E_a + B^c B_c) \tag{2.21}$$

and the *evolution equations*

$$\frac{\partial}{\partial t} S_{ab} = 2S_a^c S_{cb} - SS_{ab} - {}^{(3)}R_{ab} - 2(E_a E_b + B_a B_b) + g_{ab}(E^2 + B^2) \tag{2.22}$$

Here, the three components of the electric and magnetic field vectors are as usually defined by

$$E_b \equiv F_{b0}, \qquad B_a \equiv \tfrac{1}{2} \epsilon_{abc} F^{bc} \tag{2.23}$$

In the case of zero electric and magnetic field the natural metric on $W^3 = S^1 \times S^2$

$$dl_W^2 = d\mu^2 + d\Omega^2, \qquad -\pi < \mu \leq \pi \tag{2.24}$$

where

$$d\Omega^2 \equiv d\theta^2 + \sin^2 \theta d\sigma^2 \tag{2.25}$$

in the standard metric on S^2, may be conformally deformed to

$$d\bar{l}_W^2 = \phi^4(\mu, \theta) dl_W^2 \tag{2.26}$$

in order to satisfy the initial value equations [36]. Since the wormhole is assumed to be initially at rest, i.e., $S_{ab} = 0$, the problem is reduced to Brill's wave equation

$$\Delta_W \phi - \tfrac{1}{8} {}^{(3)}R_W \phi = 0 \tag{2.27}$$

for which solutions are known on compact manifolds $M_c^3 = W^3$.

Another realization [37] of that topology is the Schwarzschild geometry

$$ds^2 = -\left(1 - \frac{2m}{r}\right)dt^2 + \left(1 - \frac{2m}{r}\right)^{-1}dr^2 + r^2 d\Omega^2 \qquad (2.28)$$

where the two asymptotically flat regions are identified at a distance far away from the throat (Einstein–Rosen bridge). In order to visualize this, the θ rotational degree of freedom will be suppressed and a parameter $z(r)$ for a trivial embedding into the Euclidean space

$$de^2 = dz^2 + dr^2 + r^2 d\sigma^2 \qquad (2.29)$$

can be calculated according to the formula [39]

$$z(r) = \int_{r_0}^{r} \left[\frac{r}{2m(r)} - 1\right]^{-1/2} dr \qquad (2.30)$$

Thus the Schwarzschild wormhole can be portrayed (Figure 1) by the paraboloid of revolution

$$r = 2m + z^2/8m \qquad (2.31)$$

Since the creation and pinch-off of the throat occurs very rapidly within the time $T = 2\pi m$, no light signals can be sent between two asymptotically flat regions; i.e., the Schwarzschild wormhole is a globally causal space-time [12]. If the center of attraction apparently carries a charge q in addition to mass m, such that the dual of the electric field is given by

$$*\omega = q \sin \theta d\theta \wedge d\sigma \qquad (2.32)$$

the Reissner–Nordstrøm metric will be obtained:

$$ds^2 = -\left(1 - \frac{2m}{r} + \frac{q^2}{r^2}\right)dt^2 + \left(1 - \frac{2m}{r} + \frac{q^2}{r^2}\right)^{-1}dr^2 + r^2 d\Omega^2 \qquad (2.33)$$

The embedding function for the case of maximal charge $q = \pm m$ is

$$z_{RN} = 2(\sqrt{2mr - m^2} - m) - 2m\,\text{argtanh}\,(\sqrt{2mr - m^2}/m - 1) \qquad (2.34)$$

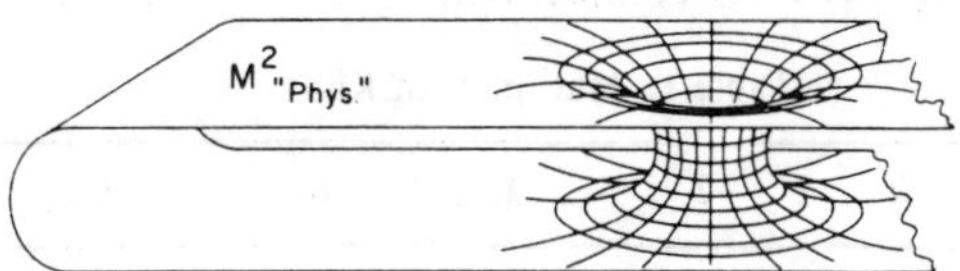

Fig. 1. Schwarzschild wormhole (adapted from MTW).

182   MIELKE

In the case where the mass exceeds the charge, $m^2 > q^2$, the extremal radius of the throat oscillates between

$$r_{\substack{min \\ max}} = m \mp \sqrt{m^2 - q^2} \qquad (2.35)$$

within a charge-independent period of $T = 2\pi m$. As shown by Graves and Brill [20] this is due to the dynamical balance of gravitational pull and Maxwell pressure. Thus the Reissner–Nordstrøm wormhole can be considered as the realization of a topological model of classical, i.e., unquantized charge within the context of geometrodynamics.

§(3): *Knot Wormholes and Links*

In order to present the placement problem of closed hypersurfaces M_c^3 in space-time from a more intuitive viewpoint, a 2 + 1 space-time will be discussed first, considering the fact that in this model the Einstein equations (2.17) in vacuum have only flat 3-metrics as solutions.

According to Crowell and Fox [6], a *knotted n-sphere* (a *knot* for $n = 1$) is defined as the image of a homeormorphism h of the unit sphere S^n into R^{n+2}

$$S^n \xrightarrow{\ h\ } K^n \subset R^{n+2} \qquad (3.1)$$

As a two-dimensional *knot wormhole* the topological product

$$\tilde{W}^2 \equiv S^1 \times K^1 \qquad (3.2)$$

will be understood, where the first factor S^1 is assumed to be trivially embedded, i.e., unknotted (in accord with a result in 3 + 1 dimensions). Only tame or equivalently PL (piece-wise linear) knots will be considered in what follows. $\tilde{W}^2$ is equivalent to the boundary of a "cube with knotted holes" [2]. For this reason, the classification problem of $\tilde{W}^2$ is the same as that of ordinary knot theory. It is known from the Alexander–Briggs table [44] of all *prime* knots (i.e., knots that cannot be represented as the connected sum $K_i \# K_j$ of other knots) that there are nonequivalent knots for crossing numbers $\geqslant 5$. Following the extension of this list given by Conway [5], the multiplets of knots for given crossing number are given in Table I. It is, therefore, necessary to distinguish knots by invariant concepts most of which are derived from its *knot group* $\pi_1(R^3 - K^1)$, i.e., the fundamental group of the knot complement.

Table 1. Multiplets of Knots

Crossing number	0	1	2	3	4	5	6	7	8	9	10	11
Number of different knot types	1	0	0	1	1	2	3	7	21	49	166	549

More generally, *links* of multiplicity μ

$$\tilde{L}_\mu(S^n) \equiv K_1^n \cup \cdots \cup K_i^n \cup K_j^n \cup \cdots \cup K_\mu^n \subset R^{n+2} \qquad (3.3)$$

i.e., the smooth (or PL) embedding of the disjoint union of μ copies of S^n in R^{n+2} will be considered. $\tilde{L}_\mu(S^n)$ is said to be trivial if it extends to a smooth embedding of the disjoint union of μ disks D^{n+1}. A characteristic invariant is the *linking number* $l(K_i^n, K_j^n)$. In the case of all components of $\tilde{L}_\mu(S^1)$ being torus knots with *winding number* (α, β) around the meridian or longitudinal direction of the torus, the relationship

$$l(K_i^1, K_j^1) = \alpha\beta(1 - \delta_{ij}) \qquad (3.4)$$

can be shown to be true [28].

The extension to *links of wormholes* will be denoted by

$$\tilde{L}_\mu(W^n) \equiv \tilde{W}_1^n \cup \cdots \cup \tilde{W}_\mu^n \subset R^{n+1}$$
$$= S^1 \times \tilde{L}_\mu(S^{n-1}) \qquad \text{if } n > 2 \qquad (3.5)$$

where the second identity can be inferred from a result of Gluck [17].

As examples a link $\tilde{L}_2(W^2)$ of a trivial wormhole with a "trefoil" wormhole is drawn in Figure 2e, whereas links of unknotted wormholes are shown in Figure 2d and on p. 1210 of MTW [39] [a link L "$_{10,000}$" (W^2)!]. Links and knots can exhibit important discrete symmetries, which likewise will appear at least in wormhole links $\tilde{L}_\mu(W^2)$.

Consider the following inversions of knots:

(a) Reversing the string orientation: $\vec{K} \to \overleftarrow{K}$

(b) Taking the mirror image: $K \to \bar{K} \equiv \{(x^1, x^2, -x^3) | x^a \in K\}$

Knots obeying the following equivalence relations are called

$\vec{K} = \vec{\bar{K}}$: amphicheiral

$\vec{K} = \overleftarrow{K}$: reversible

$\vec{K} = \overleftarrow{\bar{K}}$: involutory

As an example, the figure-eight handle (Figure 2c) enjoys all these symmetries, whereas the trefoil τ is known to be only reversible (see Conway [5] for a complete list of knot symmetries up to 10 crossings). Left- and right-handed trefoil wormholes are shown in Figures 2b and 2$\bar{\text{b}}$, respectively.

Of similar interest is the more general case [18] where an oriented link admits a rotational symmetry around a disjoint axis, which is periodical of degree p. Especially the complexity number of a complete symmetric link deserves further attention in future topological models since this invariant measures the number of consecutive projections to orbit spaces under the symmetry that are necessary to get the trivially knotted S^1.

With regard to the embedding problem of wormholes W^3 in the "physical" space-time that will be assumed locally as R^4, much less is known. Since a locally

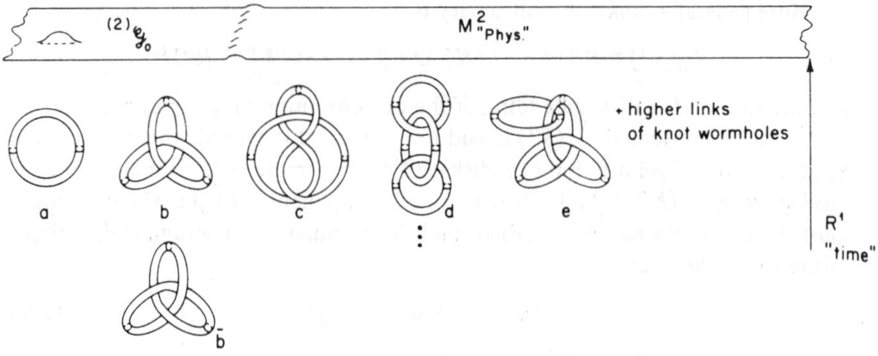

Fig. 2. "Vacuum sea" of "virtual" knot wormholes.

tame embedded S^1 is not knottable [17] in R^4, the problem is reduced to the analysis of knotted 2-spheres K^2. In order to avoid further complications, only locally flat spheres (without singularities) will be studied. Taking into account some additional conditions, S^2 can be smoothly embedded in R^4 with its regular neighborhood homeomorphic [16] to $\mathring{D}^2 \times S^2$. A subclass of knotted 2-spheres can be produced by spinning a knotted arc in half-3-space R^3_+ about a plane in 4-space [11]. Since

$$\pi_1(R^4 - K_s^2) = \pi_1(R^3_+ - K^1) \tag{3.6}$$

the classification problem of *spun knots* K_s^2 is the same as that of knots. Another concept that will be of use for the physical models analyzed here is that of a *ribbon 2-knot* $\overline{\overline{K}}^2$ or "simple-knotted 2-sphere." Essential is the fact [62] that $\overline{\overline{K}}^2$ bounds a 3-manifold that is homeomorphic either to a closed 3-ball D^3 or to the manifold obtained by taking the connected sum of "virtual" wormholes W^3 and then removing from this "hollow pretzel" [53] the interior $\mathring{D}^3$ of a 3-ball:

$$\overline{\overline{K}}^2 = \partial(\approx D^3)$$

or

$$\overline{\overline{K}}^2 = \partial(\approx(\#W^3 - \mathring{D}^3)) \tag{3.7}$$

In the second case, it can be shown that $\overline{\overline{K}}^2$ is the fusion of a split link of a trivial system of spheres S^2 in R^4. Furthermore, $\overline{\overline{K}}^2$ is unknotted in the (compactified) space-time S^4 if and only if [63]

$$\pi_1(S^4 - \overline{\overline{K}}^2) = Z \tag{3.8}$$

Like spun knots, ribbon 2-knots, too, do not exhaust all possible knotted spheres K^2. Even less is known about the classification of all (finite) collections

$\{\tilde{L}_{\mu_k}(W^3)\}$ of links of knotted wormholes. In spite of this, they will be regarded as the most general class of hypersurfaces out of which the "geometrodynamical vacuum" is supposed to be constructed alongside $M^3_{\text{"phys"}} = S^3$.

The crucial problem of this work is that if such *knotted* manifolds can be realized as hypersurfaces in space-time, which ones are compatible with the "embeddability equations" (2.20) and (2.21) of geometrodynamics? Since there does not seem to be any proof on this topic, one has to rely more on heuristic arguments:

First, the 2 + 1 space-time model will be considered in order to develop an idea how this issue can be tackled. Guided by the piece-wise linear viewpoint of topologists, the first step is to realize that disjoint tubes $S^1 \times [a, b]$ are admitted by the initial value equations as hypersurfaces [suppress θ in the considerations following (2.23)]. In order to construct, for example, a "PL"-trefoil wormhole $S^1 \times \tau$ out of six such tubes (Figure 3), three of them have to be placed in (flat) 2 + 1 space-time in such a position as to allow for one over- and one underpass, respectively. Their ends will be joined together by the other tubes such that the corners have been smoothed out in accordance with the familiar matching conditions in the theory of partial differential equations on manifolds. (The method above is not new in general relativity; for example, it has been used to construct "lattice universes" [31].)

For a transfer of such PL-arguments to the knotting problem of the "real" 3 + 1 space-time it should be very economical to apply the method of hyperplane cross sections [11] to the knotted 2-spheres K^2 of the $\tilde{L}_{\mu}(W^3)$ in question. Although it is highly nontrivial to extend these topological devices to the solutions of the PDE's (2.20) and (2.21), it seems to be reasonable to conjecture that, at least for spun knots K_s^2 and ribbon knots $\overline{\overline{K}}^2$, the arguments of the 2 + 1 case can be extended to the physical space-time in order to prove the existence of links of knotted wormholes in GMD. At present, the more far-reaching question seems to be totally out of sight, namely, how these knotted hypersurfaces develop in "time" due to the evolution equations (2.22). Do these equations constitute restrictions on the otherwise infinite number of knotted embeddings that form invertible knot cobordisms [51]?

Fig. 3. "Piecewise linear" trefoil wormhole.

186   MIELKE

§(4): *Wormhole Scattering*

In the model discussed so far, the "physical" manifold $M^3_{\text{"phys"}}$ corresponds to an empty space and is disconnected from the "vacuum sea" of "*virtual*" knot wormholes and links (Figure 2). In order to introduce mass and charge into this space, the global topology of $M^3_{\text{"phys"}}$ has to be changed.

First those consequences of *topology change* will be studied that do not depend on the issue of its compatability with GMD: The *connected sum* $M^n \# N^n$ of two (not necessarily distinct) orientable manifolds M^n and N^n of dimension n will be obtained by (a) removing open balls $\mathring{D}^n$ from each by surgery and then (b) identifying the resulting boundary spheres $\partial D^n = S^{n-1}$ by an orientation reversing diffeomorphism in order to obtain a new orientable manifold.

Figure 4 shows intuitively how in a 2 + 1 space-time a hypersurface of the wormhole type W^2 can be "created" by this process (the arrow indicates an ambient isotopy). Indeed, if the "physical" manifold is assumed to be closed, i.e., $M^n_{\text{"phys"}} = S^n$, the result is

$$M^n_{\text{"phys"}} \# W^n = S^n \# (S^1 \times S^{n-1}) = W^n \tag{4.1}$$

since $\#$ is commutative, associative, and has S^n as the zero element. There are no obstructions against taking the connected sum of a manifold with itself ("adding handles")

$$M^n \#^\nu W^n \equiv \begin{cases} M^n \# W^n_1 \# \cdots \# W^n_\nu, & \nu > 0 \\ M^n \cup W^n, & \nu = 0. \end{cases} \tag{4.2}$$

For 3-manifolds, the decomposition (4.2) is even unique [33].

The process of wormhole scattering has more far-reaching consequences if links of knotted handles are considered. The link $\tilde{L}_2(W^2) = S^1 \times S^1 \cup S^1 \times \tau$ in 2 + 1 dimensions (Figure 2e) may serve as an illustration. If it is connected once with $M^2_{\text{"phys"}}$, one knot wormhole tied to an unknotted torus or vice versa will be obtained depending on the location of the connection. If this is appropriately repeated at the other component, there is the surprise:

$$M^2_{\text{"phys"}} \#^2 \tilde{L}_2(W^2) = W^2_1 \# W^2_2 \tag{4.3}$$

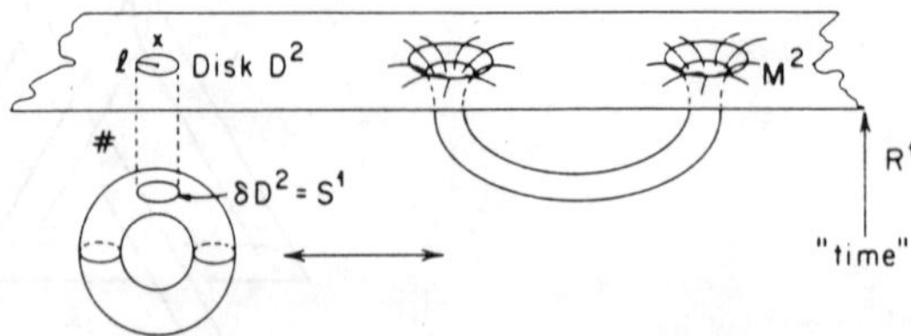

Fig. 4. Creation of wormholes by means of the connected sum $M^2 \# (S^1 \times S^1)$.

i.e., two unknotted and unlinked handles have been created. This becomes evident from the sequence of isotopies indicated in Figure 5, which forms the clue to this famous example [1]. The unknotting can be prevented, if, for example, another link component is inserted (Figure 6, W_6).

Denoting a *generalized wormhole* of order (μ, c) by

$$\tilde{W}_{\mu,c} \equiv M^n_{\text{"phys"}} \#^c \tilde{L}_\mu(W^n) \tag{4.4}$$

in accordance with the above example the process of *"wormhole scattering"* will be more generally introduced by

$$\tilde{W}_{\mu_1,c_1} \cup \tilde{W}_{\mu_2,c_2} \longleftrightarrow \tilde{W}_{\mu_1,c_1} \#^\nu \tilde{W}_{\mu_2,c_2} = \bigcup_{i=1}^{m} \tilde{W}_{\kappa_i,e_i}, \quad 0 < c_1 + c_2 + \nu \tag{4.5}$$

(Notice that ν may become negative because some of the already existing connections are allowed to be "killed" again in this scheme.) As a result, the new links $\tilde{W}_{\kappa_i,e_i}$ will be differently knotted and tied together. The total connection number

$$c_{\text{tot}} = c_1 + c_2 - \sum_{i=1}^{m} e_i \tag{4.6}$$

will be conserved if the connections have only been rearranged.

With respect to the process (4.5) a basic invariant is the minimal number c^0 of connections that are necessary in order to split a given link in unknotted and unlinked wormholes, i.e., in the "basic constituents" of this theory:

$$\tilde{W}_{\mu,c^0} = \#^{c^0} W^n \tag{4.7}$$

In two dimensions, the minimal bridge number [47] b will be an upper bound:

$$c^0 \leqslant b \tag{4.8}$$

(Schubert, private communication 1974). Indeed, if a bridge presentation of $\tilde{L}_\mu(W^2)$ characterized by the torsion number α and the crossing class β has been chosen, and if the "surgery" is performed at each bridge once, the link will split into (4.7). A computation of b has been achieved only for a few cases. It should

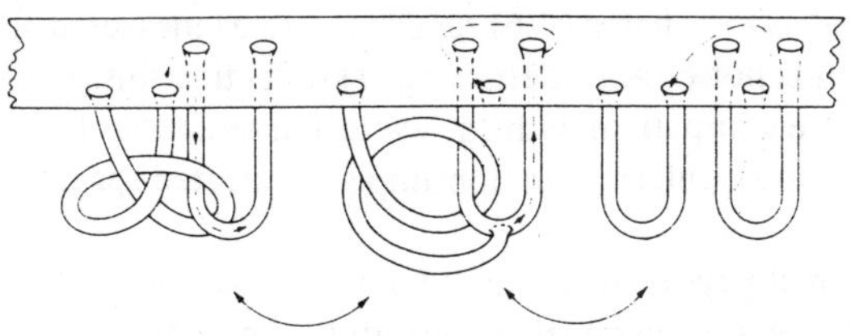

Fig. 5. Bing's unknotting example of a wormhole link.

be noticed that in the case of knots with two bridges (4.8) can be sharpened to $c^0 = b = 2$.

In $3 + 1$ dimensions it is even more difficult to identify c^0 with any known knot invariant. Because of (3.6), there are, for links $\tilde{L}_\mu(W^3)$ constructed from spun knots K_s^2, indications from unlinking theorems [52] that

$$c^0 = \mu_0 + 1 \tag{4.9}$$

where μ_0 is the Gordian number of $\tilde{L}_\mu(S^2)$. In this case, the bridge number could again form an upper bound. To see this, a rectangular bridge presentation with unequally distributed, parallel bridges can be spun such as to produce b nonintersecting "bridge tubes" in K_s^2. Doing surgery at these tubes by appropriate connections should yield the unknotting.

The question of whether topology change can actually occur in pseudo-Riemannian space-times carrying a metric of Lorentzian signature has been answered positively by introducing the concept of *Lorentz cobordism*. It consists, together with a vector field u, of a connected manifold C^4 whose boundary ∂C^4 is the disjoint union of two hypersurfaces M_{in}^3 and M_{out}^3 that have different global topologies as a result of wormhole scattering. It is essential that the *cobordism manifold* C^4 is not compact as has been required before [14]. Actually, the critical points have been removed [64] in order to show that an *open* C^4 (a) is stable causal, (b) is future causal geodesic complete, and (c) satisfies the weak energy condition

$$R_{\mu\nu}u^\mu u^\nu + \tfrac{1}{2}R > 0 \tag{4.10}$$

with respect to a timelike unit vector u.

Of course, that does not guarantee that the metrics (2.18) on C^4 satisfy the geometrodynamical equations (2.20)–(2.22). In the case in which a space-time admits two-parameter isometry groups (as for example an appropriate metric on $R \times W^3$ does), Gowdy [19] has deduced from the vacuum Einstein equations that a solution on C^4 is in conflict with his "corner theorem" if topology change occurs. (Whether this statement is true in less symmetric cases, as considered here, has not yet been analyzed.)

With respect to this issue Komorowski [30] has conjectured that there exists a continuous curve of metrics $g_{ab}(t)$ on M^3 (contained in a superspace endowed with a weak enough topology) such that the "radius" l of the connected sum (Figure 4) as well as the charge (2.6) tends to zero as measured from M_{phys}^3". However, again it is not clear if metrics $g_{ab}(t)$ reflecting this process of "handle disappearance" are compatible with the evolution equations (2.22) of GMD. For a recent account on singularities in nonsimply connected space-times see Gannon [13].

Thus, as the real payoff of this discussion one may regard the claim that one is forced to consider a probabilistic description of manifolds in order to resolve the dilemma of topology change.

§(5): *Quantum Geometrodynamics*

In a fully quantized version of geometrodynamics a tunneling through the classical "barrier" against topology change should be expected. The conceptually most concise framework in which this issue can analyzed is the superspace approach [57] to quantization.

In order to include the possibility of topological scattering processes, a "grand superspace" $S^{\#}$ (compare Brill [3]) may be constructed in the following manner: Consider the space $\mathfrak{M}^4$ of all pseudo-Riemannian metrics with Lorentzian signature on open cobordism manifolds C^4 and identify metrics that are equivalent with respect to the group $\mathfrak{D}^4$ of diffeomorphisms of C^4:

$$S(C^4) \equiv \mathfrak{M}^4/\mathfrak{D}^4 \tag{5.1}$$

The restriction on the domain

$$S^{\#} \equiv S(C^4)|_{\partial C^4} \tag{5.2}$$

is a superspace on 3-manifolds in the sense of Fischer (1970), but it is "grand" enough to allow for topology change. Moreover, if one had started from the manifold $S^{\text{ext}}(C^4)$, a supermetric [7]

$$G_{abce} \equiv \tfrac{1}{2} g^{-1/2}(g_{ac}g_{be} + g_{ae}g_{bc} - g_{ab}g_{ce}) \tag{5.3}$$

would have been induced on $S^{\#\,\text{ext}}$, which could account for the "closeness" of changing topologies. On "grand" superspace a probability amplitude function

$$\psi \equiv \psi({}^{(3)}G_T;\, \widetilde{W}_{\mu_1,c_1};\, \ldots;\, \widetilde{W}_{\mu_k,c_k}) \tag{5.4}$$

can be properly defined, where ${}^{(3)}G_T \subset S^{\#}$ is a one-parameter curve of geometries. For the sake of distinctness, the dependence of ψ upon the global topology and the embedding of the underlying 3-manifold has been indicated explicitly (Figure 6 visualizes this in a lower-dimensional space-time). The Hamiltonian constraint in QGMD corresponding to the initial value equation (2.21) of empty space-time is the Wheeler–DeWitt equation:

$$\mathcal{H}\psi \equiv \left(G_{abce}\, \frac{\delta^2}{\delta g_{ab}\delta g_{ce}} - \sqrt{g}\,{}^{(3)}R \right)\psi = 0 \tag{5.5}$$

Electric and magnetic fields that are derived from a vector potential A can be included in this framework [4]. The extension of (5.5) to a "time"-dependent Einstein–Schrödinger equation

$$i(\partial/\partial \tau)\psi = \mathcal{H}'\psi \tag{5.6}$$

seems to be feasible with respect to superspace defined on asymptotically flat spaces [43].

In order to develop a tentative concept of the quantum statistics of worm-

190   MIELKE

holes, the *"geometrodynamical vacuum"*

$$H_0 \equiv \{\psi(^{(3)}\mathcal{G}_T)|\psi \in H^{\Phi}\} \qquad (5.7)$$

will be defined. It consists of the probability amplitudes of all metrics on the simply connected "enveloping" space M_0^3, which has no wormholes attached (however, the "vacuum sea" of virtual wormholes is present!). From the space

$$H_1 \equiv \{\psi(^{(3)}\mathcal{G}_T; W_1)|\psi \in H^{\Phi}\} \qquad (5.8)$$

of *"one-wormhole-states,"* an exterior algebra [22] over H_1 can be built up:

$$H^{\Phi} = \sum_{k=0}^{\infty} \wedge^k H_1 \subset {}``H(\delta^{\#})" \qquad (5.9)$$

Although it is not known whether this space admits a Hilbert-space-like structure, it will be called the *antisymmetric $\Phi 0\kappa$-space of wormholes*. Using the notation of the occupation-number formalism of quantum statistics

$$|0\rangle \equiv \psi(^{(3)}\mathcal{G}_T)$$
$$\qquad (5.10)$$
$$|w_1, \ldots, w_k, \ldots\rangle \equiv \psi(^{(3)}\mathcal{G}_T; W_1; \ldots; W_k; \ldots)$$

where $w_k = 0, 1$ are the occupation numbers of (trivially embedded) wormholes W_k, the topology change of the underlying manifold may be represented by *wormhole creation operators*

$$C_\alpha|w_1, \ldots, w_\alpha, \ldots\rangle \equiv (-1)^{P_\alpha}(1 - w_\alpha)|w_1, \ldots, w_\alpha + 1, \ldots\rangle \qquad (5.11)$$

where

$$P_\alpha = \sum_{k=1}^{\alpha-1} w_k \qquad (5.12)$$

is equal to the number of occupied holes up to the αth order, and by *wormhole annihilation operators*

$$A_\beta|w_1, \ldots, w_\beta, \ldots\rangle \equiv (-1)^{P_\beta} w_\beta|w_1, \ldots, w_\beta - 1, \ldots\rangle \qquad (5.13)$$

The algebra of these operators

$$[A_\alpha, A_\beta]_+ = [C_\alpha, C_\beta]_+ = 0$$
$$\qquad (5.14)$$
$$[A_\alpha, C_\beta]_+ = \delta_{\alpha\beta}$$

is of course that of the CAR. Since (5.14) can be regarded [8] as derived from the antisymmetrization postulate (5.9), it should be analyzed if and how the latter can be abstracted from topology.

The most promising line of approach [56] is to introduce a spin structure with group SU_2 on the tangent bundle of $M^3 = \partial C^4$ and to employ the concept

KNOT WORMHOLES IN GEOMETRODYNAMICS?   191

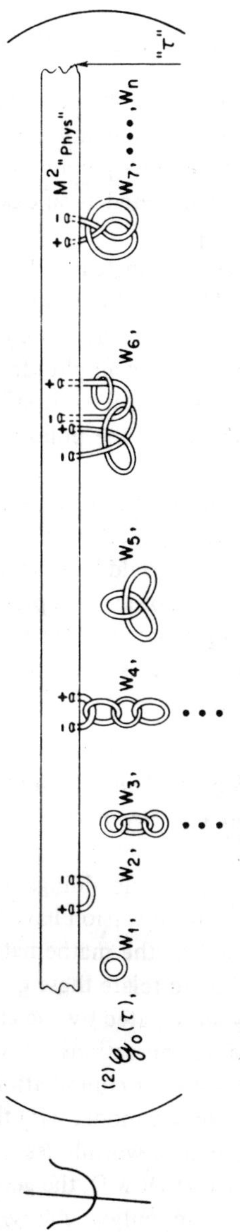

Fig. 6. Wormhole probability amplitude function.

of spin cobordism [34, 61] (compare Kiehn [29] for a related point of view).
This is possible in exactly $s = \dim H^1(M^3, Z/2)$ ways if the Stiefel–Whitney class
ω_2 of M^3 is zero. As an example, one wormhole W^3 will carry

$$s(W^3) = \dim H^1(S^1, Z/2) \otimes H^1(S^2, Z/2) = 2 \tag{5.15}$$

different spin structures. By introducing the occupation numbers $w_k^\uparrow, w_k^\downarrow = 0, 1$,
appropriate generalized creation and annihilation operators will account for that.
If this vague argument can be made more precise in order to yield a derivation
of (5.14), this equation may be considered as an example of "pregeometry as the
calculus of propositions" [39, 40].

Furthermore, if the generalized Einstein–Schrödinger equation (5.6) of
Christodoulou [4] can be solved by the "time"-dependent creation and annihilation operators of the interaction picture, the route of quantum statistics is open
to a formal (!) construction of scattering amplitudes, Green's, correlation, and
density functions. A "geometrodynamical exciton" [56], the quasiparticle of
this theory, would be characterized by the poles in the wormhole Green's
function.

Admittedly, only a preliminary step towards a more rigorous foundation
of a probabilistic theory of "topology scattering" has been achieved. Much beyond the scope of the present work would be to include the scattering of generalized wormholes in this scheme in order to represent the change of knot invariants during this process by algebraic relations of the wormhole operators.
An S-matrix theory of knot wormholes is as yet not known! (See Milnor [35]
for "statistical" results in knot theory.)

§(6): *Outlook: Topological Model of Elementary Particles in the Framework of Geometrodynamics?*

So far, the varieties of topologies, embeddings, and their changes that are
compatible with GMD or its quantized version have been analyzed without any
need for a physical interpretation of the mathematical phenomena occurring.
However, the ultimate goal will be to relate these geometric effects to elementary particle physics, as already anticipated by the choice of a suggestive physical
terminology for some mathematical operations.

Essentially three schemes (or proper combinations of them) seem to be the
most promising: (I) Wheeler's exciton model, (II) the f-wormhole model of
hadrons, and (III) flux quantization of wormholes.

(I) In the exciton model [56] of QGMD, the geometry of the enveloping
manifold is on the average Euclidean, only at a closer look will the foam-like
structure due to "birth and death" of generalized wormholes be revealed. According to the theory, this is expected to happen at spacelike separations com-

KNOT WORMHOLES IN GEOMETRODYNAMICS? 193

parable to the Planck length

$$L^* = \sqrt{\hbar G/c^3} = 1.616 \times 10^{-33} \text{ cm} \qquad (6.1)$$

For effective charges of the order

$$q = \sqrt{\hbar c} \sim 12e \qquad (6.2)$$

the electric mass-energy of the wormhole will be

$$m_W(G) = \hbar/cL^* = \sqrt{\hbar c/G} \sim 10^{28} \text{ eV} \qquad (6.3)$$

However, the gravitational self-energy

$$E_g \sim - Gm_w^2/L^* \qquad (6.4)$$

of nearby holes will "renormalize" the "bare" wormhole mass m_W by a mass shift

$$\delta m_W \sim E_g/c^2 = m_W \qquad (6.5)$$

which is of the same order of magnitude. This makes it possible to think of a particle as a weak alteration of the pattern of bare wormholes over a region of 10^{-14} cm in $M_{\text{«phys»}}$, i.e., as a geometrodynamical exciton following the viewpoint devised at the end of Section 5.

(II) For a direct particle interpretation of a handle, its mass has to be scaled to that of, for example, a proton

$$m_W(G_f) = m_P \qquad (6.6)$$

by an appropriate choice of a new gravitational constant G_f. Exactly this suggestion has been made in (strong) f-gravity, which is characterized by a corresponding Planck length of the order

$$L_f^* = 2.103 \times 10^{-14} \text{ cm} \qquad (6.7)$$

For the aspects discussed here, this is the relevant change made in f-gravity (except for a g-f mixing term in the Lagrangian, which, together with the "mass" of the f-meson, will be neglected). Thus, one might speculate with Salam [46] that hadronic matter is made out of generalized f-wormholes in the context of f-QGMD! Could their linkage and knotting resolve the dilemma of quark confinement inside normal hadrons?

For a further study of this issue, the interpretation of knot invariants according to the flux quantization approach [26] to elementary particles will be tentatively taken up. There, spinning magnetic flux lines are attached to the "hard core" of a particle (in sharp contrast to the concept of particles made out of "pure geometry"). Those with the shape of torus knots with winding numbers $(2, 1)$, $(3, 1)$, and $(3, 2)$ (i.e., a trefoil knot) are interpreted as n, p and λ quarks respectively. These assignments are not directly transferable to knotted $\widetilde{W}^3$

since the spun knots K_s^2 corresponding to the torus knots $(2, 1)$ and $(3, 1)$ are trivially embedded because of (3.6). A further transfer of Jehle's model to f-geometrodynamics would yield the result that "scattering" of f-wormholes corresponds to strong interaction, whereas weak interaction is associated with a "merging process" (4.5) with c_{tot} = const accompanied by an unknotting. Consequently, the minimal connection number c_0 would be related to the strangeness of hadrons. However, it is not at all clear whether this is the most natural interpretation in f-QGMD.

(III) With respect to wormholes with quantized flux, the phase function θ of the geometrodynamical amplitude ψ will be assumed to be single-valued modulo $2\pi n$ on closed loops on M_c^3. Following Jehle, the flux is quantized through the following choice of a pseudo-gauge-transformation:

$$A_\mu = \hbar c/e\, \partial_\mu \theta \tag{6.8}$$

This *singular* vector potential leads, by means of (2.6), to the result that the electric charge is quantized through

$$q = ne \tag{6.9}$$

without having the need to introduce magnetic monopoles into the theory (compare references 49 and 23 for an elaborate discussion of this subject).

Most of the other interpretations of knots are inherent in Jehle's model and cannot be used in GMD with one exception, namely, that the existence of parity violating neutrinos and antineutrinos may be a reflection of nonamphicheiral embeddings of the underlying space (consider, for example, the left- and right-handed trefoil wormholes $\tilde{W}^2$ in Figures 2b and 2$\bar{\text{b}}$). The idea that such topological asymmetries can be the "metaphysical" reason for the observed P- or T-violation together with CPT invariance of elementary particles has also been stressed by Sakharov [45].

Thus, there seems to be a promising program for future work that must be done before any ultimate conclusions [50] may be drawn on the view: "Everything is Geometry" [56]!

Acknowledgments

The author would like to express his sincere gratitude to Professor J. A. Wheeler for the encouragement to study within the stimulating atmosphere of the Princeton Relativity Group during the academic year 1973-74 and for his continuous support. Furthermore, the very inspiring lecture of Professor R. Fox in the Fall of 1973 was very important for this work. Much appreciation is expressed to Professors D. Brill, R. Fox, F. W. Hehl, C. D. Papakyriakopoulos, D. W. Sumners, J. A. Wheeler, Dr. W. Deppert, and D. Četković for making valuable reading suggestions and useful hints in the course of the preparation of this paper. The author is also indebted to D. Ćetković for providing him with a translation of the Russian original of reference 45.

References

1. Bing, R. H. (1965). In *Topology Seminar, Wisconsin*, eds. Bing, R. H., and Bean, R. J. (Princeton University Press, Princeton, N.J.), p. 89.
2. Bing, R. H., and Martin, J. M. (1971). *Trans. Am. Math. Soc.*, **155**, 217.
3. Brill, D. (1972). In *Magic without Magic: John Archibald Wheeler*, ed. Klauder, J. R. (W. H. Freeman and Co., San Francisco), p. 309.
4. Christodoulou, D. (1975). *Nuovo Cimento*, **26B**, 335.
5. Conway, J. H. (1970). In *Computational Problems in Abstract Algebra*, ed. Leech, J. (Pergamon Press, New York), p. 329.
6. Crowell, R. H., and Fox, R. H. (1963). *Introduction to Knot Theory*. (Ginn and Co., Boston).
7. DeWitt, B. S. (1970). In *Relativity*, eds. Carmeli, M., Fickler, S. I., and Witten, L. (Plenum Press, New York), p. 359.
8. Feynman, R. P. (1972). *Statistical Mechanics*. (W. A. Benjamin, New York).
9. Finkelstein, D., and Rubinstein, J. (1968). *J. Math. Phys.*, **9**, 1762.
10. Fischer, A. E. (1970). In *Relativity*, eds. Carmeli, M., Fickler, S. I., and Witten, L. (Plenum Press, New York), p. 303.
11. Fox, R. H. (1962). In *Topology of 3-Manifolds*, ed. Fort, M. K., Jr. (Prentice-Hall, Englewood Cliffs, N.J.), p. 120.
12. Fuller, R. W., and Wheeler, J. A. (1962). *Phys. Rev.*, **128**, 919.
13. Gannon, D. (1975). *J. Math. Phys.*, **16**, 2364.
14. Geroch, R. P. (1967). *J. Math. Phys.*, **8**, 782.
15. Geroch, R. (1966). *Ann. Phys. (N.Y.)*, **36**, 147.
16. Gluck, H. (1961). *Bull. Am. Math. Soc.*, **67**, 586.
17. Gluck, H. (1965). *Ann. Math.*, **81**, 195.
18. Goldsmith, D. L. (1975). In *Knots, Groups and 3-Manifolds*, Papers Dedicated to the Memory of R. H. Fox, ed. Neuwirth, L. P. (Princeton Univ. Press, Princeton, N.J.), p. 3.
19. Gowdy, R. H. (1974). *Ann. Phys. N.Y.*, **83**, 203.
20. Graves, J. C., and Brill, D. R. (1960). *Phys. Rev.*, **120**, 1507.
21. Graves, J. C. (1971). *The Conceptual Foundation of Contemporary Relativity Theory*. (MIT Press, Cambridge).
22. Greub, W., Halperin, S., and Vanstone, R. (1972). *Connections, Curvature, and Cohomology*, Vol. I. (Academic Press, New York).
23. Greub, W., and Petry, H.-R. (1975). *J. Math. Phys.*, **16**, 1347.
24. Grünbaum, A. (1973). *J. Phil.*, **70**, 775.
25. Hu, S.-T. (1959). *Homotopy Theory*. (Academic Press, New York).
26. Jehle, H. (1975). *Phys. Rev. D*, **11**, 2147.
27. Kanitscheider, B. (1971). *Geometrie und Wirklichkeit*, Erfahrung und Denken Band 36. (Duncker & Humblot, Berlin).
28. Kauffman, L. H. (1974). *Michigan Math. J.*, **21**, 33.
29. Kiehn, R. M. (1975). *Nuovo Cimento Lett.*, **12**, 300.
30. Komorowski, J. (1973). In: The Greek Mathematical Society C. Carathéodory Symposium, p. 318.
31. Lindquist, R. W., and Wheeler, J. A. (1957). *Rev. Mod. Phys.*, **29**, 432.
32. Mielke, E. W. (1974). *Bull. Am. Phys. Soc.*, **19**, 508.
33. Milnor, J. (1962). *Am. J. Math.*, **84**, 1.
34. Milnor, J. (1963). *L'Enseignment Mathematique*, **9**, 198.
35. Milnor, J. (1964). *Fundam. Math.*, **54**, 335.
36. Misner, C. W. (1960). *Phys. Rev.*, **118**, 1110.
37. Misner, C. W. (1963). *Ann. Phys. (N.Y.)*, **24**, 102.

38. Misner, C. W., and Wheeler, J. A. (1957). *Ann. Phys. (N.Y.)*, **2**, 525, reprinted in Wheeler, J. A. (1962). *Geometrodynamics*. (Academic Press, New York).
39. Misner, C. W., Thorne, K. S., and Wheeler, J. A. (1973). *Gravitation*. (A. W. Freeman and Co., San Francisco) (quoted as MTW).
40. Patton, C. M., and Wheeler, J. A. (1975). In *Quantum Gravity*, eds. Isham, C. J., Penrose, R., and Sciama, D. W. (Oxford University Press, London), p. 538.
41. Penrose, R. (1975). In *Quantum Gravity*, eds. Isham, C. J., Penrose, R., and Sciama, D. W. (Oxford University Press, London), p. 268.
42. Price, P. B., Shirk, E. K., Osborne, W. Z., and Pinsky, L. S. (1975). *Phys. Rev. Lett.*, **35**, 487.
43. Regge, T., and Teitelboim, C. (1974). *Ann. Phys. N.Y.*, **88**, 286.
44. Reidemeister, K. (1932). *Knotentheorie*, in Ergebnisse der Mathematik, Vol. 1. (Chelsea Publishing Company, New York).
45. Sakharov, A. D. (1972). "The Topological Structure of Elementary Charges and CPT-Symmetry," in *"Problems of Theoretical Physics, A Memorial Volume to Igor E. Tamm."* ("NAUKA", Moscow, 1972), p. 243.
46. Salam, A. (1975). In *Quantum Gravity*, eds. Isham, C. J., Penrose, R., and Sciama, D. W. (Oxford University Press, London), p. 500.
47. Schubert, H. (1956). *Math. Z.*, **65**, 133.
48. Seifert, H., and Threlfall, W. (1934). *Lehrbuch der Topology*. (Chelsea Publishing Company, New York).
49. Strazhev, V. I., and Tomil'chik, L. M. (1973). *Sov. J. Part. Nucl.*, **4**, 78.
50. Stachel, J. (1974). in *Boston Studies in the Philosophy of Science*, Volume XX, Schaffner, K. F., and Cohen, R. S., eds. (D. Reidel Publishing Company, Dordrecht-Holland), p. 31.
51. Sumners, D. W. (1971). *Comm. Math. Helv.*, **46**, 240.
52. Sumners, D. W. (1972). *Proc. Cambridge Phil. Soc.*, **71**, 1.
53. Waldhausen, F. (1968). *Topology*, **7**, 195.
54. Westenholz, C. (1971). *Ann. Inst. Henri Poincaré*, **XV**, 182.
55. Wheeler, J. (1962). In *Logic, Methodology and Philosophy of Science: Proceedings of the 1960 International Congress*, eds. Nagel, E., Suppes, P., and Tarski, A. (Stanford University Press, Stanford, Calif.), p. 361.
56. Wheeler, J. A. (1968). *Einsteins Vision*, (Springer-Verlag, Berlin) contained in part in *Battelle Rencontres: 1967 Lectures in Mathematics and Physics*, DeWitt, C., and Wheeler, J. A., eds. (W. A. Benjamin, New York).
57. Wheeler, J. A. (1970a). In *Analytic Methods in Mathematical Physics*, eds. Gilbert, R. P., and Newton, R. (Gordon and Breach, New York), p. 335.
58. Wheeler, J. A. (1970b). In *Relativity*, eds. Carmeli, M., Fickler, S. I., and Witten, L. (Plenum Press, New York), p. 31.
59. Wheeler, J. A. (1973). In *The Physicist's Conception of Nature*, ed. Mehra, J., (D. Reidel Publishing Company, Dordrecht-Holland), p. 202.
60. Whiston, G. S. (1973). *Int. J. Theor. Phys.*, **3**, 99.
61. Whiston, G. S. (1975). *Int. J. Theor. Phys.*, **12**, 225.
62. Yanagawa, T. (1969). *Osaka J. Math.*, **6**, 447.
63. Yanagawa, T. (1970). *Osaka J. Math.*, **7**, 165.
64. Yodzis, P. (1973). *Gen. Rel. Grav.*, **4**, 299.

Helicity and the Călugăreanu invariant†

By H. K. Moffatt and Renzo L. Ricca‡

Department of Applied Mathematics and Theoretical Physics, Silver Street,
Cambridge CB3 9EW, U.K.

The helicity of a localized solenoidal vector field (i.e. the integrated scalar product of the field and its vector potential) is known to be a conserved quantity under 'frozen field' distortion of the ambient medium. In this paper we present a number of results concerning the helicity of linked and knotted flux tubes, particularly as regards the topological interpretation of helicity in terms of the Gauss linking number and its limiting form (the Călugăreanu invariant). The helicity of a single knotted flux tube is shown to be intimately related to the Călugăreanu invariant and a new and direct derivation of this topological invariant from the invariance of helicity is given. Helicity is decomposed into writhe and twist contributions, the writhe contribution involving the Gauss integral (for definition, see equation (4.8)), which admits interpretation in terms of the sum of signed crossings of the knot, averaged over all projections. Part of the twist contribution is shown to be associated with the torsion of the knot and part with what may be described as 'intrinsic twist' of the field lines in the flux tube around the knot (see equations (5.13) and (5.15)). The generic behaviour associated with the deformation of the knot through a configuration with points of inflexion (points at which the curvature vanishes) is analysed and the role of the twist parameter is discussed. The derivation of the Călugăreanu invariant from first principles of fluid mechanics provides a good demonstration of the relevance of fluid dynamical techniques to topological problems.

1. Introduction

The purpose of this paper is to gather together a number of results concerning the *helicity* $\mathscr{H}$ of an arbitrary solenoidal vector field $B(x)$ confined to knotted or linked tube-like structures, particularly as regards its topological interpretation. This interpretation is straightforward when the field B is confined to two linked oriented flux tubes carrying fluxes Φ_1 and Φ_2: provided each tube is unknotted and the field lines within each tube are unlinked closed curves circulating once parallel to the tube axis, the helicity is given by

$$\mathscr{H} = 2n\Phi_1\Phi_2,\qquad(1.1)$$

where n (an integer, positive, negative or zero) is the (Gauss) linking number of the two tubes (Moreau 1961; Moffatt 1969). If, however, B is confined to a single knotted

---

† This paper was accepted as a rapid communication.

‡ Present address: Department of Mathematics, University College London, Gower Street, London WC1E 6BT, U.K.

Proc. R. Soc. Lond. A (1992) **439**, 411–429

© 1992 The Royal Society

Printed in Great Britain 411

412   *H. K. Moffatt and R. L. Ricca*

flux tube with flux Φ, then the helicity is related to the topology of the knot in a more subtle way. On purely dimensional grounds, a result of the form

$$\mathscr{H} = h\Phi^2, \tag{1.2}$$

where h is a real number determined partly by the topology of the knot and partly by the twist of the field B within the knot tube, is to be expected. Since this twist can be arbitrarily prescribed (equivalent to an arbitrary 'framing' of the knot), the number h can take any value, positive or negative. Nevertheless, the fact that h is then invariant under topological deformation of the knot tube and the field within it, does carry some important information about the knot itself.

This type of problem appears to have been first addressed by Călugăreanu (1959, 1961; hereafter referred to as C59 and C61). Călugăreanu considered two neighbouring closed curves C, C^* forming the boundaries of a (possibly knotted) ribbon of small spanwise width ϵ, and showed that the linking number n of C and C^* can be expressed in the form

$$n = \mathscr{W} + \mathscr{T} + \mathscr{N} \tag{1.3}$$

(this equation is given at the end of p. 613 of C61), where $\mathscr{W}$ and $\mathscr{T}$ are respectively the writhe and the normalized total torsion of C (for definition of these quantities, see (4.8) and (5.13) below), and $\mathscr{N}$ is an integer representing the number of rotations of the unit spanwise vector N on the ribbon relative to the Frenet pair (n, b) (unit principal normal and unit binormal) in one passage round C. For the moment, we simply note that $\mathscr{T}$ and $\mathscr{N}$ are well defined only if C has no point of inflexion (i.e. no point at which the curvature vanishes). If C is continuously deformed through an inflexional configuration (i.e. through a state that does contain a point of inflexion) then, as will be shown in §6 below, $\mathscr{T}$ is discontinuous by ± 1, but $\mathscr{N}$ is simultaneously discontinuous by an equal and opposite amount ∓ 1 as a result of the discontinuous behaviour of the Frenet pair (n, b) in going through the inflexion, so that the sum $\mathscr{T} + \mathscr{N}$ does vary continuously.

The difficulty associated with inflexion points was recognized by Călugăreanu (see the long footnote on p. 8 of C59) and was to some extent resolved through consideration of a particular example of deformation through an inflexional configuration in C61 (pp. 616–617). Deformations of this type were described as 'degenerate' by Pohl (1968, p. 83); in fact, as pointed out by Ricca & Moffatt (1992), any deformation whose projection on any plane involves a type I Reidemeister move (see, for example, Kauffman 1991, p. 16) must involve passage through an inflexional configuration. General deformations (or 'ambient isotopies') do therefore typically involve such passage and consideration of the associated behaviour of $\mathscr{T}$ and $\mathscr{N}$ cannot be avoided.

The concept of the self-linking number SL of a closed curve C having no points of vanishing curvature was introduced by Pohl (1968). SL is defined as half the sum of the indices of the cross-tangents of C (i.e. the tangents of C which intersect C in a point distinct from the point of tangency). Pohl showed that SL is an integer, and he proved that

$$SL = \mathscr{W} + \mathscr{T}. \tag{1.4}$$

Under *regular isotopy* (i.e. continuous deformation of C not passing through any inflexional configuration), SL is invariant so that Pohl's result provides an alternative proof of the invariance under regular isotopy of $\mathscr{W} + \mathscr{T}$, as proved in C59. Pohl's work was extended to higher dimensions by White (1969), who again restricted

Helicity and the Călugăreanu invariant **413**

consideration to regular isotopy (which he described (p. 179) as 'non-degenerate isotopy').

A third strand of inquiry was introduced by Fuller (1971) who defined the total twist number Tw for a ribbon by

$$Tw = \frac{1}{2\pi} \oint_C (N' \times N) \cdot t\, ds, \tag{1.5}$$

where $N' = dN/ds$. It is easy to show (see §5 below) that

$$Tw = \mathcal{T} + \mathcal{N}, \tag{1.6}$$

(a result nowhere actually stated by Fuller). Fuller then *defines* the writhe $\mathcal{W}$ through

$$\mathcal{W} = n - Tw, \tag{1.7}$$

where n is the linking number of the curves C, C^* bounding the ribbon, but nowhere does he prove that this $\mathcal{W}$ is the same as that defined by the Gauss integral (equation (4.8) below). This identification is however established by Călugăreanu's result (1.3) in conjunction with (1.6). The definition (1.5) of Tw provides a quantity which evidently varies continuously under all continuous deformations of the ribbon (i.e. under ambient isotopy).

Curiously, Fuller (1972), in a paper dedicated to Călugăreanu (on his 70th anniversary), gives White (1969) the credit for the result $n = \mathcal{W} + Tw$, although this result (with $Tw = \mathcal{T} + \mathcal{N}$) can be found clearly stated, and coupled with a tentative discussion of the role of inflexion points, in C61. White's achievement was to place this result in the wider context of differentiable manifolds of arbitrary dimension; but the theorem in the form (1.3), or in the equivalent form

$$n = \mathcal{W} + Tw \tag{1.8}$$

should surely be described as Călugăreanu's theorem.

We feel it necessary to emphasize this point because in some more recent papers and text books, Călugăreanu is given less than due credit for his achievement. Thus, for example, Pohl (1980) describes (1.8) as 'White's formula', and only rather grudgingly states that 'White's formula was actually put forward by Georges Călugăreanu (1961), originally, for curves C having nowhere vanishing curvature. This proof was very complicated and his formulation confusing...'. We question this judgement and would simply reiterate that Călugăreanu (1961) explicitly considers the zero curvature, or inflexional, problem, whereas White (1969) explicitly excludes such considerations. A general misunderstanding of Călugăreanu's contribution has gradually led people to refer to equation (1.8) as 'White's theorem', so that even in text books (e.g. Kauffman 1987, p. 18; 1991, p. 489), references to C59 and C61 have gradually disappeared.

Our aim in the present paper is to show that all of these results can be obtained in a straightforward manner starting from the helicity invariant of classical fluid dynamics. The link between helicity and the Călugăreanu invariant was conjectured by Moffatt (1981) and was developed on the basis of the result (1.8) by Berger & Field (1984). However, a direct derivation of (1.8) from the invariance of helicity has not previously been given. We provide such a direct proof in §§2–6 of this paper. First, in §2, basic results concerning the helicity of linked flux tubes are stated. Then in §3, it is shown that for a knotted flux tube constructed in such a way that the field lines are closed satellites of C, each pair of field lines having linking number n, the helicity

                         H. K. Moffatt and R. L. Ricca

is given by $\mathcal{H} = n\Phi^2$ (the proof was given by Ricca & Moffatt (1992), but is repeated here for completeness). In §4, the helicity is decomposed into writhe and twist contributions, the writhe contribution involving the Gauss integral (4.8), which admits interpretation in terms of the sum of signed crossings of C averaged over all projections. In §5, the twist contribution is considered, part of this being associated with torsion of C and part with what may be described as 'intrinsic twist' of the field lines in the flux tube around C. In §6 the generic behaviour associated with inflexional configurations is analysed and the role of the twist parameter is discussed in §7. Finally in §8 we summarize the conclusions. Our hope is that the alternative proof presented in this paper and the associated discussion may help to demonstrate the relevance of fluid dynamical techniques to topological problems.

2. The helicity of linked flux tubes

Consider an arbitrary solenoidal vector field $\boldsymbol{B}(\boldsymbol{x}) = \nabla \times \boldsymbol{A}(\boldsymbol{x})$ of compact support $\mathcal{D}$ in $\mathbb{R}^3$. We suppose that $\boldsymbol{n} \cdot \boldsymbol{B} = 0$ on $\partial\mathcal{D}$, the boundary of $\mathcal{D}$. The helicity $\mathcal{H}$ of $\boldsymbol{B}$ is then the pseudo-scalar quantity defined by

$$\mathcal{H} = \int_{\mathcal{D}} \boldsymbol{A} \cdot \boldsymbol{B} \, \mathrm{d}V, \tag{2.1}$$

where $\mathrm{d}V$ is the volume element d^3x. Note immediately that $\mathcal{H}$ does not depend on the gauge of $\boldsymbol{A}$; for if $\boldsymbol{A}$ is replaced by $\boldsymbol{A} + \nabla\psi$, then $\mathcal{H}$ is unchanged since

$$\int_{\mathcal{D}} \boldsymbol{B} \cdot \nabla\psi \, \mathrm{d}V = \int_{\partial\mathcal{D}} \boldsymbol{n} \cdot \boldsymbol{B}\psi \, \mathrm{d}S = 0. \tag{2.2}$$

If we adopt the Coulomb gauge for $\boldsymbol{A}$ (i.e. $\nabla \cdot \boldsymbol{A} = 0$) and impose the further condition $\boldsymbol{A} = O(|\boldsymbol{x}|^{-3})$ as $|\boldsymbol{x}| \to \infty$, then $\boldsymbol{A}(\boldsymbol{x})$ is given by the Biot–Savart law:

$$\boldsymbol{A}(\boldsymbol{x}) = \frac{1}{4\pi} \int \frac{\boldsymbol{B}(\boldsymbol{x}^*) \times (\boldsymbol{x} - \boldsymbol{x}^*)}{|\boldsymbol{x} - \boldsymbol{x}^*|^3} \, \mathrm{d}V^*, \tag{2.3}$$

so that, from (2.1),

$$\mathcal{H} = \frac{1}{4\pi} \iint \frac{[\boldsymbol{B}(\boldsymbol{x}) \times \boldsymbol{B}(\boldsymbol{x}^*)] \cdot (\boldsymbol{x} - \boldsymbol{x}^*)}{|\boldsymbol{x} - \boldsymbol{x}^*|^3} \, \mathrm{d}V \, \mathrm{d}V^*. \tag{2.4}$$

Consider now the special situation in which $\boldsymbol{B}$ is zero except in two flux filaments centred on two unknotted oriented closed curves C_1, C_2 which may be linked (figure 1). We may suppose that the cross sections of the filaments are small, and that they carry fluxes $\delta\Phi_1$, $\delta\Phi_2$. We suppose further that, within each filament, the $\boldsymbol{B}$-lines are themselves unlinked curves which close on themselves after just one passage round the filament, running 'parallel' to C_1, C_2 respectively. In these circumstances, $\mathcal{H}$ may be evaluated directly from (2.1): integrating first over the cross section, $\boldsymbol{B}\,\mathrm{d}V \to \delta\Phi_1 \, \mathrm{d}\boldsymbol{x}_1$, $\delta\Phi_2 \, \mathrm{d}\boldsymbol{x}_2$ on C_1, C_2 respectively, so that

$$\mathcal{H} = \delta\Phi_1 \oint_{C_1} \boldsymbol{A} \cdot \mathrm{d}\boldsymbol{x}_1 + \delta\Phi_2 \oint_{C_2} \boldsymbol{A} \cdot \mathrm{d}\boldsymbol{x}_2. \tag{2.5}$$

Now

$$\oint_{C_1} \boldsymbol{A} \cdot \mathrm{d}\boldsymbol{x}_1 = \int_{D_1} \boldsymbol{B} \cdot \boldsymbol{n} \, \mathrm{d}S, \tag{2.6}$$

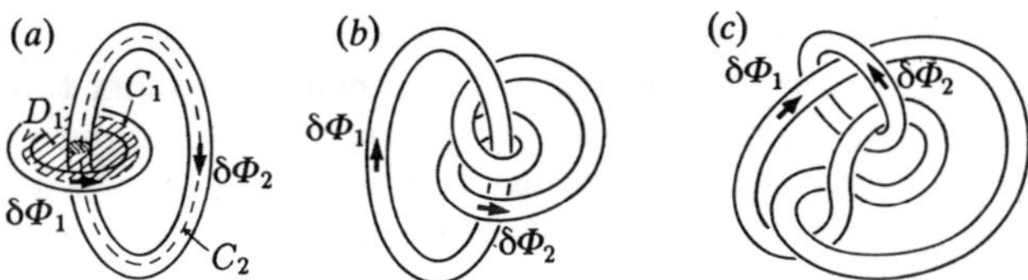

Figure 1. Linked, oriented and unknotted flux tubes with no internal contribution to helicity. In this case $\mathscr{H} = 2n\Phi_1\Phi_2$, where n is the (Gauss) linking number of the two tube axes. (a) $n = +1$; (b) $n = -2$; (c) $n = 0$.

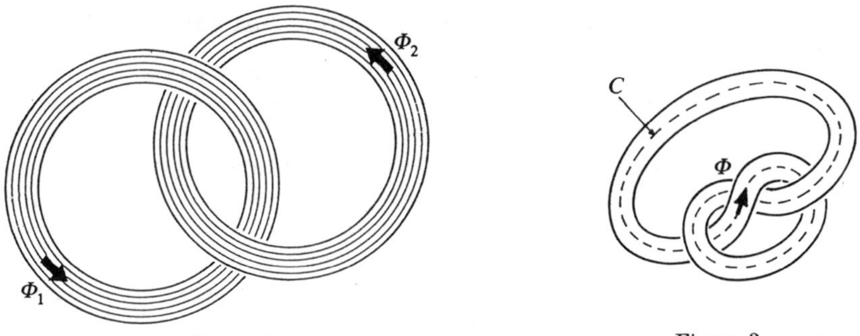

Figure 2   Figure 3

Figure 2. Two linked oriented flux tubes, each one of which is made up of a large number of filaments of small cross section. Each pair of filaments (one from each tube) makes a contribution $2n\,\delta\Phi_1\,\delta\Phi_2$ to the total helicity, and the total helicity is $2n\Phi_1\Phi_2$.

Figure 3. Knotted flux tube whose tube axis is a trefoil knot.

where D_1 is the open disc surface spanning C_1. Moreover

$$\int_{D_1} \boldsymbol{B}\cdot\boldsymbol{n}\,\mathrm{d}S = n\,\delta\Phi_2, \tag{2.7}$$

where n is the (Gauss) linking number of $\{C_1, C_2\}$, i.e. the algebraic number of times that C_2 crosses D_1 (allowing for direction of crossing). Three examples are shown in figure 1. Similarly,

$$\oint_{C_2} \boldsymbol{A}\cdot\mathrm{d}\boldsymbol{x}_2 = n\,\delta\Phi_1, \tag{2.8}$$

and hence, from (2.5),

$$\mathscr{H} = 2n\,\delta\Phi_1\,\delta\Phi_2. \tag{2.9}$$

Thus $\mathscr{H}$ is determined solely by the two fluxes and the linking number of the two filaments.

In this derivation, it is essential that each flux tube should by itself have zero helicity and this is ensured by the above assumption that the $\boldsymbol{B}$-lines within either tube on its own are unlinked closed curves. In these circumstances also, the value of n is given from (2.4) by integrating over the two cross sections: allowing for the fact that we may have $\boldsymbol{x}\in C_1$, $\boldsymbol{x}^*\in C_2$ or $\boldsymbol{x}\in C_2$, $\boldsymbol{x}^*\in C_1$, we find again $\mathscr{H} = 2n\,\delta\Phi_1\,\delta\Phi_2$ with

$$n = \frac{1}{4\pi}\oint_{C_1}\oint_{C_2} \frac{(\mathrm{d}\boldsymbol{x}\times\mathrm{d}\boldsymbol{x}^*)\cdot(\boldsymbol{x}-\boldsymbol{x}^*)}{|\boldsymbol{x}-\boldsymbol{x}^*|^3}. \tag{2.10}$$

This is the well-known Gauss formula for n.

                    H. K. Moffatt and R. L. Ricca

The restriction to small cross sections of the two filaments is easily lifted. In the situation indicated by figure 2 in which $\boldsymbol{B}$ is confined to two flux tubes, in each one of which the $\boldsymbol{B}$-lines are again unlinked closed curves passing once (the long way) round the torus, we may regard each flux tube as made up of a large number of filaments of small cross section. Each pair of filaments (one from each tube) make a contribution $2n\,\delta\Phi_1\,\delta\Phi_2$ to the total helicity, so that summing over all such pairs, this is now given by

$$\mathcal{H} = 2n\Phi_1\,\Phi_2. \tag{2.11}$$

3. The helicity of a single knotted flux tube

If a flux tube is knotted (as for example in figure 3) then its axis C is necessarily a twisted closed curve in $\mathbb{R}^3$, and we cannot avoid consideration of the twist of $\boldsymbol{B}$ within the tube itself (which, as will become clear, may change as the flux tube is continuously deformed). It is useful first to define a standard procedure for the construction of a flux tube of prescribed helicity around any given knot K. Here we adopt the procedure of Moffatt (1990).

Suppose we deform the knot continuously to lie entirely in the (x, y) plane except at crossing points (the knot being viewed in projection) where we must allow indentations into $z > 0$ or $z < 0$ (figure 4). The crossings are labelled $+$ or $-$ according as the overpass must be rotated anticlockwise or clockwise to bring it into coincidence (complete with arrow) with the underpass. By a finite number of crossing switches (i.e. reflections of indentations), it is always possible to convert K to the unknot K_0 which may then be continuously deformed to the circle $C_0 : x^2 + y^2 = R^2$. Conversely, the circle C_0 may be converted to K by simply reversing these steps (i.e. deformation to K_0 followed by appropriate crossing switches).

Suppose then that we start with the circle C_0 and place around this a tubular neighbourhood of small cross section

$$T_0 : (r - R)^2 + z^2 < (\epsilon R)^2,$$

where $r = (x^2 + y^2)^{\frac{1}{2}}$. Within T_0, we now define in cylindrical polar coordinates (r, θ, z), a field

$$\boldsymbol{B}_0 = (0, 2\pi r\Phi/V, 0), \tag{3.1}$$

where $V = 2\pi^2\epsilon^2 R^3$ is the volume of T_0, and Φ (as may be easily verified) is the flux of $\boldsymbol{B}_0$ through any section of the tube. The field lines of $\boldsymbol{B}_0$ are thus unlinked circles near $r = R$. The helicity of the field is clearly zero.

We may now inject helicity (figure 5) by 'Dehn surgery', i.e. by cutting the tube at a section $\theta = $ const., twisting the free ends through a relative angle $2\pi h$, and reconnecting. We may suppose that the resulting twist is uniformly distributed round the tube. If h is an integer n_0 say, then each $\boldsymbol{B}$-line in the new flux tube is a closed curve in the form of a helix with axis the circle $r = R$, and each pair of $\boldsymbol{B}$-lines has linking number n_0. The helicity thus generated is given by

$$\mathcal{H}_0 = \int_0^\Phi 2n_0\,\phi\,\mathrm{d}\phi = n_0\,\Phi^2, \tag{3.2}$$

since we may build up the tube by increments $\mathrm{d}\phi$, the increment in helicity at each stage being $2n_0\,\phi\,\mathrm{d}\phi$, from (2.11).

We now propose to distort C_0 to the curve K_0 defined above, carrying the tube T_0

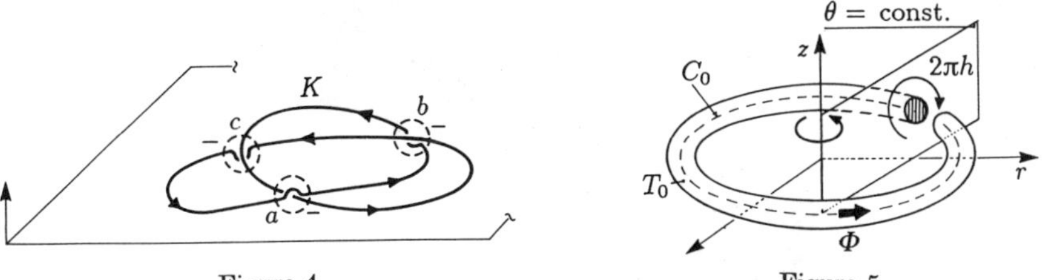

Figure 4   Figure 5

Figure 4. Oriented trefoil knot in a projection plane: at crossing points we must allow indentations into $z > 0$ (as at the point (a) in the figure) or $z < 0$ (as at points (b) and (c) in the figure).

Figure 5. Dehn surgery: 'cut' the tube at a section $\theta = $ const., 'twist' the free ends through a relative angle $2\pi h$ and then 'reconnect'.

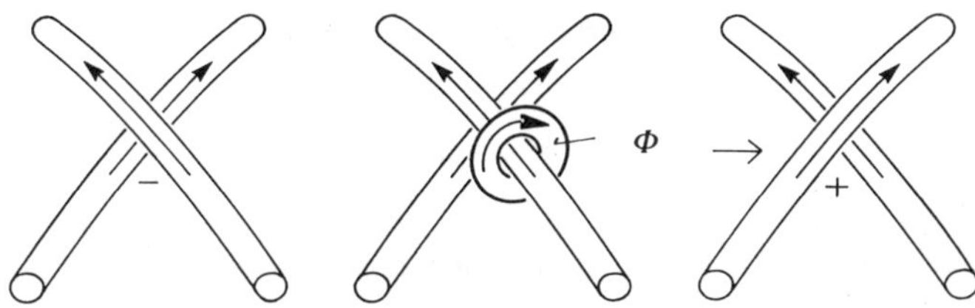

Figure 6. Each negative (positive) switch is equivalent to the insertion of a small loop of flux Φ annihilating flux on one side of the crossing and creating flux on the other side. In the figure the net increment of helicity is $+2\Phi^2$.

with it. To do this we must specify the isotopy that acts upon the field B_0. We picture the field as embedded in an incompressible fluid medium which moves with velocity $u(x,t)$ (where $\nabla \cdot u = 0$) carrying $B(x,t)$ with it according to the 'frozen field' equation

$$\partial B/\partial t = \nabla \times (u \times B). \tag{3.3}$$

It is well known that the flux of B through any material surface is conserved under this evolution, and that the helicity of the field is also conserved (Woltjer 1958; Moffatt 1969). We choose a velocity field $u(x,t)$, $t \in (-t_0, t_0)$, that brings C_0 into coincidence with K_0 and that carries T_0 into a tubular neighbourhood T_0' of K_0. The flux in this tube is then still Φ and its helicity is still $n_0 \Phi^2$.

We now convert K_0 to K by appropriate crossing switches. Suppose that N_+ positive switches (i.e. switches which create positive crossings) and N_- negative switches are needed to effect this transformation. Each positive switch is equivalent to the insertion of a small loop of flux Φ annihilating flux on one side of the crossing and creating flux on the other side (figure 6). The net increment of helicity is $2\Phi^2$. Similarly a negative switch gives a net increment of helicity $-2\Phi^2$. Hence the helicity of the field in the new tubular neighbourhood T around K is

$$\mathcal{H} = N\Phi^2, \quad N = n_0 + 2(N_+ - N_-). \tag{3.4}$$

By this construction, the B-lines within T are still clearly closed curves, all satellites of K, and each pair of B-lines having the same linking number n, since the crossing switches treat all pairs in the same way. Following the argument of Ricca & Moffatt (1992), n is in fact equal to N; for suppose we divide the flux tube into $m\,(>1)$ sub-tubes, each sub-tube carrying flux $\Phi_m = \Phi/m$ (figure 7). The helicity of a

                          H. K. Moffatt and R. L. Ricca

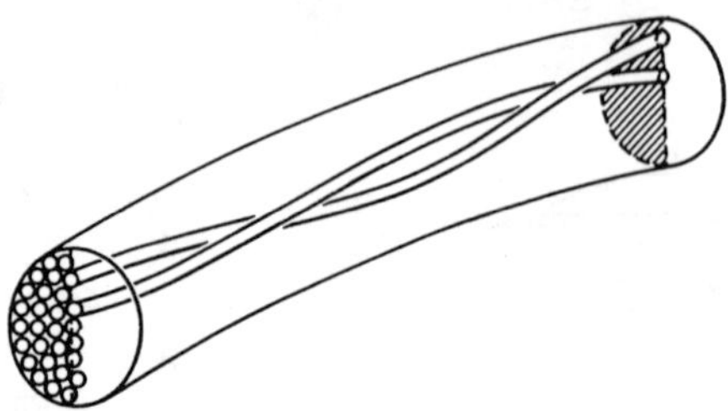

Figure 7. Subdivision of the flux tube into $m(> 1)$ sub-tubes.

sub-tube in isolation is $\mathscr{H}_m = \mathscr{H}/m^2$, since $\mathscr{H}_m$ is quadratically related to the flux Φ_m. The total helicity is therefore $m\mathscr{H}_m$ plus the sum of the interaction helicities due to linkage of pairs of sub-tubes with linking number n, i.e.

$$\begin{aligned}
\mathscr{H} &= m\mathscr{H}_m + \tfrac{1}{2}m(m-1)\cdot 2n\Phi_m^2 \\
&= \mathscr{H}/m + (m-1)\,n\Phi^2/m,
\end{aligned} \tag{3.5}$$

whence
$$\mathscr{H} = n\Phi^2, \tag{3.6}$$

so that $n = N$ as asserted.

It is obvious now that the linking number n may be given any desired value by suitable (retrospective) choice of n_0:

$$n_0 = n - 2(N_+ - N_-). \tag{3.7}$$

In particular, the choice $n_0 = -2(N_+ - N_-)$ makes $n = 0$, so that the linking number of every pair of $\boldsymbol{B}$-lines in the knotted flux tube is zero. Consideration of the example of figure $1c$ shows that this does not in general imply that the $\boldsymbol{B}$-lines are unlinked!

4. The writhe contribution to helicity

Suppose now that the knot K is in the form of a curve C which has no inflexion points (i.e. points of zero curvature). Let s be arc length on C from some origin O, and let the parametric equations of C be $\boldsymbol{x} = \boldsymbol{X}(s)$, where $\boldsymbol{X}(s)$ is periodic with period L, the length of C. The unit tangent vector is $\boldsymbol{t} = \mathrm{d}\boldsymbol{X}/\mathrm{d}s$, and the unit principal normal $\boldsymbol{n}$ and binormal $\boldsymbol{b} = \boldsymbol{t} \times \boldsymbol{n}$ then satisfy the Frenet equations

$$\mathrm{d}\boldsymbol{t}/\mathrm{d}s = c\boldsymbol{n}, \quad \mathrm{d}\boldsymbol{n}/\mathrm{d}s = -c\boldsymbol{t} + \tau\boldsymbol{b}, \quad \mathrm{d}\boldsymbol{b}/\mathrm{d}s = -\tau\boldsymbol{n}, \tag{4.1}$$

where $c(s)$ is the curvature and $\tau(s)$ the torsion of C at position s. (Note that $\boldsymbol{n}, \boldsymbol{b}$ and τ would not be defined at an inflexion point where $c = 0$; problems associated with deformation through inflexional configuration will be treated in §6 below.)

We now seek to obtain an alternative expression for the helicity in the flux tube T constructed around K, directly from the formula (2.1) by considering its limiting behaviour as the cross section of T tends to zero. The field $\boldsymbol{B}$ may be decomposed into the sum of two parts:

$$\boldsymbol{B} = \boldsymbol{B}_\mathrm{a} + \boldsymbol{B}_\mathrm{m}, \tag{4.2}$$

where $\boldsymbol{B}_\mathrm{a}$ is the axial field parallel to the tube axis and $\boldsymbol{B}_\mathrm{m}$ is the meridional field in meridian planes perpendicular to the tube axis. When the cross section of the tube is small, we may adopt a local cylindrical coordinate system (r, θ, z) and suppose that

$$\boldsymbol{B}_\mathrm{a} = (0, 0, B_z(r)), \quad \boldsymbol{B}_\mathrm{m} = (0, B_\theta(r), 0). \tag{4.3}$$

Evidently $\nabla \cdot \boldsymbol{B}_a = 0$ and $\nabla \cdot \boldsymbol{B}_m = 0$, so that we may introduce separate vector potentials:

$$\boldsymbol{B}_a = \nabla \times \boldsymbol{A}_a, \quad \boldsymbol{B}_m = \nabla \times \boldsymbol{A}_m, \tag{4.4}$$

with $\nabla \cdot \boldsymbol{A}_a = 0$ and $\nabla \cdot \boldsymbol{A}_m = 0$. The lines of force of the $\boldsymbol{B}_m$-field are unlinked circles, so that

$$\int_T \boldsymbol{A}_m \cdot \boldsymbol{B}_m \, \mathrm{d}V = 0. \tag{4.5}$$

Hence the total field helicity is given by

$$\mathcal{H} = \int_T \boldsymbol{A}_a \cdot \boldsymbol{B}_a \, \mathrm{d}V + \int_T \boldsymbol{A}_a \cdot \boldsymbol{B}_m \, \mathrm{d}V + \int_T \boldsymbol{A}_m \cdot \boldsymbol{B}_a \, \mathrm{d}V$$

$$= \int_T \boldsymbol{A}_a \cdot \boldsymbol{B}_a \, \mathrm{d}V + 2 \int_T \boldsymbol{A}_a \cdot \boldsymbol{B}_m \, \mathrm{d}V \tag{4.6}$$

(using integration by parts and the divergence theorem) the integrals in each case being over the tube T.

Consider first the axial contribution $\mathcal{H}_a = \int_T \boldsymbol{A}_a \cdot \boldsymbol{B}_a \, \mathrm{d}V$. Here we may use the Biot–Savart expression (2.3) in the limiting form

$$\boldsymbol{A}(\boldsymbol{x}) = \frac{1}{4\pi} \Phi \oint_C \frac{\mathrm{d}\boldsymbol{x}^* \times (\boldsymbol{x} - \boldsymbol{x}^*)}{|\boldsymbol{x} - \boldsymbol{x}^*|^3}. \tag{4.7}$$

Although this expression diverges when $\boldsymbol{x} \in C$, its axial component does not diverge, and the limiting expression

$$\mathcal{H}_a = \frac{1}{4\pi} \Phi^2 \oint_C \oint_C \frac{(\mathrm{d}\boldsymbol{x} \times \mathrm{d}\boldsymbol{x}^*) \cdot (\boldsymbol{x} - \boldsymbol{x}^*)}{|\boldsymbol{x} - \boldsymbol{x}^*|^3} = \Phi^2 \mathcal{W}, \tag{4.8}$$

say, is finite. The quantity $\mathcal{W}$ is called the *writhing number* (Fuller 1971) or simply the *writhe* of C, and bears a formal similarity to the Gauss integral (2.10). However, it is important to recognize that $\mathcal{W}$ is not a topological invariant of C; in fact it changes continuously (in general) under continuous deformation of C.

The physical meaning of the writhe is as follows. Suppose we view the closed curve C projected on a plane with unit normal $\boldsymbol{v}$. We then see a number $n_+(\boldsymbol{v})$ of positive crossings and $n_-(\boldsymbol{v})$ of negative crossings. Then

$$\mathcal{W} = \langle n_+(\boldsymbol{v}) - n_-(\boldsymbol{v}) \rangle, \tag{4.9}$$

where the angular brackets denote averaging over all directions $\boldsymbol{v}$ of projection. This fact is evident from consideration of the diagram of figure 8. The elements $\mathrm{d}\boldsymbol{x}$, $\mathrm{d}\boldsymbol{x}^*$ will intersect in projection if and only if $\boldsymbol{n}$ is parallel to $\pm(\boldsymbol{r} + \lambda \, \mathrm{d}\boldsymbol{x} - \mu \, \mathrm{d}\boldsymbol{x}^*)$ where $\boldsymbol{r} = \boldsymbol{x} - \boldsymbol{x}^*$, $0 < \lambda < 1$ and $0 < \mu < 1$, i.e. only if $\boldsymbol{v}$ lies within a solid angle

$$\mathrm{d}\varpi = 2(\mathrm{d}\boldsymbol{x} \times \mathrm{d}\boldsymbol{x}^*) \cdot \boldsymbol{r}/4\pi r^3 \tag{4.10}$$

(the factor 2 allowing for the $\pm$ possibilities above). Thus when we average over all directions of $\boldsymbol{v}$, take account of crossing signs and then integrate over all pairs of elements $\mathrm{d}\boldsymbol{x}, \mathrm{d}\boldsymbol{x}^*$, we obtain

$$\mathcal{W} = \frac{1}{4\pi} \oint_C \oint_C \frac{(\mathrm{d}\boldsymbol{x} \times \mathrm{d}\boldsymbol{x}^*) \cdot \boldsymbol{r}}{r^3} = \langle n_+(\boldsymbol{v}) - n_-(\boldsymbol{v}) \rangle, \tag{4.11}$$

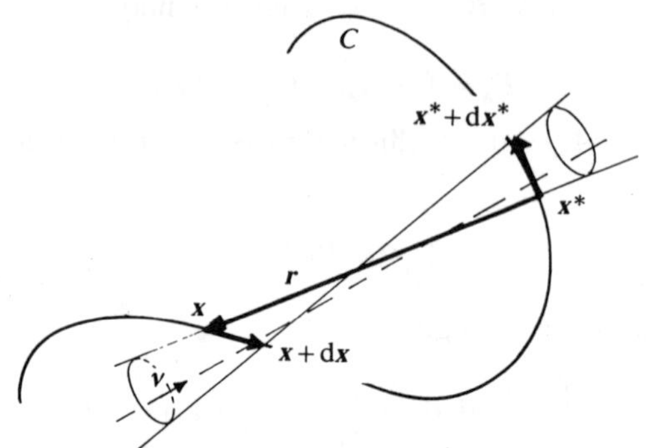

Figure 8. Contribution to the solid angle by elements $\mathrm{d}x$, $\mathrm{d}x^*$ on the curve C. The v-direction indicates a line of apparent intersection of $\mathrm{d}x$, $\mathrm{d}x^*$.

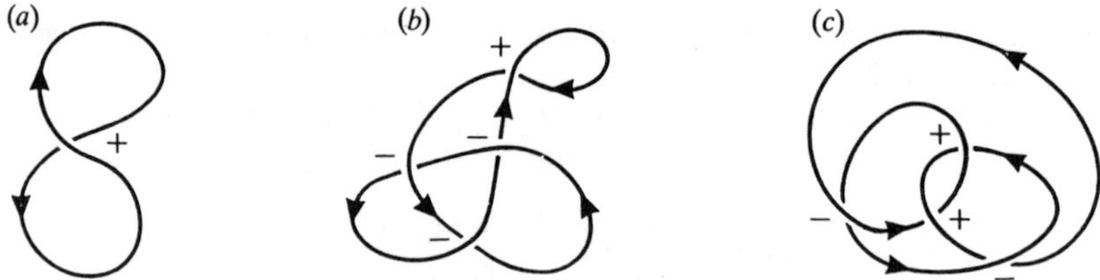

Figure 9. Values of the writhe for a number of flattened configurations. (a) $\mathcal{W} = +1$; (b) $\mathcal{W} = -2$; (c) $\mathcal{W} = 0$.

as stated. The geometric interpretation of $\mathcal{W}$ in terms of solid angle was originally discussed in C59 and later, in terms of spherical area, by Fuller (1978); its interpretation in terms of average number of apparent crossings was given by Fuller (1971). The same type of argument leads to Freedman & He's (1991) expression for the *crossing* number

$$\mathcal{C} = \frac{1}{4\pi} \oint_C \oint_C \frac{|(\mathrm{d}x \times \mathrm{d}x^*) \cdot r|}{r_3} = \langle n_+(v) + n_-(v) \rangle. \tag{4.12}$$

When the knot K is flattened onto the (x, y) plane except for indentations at the crossings, the writhe may be easily computed from the formula (4.11), since in this case $n_+(v)$ and $n_-(v)$ are independent of the viewing direction v except for a vanishingly small solid angle of directions nearly parallel to the (x, y) plane. Thus the writhe in this limiting situation is simply given by

$$\mathcal{W} \sim (n_+ - n_-), \tag{4.13}$$

and is an integer. Values of the writhe for a number of flattened configurations are shown in figure 9.

We note also that the field A_a provides a family of Seifert surfaces for the knot, as noted by Akhmet'ev & Ruzmaikin (1992). A Seifert surface is a non-self-intersecting oriented open surface bounded by the knot K. Let the cross section of the knot tube T tend to zero. Outside T, $\nabla \times A_\mathrm{a} = 0$ so there exists a scalar field (not single-valued) such that

$$A_\mathrm{a} = \nabla \Psi_\mathrm{a} \tag{4.14}$$

and, near K, $\Psi_\mathrm{a} \sim (2\pi)^{-1} \Phi \theta$ where θ is the azimuth angle used above. Thus the

Helicity and the Călugăreanu invariant 421

surfaces $\Psi_a = $ const. are all bounded by the knot, and since $\nabla\Psi_a$ is single valued the knot does not intersect such a surface at any other point. Any surface $\Psi_a = $ const. is therefore a Seifert surface.

5. Torsion and twist contributions to helicity

Consider now the second contribution in equation (4.6),

$$\mathscr{H}_m = 2\int_T \boldsymbol{A}_a \cdot \boldsymbol{B}_m \, \mathrm{d}V = 2\int_T A_\theta(r) B_\theta(r) \, \mathrm{d}V, \tag{5.1}$$

arising from the meridional component of the field $\boldsymbol{B}_m$. Note that from the first of (4.3) and the first of (4.4), $\boldsymbol{A}_a = (0, A_\theta(r), 0)$ where

$$\frac{1}{r}\frac{\mathrm{d}}{\mathrm{d}r}(rA_\theta) = B_z(r). \tag{5.2}$$

Let us consider the change in $\mathscr{H}_m$ under a virtual displacement $\delta\boldsymbol{\xi}(s)$ of the flux tube due to instantaneous changes $\delta c(s)$, $\delta\tau(s)$ in curvature and torsion of C. With plane polar coordinates (r, θ) in the cross-sectional plane at any section s of the tube T (figure 10), with θ measured from the direction of the principal normal $\boldsymbol{n}$, let

$$\left.\begin{array}{l} \boldsymbol{\xi} = r\hat{\boldsymbol{e}}_r = r(\boldsymbol{n}\cos\theta + \boldsymbol{b}\sin\theta) \\ \hat{\boldsymbol{e}}_\theta = -\boldsymbol{n}\sin\theta + \boldsymbol{b}\cos\theta, \end{array}\right\} \tag{5.3}$$

so that, assuming $\delta\boldsymbol{\xi}$ to be the same for all (r, θ),

$$\delta\boldsymbol{\xi} = r\cos\theta\,\delta\boldsymbol{n} + r\sin\theta\,\delta\boldsymbol{b}, \tag{5.4}$$

and so $$\frac{\mathrm{d}}{\mathrm{d}s}\delta\boldsymbol{\xi} = (r\cos\theta)\frac{\mathrm{d}}{\mathrm{d}s}\delta\boldsymbol{n} + (r\sin\theta)\frac{\mathrm{d}}{\mathrm{d}s}\delta\boldsymbol{b}. \tag{5.5}$$

Since it is only the variation of $\delta\boldsymbol{\xi}$ with arc length s that contributes to distortion of the field, we may suppose that at $s = s_1$, $\delta\boldsymbol{\xi}(s_1) = 0$, i.e. $\delta\boldsymbol{n}(s_1) = \delta\boldsymbol{b}(s_1) = 0$. Then, from the Frenet relations (4.1), we have

$$\left.\begin{array}{l} \dfrac{\mathrm{d}}{\mathrm{d}s}\delta\boldsymbol{n} = -\delta c\boldsymbol{t} + \delta\tau\boldsymbol{b} \\[2mm] \dfrac{\mathrm{d}}{\mathrm{d}s}\delta\boldsymbol{b} = -\delta\tau\boldsymbol{n} \end{array}\right\} \quad \text{at} \quad s = s_1. \tag{5.6}$$

Now under the assumed virtual displacement $\delta\boldsymbol{\xi}(s)$, the axial field $\boldsymbol{B}_a$ (and so $A_\theta(r)$) is unchanged, but the meridional field $\boldsymbol{B}_m$ at $s = s_1$ is changed by an amount

$$\delta\boldsymbol{B}_m = (\boldsymbol{B}_a \cdot \nabla)\,\delta\boldsymbol{\xi} = B_z(r)\frac{\mathrm{d}}{\mathrm{d}s}\delta\boldsymbol{\xi}, \tag{5.7}$$

due to the variation of $\delta\boldsymbol{\xi}$ with arc length (this is the process that in magnetohydrodynamics is known as 'generation of toroidal field by differential rotation' (see Moffatt 1978). Hence, at $s = s_1$,

$$\delta B_\theta = B_z(r)\left(\frac{\mathrm{d}}{\mathrm{d}s}\delta\boldsymbol{\xi}\right)_\theta$$

$$= B_z(r)\left[-\sin\theta\boldsymbol{n}\cdot\left(\frac{\mathrm{d}}{\mathrm{d}s}\delta\boldsymbol{\xi}\right) + \cos\theta\boldsymbol{b}\cdot\left(\frac{\mathrm{d}}{\mathrm{d}s}\delta\boldsymbol{\xi}\right)\right]. \tag{5.8}$$

H. K. Moffatt and R. L. Ricca

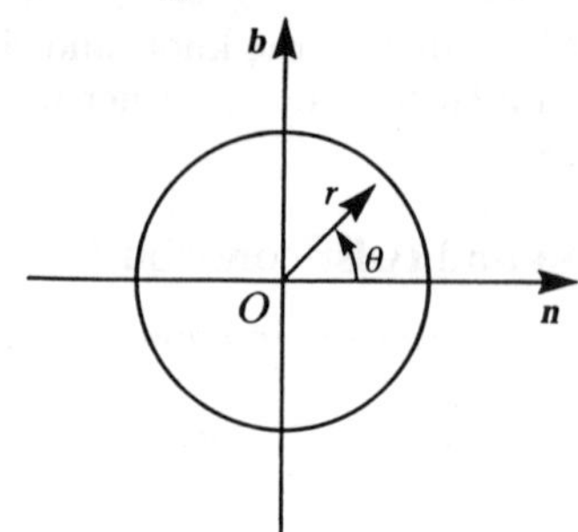

Figure 10. Cross section of the flux tube with plane polar coordinates (r, θ).

Substituting from (5.5) and (5.6), we have

$$\delta B_\theta = B_z(r)\, r\, \delta\tau(s) \quad \text{at} \quad s = s_1. \tag{5.9}$$

Since the same argument may be used at any section, (5.9) gives the field perturbation due to the virtual displacement for all s_1, and the resulting change in the $\mathscr{H}_\mathrm{m}$ is therefore

$$\delta\mathscr{H}_\mathrm{m} = 2\int_T A_\theta(r)\, \delta B_\theta(r)\, \mathrm{d}V$$

$$= 2\int_T A_\theta(r)\, B_z(r)\, r\, \delta\tau(s)\, \mathrm{d}V. \tag{5.10}$$

If we integrate first over the cross section, using (5.2) and the result

$$\int_0^\infty A_\theta \cdot \frac{1}{r}\frac{\mathrm{d}}{\mathrm{d}r}(rA_\theta) \cdot r \cdot 2\pi r\, \mathrm{d}r = \tfrac{1}{2}[(rA_\theta)^2]_0^\infty = 2\pi \cdot \tfrac{1}{2} \cdot \left(\frac{\Phi}{2\pi}\right)^2, \tag{5.11}$$

then from (5.10)
$$\delta\mathscr{H}_\mathrm{m} = \Phi^2\, \delta\mathscr{T}, \tag{5.12}$$

where
$$\mathscr{T} = \frac{1}{2\pi}\oint_C \tau(s)\, \mathrm{d}s \tag{5.13}$$

is the total torsion of C, normalized by the factor $(2\pi)^{-1}$.

It is easy to show how the total twist number Tw defined by (1.5) is related to the normalized total torsion. For this, let us take $N = n\cos\Theta + b\sin\Theta$ as the unit spanwise vector on the ribbon relative to the Frenet pair (n, b). By the Frenet equations (4.1), we have

$$N' = \mathrm{d}N/\mathrm{d}s = -c\cos\Theta\, t + (\tau + \mathrm{d}\Theta/\mathrm{d}s)\,\hat{e}_\theta, \tag{5.14}$$

where $\hat{e}_\theta = -n\sin\Theta + b\cos\Theta$. By (1.5), the total twist number for a ribbon is thus given by

$$Tw = \frac{1}{2\pi}\oint_C (N' \times N)\cdot t\, \mathrm{d}s = \frac{1}{2\pi}\oint_C \left(\tau + \frac{\mathrm{d}\Theta}{\mathrm{d}s}\right)\mathrm{d}s = \mathscr{T} + \frac{1}{2\pi}[\Theta]_C \tag{5.15}$$

and we identify $(1/2\pi)\,[\Theta]_C \equiv \mathscr{N}$. As was pointed out by Banchoff & White (1975), the total twist number Tw depends on the choice of the vector field N.

Now, if we consider a time-dependent deformation of C which does not pass through any inflexional configuration, then (5.12) may be written

$$\mathrm{d}\mathscr{H}_\mathrm{m}/\mathrm{d}t = \Phi^2\, \mathrm{d}\mathscr{T}/\mathrm{d}t, \tag{5.16}$$

or equivalently
$$\mathscr{H}_{\mathrm{m}} = \Phi^2(\mathscr{T} + \mathscr{T}_0), \tag{5.17}$$

where $\mathscr{T}_0$ is a constant. In fact $\mathscr{T}_0 = \mathscr{N}$; however, to establish this point, we have to consider the behaviour when C *does* pass through an inflexional configuration.

6. Generic behaviour associated with inflexion points

If a curve $x = X(s)$ has an inflexion point at $s = s_{\mathrm{c}}$, then $t' = \mathrm{d}t/\mathrm{d}s = \mathrm{d}^2X/\mathrm{d}s^2 = 0$ at $s = s_{\mathrm{c}}$, so that near $s = s_{\mathrm{c}}$ we have the Taylor expansions

$$t(s) = t_{\mathrm{c}} + \tfrac{1}{2}(s - s_{\mathrm{c}})^2 t''_{\mathrm{c}} + \dots, \tag{6.1}$$

$$X(s) = X_{\mathrm{c}} + (s - s_{\mathrm{c}})\, t_{\mathrm{c}} + \tfrac{1}{6}(s - s_{\mathrm{c}})^3 t''_{\mathrm{c}} + \dots. \tag{6.2}$$

Moreover, since $|t| = 1$,

$$(t'' \cdot t)_{s=s_{\mathrm{c}}} = \frac{\mathrm{d}^2}{\mathrm{d}s^2} t^2 \bigg|_{s=s_{\mathrm{c}}} = 0, \tag{6.3}$$

so that t''_{c} is perpendicular to t_{c}. We may therefore choose origin at the inflexion point ($X_{\mathrm{c}} = 0$, $s_{\mathrm{c}} = 0$) and axes $Oxyz$ with Ox parallel to t_{c} and Oz parallel to t''_{c}. The form of the curve near the inflexion point is then given by

$$X(s) = (s, 0, \alpha s^3), \tag{6.4}$$

where $\alpha = \tfrac{1}{6}|t''_{\mathrm{c}}|$, i.e. it is the plane cubic curve $y = 0$, $z = \alpha x^3$. By simple rescaling, we may take $\alpha = 1$.

We now wish to consider a time-dependent curve $x = X(s,t)$ passing through the inflexional configuration (6.4) at $t = 0$, but having $\partial t/\partial s \neq 0$ when $t \neq 0$. Since

$$t' \cdot t = \tfrac{1}{2}\partial(t^2)/\partial s = 0, \tag{6.5}$$

we may always, by rigid rotation, ensure that at $s = 0$, t remains parallel to Ox and t' remains parallel to Oy. These conditions are satisfied by the time-dependent twisted cubic

$$X(s,t) = (s - \tfrac{2}{3}t^2 s^3, ts^2, s^3), \tag{6.6}$$

for which
$$t = \partial X/\partial s = (1 - 2t^2 s^2, 2ts, 3s^2) \tag{6.7}$$

and
$$|t| = 1 + O(s^4), \tag{6.8}$$

so that, near $s = 0$, t is indeed the unit tangent vector. Figure 11 shows this family of curves and their projections on the three coordinate planes.

From (6.7), to leading order in $|t|$ and $|s|$,

$$\partial t/\partial s \sim 2(0, t, 3s), \tag{6.9}$$

so that
$$c(s,t) = |\partial t/\partial s| \sim 2(t^2 + 9s^2)^{\frac{1}{2}} \tag{6.10}$$

and
$$n(s,t) = \frac{1}{c}\frac{\partial t}{\partial s} \sim \frac{(0, t, 3s)}{(t^2 + 9s^2)^{\frac{1}{2}}}. \tag{6.11}$$

Note here that for very small t, n rotates through an angle π about the direction $t_{\mathrm{c}} = (1, 0, 0)$ as s increases from $-s_0$ to $+s_0$ where $s_0 \gg |t|$; and that this rotation is clockwise (right-handed) for $t < 0$, and anticlockwise (left-handed) for $t > 0$; thus the number of rotations of the pair (n, b) about the tangent direction t in the anticlockwise sense increases by $+1$ as t increases through zero (at the instant $t = 0$, this number is undefined).

Now the binormal is given by $b = t \times n$, and the torsion is obtained from (4.1): for $|t|$ and $|s|$ small,

$$\tau(s,t) \sim 3t/(t^2 + 9s^2). \tag{6.12}$$

H. K. Moffatt and R. L. Ricca

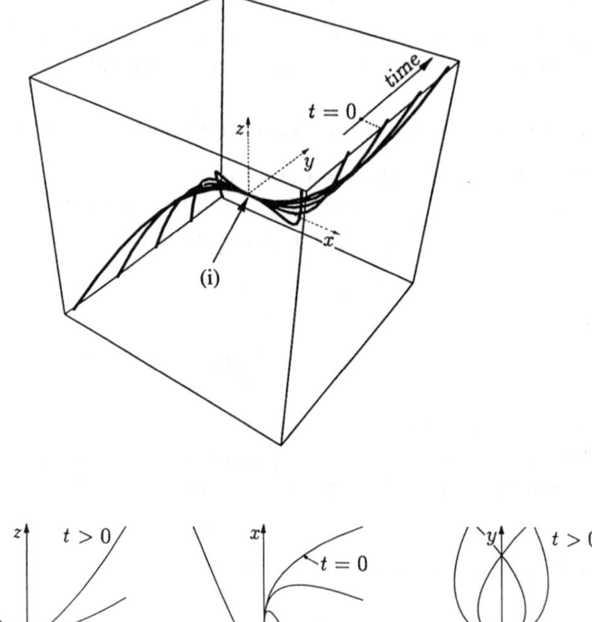

Figure 11. The twisted cubic (6.6) for $-1 \leqslant s \leqslant 1$ and for various values of t. The curve contains an inflexion point at (i) $s = 0$ when $t = 0$. Plane projections are shown below.

As expected, c vanishes only at $t = s = 0$, and τ is singular at this inflexion. However, the singularity is integrable; the contribution to the normalized total torsion $\mathcal{T}$ from any small interval $[-s_0, s_0]$ is

$$\frac{1}{2\pi} \int_{-s_0}^{s_0} \tau(s, t)\,\mathrm{d}s = \frac{1}{\pi} \int_0^{s_0} \frac{3t}{t^2 + 9s^2}\,\mathrm{d}s = \frac{1}{\pi} \arctan\left(\frac{3s_0}{t}\right), \tag{6.13}$$

and, irrespective of the value of s_0, this jumps from $-\frac{1}{2}$ to $+\frac{1}{2}$ as t increases through zero, i.e. as the curve passes through the inflexional configuration. Hence $\mathcal{T}$ is discontinuous as C passes through the inflexion, with discontinuity $[\mathcal{T}] = +1$. The reverse passage (or equivalently replacement of t by $-t$ in (6.6)) gives a jump $[\mathcal{T}] = -1$. This behaviour, recognized by Călugăreanu (1961) for a particular example, appears to be generic.

7. Role of the twist parameter

We have seen from §§4 and 5 above that the helicity of a twisted flux tube with axis C can be expressed in the form

$$\mathcal{H} = \Phi^2 h = \Phi^2(\mathcal{W} + \mathcal{T} + \mathcal{T}_0). \tag{7.1}$$

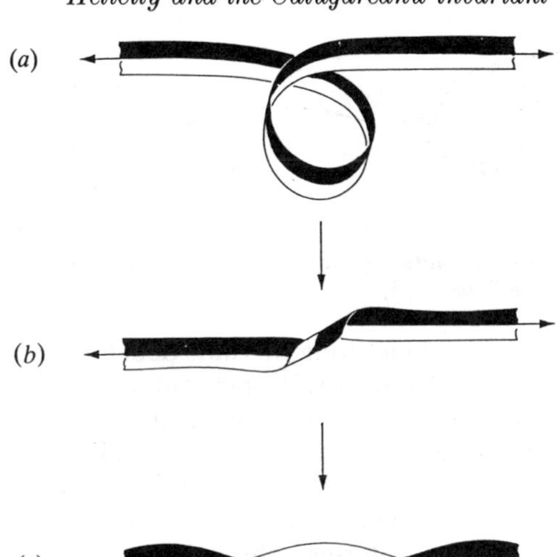

Figure 12. (*a*) Writhe, (*b*) torsion and (*c*) twist contributions of a ribbon to the Cǎlugǎreanu invariant. If a coiled ribbon is stretched so that its centre-line becomes straight, then the initial torsion of the centre-line is converted to the final twist of the ribbon about its centre-line.

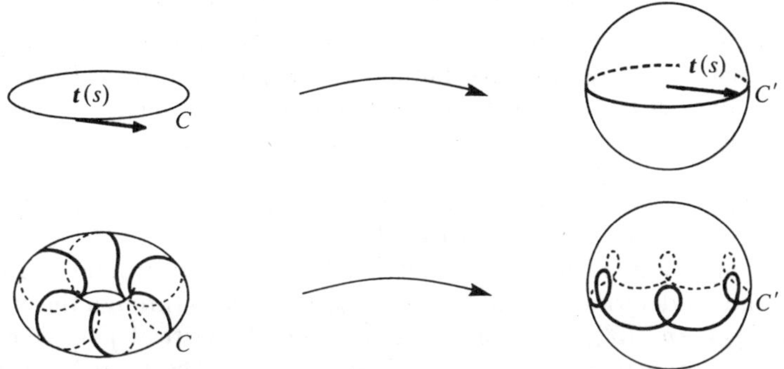

Figure 13. Mapping of closed curves C to their images C' on the unit sphere.

Under continuous deformation of the flux tube, $\mathcal{H}$ is conserved and hence

$$h = \mathcal{W} + \mathcal{T} + \mathcal{T}_0 = \text{const.} \tag{7.2}$$

The writhe $\mathcal{W}$ defined by (4.8) varies continuously as C is deformed continuously; however, if C passes through an inflexional configuration, then $\mathcal{T}$ jumps by ± 1. Hence, as $\mathcal{T}$ jumps by ± 1, the term $\mathcal{T}_0$ must jump by a compensating amount ∓ 1, to maintain the overriding invariance of helicity.

The equal and opposite jumps in $(\mathcal{W} + \mathcal{T})$ and $\mathcal{T}_0$ may be understood with reference to the simple example illustrated in figure 12. If a coiled ribbon is stretched so that its centre-line becomes straight (at which stage $\mathrm{d}t/\mathrm{d}s \equiv 0$ on the centre-line!), then the initial writhe of the centre-line is converted to the final twist of the ribbon about its centre-line. This example is not generic since it involves the appearance of a continuum of inflexion points. However, it captures the essence of the nature of the interchange between $(\mathcal{W} + \mathcal{T})$ and $\mathcal{T}_0$: $\mathcal{T}_0$ represents the intrinsic twist of the ribbon about its centre-line, and this in general jumps by ∓ 1 when the centre line is deformed through an isolated inflexion point.

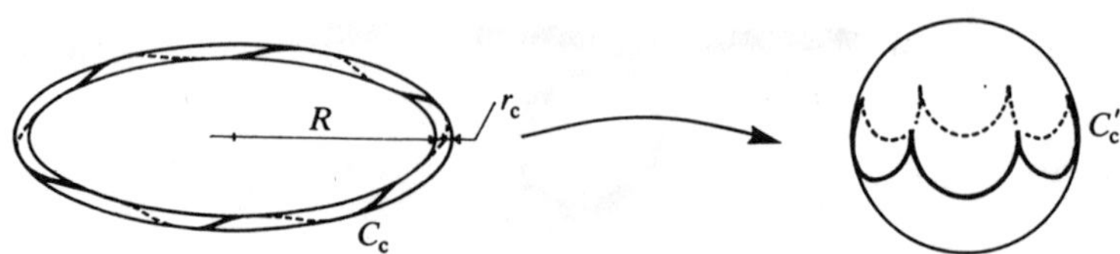

Figure 14. Mapping of the curve C_c on a 'critical' torus to its image C_c' on the unit sphere. At the critical value $r_c = R/m^2$, inflexion points appear on C_c and they are mapped to the m cusps on the unit sphere ($m = 6$ in the case illustrated).

There is a further useful way of picturing this inflexional behaviour (see figure 13); the set of unit vectors $t(s)$ on a closed curve C trace out a corresponding closed curve C' on the unit sphere. For example a circle C corresponds to an equatorial circle C'; a helix wound around a fat torus corresponds to an epicyclic curve with a number of double points, and so on. If C is continuously deformed, then its image C' is continuously deformed also, and the number of double points of C' may change. In fact this does happen when C is deformed through an inflexion (at $s = s_1$ say) at which $dt/ds = 0$. When C has an inflexion, C' has a cusp at the corresponding point on the sphere.

The case of a helix wound around a torus,

$$X(s) = [(R + r_0 \cos m\phi) \cos \phi, (R + r_0 \cos m\phi) \sin \phi, r_0 \sin m\phi], \qquad (7.3)$$

(where m is an integer) is particularly interesting in this respect (figure 14). If $r_0 = 0$, this is a circle, the principal normal $n = n_0$ pointing towards its centre. For very small values of r_0, the direction of n simply oscillates about the position n_0 as we move round C. As r_0 increases further the amplitude of these oscillations increases, until at a critical value $r_c (= R/m^2)$ inflexion points appear on C at the points where $\cos m\phi = -1$; for $r_0 > r_c$ the principal normal n makes m complete rotations around the axis of the torus in one passage round C.

If we now place a flux tube of cross-sectional radius $\epsilon \ll r_c$ around C, and consider a time-dependent deformation in which $r_0 = r_0(t)$ decreases through the critical value r_c, then $[\mathcal{T}] = -m$, $[\mathcal{T}_0] = +m$ in going through the critical point, i.e. torsional helicity is instantaneously converted to twist helicity, the total helicity being of course conserved.

The fact that $\mathcal{T}_0$ jumps by ± 1 whenever C passes through a single inflexion suggests the interpretation that it represents, in some sense, the number of rotations of the flux tube (or of the associated set of ribbons) about its axis in one passage around C. This concept is, however, quite elusive, because one must specify carefully the frame of reference with respect to which the flux tube rotates. There is no difficulty in this when C is not in an inflexional configuration, for then we may use the Frenet frame (t, n, b). Let C, C^* be two neighbouring B-lines in the flux tube (the boundaries of a ribbon), and, as before, let $N(s)$ be the spanwise vector from C to C^* on this ribbon. Let $\mathcal{N}$ be the (integer) number of rotations of $N(s)$ about t with respect to the Frenet frame (as defined in §5). We shall show that in fact $\mathcal{T}_0 = \mathcal{N}$.

Under arbitrary continuous deformations of the flux tube, $N(s, t)$ is a continuous vector function of (s, t), and its components with respect to a fixed cartesian frame of reference are also continuous. However, if C passes through an inflexion at s_c, then, as we have seen in §6, the number of rotations of the Frenet pair (n, b) about the tangent vector t in one passage round C changes by ± 1; hence the number $\mathcal{N}$ of rotations of $N(s, t)$ relative to the Frenet frame changes by ± 1.

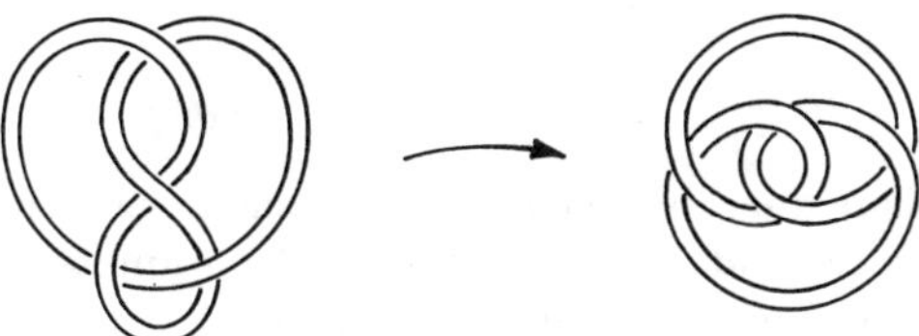

Figure 15. A plane projection of a knot may always be arranged so to have no inflexion points. Here the figure-of-eight knot, with two inflexions, is isotoped to a configuration with no inflexion points.

The knot K and its associated flux tube may always be deformed to lie nearly in the (x, y) plane (as in §3) so that the write $\mathscr{W}$ is (in the limit) an integer $(n_+ - n_-)$. We may also always arrange that this plane projection of K has no inflexion points; for example the figure-of-eight, containing two prototype inflexions, can be isotoped (figure 15) to a curve of non-vanishing curvature. In such a 'standard' configuration, the torsion $\tau(s)$ is zero everywhere except in the small indentations where it is small; hence in the plane limit, $\mathscr{T} = 0$. In this configuration therefore the helicity is given by

$$\mathscr{H} = n\Phi^2 = (n_+ - n_-)\,\Phi^2 + \mathscr{T}_0\,\Phi^2, \tag{7.4}$$

so that

$$\mathscr{T}_0 = n - (n_+ - n_-) \tag{7.5}$$

is an integer. Consideration of the special case of a circle $(n_+ = n_- = 0)$ shows that $\mathscr{T}_0$ is indeed the number of twists of the ribbon (unambiguous in the plane configuration) about its centre-line. We thus identify $\mathscr{T}_0$ and $\mathscr{N}$ in this standard configuration, and hence in every configuration of the knot.

8. Summary

In this paper we have discussed several properties of the helicity of linked and knotted flux tubes. Different contributions to helicity have been analysed in terms of the Gauss linking number (§2) and in terms of the Călugăreanu invariant (1.3). If the field lines in a single knotted flux tube are twisted closed curves which close on themselves after one passage around the tube, then the helicity of the flux tube is given by

$$\mathscr{H} = \int_T A \cdot B \, \mathrm{d}V = n\Phi^2, \tag{8.1}$$

where Φ is the flux associated with the tube (§3). The integer n is an invariant under frozen field distortion of the tube, and is identified with the Călugăreanu invariant (1.3). We have demonstrated this invariance by starting from the known invariance of helicity. The helicity has been decomposed into writhe and twist contributions, the writhe contribution involving the Gauss integral, which admits interpretation in terms of the sum of signed crossings of the knot averaged over all projections (§4). Part of the twist contribution is shown to be associated with the torsion of the knot and part with what may be described as 'intrinsic twist' of the field lines in the flux tube around the knot (§5). The generic behaviour associated with the deformation of the knot through a configuration with points of inflexion (points at which the curvature vanishes) has been analysed (§6) and the role of the twist parameter has been discussed (§7).

428   *H. K. Moffatt and R. L. Ricca*

In deriving the relation

$$\mathscr{H} = n\Phi^2 = (\mathscr{W} + \mathscr{T} + \mathscr{N})\,\Phi^2, \tag{8.2}$$

where the writhe $\mathscr{W}$ is defined by equation (4.8), the normalized total torsion $\mathscr{T}$ is defined by equation (5.13) and $\mathscr{N} = \mathscr{T}_0$ is the twist parameter, we have shown that (generically) $(\mathscr{W} + \mathscr{T})$ jumps discontinuously through ± 1 as C passes through an inflexional configuration and that by virtue of the invariance of $\mathscr{H}$, there is then a compensating jump of ∓ 1 in $\mathscr{N}$. This behaviour has been previously recognized by Ricca & Moffatt (1992) and is associated with the classical type I Reidemeister move of ambient isotopies.

The Călugăreanu invariant is fundamental in relation to many problems that involve continuous deformation of tube-like structures. Examples in the literature range from the theory of dynamical systems (Uezu 1990) to the biochemistry of excitable media (Winfree 1990), from the quantum field theory of string-like objects (Tze & Nam 1989) to studies of DNA coiling (Tsuru & Wadati 1986), from the theory of propagation of spinning particles (Jaroszewicz & Kurzepa 1991) to the general problem of protein folding (De Santis *et al.* 1986). Fundamental topological aspects of these phenomena can be successfully described in terms of the Călugăreanu invariant. In this paper we hope that the direct derivation of the Călugăreanu invariant from first principles of fluid mechanics together with the discussion of the generic behaviour associated with inflexional configurations, which are ubiquitous in many processes of continuous deformation of field structures, provides a good demonstration of the relevance of fluid dynamical techniques to topological problems.

We thank Dr J. S. Langer, Director of the Institute for Theoretical Physics at UCSB (which is supported by NSF Grant PHY89-04035), for his kind hospitality at the Institute, where this work was initiated. One of us (R. L. R.) acknowledges the financial support from Associazione Sviluppo Scientifico e Tecnologico del Piemonte (Turin, Italy).

References

Akhmet'ev, P. & Ruzmaikin, A. 1992 Borromeanism and bordism. In *Topological aspects of the dynamics of fluids and plasmas* (ed. H.K. Moffatt *et al.*) pp. 249–264. Dordrecht: Kluwer.

Banchoff, T. F. & White, J. H. 1975 The behaviour of the total twist and self-linking number of a closed space curve under inversions. *Math. Scand.* **36**, 254–262.

Berger, M. A. & Field, G. B. 1984 The topological properties of magnetic helicity. *J. Fluid Mech.* **147**, 133–148.

Călugăreanu, G. 1959 L'intégral de Gauss et l'analyse des noeuds tridimensionnels. *Rev. Math. pures appl.* **4**, 5–20. (C59 in text.)

Călugăreanu, G. 1961 Sur les classes d'isotopie des noeuds tridimensionnels et leurs invariants. *Czechoslovak Math. J.* **11**, 588–625. (C61 in text.)

De Santis, P., Palleschi, A. & Chiavarini, S. 1986 Topological approach to protein folding. *Gazz. Chim. It.* **116**, 561–567.

Freedman, M. H. & He, Z.-X. 1991 Divergence-free fields: Energy and asymptotic crossing number. *Ann. Math.* **134**, 189–229.

Fuller, F. B. 1971 The writhing number of a space curve. *Proc. natn. Acad. Sci. U.S.A.* **68**, 815–819.

Fuller, F. B. 1972 How the writhing number of a curve depends on the curve. *Rev. Roum. Math. pures appl.* **17**, 1329–1334.

Fuller, F. B. 1978 Decomposition of the linking of a closed ribbon: a problem from molecular biology. *Proc. natn. Acad. Sci. U.S.A.* **75**, 3557–3561.

Helicity and the Călugăreanu invariant   429

Jaroszewicz, T. & Kurzepa, P. S. 1991 Spin, statistics and geometry of random walks. *Ann. Phys.* **210**, 255–322.

Kauffman, L. H. 1987 *On knots*. Princeton University Press.

Kauffman, L. H. 1991 *Knots and physics*. World Scientific Publishing.

Moffatt, H. K. 1969 The degree of knottedness of tangled vortex lines. *J. Fluid Mech.* **35**, 117–129.

Moffatt, H. K. 1978 *Magnetic field generation in electrically conducting fluids*. Cambridge University Press.

Moffatt, H. K. 1981 Some developments in the theory of turbulence. *J. Fluid Mech.* **106**, 27–47.

Moffatt, H. K. 1990 The energy spectrum of knots and links. *Nature, Lond.* **347**, 367–369.

Moreau, J. J. 1961 Constantes d'un îlot tourbillonnaire en fluid parfait barotrope. *C.r. hebd. Séanc. Acad. Sci., Paris* **252**, 2810–2812.

Pohl, W. F. 1968 The self-linking number of a closed space curve. *J. Math. Mech.* **17**, 975–985.

Pohl, W. F. 1980 DNA and differential geometry. *Math. Intelligencer* **3**, 20–27.

Ricca, R. L. & Moffatt, H. K. 1992 The helicity of a knotted vortex filament. In *Topological aspects of the dynamics of fluids and plasmas* (ed. H. K. Moffatt *et al.*) pp. 225–236. Dordrecht: Kluwer.

Tsuru, H. & Wadati, M. 1986 Elastic model of highly supercoiled DNA. *Biopolymers* **25**, 2083–2096.

Tze, C.-H. & Nam, S. 1989 Topological phase entanglements of membrane solitons in division algebra sigma models with a Hopf term. *Ann. Phys.* **193**, 419–471.

Uezu, T. 1990 Topological structure in flow systems. *Prog. theor. Phys.* **83**, 850–874.

White, J. H. 1969 Self-linking and the Gauss integral in higher dimensions. *Am. J. Math.* **91**, 693–728.

Winfree, A. T. 1990 Stable particle-like solutions to the nonlinear wave equations of three-dimensional excitable media. *SIAM Rev.* **32**, 1–53.

Woltjer, L. 1958 A theorem on force-free magnetic fields. *Proc. natn. Acad. Sci. U.S.A.* **44**, 489–491.

Received 20 July 1992; accepted 24 August 1992

Witten's Invariant of 3-Dimensional Manifolds: Loop Expansion and Surgery Calculus[1]

L. Rozansky[2]

Theory Group, Department of Physics, University of Texas at Austin

Austin, TX 78712-1081, U.S.A.

Abstract

We review two different methods of calculating Witten's invariant: a stationary phase approximation and a surgery calculus. We give a detailed description of the 1-loop approximation formula for Witten's invariant and of the technics involved in deriving its exact value through a surgery construction of a manifold. Finally we compare the formulas produced by both methods for a 3-dimensional sphere S^3 and a lens space $L(p,1)$.

A substantial understanding of quantum theory is *not* a prerequisite for reading this paper.

---

[1]to be published in the volume *Knots and Applications*

[2]Work supported by NSF Grant 9009850 and R. A. Welch Foundation.

Contents

1 Introduction

A quantum field theory based on Chern-Simons action has been developed by E. Witten in his paper [1]. Consider a connection A_μ of a G bundle E on a 3-dimensional manifold M, G being a simple Lie group. If the bundle is trivial, then an integral

$$S_{CS} = \frac{1}{2}\epsilon^{\mu\nu\rho}\mathrm{Tr}\int_M (A_\mu\partial_\nu A_\rho + \frac{2}{3}A_\mu A_\nu A_\rho)\, d^3x. \tag{1}$$

defines a Chern-Simons action as a function of A_μ. A manifold invariant $Z(M)$ is a path integral

$$Z(M,k) = \int [\mathcal{D}A_\mu]e^{\frac{i}{\hbar}S_{CS}[A_\mu]}, \quad \hbar = \frac{\pi}{k}, \tag{2}$$

here $k \in \mathbf{Z}$, and the brackets in $[\mathcal{D}A_\mu]$ mean that we integrate over the gauge equivalence classes of connections. The action (1) does not depend on the choice of local coordinates on M, neither does it depend on the metric of the manifold M. Therefore the integral $Z(M,k)$, also known to physicists as a partition function, is a topological invariant of the manifold (modulo the possible metric dependence of the integration measure $[\mathcal{D}A_\mu]$).

Witten considered two different methods of calculating $Z(M,k)$. He first applied a stationary phase approximation to the integral (2). This is a standard method of quantum field theory. It expresses $Z(M,k)$ as asymptotic series in k^{-1}. The first term in this series contains such ingredients as Chern-Simons action of flat connections and Reidemeister torsion. The other method of calculating the invariant, which we call "surgery calculus" is based upon a construction of M as a surgery on a link in S^3 (or in $S^1 \times S^2$). It presents $Z(M,k)$ as a finite sum, however the number of terms in it grows as a power of k. Reshetikhin and Turaev used the surgery calculus formula in [2] as a definition of Witten's invariant and proved its invariance (i.e. independence of the choice of surgery to construct a given manifold M) without referring to the path integral (2).

A systematic comparison between both methods of calculating Witten's invariant has been initiated in [3]. D. Freed and R. Gompf compared the numeric values of the invarints of some lens spaces and homology spheres for large values of k as given by the two methods. The full analytic comparison has been carried out in [4] and [5] for lens spaces and mapping tori. It was extended further to Seifert manifolds in [6]. A complete agreement between the stationary phase approximation and surgery calculus has been found in all these papers.

In this paper we will review both methods of calculating Witten's invariant and compare their results. In section 2 we explain the stationary phase approximation method. Section 3 contains the basics of surgery calculus. In section 4 we apply both methods to the calculation of Witten's invariant of the sphere S^3 and lens space $L(p,1)$.

2 Stationary Phase Approximation

2.1 Finite Dimensional Integrals

Let us start with a simple example of the stationary phase approximation. Consider a finite dimensional integral

$$Z(\hbar) = \int \frac{d^n X}{(2\pi\hbar)^{n/2}} \exp\left[\frac{i}{\hbar} S(X_1,\ldots,X_n)\right] \tag{3}$$

for some function S. Here $\hbar$ is an arbitrary small constant, called Planck's constant in quantum theory. The integral (3) is a finite dimensional version of the path integral (2). Note that a path integral measure $[\mathcal{D}A_\mu]$ includes implicitly a factor $(2\pi\hbar)^{-1/2} = \pi^{-1}(k/2)^{1/2}$ for each of the one-dimensional integrals comprising the full path integral.

In the limit of small $\hbar$ the dominant contribution to $Z(\hbar)$ comes from the extrema of S, i.e. from the points X_i^a such that

$$\left.\frac{\partial S}{\partial X_i}\right|_{X_i=X_i^{(a)}} = 0, \quad 1 \le i \le n \tag{4}$$

If we retain only the quadratic terms in Taylor expansion of S in the vicinity of these points, then

$$\begin{aligned}
Z(\hbar) &= \sum_a e^{\frac{i}{\hbar}S(X^{(a)})} \int d^n x \exp\left[i\pi \sum_{i,j=1}^n x_i x_j \left.\frac{\partial^2 S}{\partial X_i \partial X_j}\right|_{X_i=X_i^{(a)}}\right] \\
&= \sum_a e^{\frac{i}{\hbar}S(X^{(a)})} \det^{-1/2}\left(-i\left.\frac{\partial^2 S}{\partial X_i \partial X_j}\right|_{X_i=X_i^{(a)}}\right).
\end{aligned} \tag{5}$$

A phase of this expression requires extra care. The matrix $\frac{\partial^2 S}{\partial X_i \partial X_j}$ is hermitian. It has only real eigenvalues λ, but they can be both positive and negative. Each positive eigenvalue contributes a phase factor $(-i)^{-1/2} = e^{i\pi/4}$ to the inverse square root of the determinant in eq. (5), while each negative eigenvalue contributes $i^{-1/2} = e^{-i\pi/4}$. Therefore a refined version of the formula (5) is

$$Z(\hbar) = \sum_a e^{\frac{i}{\hbar} S(X^{(a)})} e^{i\frac{\pi}{4}\eta_a} \left| \det \left(\left. \frac{\partial^2 S}{\partial X_i \partial X_j} \right|_{X_i = X_i^{(a)}} \right) \right|^{-1/2}, \tag{6}$$

here

$$\eta^a = \#\,\text{positive}\,\lambda - \#\,\text{negative}\,\lambda \tag{7}$$

In the context of quantum field theory the integral (3) becomes an infinite dimensional path integral, however the stationary phase approximation method remains the same if we can make sense of infinite dimensional determinants. Physicists call the formula (5) a 1-loop approximation, because it can be derived by summing up all 1-loop Feynman diagrams.

2.2 Gauge Invariant Theories

The integral (2) presents a special challenge, because the action (1) is invariant under a gauge transformation (i.e. under a local change of basis in the fibers)

$$A_\mu \to A_\mu^g = g^{-1} A_\mu g + g^{-1} \partial_\mu g. \tag{8}$$

The integral over the gauge equivalence classes of connections is equal to the integral over all connections divided by the volume of the group of gauge transformations. However the latter integral can not be calculated through eq. (6), because its stationary phase points are not isolated. They form the orbits of the gauge action (8). Therefore we should rather integrate over the submanifold in the space of all connections, which is transversal to gauge orbits, multiply the terms in the sum (6) by the volumes of those orbits and divide the whole sum by the volume of the group of gauge transformations.

The problem of reducing an integral of a function invariant under the action of a group, to an integral over a factor manifold, is not unfamiliar to mathematicians. For example, an integral of the product of characters over a simple Lie group can be reduced to its maximal torus at a price of adding an extra factor which accounts for the volume of the orbits of adjoint action. This factor is equal to the square of denominator in the Weyl character formula and appears as a Jacobian of a certain coordinate transformation.

A similar trick was developed for gauge invariant path integrals by Faddeev and Popov. Consider a Lie algebra valued functional $\Phi[A_\mu]$ such that each gauge orbit intersects transversally the set of its zeros

$$\Phi[A_\mu] = 0 \tag{9}$$

For a constant function $g(x) = g = \text{const}$, the second term in eq. (8) vanishes. If g also belongs to the center $Z(G)$ of G, then, obviously, $A^g = A$. Therefore a general gauge orbit intersects the set (9) at the same point $\text{Vol}(Z(G))$ times. $\text{Vol}(Z(G))$ denotes the number of elements in $Z(G)$. We use this notation to make connection with the formula (35), which, as we will see, works also for the case when the tangent spaces of the manifold (9) and a gauge orbit intersect along a finite dimensional space.

A path integral generalization of a simple formula

$$\int_{-\infty}^{+\infty} \delta(f(x))\frac{dx}{\sqrt{2\pi\hbar}} = \sum_{x_i:f(x_i)=0} \left|\sqrt{2\pi\hbar}f'(x_i)\right|^{-1} \tag{10}$$

can be used to derive the following identity:

$$1 = \frac{1}{\text{Vol}(Z(G))} \left|\det\left(\frac{\delta(\sqrt{2\pi\hbar}\Phi[A^g])}{\delta g}\bigg|_{\Phi[A^g]=0}\right)\right| \int \mathcal{D}g\, \delta(\Phi[A^g]). \tag{11}$$

A path integral δ-function which will reduce the integral over all connections to a submanifold (9) is called "gauge fixing".

A multiplication of the integral (2) by the r.h.s. of the identity (11) and a subsequent change of variables $A_\mu^g \to A_\mu$ allows us to factor the volume of the group of gauge transformations out of the integral over all gauge connections:

$$
\begin{aligned}
Z(\hbar) &= (\mathcal{D}g)^{-1} \int \mathcal{D}A_\mu\, e^{\frac{i}{\hbar}S_{CS}[A_\mu]} \\
&= (\mathcal{D}g)^{-1} \frac{1}{\text{Vol}(Z(G))} \int \mathcal{D}A_\mu\, e^{\frac{i}{\hbar}S_{CS}[A_\mu]} \left|\det\left(\frac{\delta(\sqrt{2\pi\hbar}\Phi[A^g])}{\delta g}\bigg|_{\Phi[A^g]=0}\right)\right| \int \mathcal{D}g\, \delta(\Phi[A^g])
\end{aligned}
$$

$$= \frac{1}{\text{Vol}(Z(G))} \int \mathcal{D}A_\mu \, e^{\frac{i}{\hbar}S_{CS}[A_\mu]} \delta(\Phi[A]) \left| \det \left(\left. \frac{\delta(\sqrt{2\pi\hbar}\Phi[A^g])}{\delta g} \right|_{g=1} \right) \right|. \tag{12}$$

2.3 Chern-Simons Path Integral

Let us apply a stationary phase approximation to the integral (12). We first look for the stationary phase points[3]

$$\frac{\delta S}{\delta A_\mu} \sim \epsilon^{\mu\nu\rho} F_{\nu\rho} = 0 \tag{13}$$

These points are flat connections, i.e. connections for which $F_{\mu\nu} = 0$. The gauge equivalence classes of flat connections are in one-to-one correspondence with the homomorphisms

$$\pi_1(\mathcal{M}) \overset{A}{\to} G, \quad A : x \mapsto g(x) \in G \tag{14}$$

up to a conjugacy, that is, the homomorphisms $x \to g(x)$ and $x \to h^{-1}g(x)h$ are considered equivalent.

The next step is to expand the action (1) up to the terms quadratic in gauge field variation a_μ around a particular flat connection $A_\mu^{(a)}$:

$$S_{CS}[A_\mu^{(a)} + \pi\sqrt{\frac{2}{k}}a_\mu] \approx S_{CS}[A_\mu^{(a)}] + \frac{\pi^2}{k}\epsilon^{\mu\nu\rho}\text{Tr}\int a_\mu D_\nu a_\rho \, d^3x. \tag{15}$$

Here D_ν is a covariant derivative with respect to the "background field" $A_\mu^{(a)}$:

$$D_\nu = \partial_\nu + [A_\nu^{(a)}, *]. \tag{16}$$

---

[3]Eq.(13) can be used to verify the gauge invariance of the action (1) under small gauge transformations. The infinitesimal version of eq. (8) is

$$\delta_\omega A_\mu = D_\mu \omega,$$

so that

$$\begin{aligned}
\delta_\omega S &= \int \frac{\partial S}{\partial A_\mu} D_\mu \omega \, d^3x \sim \epsilon^{\mu\nu\rho} \text{Tr} \int F_{\nu\rho} D_\mu \omega \, d^3x \\
&= -\text{Tr} \int \omega \epsilon^{\mu\nu\rho} D_\mu F_{\nu\rho} \, d^3x = 0
\end{aligned}$$

because of Bianchi identity.

A gauge fixing condition should be imposed on the fluctuation field a_μ. Witten suggested a covariant (with respect to $A_\mu^{(a)}$) choice[4]:

$$\Phi[a_\mu] = D_\mu a_\mu. \tag{17}$$

According to eq. (8), a change of a_μ under an infinitesimal gauge transformation $g(x) \approx 1 + \omega(x)$ is

$$\delta_\omega a_\mu = \frac{1}{\pi}\sqrt{\frac{k}{2}}\,D_\mu\omega. \tag{18}$$

Therefore the operator $\delta(\sqrt{2\pi\hbar}\Phi[A^g])/\delta g\big|_{g=1}$ is a covariant Laplacian $\Delta = D_\mu D_\mu$ acting on 0-forms. As for the δ-function of $\Phi[a_\mu]$, it can be presented as a path integral over a Lie algebra valued scalar field:

$$\delta(\Phi[a_\mu]) = \int \mathcal{D}\phi \exp\left[2\pi i\,\mathrm{Tr}\int \phi D_\mu a_\mu\, d^3x\right]. \tag{19}$$

As a result, at the 1-loop level

$$\begin{aligned}
Z(M,k) &\approx \frac{1}{\mathrm{Vol}\,(Z(G))}\sum_a e^{i\frac{k}{\pi}S_{CS}[A_\mu^{(a)}]}|\det\Delta| \\
&\quad \times \int \mathcal{D}a_\mu \mathcal{D}\phi \exp\left[i\pi\,\mathrm{Tr}\int d^3x\,(\epsilon^{\mu\nu\rho}a_\mu D_\nu a_\rho + 2\phi D_\mu a_\mu)\right] \\
&= \frac{1}{\mathrm{Vol}\,(Z(G))}\sum_a e^{i\frac{k}{\pi}S_{CS}[A_\mu^{(a)}]}\frac{|\det\Delta|}{(\det L_-)^{1/2}}.
\end{aligned} \tag{20}$$

Here L_- is the operator of the quadratic form in the exponential of the path integral. L_- acts on the direct sum of 0-forms and 1-forms on M:

$$L_-(\phi, a_\mu) = (D_\mu a_\mu, \epsilon_{\mu\nu\rho}D_\nu a_\rho - D_\mu\phi). \tag{21}$$

If we use 3-forms instead of 0-forms, then $L_- = \star D + D\star$, $\star$ is the Hodge operator. Note that the integration measures of the fluctuation fields $\mathcal{D}\phi$ and $\mathcal{D}a_\mu$ do not contain any implicit factors in contrast to $\mathcal{D}g$ and $\mathcal{D}A_\mu$.

---

[4]Note that $\Phi[a_\mu]$ depends on the metric of M. This dependence ultimately results in a framing dependence of $Z(M,k)$.

2.4 η-Invariant

A. Schwartz observed in [7] that the absolute value of the ratio of determinants in eq. (20) was equal to the square root of the Reidemeister-Ray-Singer analytic torsion. The phase of the ratio is equal to the η-invariant of Atiyah, Patodi and Singer. Similarly to eq. (7) it is a difference between the number of positive and negative eigenvalues of L_-. Thus the formula for the ratio of determinants is

$$\frac{|\det \Delta|}{(\det L_-)^{1/2}} = \tau_R^{1/2} e^{i\frac{\pi}{4}\eta}. \tag{22}$$

Actually L_- has infinitely many eigenvalues, so a regularization is needed to define η. To get some idea of how η might depend on the background connection $A_\mu^{(a)}$ consider the following simple problem. Let the eigenvalues be $\lambda_m = m + a$, $m \in \mathbf{Z}$ and let us calculate the number of positive λ_m minus the number of negative ones as a function of a. The simplest regularization is

$$\eta_a = \lim_{\epsilon \to 0} \left[\sum_{\lambda_m > 0} e^{-\lambda_m \epsilon} - \sum_{\lambda_m < 0} e^{\lambda_m \epsilon} \right]. \tag{23}$$

Suppose that $0 < a < 1$, then

$$\eta_a = \lim_{\epsilon \to 0} \left[\frac{e^{-a\epsilon}}{1 - e^{-\epsilon}} - \frac{e^{-(1-a)\epsilon}}{1 - e^{-\epsilon}} \right] = 1 - 2a. \tag{24}$$

In particular $\eta_{1/2} = 0$ because of the symmetry between the positive and negative λ for $a = 1/2$. A dependence of η_a on a is a nontrivial consequence of the infinity of the number of eigenvalues, since naively the number of positive and negative λ is the same for any $a \in (0, 1)$.

Obviously, η_a is a periodic function of a: $\eta_{a+n} = \eta_a$, $n \in \mathbf{Z}$, because a and $a + n$ define the same set of eigenvalues λ. Therefore eq. (24) requires a modification to work for all a. Indeed, when a moves through an integer number n, an eigenvalue λ_{-n} changes the sign and the value of η_a jumps by two units. Define I_a to be a number of positive eigenvalues becoming negative minus a number of negative eigenvalues becoming positive when the paramemter goes from $1/2$ to a. Then

$$\eta_a = 1 - 2a - 2I_a. \tag{25}$$

If we define η_0 to be equal to 1 as if $\lambda_0 = 0$ is counted as positive, then we arriive to the formula

$$\eta_a = \eta_0 - 1 + (1 - 2a) - 2I_a. \tag{26}$$

A similar formula for the η-invariant of L_- was derived in [3]:

$$\eta_a = \eta_0 - (1 + b^1(M))\dim G + \frac{4}{\pi^2} c_V S_{CS}[A_\mu^{(a)}] - 2I_a, \tag{27}$$

here η_0 is the η-invariant of the trivial connection, $b^1(M)$ is the first Betti number of M, I_a is a spectral flow of L_- and c_V is a dual Coxeter number of the group G (e.g. $c_V = N$ for $SU(N)$). The operator L_- for a trivial connection has $\dim G$ 0-form zero modes which are constant Lie algebra valued functions and $b^1(M)\dim G$ 1-form zero modes which are Lie algebra valued closed 1-forms. All these modes are counted as positive in η_0, hence the term $(1 + b^1(M))\dim G$. The role of the smooth function $1 - 2a$ is played by $\frac{4}{\pi^2}c_V S_{CS}[A_\mu^{(a)}]$.

The metric of M enters the gauge fixing functional (17) as well as the operators Δ and L_-. We could naively assume that this dependence would cancel out from the ratio of determinants in eq. (22). However the phase η has an "anomalous" dependence on the metric. It can be compensated by multiplying $Z(M, k)$ by an extra phase factor

$$\exp\left[-i\dim G\frac{1}{96\pi} \int_M \mathrm{Tr}(\omega \wedge d\omega + \frac{2}{3}\omega \wedge \omega \wedge \omega)d^3x\right], \tag{28}$$

here ω is a Levi-Chivita connection on M and the integral in the exponent is its Chern-Simons invariant. This invariant is defined relative to the choice of basis in the tangent space at each point of M. The local change in that basis is the analog of the gauge transformation. The exponent of eq. (28) is invariant under the transformations which are homotopic to identity. The choice of basis modulo such transformations is called "framing". The change in framing by n units shifts the phase of the factor (28) by $\pi n \dim G/12$. Actually the whole invariant (3) with a compensated metric dependence would be multiplied by a factor

$$\exp\left[i\frac{\pi}{12}n \dim G\frac{k}{k + c_V}\right]. \tag{29}$$

Physicists call the exponent of eq. (28) a 1-loop counterterm. It converts the metric dependence of η into a framing dependence of the invariant $Z(M, k)$. According to [3],

$$\eta_0 = 0 \tag{30}$$

in the special framing of M called canonical.

2.5 Zero Modes

To complete the study of the stationary phase approximation we have to consider the flat connections for which the operators L_- and Δ have zero modes. The 0-form zero modes of these operators satisfy the same equation

$$D_\mu\omega = 0, \tag{31}$$

so they are the elements of a cohomology H_a^0 built upon a covariant dreivative (16). A 1-form a_μ which is a zero mode of L_-, satisfies two equations

$$\epsilon^{\mu\nu\rho}D_\nu a_\rho = 0, \quad D_\mu a_\mu = 0 \tag{32}$$

The first equation means that a_μ is a closed form with respect to D, the second one means that it is not exact: if $a_\mu = D_\mu\omega$, then $D_\mu^2\omega = 0$, hence $D_\mu\omega = 0$. As a result, the 1-form zero modes are the elements of the cohomology H_a^1.

Let us remove the zero modes from the operators L_- and Δ. The absolute value of the ratio of their determinants is still equal to the square root of the Reidemeister torsion, which, as noted in [4], becomes an element of $\Lambda^{\max}H_a^0 \otimes (\Lambda^{\max}H_a^1)^*$. As for the phase η, it can be obtained by a simple correction of eq. (27) presented in [3]:

$$\eta_a = \eta_0 - (1 + b^1(M))\dim G - (\dim H_a^0 + \dim H_a^1) + \frac{4}{\pi^2}c_V S_{CS}[A_\mu^{(a)}] - 2I_a. \tag{33}$$

The zero modes of L_- are counted as positive in the spectral flow I_a. Therefore their number had to be subtracted from η_a since they are removed from the l.h.s. of eq. (22) and do not affect its phase.

According to eq. (31), the 0-form zero modes are the infinitesimal gauge transformations that do not change the background field $A_\mu^{(a)}$. The group of gauge transformations which is a symmetry of $A_\mu^{(a)}$, is isomorphic to a subgroup $H_a \subset G$ which commutes with the image of the homomorphism (14). Therefore H_a^0 is isomorphic to a Lie algebra of H_a.

The 1-form zero modes a_μ are the deformations of a flat connection $A_\mu^{(a)}$ which preserve its flatness in the linear order in a_μ:

$$F_{\mu\nu}[A_\rho^{(a)} + a_\rho] \approx \epsilon_{\mu\nu\lambda}\epsilon^{\lambda\sigma\rho}D_\sigma a_\rho = 0. \tag{34}$$

In most cases these infinitesimal deformations can be extended up to the finite flatness preserving deformations. Then H_a^1 is a tangent space of the moduli space $\mathcal{M}_a$ of flat connections at the point $A_\mu^{(a)}$.

A removal of the 0-form zero modes from the determinants of the r.h.s. of eq. (22) amounts to "forgetting" about H_a as a part of the group of gauge transformations. In other words, the symmetry under the global H_a gauge transformations[5] is not fixed, as it is demonstrated on a simple finite dimensional example in the Appendix of [6]. As a result, we have to divide the integrals of eq. (12) by the volume of H_a "by hands". A square root of the Reidemeister torsion as an element of $\Lambda^{\max}H_a^0 \otimes (\Lambda^{\max}H_a^1)^*$ defines a "ratio" of the volume forms on $\mathcal{M}_a$ and H_a. Therefore $\sqrt{\tau_R}/\mathrm{Vol}(H_a)$ is a volume form on $\mathcal{M}_a$ and it is quite natural to supplement a sum in eq. (20) by an integral over the components of the moduli space.

The volume forms for H_a and $\mathcal{M}_a$ being the part of the path integral measure, contain the factors $(2\pi\hbar)^{-1/2} = \pi(k/2)^{1/2}$. After extracting these factors from the integration measures we obtain the following 1-loop formula:

$$\begin{aligned}
Z(M,k) &= \sum_a e^{\frac{i}{\pi}(k+c_V)S_{CS}[A_\mu^{(a)}]}e^{-i\frac{\pi}{4}[(1+b^1(M))\dim G+\dim H_a^0+\dim H_a^1+2I_a]}\left(\frac{k}{2\pi^2}\right)^{\frac{\dim H_a^1-\dim H_a^0}{2}} \\
&\quad \times\frac{1}{\mathrm{Vol}(H_a)}\int_{\mathcal{M}_a}\tau_R^{1/2}.
\end{aligned} \tag{35}$$

The sum goes over the connected components of the moduli space of flat connections on M. The Chern-Simons action $S_{CS}[A_\mu^{(a)}]$ is constant within those components, because its derivative is zero due to eq. (13).

The formula (35) is not the end of the story, because sometimes $\dim\mathcal{M}_a < \dim H_a^1$. In other words, not all the 1-form zero modes of L_- can be extended to finite deformations of

---

[5] a gauge transformation (8) is called global if the transformation parameter $g(x)$ is (covariantly) constant.

the flat connection $A_\mu^{(a)}$. This means that the Chern-Simons action is not constant in the direction of these modes, rather its expansion around $A_\mu^{(a)}$ starts with the terms of order $m > 2$. The corresponding piece of the path integral has a form

$$\int d^n x \, \exp\left[2\pi i (2\pi\hbar)^{\frac{m-2}{2}} \frac{1}{m!} \frac{\partial^{(m)} S}{\partial X_{i_1} \ldots \partial X_{i_m}}\bigg|_{X_i = X_i^{(a)}} x_{i_1} \cdot \ldots \cdot x_{i_m}\right] \sim e^{i\pi \frac{n}{m}} (2\pi\hbar)^{\frac{n(2-m)}{2m}}, \quad (36)$$

here $n = \dim H_a^1 - \dim \mathcal{M}_a$ and $\hbar = \pi/k$ as defined in eq. (2). Therefore if $\dim \mathcal{M}_a < \dim H_a^1$, then we should substitute $\dim \mathcal{M}_a$ instead of $\dim H_a^1$ in the r.h.s. of eq. (35) and multiply it by the factor (36).

3 Surgery Calculus

3.1 Multiplicativity in Quantum Theory

The surgery calculus uses one of the basic principles of quantum field theory: a multiplicativity of the path integral (2). We are going to describe briefly what this multiplicativity means. Suppose that a 3-dimensional manifold M has a boundary ∂M. Let us impose a boundary condition on a connection A_μ. For example, we choose a tangent vector field v_μ omn ∂M and demand that $A_\mu v_\mu$ is equal to some fixed function A on ∂M:

$$A_\mu v_\mu = A. \quad (37)$$

Then a path integral (2) taken over all the connections on M satisfying this condition becomes a functional $\Psi[A]$. Such functional is called a wave function or a state in quantum theory. All possible functionals $\Psi[A]$ for a given manifold ∂M form a Hilbert space $\mathcal{H}_{\partial M}$. A scalar product in it is defined by a path integral over all functions A on ∂M:

$$\langle \Psi_2 | \Psi_1 \rangle = \int \bar{\Psi}_2[A] \Psi_1[A] \mathcal{D}A. \quad (38)$$

Suppose now that two manifolds M_1 and M_2 have diffeomorphic boundaries (with opposite orientations): $\partial M_1 = \partial M_2$. We can glue them together to form a single manifold M.

An integration over connections A_μ on M can be split into an integration over connections $A_\mu^{(1)}$ on M_1 and connections $A_\mu^{(2)}$ on M_2 satisfying the same condition (37) and an integration over all boundary conditions A. If $A_\mu^{(1)} v_\mu\big|_{\partial M_1} = A_\mu^{(2)} v_\mu\big|_{\partial M_2}$, then the Chern-Simons action is additive[6]:

$$S_M[A_\mu] = S_{M_1}[A_\mu^{(1)}] + S_{M_2}[A_\mu^{(2)}]. \tag{39}$$

Since the exponential is multiplicative, the integrals over $A_\mu^{(1)}$ and $A_\mu^{(2)}$ can be calculated separately yielding the wave functions $\Psi_{1,2}[A]$. The whole integral is a product $\Psi_1[a]\bar{\Psi}_2[A]$ (Ψ_2 is complex conjugated because ∂M_1 and ∂M_2 have opposite orientations). The final integral over A gives a scalar product:

$$Z(M,k) = \langle \Psi_2 | \Psi_1 \rangle. \tag{40}$$

To summarize, multiplicativity means that gluing the manifolds is achieved by taking a scalar product of the states appearing on their boundaries.

We adopt the strategy of [2]. Each 3-dimensional manifold can be constructed by a surgery on a link in S^3. The tubular neighborhoods of the link components are cut out, the modular transformations on their boundaries are performed and then they are glued back. So if we find the wave functions on both sides of the boundaries of tubular neighborhoods, then we can use eq. (40) to find Witten's invariant. The boundaries of the tubular neighborhoods are 2-dimensional tori T^2, so we start by describing the Hilbert space $\mathcal{H}_{T^2}$. We use canonical quantization as described in [8], where it was called "first constraining, then quantizing".

3.2 Canonical Quantization

Consider a manifold $M = \mathrm{R}^1 \times T^2 = \mathrm{R}^1 \times S^1 \times S^1$ with coordinates t along R^1 and $x_{1,2}$ along both circles, $0 \le x_{1,2} < 1$. The Chern-Simons action (1) can be cast in the form (up

---

[6]In fact, the action (1) on a manifold with a boundary should be corrected by a certain boundary term which guarantees that a derivative transversal to ∂M does not act on a tangential component of A_μ which is not fixed by condition (37) and hence is not necessarily continuous after the gluing.

to a total derivative in t that can be removed by adding appropriate boundary terms):

$$S_{CS} = \text{Tr} \int dt\, d^2x\, (A_2 \partial_0 A_1 + A_0 F_{12}). \tag{41}$$

The 1-form A_μ takes values in the Lie algebra of G. Lie algebra elements are antihermitian matrices in the adjoint representation. Quantum field theory deals usually with hermitian objects, so we introduce hermitian forms

$$\tilde{A}_\mu = -iA_\mu, \quad \tilde{F}_{\mu\nu} = -iF_{\mu\nu}. \tag{42}$$

Now

$$S_{CS} = -\text{Tr} \int dt\, d^2x\, (\tilde{A}_2 \partial_0 \tilde{A}_1 + \tilde{A}_0 \tilde{F}_{12}). \tag{43}$$

Compare this with the action of a constrained mechanical system

$$S = \int dt\, [p_i \dot{q}_i + h(p_i, q_i) + \lambda_\alpha \phi_\alpha(p_i, q_i)], \tag{44}$$

here q_i are coordinates, p_i are conjugate momenta, $h(p_i, q_i)$ is a hamiltonian, $\phi_\alpha(p_i, q_i)$ are constraints and λ_α are Lagrange multipliers. We see that $\tilde{A}_1$ and $-\tilde{A}_2$ are conjugate coordinates and momenta. The hamiltonian is zero as it happens in diffeomorphism invariant theories.

A path integral over A_0 in eq. (2) produces a δ-function of the constraint F_{12}, so we should, in fact, study only flat 2-dimensional connections as coordinates in the phase space. A gauge transformation can make both A_1 and A_2 constant. Moreover, since $\pi_1(T^2)$ is commutative, A_1 and A_2 will belong to the same Cartan subalgebra (e.g. they will be made diagonal simultaneously for $G = SU(N)$). The action (43) becomes simply

$$S_{CS} = \int dt\, \tilde{A}_2^a \dot{\tilde{A}}_1^a, \tag{45}$$

an index a runs over the orthonormal basis of Cartan subalgebra. After a quantization the fields $\tilde{A}_i^a$ become hermitian operators $\hat{\tilde{A}}_i^a$ satifying the Heisenberg commutation relation:

$$[\hat{\tilde{A}}_2^a, \hat{\tilde{A}}_1^b] = i\hbar \delta^{ab} \equiv i\frac{\pi}{k}\delta^{ab}. \tag{46}$$

This algebra can be represented in a space of functions $\psi(\tilde{A}_1^a)$:

$$\hat{\tilde{A}}_1^a \psi(\tilde{A}_1^b) = \tilde{A}_1^a \psi(\tilde{A}_1^b), \quad \hat{\tilde{A}}_2^a \psi(\tilde{A}_1^b) = i\hbar \frac{\partial}{\partial \tilde{A}_1^a} \psi(\tilde{A}_1^b). \tag{47}$$

The eigenfunctions of $\hat{\tilde{A}}_1^a$ are δ-functions, while the eigenfunctions of $\hat{\tilde{A}}_2^a$ are exponentials

$$|\alpha_a; 2\rangle \sim e^{i\alpha_a \tilde{A}_1^a}, \tag{48}$$

here we use a standard quantum mechanical notation for eigenstates:

$$\hat{\tilde{A}}_i^a |\alpha_b; i\rangle = \hbar \alpha^a |\alpha_b; i\rangle. \tag{49}$$

However a construction of a representation for the algebra $\hat{\tilde{A}}_i^a$ should reflect the fact that a Cartan subalgebra is not an appropriate configuration space for the torus T^2.

3.3 $U(1)$ Theory

Let us study carefully the simplest case of $G = U(1)$. The constant field components appearing in the action (45) are equal to the contour integrals along the periods $C_{1,2}$ of T^2

$$\tilde{A}_i = \oint_{C_i} \tilde{A}_j(x) \, dx^j \tag{50}$$

of any connection $\tilde{A}_j(x)$ which can be reduced to a constant one by a homotopically trivial gauge transformation. A homotopically nontrivial gauge transformation

$$g(x) = e^{2\pi i (m_1 x_1 + m_2 x_2)} \tag{51}$$

is well defined if $m_{1,2} \in \mathbf{Z}$. Eq. (8) shows that $\tilde{A}_1$ and $\tilde{A}_2$ remain constant under this transformation, but their values are shifted:

$$\tilde{A}_i \to \tilde{A}_i + 2\pi m_i. \tag{52}$$

Thus both coordinate $\tilde{A}_1$ and momentum $\tilde{A}_2$ are periodic with a period of 2π. The phase space is compact (it is $S^1 \times S^1$), its volume is $(2\pi)^2$ and, according to the WKB approximation, the dimension of the Hilbert space should be approximately

$$\dim \mathcal{H}_{T^2}^{U(1)} \approx \frac{(2\pi)^2}{2\pi \hbar} \equiv 2k \tag{53}$$

in the limit of large k. In fact, as we will see, eq. (53) is exact.

A periodicity in $\tilde{A}_1$ leads to a quantization of the eigenvalues of $\hat{\tilde{A}}_2$: α should be integer to make the eigenfunctions (48) periodic. On the other hand, since $\tilde{A}_2$ is also periodic, we should limit the number of independent values of α, e.g.

$$- k \leq \alpha < k, \ \ \alpha \in \mathbf{Z} \tag{54}$$

This procedure is self-consistent, because for integer k the period of $\tilde{A}_2$ (2π) is a multiple of the spacing of its eigenvalues $(\hbar = \pi/k)$.

The $2k$ values of α determine the momentum eigenstates $|\alpha; 2\rangle$, which form an orthonormal basis of $\mathcal{H}_{T^2}^{U(1)}$. Another basis is formed by the coordinate eigenstates $|\alpha; 1\rangle$ with the same range (54) of possible values of α. These two bases are related by a finite dimensional version of the Fourier transform (which also provides a relation between coordinate and momentum eigenstates in quantum mechanics of a particle on a line):

$$|\alpha; 2\rangle = \frac{1}{\sqrt{2k}} \sum_{\beta=-k}^{k-1} e^{i\frac{\pi}{k}\alpha\beta} |\beta; 1\rangle. \tag{55}$$

3.4 Modular Transformations

A unimodular transformation of cycles $C_{1,2}$ in eq. (50) generates a canonical transformation of our system:

$$C_i \overset{U}{\mapsto} C_i' = U_{ij}C_j, \ \tilde{A}_i \overset{U}{\mapsto} \tilde{A}_i' = U_{ij}\tilde{A}_j, \ U \in SL(2, \mathbf{Z}). \tag{56}$$

Therefore $SL(2, \mathbf{Z})$ can be represented in $\mathcal{H}_{T^2}^{U(1)}$. This group is generated by two elements

$$S = \begin{pmatrix} 0 & -1 \\ 1 & 0 \end{pmatrix}, \ T = \begin{pmatrix} 1 & 1 \\ 0 & 1 \end{pmatrix} \tag{57}$$

satisfying a relation

$$(ST)^3 = S^2 \tag{58}$$

Each matrix

$$U^{(p,q)} = \begin{pmatrix} p & r \\ q & s \end{pmatrix} \in SL(2, \mathbf{Z}) \tag{59}$$

can be presented as their product

$$U^{(p,q)} = T^{a_t} S \ldots T^{a_1} S. \tag{60}$$

The integer numbers a_i form a continued fraction expansion of p/q:

$$\frac{p}{q} = a_t - \cfrac{1}{a_{t-1} - \cfrac{1}{\cdots - \frac{1}{a_1}}}. \tag{61}$$

We denote as $\hat{U}^{(p,q)}$ an action of $U^{(p,q)}$ in $\mathcal{H}_{T^2}^{U(1)}$. According to eq. (60), it is determined by choosing $\hat{S}$ and $\hat{T}$.

The matrix S interchanges coordinate and momentum operators:

$$\hat{S}\hat{\tilde{A}}_1\hat{S}^{-1} = \hat{\tilde{A}}_2, \quad \hat{S}\hat{\tilde{A}}_2\hat{S}^{-1} = -\hat{\tilde{A}}_1. \tag{62}$$

The same is achieved by the matrix of eq. (55), so

$$\hat{S}_{\alpha\beta} = \frac{1}{\sqrt{2k}} e^{i\frac{\pi}{k}\alpha\beta} \tag{63}$$

in the coordinate basis $|\alpha; 1\rangle$. We use a formula

$$\exp\left[\frac{ik}{2\pi}\hat{\tilde{A}}_1^2\right] \hat{\tilde{A}}_2 \exp\left[-\frac{ik}{2\pi}\hat{\tilde{A}}_1^2\right] = \hat{\tilde{A}}_2 + \hat{\tilde{A}}_1, \tag{64}$$

which is easy to check by using a representation (47), in order to find $\hat{T}$ in the coordinate basis:

$$\hat{T}_{\alpha\beta} = e^{-i\frac{\pi}{12}} e^{\frac{i\pi}{2k}\alpha^2} \delta_{\alpha\beta} \tag{65}$$

The phase of $\hat{T}$ is chosen to comply with eq. (58).

3.5 $SU(2)$ Theory

Let us turn now to the case of $G = SU(2)$. Its Cartan subalgebra (an algebra of diagonal traceless antihermitian 2×2 matrices) is isomorphic to that of $U(1)$. A new feature is the Weyl reflection. A global gauge transformation

$$g = \begin{pmatrix} 0 & 1 \\ -1 & 0 \end{pmatrix} \tag{66}$$

changes the signs of $\tilde{A}_1$ and $\tilde{A}_2$:

$$g\tilde{A}_i g^{-1} = -\tilde{A}_i, \quad i = 1, 2. \tag{67}$$

The phase space should be factored by this transformation, its volume becoming half of that for $U(1)$. Therefore we expect a dimension of the Hilbert space $\mathcal{H}_{T^2}^{SU(2)}$ also to be approximately half of that of $\mathcal{H}_{T^2}^{U(1)}$.

The space $\mathcal{H}_{T^2}^{SU(2)}$ is isomorphic to a subspace of $\mathcal{H}_{T^2}^{U(1)}$ which is antisymmetric under the Weyl reflection (67), if we make an identification

$$k_{U(1)} = k_{SU(2)} + 2 = K_{SU(2)}. \tag{68}$$

In other words, the basis of $\mathcal{H}_{T^2}^{SU(2)}$ is formed by the antisymmetric combinations

$$|\alpha; i\rangle_{SU(2)} = \frac{1}{\sqrt{2}}\left(|\alpha; i\rangle_{U(1)} - |-\alpha; i\rangle_{U(1)}\right), \; 0 < \alpha < K, \quad \dim \mathcal{H}_{T^2}^{SU(2)} = K - 1. \tag{69}$$

As a result, the matrices $\hat{S}_{\alpha\beta}$ and $\hat{T}_{\alpha\beta}$ for $SU(2)$ are obtained (after a minor change in phase factors) by restricting the matrices (63) and (65) to the subspace (69):

$$\hat{S}_{\alpha\beta} = \sqrt{\frac{2}{K}}\sin\frac{\pi\alpha\beta}{K}, \quad \hat{T}_{\alpha\beta} = e^{-\frac{i\pi}{4}}e^{\frac{i\pi}{2K}\alpha^2}\delta_{\alpha\beta} \tag{70}$$

Eq. (60) was used in [4] in order to derive a formula for $M^{(p,q)}$:

$$\hat{U}_{\alpha\beta}^{(p,q)} = -i\frac{\text{sign}(q)}{\sqrt{2K|q|}}e^{-\frac{i\pi}{4}\Phi(M^{(p,q)})}\sum_{\mu=\pm 1}\sum_{n=0}^{q-1}\mu\exp\frac{i\pi}{2Kq}\left[p\alpha^2 + 2\mu\alpha(\beta + 2Kn) + s(\beta + 2Kn)^2\right] \tag{71}$$

Here $\Phi(M)$ is a Rademacher phi function defined as follows

$$\Phi\begin{bmatrix} p & r \\ q & s \end{bmatrix} = \begin{cases} \frac{p+s}{q} - 12s(s,q) & \text{if} \quad q \neq 0 \\ \frac{r}{s} & \text{if} \quad q = 0 \end{cases}, \tag{72}$$

a function $s(s,q)$ being a Dedekind sum:

$$s(m,n) = \frac{1}{4n}\sum_{j=1}^{n-1}\cot\frac{\pi j}{n}\cot\frac{\pi m j}{n}. \tag{73}$$

3.6   A General Simple Lie Group

Consider now a general simple Lie group G. A gauge transfomation can make the fields A_i constant and belonging to a Cartan subalgebra of the Lie algebra associated with G. The homotopically nontrivial gauge transformations like (51) force the eigenvalues α of the eigenvectors $|\alpha; i\rangle$ of $\hat{A}_i$ to belong to the weight lattice Λ_w of G factored by the root lattice Λ_R magnified K times (here $K = k + c_V$). The Weyl reflections similar to (67) require us to take only the Weyl antisymmetric combinations

$$\sum_{w \in W} (-1)^{|w|} |w(\alpha); i\rangle, \tag{74}$$

here W is the Weyl group and $|w|$ denotes a determinant of the transformation w. As a result, the basis of $\mathcal{H}_{T^2}^G$ is formed by the states (74) (i is either 1 or 2), for which $\alpha \in \Lambda_w$ belongs to the fundamental domain of the affine Weyl group $\tilde{W}$. This group is a semidirect product of the Weyl group W and a group of translations by the elements of the lattice $K\Lambda_R$. The walls of the fundamental domain should be excluded, that is, we require $\tilde{w}(\alpha) \neq \alpha$ for any $\tilde{w} \in \tilde{W}$.

A scalar product $\langle \alpha; i | \beta; i \rangle$ of the basis elements of $\mathcal{H}_{T^2}^G$ is equal to 1 if α and β are the shifted highest weights of conjugate representations, otherwise it is zero.

The formulas for $\hat{S}$ and $\hat{T}$ matrices of the simply laced Lie group G, as presented in [4] (see also [9]), are

$$\hat{S}_{\alpha\beta} = i^{|\Delta_+|} \left| \frac{\text{Vol } \Lambda_w}{\text{Vol } K \Lambda_R} \right|^{1/2} \sum_{w \in W} (-1)^{|w|} \exp \left(-\frac{2\pi i}{K} \langle w(\alpha), \beta \rangle \right), \tag{75}$$

$$\hat{T}_{\alpha\beta} = \delta_{\alpha\beta} \exp \left(\frac{i\pi}{K} \langle \alpha, \alpha \rangle - \frac{i\pi}{c_V} \langle \rho, \rho \rangle \right), \tag{76}$$

here Δ_+ is a set of positive roots of G, $|\Delta_+|$ is their number, $\langle \ , \ \rangle$ is a Cartan scalar product normalized so that the length of all roots of G is equal to $\sqrt{2}$, $\rho = \frac{1}{2} \sum_{\alpha \in \Delta_+} \alpha$. Note that $\hat{S}_{\alpha\beta}$ is proportional to the numerator of the Weyl character formula.

Now we should explain how to produce the states $|\alpha; i\rangle$ by taking a path integral over a 3-dimensional manifold M with a boundary $\partial M = T^2$. First we extend the path integral (2) by adding the so-called Wilson lines. Consider a closed manifold M with a link L inside it. Let us attach representations V_{α_i} of G to its components L_i. Here V_α denotes a representation of G with the shifted highest weight α (i.e. the highest weight of V_α is $\alpha - \rho$). For any connection A_μ on M the trace of its holonomy

$$\mathcal{O}_i = \mathrm{Tr}_{V_{\alpha_i}} \mathrm{P} \exp \oint A_\mu \, dx^\mu \tag{77}$$

is invariant under gauge and coordinate transformations. Therefore the path integral

$$Z^{\alpha_1,\dots,\alpha_n}(M, L, k) = \int [\mathcal{D}A_\mu] e^{i\frac{k}{\pi} S_{CS}(A_\mu)} \prod_i \mathcal{O}_i \tag{78}$$

is an invariant of the link L in M. Witten used the methods of conformal field theory to prove that it satisfies skein relations. Thus he proved that this invariant is equal to the Jones polynomial up to a normalization constant.

Take the integral (78) for a solid torus $M = S^1 \times D^2$ $(\partial M = S^1 \times \partial D^2 = T^2)$ and L consisting of one component $S^1 \times P$, P being the center of the disk D^2. Let C_1 be the cycle which is contractible through M. Witten claimed in [1], that if we attach a representation V_α to the only component of L, then the integral (78) produces a state $|\alpha; 1\rangle$ in $\mathcal{H}^G_{T^2}$. In particular, a solid torus without any link inside it is equivalent to a torus with a link carrying a trivial representation, hence it has a state $|\rho; 1\rangle$ on its boundary.

Let us cut out a tubular neighborhood of an n-component link L in S^3. A remaining piece $S^3 \setminus L$ has a boundary $(T^2)^n$. Therefore a path integral (2) taken over $S^3 \setminus L$ produces a state $|L\rangle$ in $(\mathcal{H}^G_{T^2})^{\otimes n}$. Suppose that we glue a tubular neighborhood back after putting a Wilson line (77) inside each of its components. Then according to the multiplicativity law (40),

$$Z^{\alpha_1,\dots,\alpha_n}(M, L, k) = \langle L| \bigotimes_{i=1}^{n} |\alpha_i; 1\rangle \tag{79}$$

Therefore the state $|L\rangle$ can be identified with the tensor $Z^{\alpha_1,\dots,\alpha_n}$ belonging to the dual Hilbert space $\left[(\mathcal{H}^G_{T^2})^{\otimes n}\right]^*$. The latter can be calculated by using cabling (for the link components wich carry representations of G other than the fundamental one) and skein relations.

Let us glue the components of the tubular neighborhood of L back after performing modular transformations $U^{(p_i, q_i)}$ on their boundaries. Any 3-dimensional manifold M can be constructed in this way. A multiplicativity law (40) leads to the following expression for its invariant:

$$Z(M,K) = \sum_{\alpha_1,\ldots,\alpha_n} Z^{\alpha_1,\ldots,\alpha_n}(S^3,L,k)\hat{U}^{(p_1,q_1)}_{\alpha_1\rho}\ldots\hat{U}^{(p_n,q_n)}_{\alpha_n\rho}. \tag{80}$$

The sum here goes, of course, over α_i belonging to the fundamental domain of the affine Weyl group $\tilde{W}$. Different surgeries on different knots in S^3 can produce the same manifold M. Reshetikhin and Turaev proved in [2] that the value of the r.h.s. of eq. (80) is the same for all such surgeries.

4 Some Examples

4.1 A Gluing Formula

We are going to use the surgery calculus in order to calculate the $U(1)$ and $SU(2)$ invariants of some simple 3-dimensional manifolds M. We will construct these manifolds by gluing together 2 solid tori after performing a modular transformation $U \in SL(2,\mathbf{Z})$ on the surface of one of them. Since neither of the tori has a Wilson line (77) inside it, then they have a state $|\rho;1_i\rangle \in \mathcal{H}^G_{T^2}$ corresponding to a trivial representation, on thier boundary. An index $i = 1,2$ refers to the fact that there are two contractible cycles $C_1^{(i)}$ on the common boundary T^2: $C_1^{(1)}$ is contractible through the first solid torus while $C_1^{(2)}$ is contractible through the second one. Let us use the basis in the Hilbert space $\mathcal{H}^G_{T^2}$ corresponding to the cycles $C_i^{(1)}$ of the first solid torus, then

$$|\rho,1_2\rangle = \sum_\alpha \hat{U}_{\rho\alpha}|\alpha;1_1\rangle. \tag{81}$$

According to the multiplicativity law (40), Witten's invariant of the manifold constructed by gluing the tori, is a scalar product

$$\langle\rho;1_1|\rho;1_2\rangle = \hat{U}_{\rho\rho}. \tag{82}$$

We will calculate the matrix element $U_{\rho\rho}$ with the help of eqs. (63), (65) and (70). The gluing induces a particular framing of the manifold, which may differ from the canonical one, so we will supplement $\hat{U}_{\rho\rho}$ with a correction factor (29). Then we will compare its large k limit with the stationary phase approximation formula (35). Since $U(1)$ is abelian, its Chern-Simons action is purely quadratic. Therefore the path integral (2) is gaussian and the formula (35) should be exact for the $U(1)$ invariant.

4.2   3-Dimensional Sphere

We start with the 3-dimensional sphere S^3. The two solid tori that form it are glued through a modular transformation S. The induced framing is canonical, so the invariants are

$$Z_{U(1)}(S^3, k) \;=\; \hat{S}_{00} = \frac{1}{\sqrt{2k}}, \tag{83}$$

$$Z_{SU(2)}(S^3, k) \;=\; \hat{S}_{11} = \sqrt{\frac{2}{K}} \sin \frac{\pi}{K} \stackrel{k\to\infty}{\longrightarrow} \sqrt{2}\pi K^{-3/2}. \tag{84}$$

To determine the invariants entering eq. (35) we note that $\pi_1(S^3)$ is trivial and so is the only flat connection on S^3. Its Chern-Simons action is zero and $\tau_R = 1$. All the phase factors of eq. (35) can be dropped due to eq. (30). For any $U(1)$ flat connection on a manifold with $b_1(M) = 0$

$$H_a = U(1), \; \dim H_a^0 = \dim H_a = 1, \; \dim H_a^1 = 0, \tag{85}$$

while for the trivial $SU(2)$ connection

$$H_a = SU(2), \; \dim H_a^0 = \dim H_a = 3, \; \dim H_a^1 = 0 \tag{86}$$

As a result eq. (83) and the r.h.s. of eq. (84) coincide with the 1-loop formula (35) if we assume that

$$\mathrm{Vol}\,(U(1)) = 2\pi, \;\; \mathrm{Vol}\,(SU(2)) = 2\pi^2. \tag{87}$$

Both volumes are perfectly consistent since $SU(2)$ is a 3-dimensional sphere and $U(1)$ is its big circle.

4.3 A Lens Space $L(p,1)$

A less trivial example of a manifold is a lens space $L(p,1)$. It is constructed by gluing two solid tori through a modular transformation

$$U^{(-p,1)} = ST^{-p}S. \tag{88}$$

According to [3] and [4], the induced framing differs from the canonical one by $p - 3$ units, so in canonical framing

$$Z_{U(1)}(L(p,1),k) = e^{-i\frac{\pi}{12}(p-3)}(\hat{S}\hat{T}^{-p}\hat{S})_{00} = e^{i\frac{\pi}{4}}\frac{1}{2k}\sum_{\alpha=0}^{2k-1}\exp\left(-\frac{i\pi}{2k}p\alpha^2\right). \tag{89}$$

We changed here the range of summation over α from (54) to an equivalent one $0 \le \alpha < 2k$.

The fundamental group of $L(p,1)$ is $\mathbf{Z}_p$, so there are p flat $U(1)$ connections corresponding to different homomorphisms (14). Therefore our objective is to transform the r.h.s. of eq. (88) into a sum of p terms. We are going to use a Poisson resummation formula, which relates a sum over integer numbers of a function and its Fourier transform:

$$\sum_{\alpha\in\mathbf{Z}} f(\alpha) = \sum_{m\in\mathbf{Z}}\int_{-\infty}^{+\infty} e^{2\pi im\alpha}f(\alpha)\,d\alpha \tag{90}$$

The sum in eq. (89) has a finite range, but we can extend it by using a periodicity of its summand as a function of integer numbers:

$$\exp\left[-\frac{i\pi}{2k}p(\alpha+2kn)^2\right] = \exp\left[-\frac{i\pi}{2k}p\alpha^2\right], \quad \text{for } \alpha,n \in \mathbf{Z}. \tag{91}$$

A "regularization" formula

$$\sum_{\alpha=0}^{T-1} f(\alpha) = \lim_{\epsilon\to0}(T\epsilon^{1/2})\sum_{\alpha\in\mathbf{Z}} e^{-\pi\epsilon\alpha^2}f(\alpha), \quad \text{if } f(\alpha+T) = f(\alpha) \text{ for } \alpha,T \in \mathbf{Z} \tag{92}$$

together with eq. (90) allow us to reexpress the invariant (88)

$$Z_{U(1)}(L(p,1),k) = \frac{e^{i\frac{\pi}{4}}}{2k}\lim_{\epsilon\to0}(2k\epsilon^{1/2})\sum_{m\in\mathbf{Z}}\int_{-\infty}^{+\infty} e^{-\pi\epsilon\alpha^2}\exp i\pi\left[-\frac{1}{2k}p\alpha^2+2m\alpha\right]. \tag{93}$$

The integral over α is gaussian, it is exactly equal to the contribution of the stationary pahse point

$$\alpha_m = 2\frac{k}{p}m \tag{94}$$

determined by eq. (5). The only effect of the prefactor $e^{-\pi\epsilon\alpha^2}$ to the leading order in ϵ is to suppress that contribution by the factor $e^{-\pi\epsilon\alpha_m^2}$:

$$Z_{U(1)}(L(p,1),k) = \lim_{\epsilon\to 0}(2k\epsilon^{1/2})\sum_{m\in\mathbf{Z}} e^{-\pi\epsilon\alpha_m^2}\frac{1}{\sqrt{2kp}}\exp\left(\frac{i\pi}{2k}p\alpha_m^2\right). \tag{95}$$

The stationary phase points α_m and their contributions exhibit the same symmetry under the action of the affine Weyl group, as the original summand in eq. (89). This means that if we add p to m, then α_m is shifted by $2k$, while the last exponential of eq. (95) remains unchanged. Therefore we can roll eq. (92) backwards in order to limit the summation range of m to its fundamental domain $0 \le m < p$:

$$Z_{U(1)}(L(p,1),k) = \sum_{m=0}^{p-1}\frac{1}{\sqrt{2kp}}\exp\left(2\pi i k\frac{m^2}{p}\right). \tag{96}$$

Thus we conclude that the 1-loop contributions of stationary points (94) appear to be exact. We just have to limit the sum to those α_m which belong to the fundamental domain of α:

$$0 \le \alpha_m < 2k. \tag{97}$$

We achieved our goal of resumming eq. (89). Now a comparison with the stationary phase formula (35) is straightforward. The Chern-Simons action of the flat $U(1)$ connection is known to be

$$S_m = 2\pi^2\frac{m^2}{p}, \tag{98}$$

its Reidemeister torsion is $1/p$. We can again drop all the phase factors of eq. (35) since in this case all $\eta_m = 0$. Eqs.(85) and (87) complete the picture: the 1-loop formula (35) is really exact.

A calculation of the $SU(2)$ invariant of the lens space $L(p,1)$ goes along the similar lines. The invariant in the canonical framing is equal to

$$Z_{SU(2)}(L(p,1),k) = \exp\left[-i\tfrac{\pi}{4}(p-3)\tfrac{K-2}{K}\right]\sum_{\alpha=1}^{K-1}\hat{S}_{1\alpha}\hat{T}_{\alpha\alpha}^{-p}\hat{S}_{\alpha 1}$$

$$= -\tfrac{1}{2K}\exp\tfrac{i\pi}{4K}[2p+3(K-2)]\sum_{\alpha=1}^{K-1}\sum_{\mu_1,\mu_2=\pm 1}\mu_1\mu_2\exp-\tfrac{i\pi}{2K}[p\alpha^2+2\alpha(\mu_1+\mu_2)]. \tag{99}$$

Again we apply a Poisson resummation formula (90). We limit the sum over the stationary phase points (94) to those which belong to the $SU(2)$ affine Weyl group fundamental domain:

$$0 \le \alpha_m \le K. \tag{100}$$

Recall that it is twice as small as that of $U(1)$, because the $SU(2)$ affine Weyl group includes a reflection $\alpha \to -\alpha$. A contribution of the stationary phase points which lie on the boundaries of the domain (100) should be cut in half (in the case of $U(1)$ we could avoid this by excluding the point $\alpha = 2K$ from the domain (97)). If p is odd, then there is only one such point $\alpha_0 = 0$, and the resummed expression (99) is

$$
\begin{aligned}
Z_{SU(2)}(L(p,1),k) \;=\; & -i\sqrt{\frac{2}{Kp}}\exp\frac{i\pi}{2K}(p-3) \\
& \times \left[\frac{1}{2}\left(e^{\frac{2\pi i}{Kp}}-1\right) + \sum_{m=1}^{\frac{p-1}{2}}\left(e^{\frac{2\pi i}{Kp}}\cos\frac{4\pi m}{p}-1\right)\exp\left(2\pi i K\frac{m^2}{p}\right)\right] \\
\xrightarrow{k\to\infty}\; & \sqrt{2}\pi(Kp)^{-3/2} + \sum_{m=1}^{\frac{p-1}{2}}\frac{i}{\sqrt{2Kp}}\left(2\sin\frac{2\pi m}{p}\right)^2\exp\left(2\pi i K\frac{m^2}{p}\right)\;(101)
\end{aligned}
$$

The number of terms in this equation is approximately half of that in the $U(1)$ formula (96). The number of flat $SU(2)$ connections is also approximately twice as small as that of $U(1)$, because the Weyl reflection (a conjugation by $g \in SU(2)$ of eq. (66)) makes the nontrivial homomorphisms $\pi_1(L(p,1)) = \mathbf{Z}_p \to U(1)$ pairwise equivalent.

The first term of the r.h.s. of eq. (101) is a contribution of the trivial connection. Indeed, the Reidemeister torsion of the trivial connection is p^{-3}, so we get an agreement with eq. (35). The sum in the r.h.s. of eq. (101) goes over nontrivial flat connections. Their Chern-Simons action is again given by eq. (98). Since $\pi_1(L(p,1)) = \mathbf{Z}_p$ is abelian, it is mapped by the homomorphism (14) into $U(1) \subset SU(2)$, so that its image commutes with the group $H_m = U(1)$. Therefore

$$
\dim H_m^0 = \dim H_m = 1. \tag{102}
$$

It is also known that $\dim H_m^1 = 0$. According to [3] and [4],

$$
\exp\left(-i\frac{\pi}{2}I_m\right) = -i. \tag{103}
$$

Combining all the pieces we see that the formula (35) coincides with the r.h.s. of eq. (101). The 1-loop formula is again demonstrated to work properly.

5   Discussion

Despite an obvious progress in calculating and understanding Witten's invariant, many open questions still remain. The $1/k$ expansion of knot invariants carried through the Feynman diagram technics in [10], [11] and [12] was very successful. The terms in this expansion appeared to be Vassiliev knot invariants, and they are expressed as integrals generalizing in a certain way the gaussian linking number. However, a systematic loop expansion of the manifold invariants started in [13], proved to be technically hard. At the same time, the exact resummed formulas for lens spaces (like the middle expression in eq. (101)), which are the sums over flat connections, look abnormally simple and nice from the quantum field theory point of view. Moreover, the contributions of the irreducible flat connections on Seifert manifolds, extracted from the surgery formulas in [6], are finite loop exact (that is, the corrections to the terms of eq. (35) go only up to a finite order in $1/k$ expansion). All these facts require a genuine 3-dimensional explanation.

Another possible development of the quantum Chern-Simons theory was suggested by Witten in [14]. He noted that if the fermionic gauge fields were properly added to the action (1) (in other words, if the Chern-Simons theory was based on an appropriate supergroup), then their determinant might cancel the bosonic one (i.e. a square root of the Reidemeister torsion) in eq. (35) up to a sign. A resulting invariant would be a sum over flat connections, each taken with a certain sign. Witten conjectured that Casson invariant might be obtained in that way. A Chern-Simons invariant based on a supergroup $U(1|1)$ was studied in [15]. It is related to the Alexander polynomial and also produces a "$U(1)$ Casson invariant" which is simply the order of the homology group. It is possible that a generalization of this theory to other supergroups, such as $U(2|2)$ may produce Casson invariant and provide a quantum field theory explanation for its calculation through surgery construction (see e.g. [16]).

Acknowledgements

I am thankful to C. DeWitt-Morette, D. Freed, L. Kauffman and A. Vaintrob for inviting me to give review talks at their seminars and for encouraging me to write these notes. I am also indebted to H. Saleur for many useful discussions.

References

[1] E. Witten, *Commun.Math.Phys.* **121** (1989) 351.

[2] N. Reshetikhin, V. Turaev, *Invent.Math.* **103** (1991) 547.

[3] D. Freed, R. Gompf, *Commun.Math.Phys.* **141** (1991) 79.

[4] L. Jeffrey, *Commun.Math.Phys.* **147** (1992) 563.

[5] S. Garoufalidis, *Relations Among 3-Manifold Invariants*, University of Chicago preprint, 1991.

[6] L. Rozansky, *A Large k Asymptotics of Witten's Invariant of Seifert Manifolds*, preprint UTTG-06-93, hep-th/9303099.

[7] A. Schwarz, *Lett.Math.Phys.* **2** (1978) 247.

[8] S. Elitzur, G. Moore, A. Schwimmer, N. Seiberg, *Nucl.Phys.* **B326** (1989) 108.

[9] V. Kac, *Infinite Dimensional Lie Algebras*, (Progress in Mathematics, vol.44), Basel: Birkhäuser 1983.

[10] L. Alvarez-Gaume, J. Labastida, A. Ramallo, *Nucl.Phys.* **B334** (1990) 103.

[11] E. Guadagnini, M. Martellini, M. Mintchev, *Phys.Lett.* **277B** (1989) 111.

[12] D. Bar-Natan, *Perturbative Chern-Simons Theory*, Princeton Preprint, August 23, 1990.

[13] S. Axelrod, I. Singer, *Chern-Simons Perturbation Theory*, Proceedings of XXth Conference on Differential Geometric Methods in Physics, Baruch College, C.U.N.Y., NY, NY.

[14] E. Witten, *Nucl.Phys.* **B323** (1989) 113.

[15] L. Rozansky, S. Saleur, *Nucl.Phys.* **B376** (1992) 461; *Nucl.Phys* **B389** (1993) 365; *Reidemeister Torsion, the Alexander Polynomial and $U(1,1)$ Chern-Simons Theory*, Preprint YCTP-P35-1992.

[16] K. Walker, *An extension of Casson's Invariant to Rational Homology Spheres*, preprint.

2+1 Dimensional Quantum Gravity as a Gaussian Fermionic System and the 3D-Ising Model

Maurizio Martellini[*]

I.N.F.N., sezione di Roma, I-00185 Roma, Italy

and

Mario Rasetti

Dipartimento di Fisica, Politecnico di Torino, I-10129 Torino, Italy
I.N.F.M., Unitá Torino Politecnico, I-10129 Torino, Italy

ABSTRACT

We show that 2+1-dimensional Euclidean quantum gravity is equivalent, under some mild topological assumptions, to a Gaussian fermionic system. Furthermore we argue that the corresponding 2+1-dimensional Euclidean quantum gravity partition function may be related to the partition function of the reduced 3D-lattice Ising model for finite lattices.

PACS # 04.60 ; 05.50 ; 02.40.P

---

[*] Permanent address: Dipartimento di Fisica, Universitá di Milano, I-20133 Milano, and I.N.F.N., sezione di Pavia, I-27100 Pavia, Italy

1. Introduction

A few years ago, Witten[1] showed that 2+1-dimensional quantum gravity in a first order dreibein formalism is exactly soluble at the classical and quantum levels. The key point in ref. [1] is the observation that the dreibein e_μ^a and the spin connection $\omega_\mu^a \equiv \varepsilon_{abc}\omega_\mu^{bc}$ form a gauge field of the group $ISO(2,1)$ $(ISO(3))$ in Lorentzian (Euclidean) signature. Thus, the Einstein-Hilbert action

$$
\begin{cases}
I = \dfrac{k}{2} \displaystyle\int_{M^3} \sum_a \varepsilon^{\mu\nu\rho} e_\mu^a R_{\nu\rho}^a(\omega) \\
R_{\nu\rho}^a(\omega) \equiv \partial_\nu \omega_\rho^a - \partial_\rho \omega_\nu^a + [\omega_\nu, \omega_\rho]^a
\end{cases}
\tag{1.1}
$$

becomes the non-Abelian Chern-Simons action on M^3 with gauge group $G = ISO(2,1)$ or $ISO(3)$ depending on the signature of the 3D-manifold M^3. Here, and in the following, we shall assume M^3 closed and oriented unless otherwise stated. However, in this context the meaning of solvability is quite obscure, since in Witten's approach solvability is ascribed to the fact that the Hilbert space is essentially the space of half-densities on the moduli space of flat $SO(2,1)$ $(SO(3)$, in the Euclidean case) connections on Σ_g, where Σ_g is a spacelike surface of M^3, which is a closed Riemann surface of genus g. Witten resorts to a canonical quantization scheme, which requires that M^3 is topologically $\sim \Sigma_g \times \mathbb{R}$. This result, which doesn't tell us anything about the full quantum correlation functions, is restricted to three-manifolds topologically equivalent to $\Sigma_g \times \mathbb{R}$ and essentially necessitates solving the Hamiltonian constraints of 3D-QG before quantizing, which in this particular case can be solved explicitly because they are linear in the momenta.

In this work, we shall show how for each fixed generic (closed) three-manifold M^3, the partition function $Z_{EQG}(M^3)$ of the Euclidean continued 3D-QG is equivalent (up to a normalization factor) to the partition function of a Gaussian discrete fermionic system whose action encodes the topological nature of M^3. Namely, we shall represent M^3 as the manifold obtained by Dehn surgery[2] on S^3 along a link $L \subset S^3$ and show that (under some suitable conditions)

$Z_{EQG}(M^3) = Z_{EQG}(L; S^3)$, which is a topological invariant of M^3, independent of the representative of the equivalence class of surgery presentations, is the partition function of free fermions propagating on the link diagram D_L and on its r-parallel versions. One may then study the correlation functions of 3D-Euclidean quantum gravity directly in a Gaussian fermionic formulation, rather than in the (non-linear) Chern-Simons gauge description[◊] . In particular, in the case M^3 is a hyperbolic three-manifold N^3 modelled on $S^3 - K$ for some knot K, we shall show that $Z_{EQG}(N^3)$ is equivalent to the reduced partition function of the 3D-Ising model[3] on a lattice Λ which is embeddable in N^3.[*] This will be done by resorting to the reduction of the latter[5] to the solution of the word problem[6] for a group $\Re$ related to the mapping class group of the Riemann surface $\Sigma_g \subset N^3$ in which Λ is immersed. One should point out here that the proof of equivalence does not imply the solution of either model, but suggests a very interesting possibility of cross-breeding between known results for the two models: we extensively comment on this point in the paper, in particular in the concluding remarks.

2. The Euclidean 3D-QG Partition Function and the Alexander-Conway Polynomial

Our starting point is Witten's result about the partition function of Euclidean 3D-QG. He shows[7] that if one selects a non-degenerate metric $\bar{g}_{\alpha\beta}$ on M^3 and a (background) flat $SO(3)$ spin connection $\bar{\omega}^a_{\mu(\alpha)}$, where α is a labelling index, and uses the Landau background gauge condition (no summation is understood)

$$\bar{D}^\mu_{(\alpha)} e^a_\mu = \bar{D}^\nu_{(\alpha)} \omega^a_\mu = 0 \tag{2.1}$$

where $\bar{D}^\mu_{(\alpha)} \equiv \bar{g}^{\mu\nu} \bar{D}_{\nu(\alpha)}$ is the covariant derivative with respect to the Levi-Civita connection $\bar{\nabla}_\nu$ associated to $\bar{g}_{\mu\nu}$ plus the flat connection $\bar{\omega}_{\mu(\alpha)}$ of interest, *i.e.*

◊ So far it is not well understood which kind of generalized Jones polynomials give the non-Abelian Chern-Simons theory with a non-compact gauge group $ISO(3)$ and on a generic three-manifold M^3 not homeomorphic to S^3.

* Clearly, here the 3D-Ising model is considered over a finite lattice: if the thermodynamic limit $n \to \infty$ were taken into account, one should have $g \to \infty$ (in 3D $g \propto n^3$). In our picture, this amounts to considering a sort of double scaling limit (DSL)[4] at the level of the reduced EQG-partition function on the Riemann surface $\Sigma_g \subset N^3$. Indeed, the DSL is the usual way of formulating the genus expansion, *i.e.* the sum over all genera.

$\bar{D}_{\nu(\alpha)} = \bar{\nabla}_\nu + [\bar{\omega}_{\nu(\alpha)}, \cdot]$, then the Euclidean partition function for the 3D-QG including the Faddeev-Popov ghosts reads [1]

$$\begin{cases} Z_{EQG}(M^3) = \sum_{(\alpha)} Z_{EQG(\alpha)}(M^3) \\[2mm] Z_{EQG(\alpha)}(M^3) = \dfrac{[\mathrm{Det}'(\bar{\triangle}_{0(\alpha)})]^2}{|\mathrm{Det}'(\bar{\mathcal{L}}_{(\alpha)})|} = \dfrac{[\mathrm{Det}'(\bar{\triangle}_{1(\alpha)})]^{\frac{1}{2}}}{\mathrm{Det}'(\bar{\triangle}_{2(\alpha)})}[\mathrm{Det}'(\bar{\triangle}_{3(\alpha)})]^{\frac{3}{2}} \end{cases} \qquad (2.2)$$

(since $\mathrm{Det}'(\bar{\triangle}_{k(\alpha)}) = \mathrm{Det}'(\bar{\triangle}_{(3-k)(\alpha)}))$ where $\bar{\mathcal{L}}_{(\alpha)} = *\bar{D}_{(\alpha)} + \bar{D}_{(\alpha)}*$ (here $*$ is the standard Hodge duality on forms) is the operator that maps a one-form plus three-form $(\omega^{(1)}, \omega^{(3)})$ to a one-form plus three-form $(*\bar{D}\omega^{(1)} + \bar{D}*\omega^{(3)}, \bar{D}*\omega^{(1)})$, and $\bar{\triangle}_{i(\alpha)} \equiv (\bar{D}_{\mu(\alpha)}\bar{D}^\mu_{(\alpha)})_i$ is the Laplacian operator acting on twisted i-forms. Furthermore, $\mathrm{Det}'(*)$ in (2.2) is a functional determinant, regularized, for instance, by the zeta-function technique [8] and omitting zero-modes. Equation (2.2) is derived under the following assumptions:

i) that the moduli space $\mathcal{N}$ of flat SO(3) connections modulo $SO(3)$-gauge transformations consists of finitely many points, and $\bar{\omega}_{\mu(\alpha)}$ is an arbitrary representative labelled by α of $\mathcal{N}$. If $M^3 \sim \Sigma_g \times \mathbb{R}$ is an orientable closed Riemann surface of genus g, $\mathcal{N}$ has connected components corresponding to Euler classes $2g - 2, 2g - 3,..., -(2g - 2)$. Here, the relevant component is that one, say $\bar{N} \in \mathcal{N}$, of maximal Euler class $2g - 2$ (ref.[7]) and in this case $\alpha = 1, ... , 6g - 6$;

ii) that all $\mathcal{N}$ connections are irreducible. Of course, these conditions on $\mathcal{N}$ drastically restrict the allowed topologies of M^3, however, as this set of "good topologies" is not empty, it is reasonable to work out a quantization scheme for Euclidean 3D-gravity only for this particular set of topological three manifolds. On the basis of such an assumption, we notice first that the ratio of determinants in (2.2) is in fact the Ray-Singer analytic torsion [9] (R.S.-torsion), $T_{\rho(\alpha)}(M^3)$, relative to the orthogonal representation $\rho_{(\alpha)}$: $\pi_1(M^3) \to O(m)$ (i.e. the i-forms on the universal cover of M^3 transform according to $\rho_{(\alpha)}$). This is a combinatorial invariant of M^{3*}, which labels homotopy equivalent spaces. For closed locally symmetric 3-manifolds of

---

$\star$ Therefore, it is independent of the metric used in the gauge fixing and Faddeev-Popov terms.

non-positive sectional curvature and closed hyperbolic 3-manifolds, i. e. the "geometric structures" of the interior of every compact 3-manifold, the Ray-Singer torsion is given by the zeta function associated to $\rho_{(\alpha)}$ analytically continued at zero.

Let us assume that $\rho_{(\alpha)}$ is acyclic, *i.e.* that $H^*(M^3, \rho_{(\alpha)})$ is zero. Cheeger and Müller have shown[10] that in this case the R.S.-torsion $T_{\rho(\alpha)}(M^3)$ is equivalent to the so-called Reidemeister torsion (R-torsion), $\tau_{\rho(\alpha)}(M^3) \in \mathbb{R}^+$, that we assume defined as in Cheeger[11] :

Definition: let X be a finite CW-complex associated to M^n and $\epsilon = \{\rho_{(\alpha)}\}$ a flat orthogonal bundle over X. Set $dim H^i(X, \epsilon) \equiv b_i$ and let $0 \neq \mu_i \in \Lambda^{b_i}(H^i(X, \epsilon))$ be a volume element for the cohomology group $H^i(X, \epsilon)$. Then, if we let K^* be a real cochain irreducible complex and assume that volume elements $0 \neq \omega_i \in \Lambda^{l_i}(K^i)^*$ with $l_i \equiv dim H^i$ have been chosen, the R-torsion $\tau(X, \epsilon, \mu) \equiv \tau_{\rho(\alpha)}(M^3)$ is defined by:

$$\tau(X, \epsilon, \mu) = \prod \frac{m_{2i}}{m_{2i+1}}$$

Here the scalars $m_i \neq 0$ are given by:

$$\lambda_i \wedge d_i^*(\lambda_{i+1}) \wedge \sigma_i = m_i \omega_i$$

where $d_i : K^i \to K^{i+1}$, $d_{i+1}d_i = 0, d_n = 0, \lambda_i \in \Lambda^{l_i}(K^i)^*, \sigma_i \in \Lambda^{b_i}(K^i)^*$. Moreover, the following conditions hold: the restriction of λ_i to $d_i(K^{i-1})$ is non-vanishing and the restriction of σ_i to $ker\, d_i$ is the pull-back of the volume element μ_i.

One can check that τ is independent of the particular choices of λ_i, i. e. it is a combinatorial invariant that may be computed from the twisted (by $\rho_{(\alpha)}$) cochain complex associated to M^3 by a suitable alternating product of determinants. This invariant plays a role in simple homotopy theory, where it is used to distinguish homotopy equivalent spaces.

Returning back to Eq. (2.2), therefore, one may set$^\diamond$:

$$Z_{EQG(\alpha)}(M^3) = T_{\rho(\alpha)}(M^3) = \tau_{\rho(\alpha)}(M^3). \tag{2.3}$$

$\diamond$ Fried[12] has shown that this identification survives also in the non acyclic case if $\rho_{(\alpha)}$ is orthogonal.

A few comments are in order:

i) In the definition of the Reidemeister torsion $\tau_{\rho(\alpha)}(M^3)$ one must start with a PL-manifold; but every 3-manifold may be triangulated and hence the PL-assumption is unnecessary.

ii) The representation $\rho_{(\alpha)} : \pi_1(M^3) \to O(m)$ extends to a unique ring homomorphism from the integral group ring $\mathbb{Z}(\pi_1(M^3))$ to the ring of all real $m \times m$-matrices. Now the Reidemeister torsion $\tau_{\rho(\alpha)}(M^3)$, as defined in ref. 11, is an element of the so-called Whitehead group $\bar{K}_1 M_m(\mathbb{R})$ (see below for the definition and main properties). It is known (ref. 11) that $\bar{K}_1 M_m(\mathbb{R}) \simeq \bar{K}_1 \mathbb{R}$, which, in terms of the Reidemeister torsion, is equivalent to saying that the representation of $\pi_1(M^3)$ is given by the ring homomorphism $\varphi_{(\alpha)} : \pi_1(M^3) \to F_0$, where F_0 is the commutative multiplicative group of a field, *e.g.* of the field of real numbers $\mathbb{R}$. Thus, we have:

$$\begin{cases} \tau_{\rho(\alpha)}(M^3) = \tau_{\varphi(\alpha)}(M^3) \\ \rho_{(\alpha)} : \pi_1(M^3) \to O(m) \quad . \\ \varphi_{(\alpha)} : \pi_1(M^3) \to F_0(\mathbb{R}) \end{cases} \qquad (2.4)$$

Our next step will be connecting Eq. (2.4) to an appropriate Alexander polynomial[13] $\triangle_L$. For this purpose, we need the general definition of Dehn surgery on a 3-manifold.

Definition: suppose that the following data are given: i) a 3-manifold M^3; ii) a link $L = K_1 \cup ... \cup K_n$ in the interior M^{o3} of M^3; iii) disjoint closed tubular neighbourhoods N_i^o of the K_i in M^{o3}; iv) a specified simple closed curve J_i in each ∂N_i. Then we may construct the 3-manifold

$$M'^3 = [M^3 - \cup_i N_i^o] \underset{h}{\bigcup} [\cup_i N_i] \quad ,$$

where h is the union of homeomorphisms $h_i : \partial N_i \to \partial N_i \subset M^3$, each of which take a meridian curve μ_i of N_i onto the given J_i. Thus the manifold M'^3, which depends only on the homotopy class of the J_i in ∂N_i, is said to be obtained by a "Dehn surgery on M^3 along the link L" with surgery instructions (iii) and (iv). Equivalently one says that M'^3 has Dehn surgery presentation (M^3, L) with instructions (iii) and (iv).

Following Lickorish,[14] we may always construct M^3 by Dehn surgery along a link $L = K_1 \cup \ldots \cup K_n$ in S^3 in the following way

$$M^3 = [S^3 - (K^{\circ}_{f_1} \cup \ldots \cup K^{\circ}_{f_n})] \cup_h (K_{f_1} \cup \ldots \cup K_{f_n}) \equiv (S^3 - L_f) \cup_h L_f \quad (2.5)$$

where $K^{\circ}_{f_i}$ is the interior of K_{f_i} and K_{f_i} is the *preferred framing* f_i of each component K_i of $L \subset S^3$, *i.e.* the map $K_i \to K_{f_i} \sim S^1 \times D^2$ in which the longitude λ_i is oriented in the same way as K_i and the meridian μ_i has linking number $+1$ with K_i itself. In (2.5), h is the union of homomorphisms $h_i : \partial K_{f_i} \to \partial K_{f_i} \subset M^3$ defined by $h_*(\mu_i) = [J_i] = a_i\lambda_i + b_i\mu_i$, where b_i is the linking number between K_i and J_i, whereas J_i is a specified fixed simple closed curve in each ∂K_{f_i} (clearly $a_i, b_i \in \mathbb{Z}$). Notice that the homeomorphism type of M^3 does not depend on the choice of h. Then we have the

Fundamental Theorem (ref. [14]): every closed oriented three-manifold may be obtained by Dehn surgery on a link L in S^3 with surgery coefficients $r_i = \dfrac{b_i}{a_i}$.

The key point in the proof of the fundamental theorem and in what we shall show later about the 3D-Ising model (Sec. 4), is noticing that h, defined equivalently as $h : S^3 - \{K^{\circ}_{f_i}\} \to M^3 - \{K^{\circ}_{f_i}\}$, may be characterized by *an element τ of the genus-g mapping class group* (extensively defined in Sec. 4). We do it in the following way. Every closed, orientable three-manifold M^3 admits, apart from the Dehn surgery representation quoted previously, a Heegaard decomposition: $M^3 = H_1 \cup_{\tau} H_2$, $\partial H_1 \sim -\partial H_2 \sim \Sigma_g$, $\tau : \partial H_2 \to \partial H_1$, where H_1 and H_2 are handlebodies of genus g, *i.e.* roughly speaking, orientable three-manifolds, with boundary an algebraic curve Σ_g, which are obtained by attaching g disjoint handles $D^2 \times [-1, 1]$ to 3-balls B^3. Let us now choose a Heegaard decomposition of the same genus g for S^3 and M^3 given by: $S^3 = H_1 \cup_f H_2$ and $M^3 = H_1 \cup_{f'} H'_2$, where $f : \partial H_2 \to \partial H_1$ and $f' : \partial H'_2 \to \partial H_1$ (H_2 and H'_2 are handlebodies of the same genus g). Thus, the above homeomorphism h may be defined as $I_d \times l$, where l is the homeomorphism $l : H_2 - \{K^{\circ}_{f_i}\} \to H'_2 - \{K^{\circ}_{f_i}\}$. Now, one can show (ref. 14) that l can always be obtained as an extension of the mapping class group element $\tau = (f')^{-1}f : \partial H_2 \to \partial H'_2 \sim \partial H_2$, *i.e.* τ belongs to the group of isotopy classes of orientation preserving self-diffeomorphisms of the orientable, closed Riemann surface $\Sigma_g \sim \partial H_2$. This latter statement follows from the fact that the handlebodies H_2' and H_2 are homeomorphic since they have the same genus g (ref. 2).

As Milnor (ref. [11]) first noticed, there is a close connection between the Alexander polynomial $\triangle_L(t_1,\ldots,t_n)$ and the Reidemeister torsion $\tau_\varphi(S^3 - L_f)$. Let us recall briefly the definition of the Alexander polynomial for a link $L = K_1 \cup \ldots \cup K_n$ in S^3.

Definition: If V_L is the exterior of L, *i.e.* $V_L = S^3 - L_f$, then the homology group $H_1(V_L)$ is canonically isomorphic to a free Abelian multiplicative group with n free generators $(t_1,\ldots,t_n)$. The generator t_i corresponds to the homology class of a meridian μ_i of the preferred framing f_i of K_i, $f_i : K_i \to K_{f_i}$. Clearly, if L is a knot, that is, if $n = 1$, we simply write t_1 instead of t. The integral group ring $\mathbb{Z}[H_1(V)]$ is identified via this correspondence with the Laurent polynomial ring $\mathbb{Z}[t_1, t_1^{-1},\ldots,t_n,t_n^{-1}]$. The Alexander polynomial $\triangle_L(t_1,\ldots,t_n)$ of the link $L \subset S^3$ is this Laurent polynomial in the variables $(t_1,\ldots,t_n)$ determined up to multiplication by polynomials of the form $\pm t_1^{r_1} \cdots t_n^{r_n}$ with integral $r_1,\ldots,r_n$.

To summarize, the Alexander polynomial is a homology invariant computable from the one-dimensional homology group of the exterior of the link with appropriate twisted coefficients. In the framework of such a homological point of view, the Alexander invariant (polynomial) might be defined also for links in a closed orientable 3-manifold different from S^3. It is therefore, at this point, worth recalling another way to introduce the Alexander polynomial, by connecting it to the matrix representations of the Artins braid group B_n. The braid group B_n on n strings $(n > 3)$ is the group formed by appropriate isotopy classes of braids with the natural concatenation operation:

Definition: B_n has a presentation $< \sigma_i \,|\, \mathcal{R}_\ell >$, $i = 1,\ldots,n - 1$, where the generators σ_i's satisfy the $\frac{1}{2}(n - 1)(n - 3)$ relations

$$\mathcal{R}_\ell : \sigma_i\sigma_{i+1}\sigma_i = \sigma_{i+1}\sigma_i\sigma_{i+1} \quad , \quad \sigma_i\sigma_j = \sigma_j\sigma_i \ \ \text{if} \ |i - j| \geq 2 \ .$$

Given a braid $\alpha \in B_n$ one may always form an oriented link L by the closure, denoted $\hat{\alpha}$, of the "legs" of the braid following a certain diagram. The presentation of a link L as a closed braid is highly non-unique; however A. Markov has found purely algebraic necessary and sufficient conditions for braids $\alpha \in B_n$ and $\beta \in B_n$ to have isotopic closures. Burau[15] has obtained a (reducible) matrix representation β_i of σ_i, where β_i is the $n \times n$ matrix with: $(i, i+1)$ entry z; $(i+1, i)$ entry 1; (k, k) entry (with $k \neq i, i+1$) 1; all other entries 0. $0 \neq z \in \mathbb{C}$ is a spectral parameter, which implies that the Burau representation of B_n is a sort of deformation of the representation of the symmetric group S_n by permutations of basis vectors of an n-dimensional space.

The key feature is that $\text{Det}(Id - \psi(\alpha))$, where ψ is the *reduced* Burau representation of $\alpha \in B_n$, is $1 + z + \cdots + z^{n-1}$ times the Alexander polynomial $\Delta_L(\alpha)$ of the link $L = \hat{\alpha}$ in $S^{3\star}$. This description of the Alexander polynomial in terms of determinants of irreducible matrix representations of B_n is fundamental in obtaining a formal equivalence between the free energy of the 3D-Euclidean quantum gravity and 3D-Ising model before its thermodynamic limit, since the latter free energy admits a similar construction (see Sec. 4). However, to actually get this equivalence, it is necessary to introduce the relation existing between the Reidemeister torsion, which, as we have seen above, is related to the 3D-QG partition function, and the Alexander polynomial.

It should be noted$^\diamond$ that there is often in the mathematical literature an ambiguity connected with the combinatorial definition of R-torsion, leading to the

---

$\star$ A more conventional, yet more operative definition of Alexander polynomial is the following: one considers a generic projection Π of the knot K on an arbitrary plane. As one moves along Π (starting from a given point and in a given direction), each double point – corresponding to an intersection of Π with itself – is met twice, one as an underpass, the other as an overpass. Underpasses, which are labelled with a progressive number in the chosen direction, can be of two different types (I and II), depending on the direction (from the left or from the right) of the overpassing generator. In this picture each knot in generic position is completely characterized by a sequence of underpasses for each of which its type and number of the overpassing generator must be specified: this is the information supplied by the Alexander matrix $\mathcal{A} = \{a_{i,j}\}$. In the latter the i-th row corresponds to the i-th underpass. The only non-zero elements of such a row are (we denote by k the number of the overpassing generator):

    i) $a_{i,i} = -1$, $a_{i,i+1} = 1$, independent of the type of underpass, if $k = i$ or $k = i + 1$;

    ii) $a_{i,i} = 1$, $a_{i,i+1} = -z$, $a_{i,k} = z - 1$ for type I underpass, or $a_{i,i} = -z$, $a_{i,i+1} = 1$, $a_{i,k} = z - 1$ for type II underpass, if $k \neq i, i + 1$.

From the Alexander matrix constructed in this way the polynomial Δ_K can be obtained simply by computing any minor of order $n - 1$ (the result turns out to be independent on the choice of the minor), and multiplying it by $\pm z^{-m}$, where the sign and the integer m are chosen in such a way that the resulting polynomial in z has positive constant term and doesn't contain any negative powers of z.

$\diamond$ We are grateful to H. Saleur, D. Freed, and L. Kauffman for pointing out to us this delicate question

exchange of τ with τ^{-1}. According to the definition of Milnor[16] , torsion is the field element given by the ratio of a specific generator v_0 in the exterior power ΛC_0 of the vector space of the 0-complex C_0, and the generator $(v_1|v_2|v_3) \in \Lambda(\partial C_1)$, where the sequences

$$0 \to \partial C_{q+1} \to C_q \to \partial C_q \to 0$$

are used. In other words, torsion, in the sense of Milnor, is the inverse determinant of the map induced by ∂ in the above sequence (with $q = 0$). Such a definition leads to a torsion which is reciprocal to that defined by Franz[17] and Turaev,[18] resorting to exact sequences corresponding to the boundary homomorphism dual to that given above.

On the other hand, a recent combinatorial approach to 3D gravity[19] shows that the Einstein-Regge action can be reformulated as a spin-vector model whose hamiltonian is just the quadratic form $< v_0| \left(\hat{D}^{\mathrm{tr}}\hat{D}\right)^{-1} |v_0 >$, where $|v_0 >$ is a euclidean vector in the span of ΛC_0 and $\hat{D}$ is the orthogonal matrix representative of Milnor's torsion τ_φ. Since such a description is intrinsically bosonic, the path integral representation of the partition function, which in Hamiltonian language is nothing but the configuration sum of the Boltzmann factors associated with the above hamiltonian, generates a factor proportional to $\left[\det\left(\hat{D}^{\mathrm{tr}}\hat{D}\right)^{-1}\right]^{-\frac{1}{2}} = \det\hat{D}$ $= \tau_\varphi$, supporting our identification.

Therefore we shall not further dwell on the question of which definition of torsion one should refer to, in that the physical content of the present paper lies in proving the equivalence of two models through their free energies, as is customarily done in statistical mechanics in order to avoid normalization difficulties. Since free energy must be negative, and is defined through a logarithm, care should be exerted in adopting the definition $F \propto -|\ln\tau_\varphi|$ irrespective of the convention adopted in constructing τ_φ.

Returning to the connection between Reidemeister torsion and the Alexander polynomial, Milnor's theorem [16] states that:

Milnor Theorem (ref. [16]): the Alexander polynomial $\triangle_L$ of a link in S^3 is equal (up to a standard factor) to the Reidemeister torsion (namely Milnor's combinatorial torsion) τ_φ of the exterior of the link, *i.e.*

$$\tau_\varphi(S^3 - L_f) \simeq \triangle_L(t_1,\ldots,t_n) \tag{2.6}$$

where φ is the ring homomorphism: $\pi_1(S^3 - L_f) \to \mathbb{Z}[t_1, t_1^{-1}, \ldots, t_n, t_n^{-1}]$.

Indeed, $S^3 - K_{f_i}$ is a space with the homology of a solid torus, *i.e.* $H_{i=0,1}(S^3 - K_{f_i}) = \mathbb{Z}$ (otherwise zero), and hence in the case when $M^3 = S^3 - L_f$ we may identify the commutative <u>multiplicative</u> group F_0 defined in (2.4) with the Laurent polynomial ring $\mathbb{Z}(t_i, t_i^{-1})$. In general, if we have a set of homomorphisms $\rho_{(\alpha)} : \pi_1(S^3 - L_f) \to F_{0(\alpha)}(\mathbb{Z}) \simeq \mathbb{Z}(t_{i(\alpha)}, t_{i(\alpha)}^{-1})$, the correspondence (2.6) will take the form $\tau_{\varphi(\alpha)}(S^3 - L_f) \simeq \triangle_{L(\alpha)}(t_i)$, where $\triangle_{L(\alpha)}(t_i) \equiv \triangle_L(t_{i(\alpha)})$. However, the sums (in (α)) over $\tau_{\varphi(\alpha)}$ will become products (in (α)) over $\triangle_{L(\alpha)}$ since $\triangle_{L(\alpha)}$ is by definition an element of the <u>multiplicative</u> group, namely of $\mathbb{Z}[H_{1(\alpha)}(S^3 - L_f)]$.

To summarize, we have shown that if M^3 is obtained by Dehn surgery along a certain link L (with n components) in S^3 with the preferred framing f, then the 3D-Euclidean quantum gravity partition function in the background Landau gauge is given by (up to some irrelevant normalization factors):

$$Z_{EQG}[M^3 = (S^3 - L_f) \cup_h L_f] = \sum_{(\alpha)} \tau_{\varphi(\alpha)}[(S^3 - L_f) \cup_h L_f], \tag{2.7}$$

where in particular

$$\sum_{(\alpha)} \tau_{\varphi(\alpha)}(S^3 - L_f) \simeq \prod_{(\alpha)} \triangle_{L(\alpha)}(t_i) \qquad i = 1,\ldots,n. \tag{2.8}$$

In the next section we shall show that the argument of the sum in (2.7) may be rewritten as the vacuum average of the *link operator* L in terms of the partition function provided by the Alexander polynomial $\triangle_L$. This will be a natural consequence of the fact that $\triangle_L$ *can be represented by a free fermionic Berezin-type path integral* and L by non-local composite free fermion operators.

It is also worth noticing in particular that if M^3 is *a priori* a fixed hyperbolic three-manifold N^3, homeomorphic by a homeomorphism δ to the exterior of a

312

knot K which is not a satellite knot nor a torus knot,[20] then equation (2.7) becomes formally

$$Z_{EQG}(N^3) = \sum_{(\alpha)} \tau_{\varphi(\alpha)}[\delta(S^3 - K_f)] \simeq \prod_{(\alpha)} \delta^* \triangle_{K(\alpha)}, \qquad (2.9)$$

here δ^* is the induced map on cohomology. Notice that $N^3 = \delta(S^3 - K_f)$ defined above is topologically $T^2 \times \mathbb{R}_+$; therefore it has a toroidal boundary and is prime, i.e. is a so-called Haken manifold.[*] Of course not every hyperbolic manifold is a knot complement, but most knot complements (e.g. the complement of the figure eight knot) are hyperbolic varieties. In Eq. (2.9) δ^* denotes the lift to $\tau_{\varphi(\alpha)}$ (and hence $\triangle_K$) of δ, i.e. formally: $\tau_{\varphi(\alpha)}[\delta(\cdot)] = \delta^* \tau_{\varphi(\alpha)}(\cdot)$. δ^* may be obtained from a matrix representation of the mapping class group $\mathcal{M}_g$ canonically associated with the Heegaard decompositions of genus g for S^3 and N^3.[◊]

Here, let us define the concept of Heegaard decomposition. For this purpose recall that a handlebody of genus g is the result of attaching g disjoint "one-handles" $D^2 \times [-1, 1]$ to a 3-ball B^3 by sewing the parts $D^2 \times \{\pm 1\}$ to $2g$ disjoint disks on ∂B^3 in such a way that the result is an orientable 3-manifold with boundary. Its boundary is an orientable Riemann surface of genus g. Then if H_1 and H_2 are two handlebodies of the same genus and $h = \partial H_2 \to \partial H_1$ is a homeomorphism, we say that a closed orientable 3-manifold M^3 admits a "Heegaard decomposition" or "diagram" (H_1, H_2, h) if the following definition holds

---

[*] A 3-manifold is a Haken manifold if it is a prime and it contains a 2-sided incompressible surface which is not a 2-sphere. One can show that a closed Haken manifold has a hyperbolic structure iff it is homotopically atoroidal (see ref. [20]).

[◊] Let us recall that under (any) framing $K_i \in L$ becomes a solid torus $T_i \equiv K_{f_i}$, $\partial T_i \neq \emptyset$, and that the set of all homomorphisms, up to isotopies, of a surface is defined as the mapping class group of that surface.[21] Thus, the proof that $\delta \in \mathcal{M}_g$ follows directly from Thurston construction (ref. 20) of hyperbolic three-manifolds N^3 as $N^3 = N'^3 - (T'_1, \ldots, T'_r)$, where N'^3 is hyperbolic and the T_i's are disjoint solid tori obtained by framing a suitable link L' with r components K_i and by the so-called Lickorish twist theorem (ref. 14). In fact one may choose Heegaard decompositions of the same genus for S^3 and N^3, i.e. $S^3 = H_1 \cup_g H_2$, $N'^3 = H'_1 \cup_{g'} H'_2$, where $g : \partial H_2 \to \partial H_1$ and $g' : \partial H'_2 \to \partial H'_1$. Here we assume that $\partial N'^3 = \emptyset$. Since all handlebodies of a given genus are homeomorphic, choose any homeomorphism $h : H_1 \to H'_1$. It follows, as a consequence of the Lickorish twist theorem, that the homeomorphism $f \equiv (g')^{-1} h g : \partial H_2 \to \partial H'_2$ belonging to the genus g mapping class group $\mathcal{M}_g$ extends to a homeomorphism $\bar{f} : H_2 - (T_1, \ldots, T_r) \to H'_2 - (T'_1, \ldots, T'_r)$. This extends the chosen $h : H_1 \to H'_1$ to a homeomorphism $\delta \equiv (h, \bar{f}) : S^3 - (T_1, \ldots, T_r) = S^3 - L_f \to N'^3 - (T'_1, \ldots, T'_r) = N'^3 - L'_f = N^3$. Thus, $\delta : S^3 - L_f \to N$ carries an action of the mapping class group $\mathcal{M}_g$.

Definition: A Heegaard diagram (H_1, H_2, h) of a closed orientable manifold M^3 is the identification space

$$M^3 = H_1 \cup_h H_2 \quad , \quad H_1 \cap H_2 = \partial H_1 = \partial H_2 \quad .$$

Two Heegaard decompositions (H_1, H_2, h) and (H_1', H_2', h') are called equivalent if there exist an orientation preserving homeomorphism φ such that $\varphi : H_1 \cup H_2 \rightarrow H_1' \cup H_2'$, $\varphi(H_{1,2}) = H_{1,2}'$. It is well known that every closed orientable 3-manifold admits a Heegaard decomposition.[22]

In Sec. 4 we shall show that Eq. (2.9) is equivalent to the reduced 3d-Ising model partition function associated to the lattice Λ which describes in the continuum limit a closed Riemann surface Σ_g of genus g embedded in the hyperbolic 3-manifold N^3 modelled over $S^3 - K_f$. In this particular case, one has three natural choices for $\Sigma_g \subset N^3$:

i) the surface F resulting from the intersection of a minimal genus orientable Seifert surface for K with K_f;

ii) the boundary $\partial N^3 \sim T_2$;

iii) the Heegaard surface given by the Heegaard presentation of the above 3-manifold N^3.

Notice that if $S^3 - K_f$ has a hyperbolic structure, then it is geometrically equivalent to the quotient of the hyperbolic 3-space H^3 by the translations $z \rightarrow z + 1$ and $z \rightarrow z + w$, where w is a complex number with $\text{Im}(\omega) \neq 0$.

Eq. (2.9) may be generalized to other hyperbolic 3-manifolds by performing a suitable Dehn surgery on S^3 along K such that $N^3 \sim S^3 - K_f$ has a hyperbolic structure, i.e. for a finite number of choices of surgery coefficients (a_i, b_i). In this case the r.h.s. of (2.9) must be understood as the Alexander polynomial for K in M^3, where M^3 is the 3-manifold obtained by the above "hyperbolic Dehn surgery". In any case, one may think to obtain by successive hyperbolic Dehn surgeries a hyperbolic 3-manifold $N^{3'}$ which is geometrically equivalent to H^3/Γ, where H^3 is the hyperbolic 3-space and Γ is a quasi-Fuchsian subgroup of PLS(2,$\mathbb{C}$). Recall that a quasi-Fuchsian group is abstractly isomorphic to a Fuchsian group, but its limit set is a Jordan curve in S^3, whereas a Fuchsian group has as limit set a round circle in S^2. The key point here is to observe that now $N^{3'} \sim H^3/\Gamma$ is topologically equivalent to $\Sigma_g \times \mathcal{I}$, where $\mathcal{I} \subseteq \mathbb{R}$ is an interval (which in our application of these notions to the Ising model will be assumed closed), and $\Sigma_g \sim \partial N^{3'}$ is a closed Riemann surface. In this way one may start directly from this generalization of Eq. (2.9), and not from (2.7), as "generic" Euclidean quantum gravity partition function (2.2). Indeed, as has been remarked in this Introduction, the canonical quantization of the 3-gravity requires that M^3 be topologically equivalent to $\Sigma_g \times \mathbb{R}$.

3. 3D-Euclidean Quantum Gravity
and 3-Manifold Topological Invariants

A recent result by Kohno[23] allows one to define topological invariants $K(M^3)$ of closed orientable three-manifolds M^3 using the representations of $\mathcal{M}_g$ in such a way that $K(M^3)$ is an invariant under the Heegaard decomposition. As noticed above, any closed oriented 3-d manifold admits a Heegaard decomposition which, via a theorem of Likorish[13], is naturally in one-to-one correspondence with an element of the mapping class group of genus equal to the Heegaard genus. Kohno's construction provides a projective linear representation of $\mathcal{M}_g$

$$\Phi_k : \mathcal{M}_g \to GL(Z_k(\Gamma))/\sim \quad , \tag{3.1}$$

where k is a positive integer labelling representations, and Z_k is a finite-dimensional complex vector space, each element of which is in one-to-one correspondence (via Kohno's $k+1$ admissible weights) with the edges of the dual graph Γ, which is a trivalent graph associated with the *pants* decomposition of the Heegaard surface Σ_g. $\sim$ denotes equivalence up to a phase factor that is a root of unity in this case. $K(M^3)$ is, up to a normalization factor, the *trace* of Φ, meant as the 00 entry of the matrix $\Phi_k(h)$, $h \in \mathcal{M}_g$ being the Heegaard glueing homeomorphism, with respect to the basis of Z_k.

The problem we are faced with here, on the other hand, is the *construction of the representations of the genus g mapping class group* $\mathcal{M}_g$ *starting from a Dehn surgery presentation for* M^3, for the following two reasons:

i) the Euclidean 3-d quantum gravity partition function (2.7) is a topological invariant by way of the Reidemeister torsion of the 3-manifold M^3, given by a Dehn surgery presentation, and it is therefore interesting to investigate the relation between such an invariant and Kohno's $K(M^3)$;

ii) in view of the features of the 3-d Ising model, whose partition function is based on the whole set of irreducible representations of the mapping class group of Σ_g (in turn presented in terms of Dehn twists), and of the expected equivalence between the two partition functions, we have to express Dehn's surgery invariants in terms of Heegaard invariants.

In other words, the problem is understanding the connection existing between $K(M^3)$ and the topological invariants $I(M^3)$ of the three-manifolds M^3 obtained[24] by performing Dehn surgery on a framed link.

We sketch here a procedure to obtain representations of the mapping class group from the Dehn surgery prescription, due to Kohno.[25]

The three-manifold can be obtained either via the above Heegard decomposition, or by Dehn surgery on links. The two points of view are closely related via a construction of link surgery data from a surface homeomorphism factored into Dehn twists. Namely, the data is a set of trivalent graphs γ_i (3-holed spheres) and a link $L_0 \in S^3$. At this point one has two options: either performing surgery on $L = (L_0, \gamma_i)$ in S^3 with a choice of framing (*e.g.* the *preferred framing* discussed below Eq. (2.5)) thus obtaining a three-manifold M^3 as shown above, or, equivalently, regarding the trinions γ_i as the complementary space of the so-called *pants decomposition* of a Riemann surface Σ_g. In other words, the γ_i's with, say, $i = 1, \ldots, n$, characterize a Riemann surface of genus $g = \frac{1}{2}(n+2)$. In this second case, the three-manifold M^3 is obtained by glueing, with a homeomorphism f, the cylinder $\Sigma_g \times I$ with another copy of Σ_g (the link L_0 is inside the cylinder). Then it turns out that f, called the *cylinder map*, belongs to the mapping class group $\mathcal{M}_g$ of Σ_g.* Two comments are now in order:

---

* This construction of $\mathcal{M}_g$ representations may be understood also in terms of the so-called *plat* representation of a link.[26] Namely, if L denotes the link carrying the Dehn surgery before the framing, we may represent it by a $(2g+2)$-plat. Recall that a $(2g+2)$-plat representation of the above link L in S^3 is a triad (S^3, Σ_0, L) where (S^3, Σ_0) is a Heegaard splitting of genus zero of S^3 which separates S^3 into 3-balls $B^{(1)}$ and $B^{(2)}$ so that $B^{(i)} \cap L$

a) f provides a representation of the Heegaard decomposition;

b) whereas the process leading from the Heegaard decomposition to the Dehn surgery (and to the mapping class group representations) is one-to-one (naturally up to Heegaard equivalence), the inverse construction leading from Dehn surgery to Heegaard decomposition (and once more to a representation of $\mathcal{M}_g$) is not necessarily one-to-one.

In other words, the surgery link L depends on f, whereas f in general does not depend on L alone.

The two invariants derived, one within the Dehn surgery scheme, the other in the Heegaard decomposition, are related. We argue that the two invariants derived one within the Dehn surgery scheme, the other in the Heegaard decomposition, should be related. To begin with, we recall that Cappell-Lee-Miller[27] have recently shown, in the framework of a conformal field theory approach to the problem of topological invariants of 3-manifolds, that the above invariants $K(M^3)$ and $I(M^3)$ are the same up to a phase factor. In our specific case, if we take $I(M^3) \equiv Z_{EQG}(M^3)$, where $Z_{EQG}(M^3)$ is given by Eq. (2.7), and recall that $Z_{EQG}(M^3)$ is also equal to the Reidemeister torsion $\tau(M^3)$, Eq. (2.3), the equivalence with Kohno's invariant $K(M^3)$ follows immediately from the fact that the Reidemeister torsion can distinguish homotopy equivalent spaces[28] just like Kohno's invariant $K(M^3)$. For example, the invariant $K(M^3)$ can distinguish the Lens spaces $L(7,1)$ and $L(7,2)$, which are non-homeomorphic three-manifolds with the same homotopy, like $\tau(M^3)^{[29]}$. Of course, the equivalence between $K(M^3)$ and $Z_{EQG}(M^3) \equiv \tau(M^3)$ is up to a suitable irrelevant phase factor which, in the physical picture of $Z_{EQG}(M^3)$ as a path integral (see next section), can always be reabsorbed into the functional measure.

---

is a collection of $g+1$ unknotted and unlinked arcs with $\partial(B^{(i)} \cap L)$ a set of $2g+2$ points on $\Sigma_0 \equiv \partial B^{(i)}$ for $i = 1,2$. The topological type of the triad (S^3, Σ_0, L) is fully described by a Heegaard sewing map φ which is required to preserve the $2g+2$ points in $\partial(B^{(i)} \cap L)$; hence, up to isotopy, it is an element of the mapping class group $\mathcal{M}_{0,2g+2} \sim B_{2g+2}$ (B_n is Artin's n-strings braid group) of the $(2g+2)$-punctured sphere. Here, $\mathcal{M}_{g,n}$ stands for the mapping class group for an n-punctured genus g Riemann surface $\Sigma_{g,n}$. Then, as is well known (ref. 26), if σ_i is the standard braid generator of B_{2g+2} which interchanges the i-th and the $(i+1)$-th points of $\partial(B^{(1)} \cap L)$, σ_i lifts to the Dehn twist τ_{C_i} ($i = 1,\ldots,2g+1$) along the non-contractible circle C_i decomposing the Heegaard surface Σ_g in a standard way. We may visualize Σ_g as the 2-fold covering of the sphere Σ_0 branched over $\partial(B^{(i)} \cap L)$, $i = 1,2$. Thus, $\mathcal{M}_g$ is *minimally* generated by a homomorphic image of B_{2g+2} and one further element (ref. 16).

Furthermore, the above problem can be seen in the broader framework of distinguishing between homology and homotopy equivalence of manifolds in 3-d. Two 3-manifolds, say M and N, are said to be *simple homotopy* equivalent if their CW complexes can be obtained one from the other by a CW deformation (namely, adding a finite sequence of cells). The question whether homotopy equivalence implies simple homotopy equivalence was answered by Whitehead[30] in a theorem stating that the obstruction to such implication is just the non-vanishing of the Witehead torsion, of which the Reidemeister torsion is a representation.

We briefly summarize a few relevant known features in connection with these points. Let Π denote a multiplicative group and $Z(\Pi)$ the corresponding group ring. Clearly, Π itself is contained in the group of units $U(Z(\Pi)) = GL(1, Z(\Pi)) \subset GL(Z(\Pi))$, where $GL(A)$, $A \equiv Z(\Pi)$, is the infinite general linear group. Thus there are natural homeomorphisms given by $\Pi \to K_1(Z(\Pi)) \to \bar{K}_1(Z(\Pi))$, where $K_1(A)$ is defined as the quotient $GL(A)/E(A)$ with $E(A)$ the (normal) commutator subgroup of $GL(A)$, and $\bar{K}_1(A)$ as its extension by the image of Π. K_1 is a covariant functor of A, i.e. any ring homomorphism $A \to A'$ gives rise to a group homomorphism $K_1 A \to K_1 A'$. The "Whitehead group" $W(\Pi)$ of Π is defined as the following covariant functor

Definition: the cokernel $\bar{K}_1(Z(\Pi))/\mathrm{Image} Z(\Pi)$ is called the Whitehead group $W(\Pi)$ of Π.

If Π is finite abelian, then $W(\Pi)$ is a free abelian group of rank $r - q$, where r (q) is the number of irreducible real (rational) representations of Π. In our contest, Π is played by the fundamental group π_1 of the CW-complex X associated to M^3. Indeed, let (X, L) be a pair consisting of a finite, connected CW-complex X and a subcomplex $L \subset X$ which is a deformation retract of X (see ref. 28). Then, one may define as before the Whitehead group $W(\Pi_1)$ associated to the fundamental group $\Pi_1 = \pi_1(X)$ of X (clearly, $\pi_1(X) \simeq \pi_1(L)$). The element $\tau \equiv \tau(x, L) \in W(\Pi_1)$, $\Pi_1 \equiv \pi_1(X)$, of $W(\Pi_1)$ is called the "Whitehead torsion". We concisely summarize here its main properties:

Combinatorial Invariance Theorem (See ref. [30])

1. The torsion $\tau(X, L)$ is invariant under subdivision of the pair (X, L).[*]

2. If each component of $X - L$, where L is a deformation retract of X, is simply connected, then $\tau(X, L) = 0$.

As a consequence of the above theorem, if f is any cellular homotopy equivalence $f : X \to Y$, where X and Y are CW-complexes associated to two 3-manifolds M^3 and $M^{3'}$ respectively, one has in general that $\tau(X, Y)$ is not always zero. If $\tau(X, Y) = 0$, then f is called a "simple homotopy equivalence" and clearly Y is a deformation retract of X.[◊] A general treatment of the Simple Homotopy theory in simplicial quantum gravity, not necessarily in dimension three, is in Carfora-Marzuoli.[31] Returning now to the Kohno invariant $K(M^3)$, as both $K(M^3)$ and our invariant (2.7) are simple homotopy invariants, yet <u>not</u> homotopy invariants, the above arguments, together with the results of Turaev and Viro[32] who succeeded in connecting the invariant constructed from the q-$6j$ symbols of the quantum group $U_q(s\ell(2))^*$ to the Kohno invariant, lead us to conjecture the equivalence of $I(M^3)$ and $K(M^3)$, that we shall henceforth assume. As a consequence of ref. 32, i.e. of the equivalence of $I(M^3)$ with the 3D-EQG partition function (2.2), one has that the R-torsion (which is proportional to $I(M^3)$) and $K(M^3)$ are 3D-*diffeomorphism invariants*, since by "construction" the 3D-EQG partition function is obtained by summing over all 3-geometries.

---

[*] As we have seen in Sec. 2, a particular representation of the Whitehead torsion is the Reidemeister torsion, that is the 3D-EQG partition function. Thus the fact that τ is a combinatorial invariant follows also from physics, when one represents the 3-gravity in simplicial form using the so-called Regge calculus.

[◊] From the Regge calculus point of view quoted in a previous footnote, the fact that $\tau(X, Y)$ is zero means that the 3D-EQG partition functions associated to the CW-complexes X and Y are connected by a simple scaling relation, or, in other words, X and Y are two fixed points of the lattice renormalization group transformations which act on the "cylinder (lattice) complex" M_f interpolating between X and Y.

[*] The Turaev-Viro lattice model partition function has continuum limit equal to the semi-classical limit $q \to 1$ of the $q - 6j$ symbols, and in this limit one gets our 3D-EQG partition function (2.2).

4. Free Fermions and the 3D-Euclidean
Quantum Gravity Partition Function

The approach to the Ising model which so far appeared to be the most promising for extension to the $d = 3$ case is that referred to as the *Pfaffian* (or *dimer*) method, whose formulation holds – to a certain extent – for any number d of dimensions. We briefly review here the formulation of a method that was proposed in [5] as a possible candidate to attack some three-dimensional cases, briefly commenting on those features most relevant to the present context.

The starting step in the approach adopted in [5] is the generalization of Kasteleyn's theorem, obtained by first applying the planar case formulation to the universal covering of one of the surfaces, say Σ_g, in which the lattice Λ_δ is embedded, and then by recovering it in a form suitable to the non planar case by summing over all 2^{2g} possible boundary conditions (spinwise) for the polygon whereby by suitable side identifications Σ_g – which can of course be thought of as a Riemann surface – is obtained. Such a sum over all possible choices of the boundary conditions is included in natural way in the configuration sum giving the model partition function. Successively, that sum is shown to be equivalent – as one would expect – to summing over all possible ways of embedding Λ_δ in a surface of genus g, and the latter in turn to be identical to a sum over all the images of Σ_g with respect to a subgroup $\Re$ of the mapping class group, i.e. essentially with respect to all PL diffeomorphisms modulo isotopy of Σ_g itself which preserve the nearest neighbouring relations of Λ . This identifies the partition function with the zeta function for the (infinite dimensional) set of flows induced by the diffeomorphisms themselves. Finally, the same partition function is expressed as a product of (theta regularized, if the thermodynamic limit is to be taken) determinants, closely reminiscent of Dirichlet-type zeta functions. These, resorting to Fried's definition,[33] can in principle be treated as the formal dynamical zeta function associated with a flow at value zero of its indeterminate.

The key feature, which has in fact only recently been proven by Moscovici and Stanton[34] , is that this zeta function coincides with the R-torsion, for Σ_g with coefficients in any flat acyclic bundle. Incidentally, there is a complete consistency between such a result, and the similarity pointed out by Milnor[35] between the algebraic formalism of R-torsion in topology and zeta functions in

the sense of Weil in dynamical systems. Therefore, the result perfectly bridges
the approach to the Ising model of ref. [5] and the present approach to quantum
gravity, with the identity exhibited by Ray and Singer in the second of refs. [9]
expressing the holomorphic analog of R-torsion for surfaces of genus > 1 in terms
of classical Selberg zeta functions, whence the conjecture of equality between R-
torsion and analytic torsion, subsequently proven by Cheeger and Müller in ref.
[10], was derived.

It is the fact that the model partition function can be expressed in the form
of a product of determinants, which turns out to be an invariant of the group $\Re$,
thought of as a group of topological transformations of a manifold (the lattice Λ,
or, more precisely, the surface Σ_g), together with the combinatorial features of
the model[*] and the known property of the 2-D model partition function of being
related to a link topological invariant[36] that allows completion of the proof.

Since one of the main ingredients of such an approach is a subgroup $\Re$ of
the mapping class group of a surface in which the lattice over which the model
is defined can be embedded (thought of as a Riemann surface), together with
its representations, we review first the basic notions to be used in the sequel
concerning the mapping class group: its definition (generators and relations),
and its representations.

The mapping class group $\mathcal{M}_g$ of an orientable 2-*manifold* Σ_g of genus g is
defined as the group of path components (*i.e.* modulo isotopy) of the group of
all orientation preserving homeomorphisms of Σ_g . Baer-Nielsen's theorems gives
us an equivalent definition : the mapping class group of a surface is isomorphic
to the outer automorphism group of its fundamental group.

It is interesting to recall here a few basic facts about the representations of
$\mathcal{M}_g$. First, one fixes the cut system $\mathcal{C}_0 \equiv \{\alpha_1, \ldots, \alpha_g\}$, namely a collection of
disjoint circles on Σ_g such that $\Sigma_g \setminus \left[\bigcup_{i=1}^{g} \alpha_i \right]$ is a connected manifold, isomorphic
to a $2g$-punctured sphere. The simplest choice is: α_1 going once around the

---

[*] In any dimension, and for any choice of boundary conditions, the partition function for the
Ising model over a *closed* lattice – *i.e.* a lattice which can be immersed in a two-dimensional
closed, compact, connected surface Σ_g – is proportional to the generating function for self-
avoiding (but possibly self-intersecting transversally) closed loops on the lattice Λ. Consequently, in any dimension evaluation of the partition function is reduced to solving the
word problem for some group: $\Re$ is such a group in dimension 3.

first handle, α_i, $i = 2, \ldots, g$ going once around the g-th handle which separates the $(i-1)$-th from the i-th hole. One introduces next the dual set of circles $\{\beta_i \mid i = 1, \ldots, g\}$: β_i goes once around the throat of the i-th hole ; for $i \leq g - 1$, β_i intersects transversally once both the circle α_i and the circle α_{i+1} ; β_g intersects transversally only α_g. One defines finally the family of closed simple curves on Σ_g, $\{\omega_{i,j} ; 1 \leq i < j \leq 2g\}$: $\omega_{i,j}$ interlaces handles i and j [more precisely, $\omega_{i,j}$ enters hole i , goes around handle i, comes out of hole $(i-1)$, enters hole j, goes around handle j, comes out of hole $(j-1)$ and closes]. Denote by $\mathcal{A}_i$, $\mathcal{B}_j$, $\mathcal{W}_{i,j}$ the Dehn twists along α_i, β_j, $\omega_{i,j}$, respectively. One defines moreover the new homeomorphisms of Σ_g : $\mathcal{P} \doteq \mathcal{A}_g \mathcal{B}_g \mathcal{A}_g$, which is a simple move permuting α_g and β_g ; $\mathcal{L} \doteq \mathcal{B}_g \mathcal{A}_g \mathcal{A}_g \mathcal{B}_g$, which reverses the orientation of α_g , and $\mathcal{T}_i \doteq \mathcal{B}_i \mathcal{A}_i \mathcal{A}_{i+1} \mathcal{B}_{i+1}$, $i = 1, \ldots, g - 1$ which permutes the circles α_i and α_{i+1}.

Theorem (Hatcher and Thurston[37] , Harer[38] , and Wajnryb[39]):

The mapping class group $\mathcal{M}_g$ is *generated* by the whole set
$$\{\mathcal{L} ; \mathcal{P} ; \mathcal{A}_i , i = 1, \ldots, g ; \mathcal{T}_j, j = 1, \ldots, g - 1 ; \mathcal{W}_{i,j} , 1 \leq i < j \leq g\} .$$

Let now $\mathcal{H}_0$ be the stabilizer subgroup, generated by $\{\mathcal{A}_i ; \mathcal{W}_{i,j}\}$, of those elements of $\mathcal{M}_g$ which leave the circles $\{\alpha_i\}$ fixed ; and $\mathcal{H}$ the subgroup, generated by $\{\mathcal{H}_0 ; \mathcal{L} ; \mathcal{T}_i\}$, of the elements which leave the cut system $\mathcal{C}_0$ invariant. $\mathcal{H}$ is defined by the exact sequences:

$$1 \longrightarrow \mathcal{H}_0 \longrightarrow \mathcal{H} \xrightarrow{\ \vartheta\ } \pm S_g \longrightarrow 1 \quad ;$$

$$1 \longrightarrow [\mathbb{Z}/2\mathbb{Z}]^g \longrightarrow \pm S_g \longrightarrow S_g \longrightarrow 1 \quad ;$$

where $\vartheta(\mathcal{L}) \in [\mathbb{Z}/2\mathbb{Z}]^g$ and $\vartheta(\mathcal{T}_i)$ is the transposition $(i, i + 1)$ in the symmetric group S_g .

The complete set of relations of $\mathcal{M}_g$, generated by $\{\mathcal{H}, \mathcal{P}\}$, realizes the following identities or properties:

    $i)$ $\mathcal{P}$ commutes with $\mathcal{H}_g$ (the subgroup of $\mathcal{H}$ whose elements leave α_g and β_g invariant);

    $ii)$ $\mathcal{P}^2 \equiv \mathcal{A}_g \mathcal{L} \mathcal{A}_g \in \mathcal{H}$;

iii) $\mathcal{P}\mathcal{F}\mathcal{P}\mathcal{F}\mathcal{P} \in \mathcal{H}$ whenever $\exists$:

iii.1) a circle γ on Σ_g which intersects once transversally both α_g and β_g and does not intersect any other α_i , $i \neq g$, and

iii.2) a map $\mathcal{F} \in \mathcal{H}$ such that $[\mathcal{P}\mathcal{F}]^{-1}\gamma\,\mathcal{P}\mathcal{F} = \beta_g$; $[\mathcal{P}\mathcal{F}]^{-1}\beta_g\,\mathcal{P}\mathcal{F} = \alpha_g$; $[\mathcal{P}\mathcal{F}]^{-1}\alpha_g\,\mathcal{P}\mathcal{F} = \gamma$.

iv) $\mathcal{P}$ commutes with $\tilde{\mathcal{F}}\mathcal{P}\tilde{\mathcal{F}}^{-1}$ where $\tilde{\mathcal{F}} \in \mathcal{H}$ maps the simple closed curve $\tilde{\beta}$ encircling holes $(g-1)$ and g onto β_g .

v) $\mathcal{P}\mathcal{F}_1\mathcal{P}\mathcal{F}_2\mathcal{P}\mathcal{F}_3\mathcal{P}\mathcal{F}_4\mathcal{P} \in \mathcal{H}$ whenever $\exists$:

v.1) a circle δ on Σ_g which intersects once transversally both α_{g-1} and β_g and does not intersect β_{g-1} nor any other α_i , $i \neq g-1$, and

v.2) the maps $\mathcal{F}_j \in \mathcal{H}$, $j = 1,\ldots,4$ satisfy – upon defining $\mathcal{E}_{(0)} \doteq \mathbb{I}$; $\mathcal{E}_{(n)} \doteq \mathcal{E}_{(n-1)}\mathcal{P}\mathcal{F}_n$ (in terms of which the element of $\mathcal{H}$ we are considering reads $\mathcal{E}_{(4)}\mathcal{P}$) – the four relations : $\mathcal{E}_{(n)}\beta_{g-1}\mathcal{E}_{(n)}^{-1} = \beta_g$; $n = 1,\ldots,4$.

When Σ_g has no punctures the isotropy subgroup $\mathcal{H}$ is an extension of the coloured braid group by the permutation group, as described by the exact sequence

$$\mathbb{Z} \longrightarrow \mathbb{Z}^g \oplus B_{2g-1} \longrightarrow \mathcal{H} \longrightarrow \pm S_g \longrightarrow 1 \; ;$$

where B_{2g-1} is the *Artin* coloured *braid group* over $(2g-1)$ strings and $\pm S_g$ is the group of signed permutations of g objects.

The above presentation allows one to derive information about faithful representations of $\mathcal{M}_g$[40] [41] For $g = 1$, $\mathcal{M}_g \sim SL(2,\mathbb{Z})$, the *classical* (as opposed to the *Teichmüller* or *many-handled*) Modular Group. The related moduli space is a space whose points correspond to conformal isomorphism classes of tori. For arbitrary $g > 1$, upon denoting by $\mathcal{I}(\Sigma_g)$ the set of isotopy classes of all the closed (non oriented) curves embedded in Σ_g , and by Φ_g any foliation whose leaves are geodesics for for some metric on Σ_g (since Σ_g has negative Euler characteristics, the metric is hyperbolic), with transverse measure $\mu_\perp$, we have the following results. $\mu_\perp(\bullet)$, which is a positive real function assigning to each arc $\sigma \in \Sigma_g$ transverse to the leaves of Φ_g and with extremal points in $\Sigma_g \setminus \Phi_g$ an invariant weight, is determined by the conditions :

a) $\mu_\perp(\sigma) = \mu_\perp(\sigma')$ if σ is homotopic to σ' through arcs transverse to Φ_g and with endpoints in $\Sigma_g \setminus \Phi_g$;

b) if $\sigma = \bigcup_i \sigma_i$; with $\sigma_i \cap \sigma_j \subset \partial\sigma_i \cap \partial\sigma_j$; then $\mu_\perp(\sigma) = \sum_i \mu_\perp(\sigma_i)$;

c) $\mu_\perp(\sigma) \neq 0$ if $\sigma \cap \Phi_g \neq \emptyset$.

The collection of all these measured geodesic foliations constitutes a space Ξ_g on which $\mathcal{M}_g$ acts in a natural way. In particular, in this (faithful) representation, the elements $m \in \mathcal{M}_g$ are classified according to the following scheme : m is said to be

periodic,   if it is of finite order in $\mathcal{M}_g$;

reducible,   if there is a point in $\mathcal{I}(\Sigma_g)$ which is invariant with respect to the element m itself ;

pseudo-Anosov,   if $\exists$ mutually transverse geodesic foliations $\Phi_g^{(s)}$, $\Phi_g^{(u)} \in \Xi_g$ (s stands for *stable* , u for *unstable*) , such that $m(\Phi_g^{(s)}) = \frac{1}{\varepsilon}\Phi_g^{(s)}$ and $m(\Phi_g^{(u)}) = \varepsilon\Phi_g^{(u)}$ for some real $\varepsilon > 1$.

In order to derive a faithful representation from our finite presentation, one should first prove that <u>no</u> normal subgroup $\mathcal{N}_{\mathcal{M}_g}$ of $\mathcal{M}_g$ can have all of its elements $\neq \mathbb{I}$ which are pseudo-Anosov, because only in this case can one identify a homeomorphism $m_o \in \mathcal{N}_{\mathcal{M}_g}$ fixing some $\iota \in \mathcal{I}(\Sigma_g)$ and then proceed in the construction of an *induced* faithful representation of $\mathcal{M}_g$ as a group of matrices (possibly with entries in a field of characteristic $\neq 0$ or of anticommuting variables) .

For example, let π be a path on Σ_g which crosses the curve α_i at a finite number ℓ of points $\{p_1^{(i)}, \ldots, p_\ell^{(i)}\}$. When we act on Σ_g with $\mathcal{A}_i$, the effect on π is that it is broken at each point $p_k^{(i)}$ and a copy of α_i is inserted at the discontinuity in such a way as to coalesce (also in orientation) with the adjacent fragments of π. Resorting to the property that on any compact surface such as Σ_g there exists at least a pair of essential simple closed curves, say γ , γ', which *fill* the surface but such that one can find another essential closed curve $\tilde{\gamma}'$, disjoint from γ', such that $\gamma \cup \tilde{\gamma}'$ does *not* fill the surface, one can show (ref. 40) that – upon denoting by $\mathcal{D}_\gamma$ the Dehn twist with respect to the curve γ – $\mathcal{D}_\gamma \mathcal{D}_{\gamma'}^{-1}$ is isotopic to a pseudo-Anosov map[42] . Then $\gamma'' \equiv \mathcal{D}_\gamma \mathcal{D}_{\gamma'}^{-1} \circ \tilde{\gamma}'$ is a curve disjoint from any essential simple curve $\tilde{\gamma}$ having no intersections with $\gamma \cup \tilde{\gamma}'$. Thus there exists a map

$$\mathcal{D}_{\tilde{\gamma}'}^{-1}\mathcal{D}_{\gamma}\mathcal{D}_{\gamma'}^{-1}\mathcal{D}_{\tilde{\gamma}'}\mathcal{D}_{\gamma'}\mathcal{D}_{\gamma}^{-1} \equiv \mathcal{D}_{\tilde{\gamma}'}^{-1}\mathcal{D}_{\gamma''}$$

which fixes $\tilde{\gamma}$ and hence is *not* pseudo-Anosov.

Considering the action of $\mathcal{M}_g$ on the projective space Ξ_g of measured geodesic foliations , Dehn twists should be treated as maps with parabolic action, since they are locally conjugate to the element $\begin{pmatrix} 1 & 1 \\ 0 & 1 \end{pmatrix} \in PSL(2,\mathbb{Z})$. Moreover, recalling the presentation of the fundamental group,

$$\pi_1(\Sigma_g) \sim < \mathcal{A}_1,\mathcal{B}_1,\ldots,\mathcal{A}_g,\mathcal{B}_g \mid \prod_{i=1}^{g} [\,\mathcal{A}_i,\mathcal{B}_i\,] >$$

where $\{\mathcal{A}_i,\mathcal{B}_i \mid 1 = 1,\ldots,g\}$ are assumed to be canonical basis for the first homology group $H_1(\Sigma_g)$, and noticing that its elements which act parabolically on the hyperbolic projective space are only those which may be freely homotoped into cusps and that just these elements are non-Anosov, all that remains to be done is to check – by using the presentation – whether $\mathcal{M}_g$ has a geometrically finite subgroup $\mathcal{S}_{\mathcal{M}_g}$ on which it acts by conjugation. Then, *unless* the normal closure in $\pi_1(\Sigma_g)$ of the elements of the action of $\mathcal{M}_g$ on $\mathcal{S}_{\mathcal{M}_g}$ excludes all the cusp generators, not all of its elements $\neq \mathbb{I}$ are pseudo-Anosov.

It is worth pointing out that this conclusion holds for $g \geq 2$, when $\mathcal{M}_g$ has a set of elementary homeomorphisms equivalent to *global braids*. The corresponding matrix representation , when it exists , is that induced from the *monodromy* representation associated with the *Lefschetz fibration*[43] of Σ_g .

The dimer approach to the Ising model is particularly effective when the lattice Λ over which it is defined is homogeneous under some finitely presented (not necessarily finite) group G , and consists of a number of steps :

1] the *decorated* lattice Λ_δ is derived from Λ following Fisher's scheme[44] ;

2] the positional degrees of freedom in Λ_δ are relabelled in terms of a set of anticommuting *Grassmann* variables η_{g_ℓ}, in one-to-one correspondence with the group elements g_ℓ of G ;

3] the group G is extended to the group $\tilde{G}$ in such a way that all the bond orientations of Λ_δ compatible with the combinatorial constraints imposed by the global generalization of *Kasteleyn*'s theorem[*] to a non-planar case (*i.e.* to one in which the lattice Λ_δ cannot be embedded into a surface of genus zero, but can – still preserving the lattice coordination – be embedded into one, say Σ_g, of genus $g \geq 0$) and only those can be obtained as the invariant (under $\tilde{G}$) set of configurations of the graph Γ covering Λ_δ 2^{2g} times.

4] The partition function of the model on Λ is then given by

$$\mathcal{Z}(\Lambda) = \prod_{\alpha=1}^{d} \{\cosh(K_\alpha)\}^{N_\alpha} \operatorname{Pf} \tilde{\Im} \quad ; \quad \prod_{\alpha=1}^{d} N_\alpha = N \quad ; \qquad (4.1)$$

where $K_\alpha = \dfrac{J_\alpha}{k_B T}$ is the coupling constant of the model in direction α, and N the total number of sites of Λ. $\tilde{\Im}$ is the incidence matrix of Λ_δ , extended with respect to $\tilde{G}$ and Pf denotes the Pfaffian (for a skew-symmetric matrix such as $\tilde{\Im}$, $\operatorname{Pf} \equiv \sqrt{\det}$) .

5] If both G and g are finite, then $\tilde{G}$ is finite, and recalling that the regular representation $\mathcal{R}$ of a finite group $\tilde{G}$ is the direct sum of its irreducible representations , labelled by an index j , each contained as many times as its dimension $dim\, j$, (4.1) reduces in a natural way to :

$$\mathcal{Z}(\Lambda) = \prod_{\alpha=1}^{d} \{\cosh(K_\alpha)\}^{N_\alpha} \prod_{j^{(F)}} \left(\det \mathcal{R}\left[\tilde{\Im}^{(j^{(F)})}\right] \right)^{\frac{1}{2} dim\, j^{(F)}} \quad ; \qquad (4.2)$$

where the extra-index F refers to *Fermionic* representations , as required by the generalized Kasteleyn theorem, and $\Im^{(j)}$ is a matrix of rank j . Notice that each factor $det^{\frac{1}{2}}$ in (4.2) can be thought of as the partition function for a 2-D model.

The formal analysis of (4.2) then proceeds in the following way. Let us recall first that the group $\mathcal{G}$ is called the *extension* of the group G by the group Π *if* having G presented as $G \approx\, < \Xi\,|\,\Omega >$, where Ξ denotes the set of *generators*

---

[*] See ref. [5] for a complete discussion of this delicate issue.

and Ω the set of *relations* , and similarly having Π presented as $\Pi \approx\; < \Upsilon \,|\, \Theta >$, we have the exact sequence

$$1 \longrightarrow G \overset{\iota}{\longrightarrow} \mathcal{G} \overset{\pi}{\longrightarrow} \Pi \longrightarrow 1 \quad ; \tag{4.2}$$

and – upon denoting by φ a mapping which is the inverse of the inclusion ι – $\varphi : \Pi \to \mathcal{G}$ with $\pi \circ \varphi = \mathbf{1}_\Pi$. Since $\Upsilon^{(\varphi)} \sim \varphi(\Upsilon)$, the restriction of the relations of Π to $\mathcal{G} - G$ is presented as :

$$\mathcal{G} \approx\; < \Xi \cup \Upsilon^{(\varphi)} \,|\, \Omega \,\cup\, \{\varphi^{-1}(v)\,\xi\,\varphi(v)\lambda_v^{-1}(\xi) : \xi \in \Xi;\, v \in \Upsilon\}\cup$$

$$\cup\{\mathcal{W}_\vartheta(\xi)\,\vartheta(\varphi(v)) : \vartheta \in \Theta\} > . \tag{4.3}$$

where $\lambda_\varpi : G \to G$ is the automorphism of G induced by the action of the element $\varpi \in \Pi$ on G : $g_\ell \mapsto \varphi^{-1}(\varpi)\,g_\ell\,\varphi(\varpi)$; and $\mathcal{W}_\vartheta$ is some suitable word (one is to be selected for each $\vartheta \in \Theta$) bringing each element of $\mathcal{G}$ into the form $\mathcal{W}(\xi) \cdot \gamma(\varphi(v))$ for some $\gamma \in \iota(G)$.

Of course, each automorphism λ_ϖ can be altered by an inner automorphism of G with no essential effect. If we factor out the group of inner automorphisms we obtain a new mapping $\kappa : \Pi \to Out\,G = Aut\,G\,/\,Inn\,G$ which is a homomorphism and is basic for the extension, in that equivalent extensions define the same homomorphism. The triple $\{\Pi, G, \kappa\}$ is called an *abstract kernel*[45] , and a group $\mathcal{G}$ together with the exact sequence (4.3) is called an extension with respect to the abstract kernel if for $\gamma \in \pi^{-1}(\varpi)$, $\varpi \in \Pi$, the automorphism of G defined by $g_\ell \mapsto \iota^{-1}[\gamma^{-1}\iota(g_\ell)\,\gamma]$ belongs to the equivalence class of $\kappa(\varpi)$.

The cases of physical interest here are those in which G is a *Fuchsian* group and Σ_g is a *factor surface* of G. The center $\aleph$ of G can therefore be considered as a Π-*module* with an operation in the equivalence class of $\kappa(\varpi)$, $\varpi \in \Pi$, if Π is identified with the fundamental group $\pi_1(\Sigma_g)$. Considering now the family of *cohomology* groups $H^n(\Pi, G)$, $n \geq 1$ of Π with coefficients in G (*i.e.* the cohomology groups of the *cochain complexes* defined by $\{C^n(\Pi, G), \partial^n\}_{n \in \mathbb{Z}}$, where $\partial^n : C^n(\Pi, G) \to C^{n+1}(\Pi, G)$ is the *boundary operator* and C^n is an n-dimensional cochain*), one notices that $C^3(\Pi, \aleph)$ – upon regarding $C^n(\Pi, \aleph)$ as an

---

$\star$ Recall that $C^n(\Pi, G)$, $n \geq 1$ is the group of all functions $f : \Pi^n \equiv \overbrace{\Pi \times \cdots \times \Pi}^{n\ \text{times}} \to G$ such that $f(\varpi_1, \ldots, \varpi_n) = 0$ if some ϖ_i, $1 \leq i \leq n$ equals $\mathbf{1}$.

abelian group whose operation we write multiplicatively – is zero (one says that there is a trivial *obstruction*). The theorem of Zieschang (ref. [45]) states then the extension $\mathcal{G}$ of the abstract kernel $\{\Pi, G, \kappa\}$ exists, and that $\mathcal{G}$ is a proper subgroup of the mapping class group $\mathcal{M}_g$.

Thus the homeomorphism $Ext : G \rightarrow \tilde{\mathcal{G}}$ required at point 3] above$^\diamond$ acts locally by attaching a generalized Kasteleyn phase (indeed an element of the non-abelian group G) to the circuits on Σ_g homotopic to zero, and globally by an extension by the fundamental group, *i.e.* mapping $\pi_1(\Sigma_g)$ to $\mathbb{Z}_2$. On the other hand, all possible surfaces in which Λ_δ can be embedded are conjugate from the combinatorial point of view, and we can restrict to one, *e.g.* by fixing a cut system on Σ_g. Moreover, as stated above, the relations of the mapping class group all follow from relations supported in certain subsurfaces of Σ_g finite in number and of genus at most 2. There follows (ref. [5]) that the most general choice for $\tilde{\mathcal{G}}$ is

$$\tilde{\mathcal{G}} = \Re \bigotimes_{wr} S_{2g} ; \tag{4.4}$$

where $\otimes_{wr}$ denotes the *wreath*-product[47] , whose elements can be taken to be all $2g \times 2g$ permutation matrices in which the non- zero elements have been replaced by elements of $\Re$; whereas $\Re = \mathcal{M}_g / \mathcal{H}$, namely the subgroup of diffeomorphisms of Σ_g which preserve the isotopy class of a maximal, unordered, nonseparating system of g disjoint, smoothly embedded cycles (noncontractible and nonisotopic), *e.g.* just the cut system $\{\alpha_i ; i = 1, \ldots, g\}$. $\Re$ is then essentially generated by the elements representing homology exchange between any pair of circles $(\alpha_i, \alpha_j) ; i, j = 1, \ldots, g$.

Eq's (3.2) and (3.5) allow us now to write the *free energy* $\mathcal{F} \equiv -\kappa_B T \ln \mathcal{Z}$ as

$$-\beta \mathcal{F} = \sum_{\alpha=1}^{d} N_\alpha \ln \cosh(\beta J_\alpha) + \frac{1}{2} \sum_{j^{(F)}} \dim j^{(F)} \, \mathrm{Tr}\left(\ln \mathcal{R}\left[\tilde{\mathfrak{F}}^{(j^{(F)})} \right] \right) ; \tag{4.5}$$

from which it appears clearly that while $\mathcal{Z}$ can be expanded in terms of *characters* of $\Re$, $\mathcal{F}$, as given by the latter equation, could be rewritten in terms of

---

$\diamond$ It should be kept in mind that maps and spaces are to be thought of in the *PL* (*piecewise-linear*) category[46] , namely all morphisms referred to in present discussion should be meant in the corresponding definition as given in ref. [46].

invariant symmetric functions for $\Re$. The coefficients of such an invariant expansion retain some of the original combinatorial flavour of the problem : they count the numbers of *words* in $\Re$ equivalent to the identity, *i.e.* provide a solution for the Dehn word problem (see ref. [6]) for the subgroup $\mathcal{H}$ of $\mathcal{M}_g$.

Another *bridge* between $Z_{EQG}(N^3)$ as given in (2.9), and $\mathcal{Z}(\Lambda)$ as given in (3.1), can be ascribed to the following argument. Recalling the representation of $\mathcal{Z}(\Lambda)$ as Grassmannian path integral, as given by Itzykson[48] , we consider for simplicity the particular case in which M^3 is obtained by Dehn surgery along a knot K in S^3. It follows, that the associated Eq. (2.7) is a special form of the generalized surgery formula for a non-Abelian 3D-Euclidean Chern-Simons gauge theory defined over a generic three-manifold $\tilde{M}^3$ suggested by Witten.[49] Witten, in ref. 49, argues that:

$$\begin{cases} Z[\tilde{M}^3] = \sum_j h_0^j Z[M^3; R_j] \\ \\ Z[M^3; R_j] \equiv \langle W_{R_j}(K) \rangle_{M^3} \end{cases} , \qquad (4.7)$$

where $\tilde{M}^3 = M^3 \cup_h K_f$, h is the glueing homeomorphism on the solid torus K_f and $Z[M^3; R_j]$ is the CS-partition function of M^3 with an extra Wilson line $W_{R_j}(K)$ in the R_j representation (of the CS-gauge group $\mathcal{G}$) included on the knot K. When the CS-coupling k is an integer, using the techniques of rational conformal field theories (see *e.g.* ref. [49]), one can show that R_j is a finite-dimensional module over the representation ring of $\mathcal{G}$ with $j < \infty$. Then it turns out that the knot diagram D_K parameterized by R_j has a nice (equivalent) interpretation[50] in terms of the so-called "r-parallel version" $C * D_K$ of D_K. That is, for any $j \in \{1, \ldots, N\}$ we associate a non-negative integer $C(j)$, called the "colouring" of D_K, from the set $\{1, 2, \ldots, n\} \in \mathbb{Z}_+$. Let $C(j) * D_K$ be the diagram which can be formed by taking $C(j)$ copies all parallel , in the plane, to D_K. In this picture Eq. (4.7) becomes:

$$Z[M^3; R_j] = \langle W_{R_j}(K) \rangle_{M^3} = \langle W_\mathcal{R}[C(j) * D_K] \rangle_{M^3} \equiv \langle C(j) * D_K \rangle_{M^3}, \qquad (4.8)$$

the symbol $W_\mathcal{R}$ denoting the Wilson line in the fundamental representation $\mathcal{R}$ of $\mathcal{G}$. Similarly, one finds that the coefficients $h_j \equiv h_0^j$ (in general complex numbers) can be written as $h_j = h_{C^{-1}(\mathbb{Z}_+)} \equiv \lambda_c$ by definition of the colouring map C. Therefore, one can also write Eq. (4.7) as (recall that K denotes a knot):

$$Z[\tilde{M}^3] = \sum_{c \in C} \lambda_c \langle c * D_K \rangle_{M^3}. \tag{4.9}$$

Eq. (4.9) has recently been rigorously stated by Lickorish (ref. [24]) in the case of the one-variable Jones polynomial for $\langle c * D_K \rangle_{M^3}$ if $M^3 = S^3$ and $\mathcal{G} = su(2)$.

It is now immediate to notice that the partition function (2.7) of 3D-Euclidean quantum gravity has the form (4.9) with $\tilde{M}^3 = (S^3 - K_f) \cup_h K_f$ and $M^3 = S^3 - K_f$, if one sets

$$\langle \ldots \rangle_{(S^3 - K_f)} \simeq \sum_{(\alpha)} \tau_{\varphi(\alpha)}(S^3 - K_f) \simeq \prod_{(\alpha)} \triangle_K(t_{(\alpha)}) \equiv \prod_{(\alpha)} \triangle_{K(\alpha)}(t), \tag{4.10}$$

where we have used Eq. (2.8) and $t_{(\alpha)} \in H_{1(\alpha)}(S^3 - K_f) \equiv \dfrac{\rho_{(\alpha)}[\pi_1(S^3 - K_f)]}{[*, *]}$, and one regards $C * D_K$ as an extra "field" on which to compute the vacuum-to-vacuum expectation value given by the "partition function" $\prod_{(\alpha)} \triangle_{K(\alpha)}$. Now, such an identification of $\triangle_{K(\alpha)}$ with a certain path integral for each (α) is just what one in fact has!

Kauffman and Saleur[51] have in fact recently shown that the Alexander-Conway polynomial of a knot K is the fermionic path integral over free fermions propagating on the knot diagram D_K. Their basic idea is to describe the tangle diagram D_K as a planar Feynman graph Γ_K for a Gaussian fermionic theory. The Feynman graph is obtained by projection of the tangle diagram on a two-dimensional planar four-valent graph. To each crossing i of an oriented tangle diagram one associates four complex Grassmannian variables ψ_i^α, $\psi_j^{\beta\dagger}$, where the labels $\alpha = \beta = $ up (u), down (d) refer to edges going up and down with respect to the direction of the crossing at the point i.

All ψ's anticommute,

$$[\psi_i^\alpha, \psi_j^\beta]_+ = 0; \quad \alpha, \beta = n \text{ or } d; \quad i \neq j,$$

and in particular $(\psi_i^\alpha)^2 = 0$. The Berezin path integral is defined as usual by the rule

$$\int \Pi_i d\psi_i^u d\psi_i^{u\dagger} d\psi_i^d d\psi_i^{d\dagger} \Pi_i \psi_i^u \psi_i^{u\dagger} \psi_i^d \psi_i^{d\dagger} = 1.$$

At the Lagrangian level, if along the link (i,j) the edge is oriented from vertex i to vertex j, the propagator is $\psi_i^{\alpha\dagger}\psi_j^\beta$ with labels $\alpha,\beta = u$ *or* d depending on the particular configuration. For instance, the tangle D_K or equivalently the associated Feynman graph Γ_K, both shown in Fig. 1, correspond to the kinetic term $\psi_i^{u\dagger}\psi_j^d$.

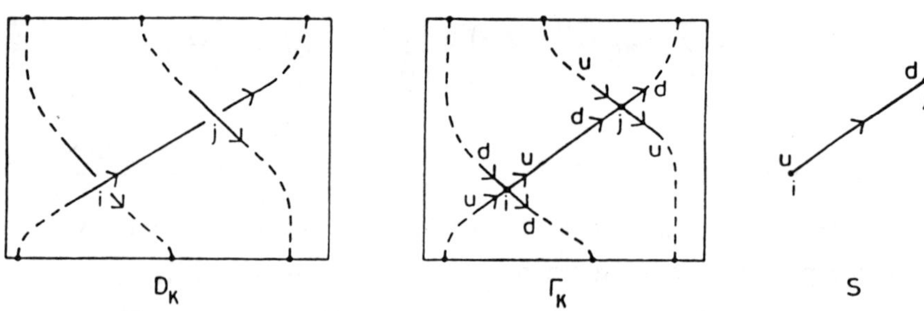

Fig. 1: Knot diagram D_K and associated Feynman rules (Γ_K, S)

Thus, the Kauffman-Saleur result is

KS-Theorem (ref. [51]): the Alexander-Conway polynomial $\nabla_K(q)^\star$ for a

---

$\star$ The Alexander-Conway polynomial for a knot is defined by the skein relation [52]

$$\nabla_{K_+}(q) - \nabla_{K_-}(q) = q\nabla_{K_0}(q)$$

and by the normalization: $\nabla_K = 0$ for $K =$ (unknot) and $\nabla_K = 1$ for $K =$ (unknotted strand).

fixed knot K has the fermionic path integral representation[◇]

$$\left\{ \begin{aligned} &\nabla_K(q) = \langle \psi_+ | \psi_- \rangle \equiv N(K;q) \int d\psi^\dagger d\psi \, \exp\left\{ \sum_{\substack{i,j \\ \alpha,\beta=u,d}} \psi_i^{\alpha\dagger} M_{\alpha,\beta}^{i,j}(K;q)\psi_j^{\beta} \right\}, \\ &N(K;q) \equiv q^{-L(K)-I(K)} \end{aligned} \right.$$

$$(4.11)$$

where $M = [M_{\alpha,\beta}^{i,j}(K;q)]$, which depends on the type of knot K selected, is a $[2 \times \#\,(crossings)]^{\otimes 2}$-matrix whose entries are ± 1 or certain rational functions of q. Furthermore, N is a normalization factor specified by the number of internal edges $I(K)$ (loops $L(K)$) of the Feynman graph Γ_K associated to D_K.

Since the usual Alexander polynomial $\triangle_K(t)$ is given by $\nabla_K(q)$ in terms of the formula (ref. [52])

$$\triangle_K(t) = \nabla_K(q \equiv \sqrt{t} - \frac{1}{\sqrt{t}}), \qquad (4.12)$$

it follows that the Gaussian Berezin path integral (4.11) also extends to $\triangle_K(t)$ and hence to $\tau_\varphi(S^3 - K_f)$ via Eq. (2.6). In our case we have a family of Alexander polynomials $\triangle_{K(\alpha)}$, thus we shall have $\triangle_{K(\alpha)} \equiv \triangle_K(t_{(\alpha)}) = \langle \psi_+ | \psi_- \rangle_{(\alpha)} \propto [Pf(M_{K(\alpha)})]^2$, where $M_{K(\alpha)}$ inherits the dependence on the labelling (α) by $t_{(\alpha)}$.

Collecting all this together, when M^3 is obtained by Dehn surgery along a knot K in S^3 we have the formula:

$$Z_{EQG}[M^3 = (S^3 - K_f) \cup_h K_f] = \sum_{c \in C} \lambda_c \prod_{(\alpha)} \langle \psi_+ | [c * D_K](\psi_i^\alpha \psi_j^{\beta\dagger}) | \psi_- \rangle_{(\alpha)}, \quad (4.13)$$

where $h \ni h_j = h_{C^{-1}(\mathbb{Z}_+)} \equiv \lambda_c$, $c \in \mathbb{Z}_+$, and $[c * D_K](\psi\psi^\dagger)$ denotes the operator associated to the c-parallel version of D_K in the Kauffman-Saleur fermionic representation. Clearly, following ref. [51], we may identify $1 * D_K = D_K$ with the action. For instance, to the trefoil diagram D_T and to the associated Feynman

---

◇ As is well known, the Berezin path integral in (4.11) just gives the square of the Pfaffian $Pf(M) \equiv \sqrt{\mathrm{Det}[M(K;q)]}$.

graph Γ_T shown in Fig.2 corresponds the matrix element $\langle \psi_+ | (\psi^\dagger M(T;t)\psi) | \psi_- \rangle$, where (ref. 51)

$$\psi^\dagger M(T;t)\psi = \psi_1^{d\dagger}\psi_2^u + \psi_1^{u\dagger}\psi_2^d + \psi_2^{d\dagger}\psi_3^u + \psi_2^{u\dagger}\psi_3^d + \psi_3^{u\dagger}\psi_1^d +$$
$$\left(\sqrt{t} - \frac{1}{\sqrt{t}}\right)\sum_{i=1}^{3}(\psi_i^{u\dagger}\psi_i^u + \psi_i^{d\dagger}\psi_i^d) + \left(t + \frac{1}{t} - 3\right)\sum_{i=1}^{3}\psi_i^{u\dagger}\psi^{di}.$$

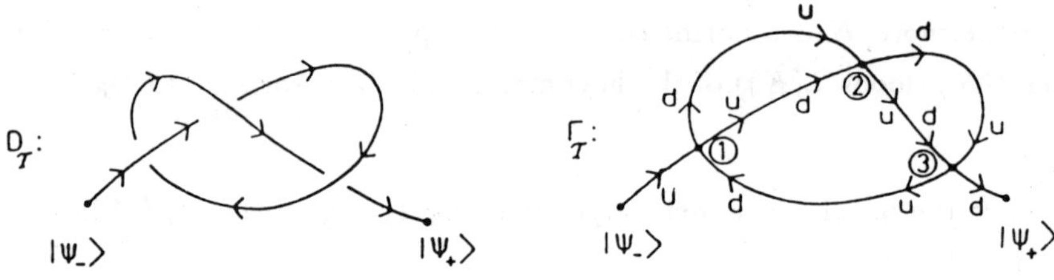

Fig. 2: Trefoil configuration and associated Feynman amplitude

As we already observed, the c-parallel version of D_K is a link analogue of the technique used to characterize the representation ring of a Lie algebra by tensor products of the fundamental representation, rather than by the irreducibles. In our fermionic picture, this corresponds to using integer powers $(\psi^\dagger M_K \psi)^{l(c)}$, $l(c) \in \mathbb{Z}_+$, of $\psi^\dagger M_K \psi$ to describe $c * D_K$. As a consequence of this description, the quantity under the product symbol in the r.h.s. of Eq. (4.13) becomes the Berezin path integral of a polynomial in $\psi_i^{\alpha\dagger}\psi_j^\beta$ and may be interpreted as the Green's function obtained from $\int d\psi^\dagger d\psi \exp(\psi^\dagger M_K \psi)$.[*]

$\star$ Clearly, one has that:

$$\langle \psi_+ | (\psi^\dagger M_K \psi)^{l(c)} | \psi_- \rangle_{(\alpha)} \sim \left(\frac{\partial}{\partial \beta}\right)^{l(c)} \int d\psi^\dagger d\psi \exp(\beta \psi^\dagger M_{K(\alpha)}\psi)|_{\beta=1}.$$

We already pointed out that it is not surprising that there exist an intimate relationship between the gaussian fermionic partition function and the topological torsion invariants which can be formally interpreted as a dynamical zeta functions (for Grassmannian stochastic systems). On one front we have representatives of Anosov flows entering the partition function, whose relative weight greatly exceeds that of closed orbits of the discrete periodic set and hence emphasizes the stochastic features of the model. On the other hand the same partition function, which describes the global dynamical (and/or thermodynamical) behaviour of the system can be viewed as the generating function of all closed loops (*i.e.* links with possibly knotted components) embedded in a Riemann surface Σ_g. V.I. Arnol'd[53] has recently studied the problem of characterizing the asymptotic images of the manifold $\mathcal{M}_{\mathcal{X} \cap \mathcal{Y}}$ obtained by intersection between two submanifolds $\mathcal{X}$ and $\mathcal{Y}$ of a given manifold $\mathcal{W}$ under iteration of the group of diffeomorphisms of $\mathcal{W}$ as, say, $\mathcal{X}$ is kept fixed ($dim\, \mathcal{W} = dim\, \mathcal{X} + dim\, \mathcal{Y}$, $dim\, \mathcal{M}_{\mathcal{X} \cap \mathcal{Y}} = |\, dim\, \mathcal{X} - dim\, \mathcal{Y}\, |$). Under an ergodic hypothesis, well motivated physically, one expects that the equilibrium features of the model we are considering are indeed controlled just by these images (where one identifies obviously $\mathcal{X}$ with Σ_g, $\mathcal{Y}$ with the isotopy-equivalents of Σ_g itself, and $\mathcal{M}_{\mathcal{X} \cap \mathcal{Y}}$ with the set of loops generated by intersection). The result of Arnol'd implies that it is just the set of topological invariants of $\mathcal{M}_{\mathcal{X} \cap \mathcal{Y}}$ and it alone which completely characterizes the asymptotic action of the group of diffeomorphisms of $\mathcal{W}$ (which in our correspondence can be thought of as a manifold of dimension $dim\, \Sigma_g + 1$, in general non-Euclidean also in the Ising case, due to the choice of boundary conditions).

A final comment is in order. The procedure described in Sect. 3, whereby we have essentially mapped the $3D$ Euclidean gravity to a free-fermion system over a lattice is an homeomorphism between the two theories. This is due to the conceptual passage through the Ising model, which allows us in principle to reconstruct from the lattice model the whole group of diffeomorphisms in $3D$, $Diff^3$. The deeper meaning of such a reconstruction can be understood in the following way: the $3D$ Euclidean quantum gravity partition function is clearly invariant with respect to $Diff^3$, as it essentially coincides with the Reidemeister torsion, which is diffeomorphism invariant by its very construction. On the other hand, at the fermionic lattice model level one has (ref. [51]) a *hidden quantum symmetry* $U_q[s\ell(1,1)]$, in other words, the link fermions ψ_i^{α}, $\psi_j^{\beta\dagger}$ have a nontrivial statistics whose *dual* is the quantum group $U_q[s\ell(1,1)]$. The key notion here is the following: the quantum group symmetry $U_q[g]$ of a lattice model appears at very short distances of the order of the lattice step; in the continuum limit, it appears at a single point, valued in the group of the Kac-Moody algebra $\hat{g}$ associated with the Lie algebra g of some finite-dimensional Lie group G.[◇] Thus we can equivalently affirm that our Gaussian fermionic system has a gauge symmetry of type $G \approx SU(1,1)$, since the corresponding Kac-Moody group just acts as a local gauge symmetry. It is known (ref. [1]) that the action in the continuum theory of this local (gauge) A_1-symmetry on the first order fields $(e_\mu, \omega_\mu) \equiv A_\mu$ describing the $3D$ gravitational field $g_{\mu\nu}$ is, on shell, equivalent to the to the action of $Diff^3$ on A_μ. This leads us to the interpretation of the presence of the group of diffeomorphisms in 3 dimensions in the continuum Einstein gravity theory as the manifestation of the *quantum* internal symmetry of the underlying lattice model (4.11).

5. Conclusions

Summarizing, we have shown that 3D-Euclidean quantum gravity in the first order dreibein formalism and in the Landau gauge, when quantized on a generic three-manifold obtained by Dehn surgery along a knot K (link L) in S^3, is equivalent to a Gaussian fermionic theory propagating on the c-parallel versions of the knot diagram D_K (link diagram D_L). In particular, we have shown in the Berezin path integral picture that the 3D-EQG partition function $Z(N^3)$ for a 3D-hyperbolic manifold N^3 is equivalent (up to some irrelevant normalization factor) to that one $\mathcal{Z}(\Lambda)$ of a 3D-Ising model on a lattice Λ embedded in $\mathbb{R}^3$.

Let us conclude with two remarks:

i) In the previous section we have shown that 3D-Euclidean quantum gravity admits a free fermion representation as well as the 3D-Ising model (ref. [48]) and that these two models seem to be in mutual relation. Furthermore, we have proved that for fixed three-dimensional topologies the 3D-EQG partition function is given by a suitable Alexander-Conway polynomial, which can explicitly be

---

◇ This observation first due to Alekseev, Faddeev and Volkov[54] in their study of the $WZNW$-model, applies here as well, with $\hat{g} = \widehat{s\ell(1,1)}$.

computed by combinatorial or (Gaussian) path integral techniques. So, thanks to the above mentioned equivalence, we also have a computable algorithm for solving the quantum 3D-Ising model before performing the thermodynamic limit.

Now, a related question is whether these are the only 3D-models which allow a free fermion description, or it is rather a general property of topological models in three dimensions. Of course any 3D-topological quantum field theory (QFT) is an "integrable" (in the sense of "exactly solvable") quantum model. If there exists some sort of "duality" between the Kac-Moody level k entering in the quantum deformation parameter q defined as $q = \exp(2\pi i/(k+2))$ (see below), and the order N of the gauge group $\mathcal{G}$ of the 3D-EQG in first-order formalism (here $\mathcal{G} = SU(N)$ with $N = 2$), i.e. $U_{q(k)}(SU(N)) \sim U_{q(N)}(SU(k))$, as happens for the WZW-models at the level of the central extended Kac-Moody algebras[55], then, perhaps, the underlying integral structure might be in correspondence with that (of KP-hierarchy type)[56] of some $k \to \infty$ limit of a $SU(k)$-Toda lattice 2D-theory. Indeed, $SU(k)$-Toda lattice models[57] have a quantum symmetry of our type $U_{q(N)}(SU(k))$, and hence the requirement $k \to \infty$ here follows from the above assumed duality and the fact that the Turaev-Viro lattice model describes the continuum 3D-EQG only in the limit $k \to \infty$, i.e. $q \to 1$ (see below).

ii) We would like to notice that an indirect hint of the possible quantum connection between the 3D-Euclidean quantum gravity partition function and the semiclassical limit of a polynomial link invariant may be found – due mainly to the recent work by Turaev – in the analysis performed by Ponzano and Regge.[58] In ref. 58 it was argued that when M^3 is "close" to S^3, the path integral of 3D-Euclidean quantum gravity in the simplicial approximation known as the Regge calculus[59] is actually proportional to the semiclassical (large angular momentum) limit of the standard $su(2)$ $6j$-symbol. We conjecture that the realization of the Regge-Ponzano program of understanding the Feynman summation of histories for the lattice 3D euclidean Einstein-Hilbert action as a sort of state model associated with the Racah coefficients, can be fully completed at the quantum level by our eq. (2.9).[60] [61]

Such a conjecture is based on the following facts:

a) it is known that the standard $su(2)$ $6j$-symbol is the "semiclassical" limit

$(q \to 1)$ of the quantum $6j$ symbol

$$\left\{ \begin{matrix} a & b & c \\ d & e & f \end{matrix} \right\}_q \equiv [D]_q \tag{4.1}$$

of the quantum group $U_q(sl(2; \mathbb{R}))$;

b) an intrinsic combinatorial approach is known which allows one to associate with the quantum $6j$ symbols of $U_q(sl(2; \mathbb{R}))$ the two-variable HOMFLY-polynomial $P_K(q, z)^{[62]}$;

c) the Alexander polynomial $\triangle_K(t)$ entering eq. (2.9), is a particular case of $P_K(q, z)$ when $q = 1^{[63]}$ and $z = \sqrt{t} - \dfrac{1}{\sqrt{t}}$.

On the other hand, the possibility of constructing directly "quantum" invariants for a closed 3-manifold M from the q-$6j$ symbols has been recently stressed by Turaev and Viro in ref. [32], supporting once more our conjecture that the Regge Ponzano idea might be extended to full quantum level. Indeed, we may recall that the Turaev-Viro 3-manifold invariant of M is given by

$$|M|_q \equiv C^{\mathcal{V}} \sum_{\{Col\}}^{<\infty} \prod_{i \in \mathcal{E}} (-1)^i [2i + 1]_q \prod_{\mathcal{T}} [D]_q \ , \tag{4.2}$$

where q is a complex root of unit of a certain degree $k \in \mathbb{Z}_+$, $k \geq 0$, C is a constant, $\mathcal{V}$, $\mathcal{E}$, $\mathcal{T}$ denote respectively the numbers of vertices, edges, and the set of tetrahedra of the simplicial complex $\mathcal{X}$, $[n]_q$ is the quantum dimension, and Col is the map which associates with the edges of $\mathcal{X}$ elements of the set $\{Col\}$ of colours $\left\{ 0, \dfrac{1}{2}, 1, \ldots, \dfrac{k}{2} \right\}$ related in standard way to the framing map introduced in (2.5).

Notice that the construction of the invariant is associated with a specific triangulation of M; however the main theorem of ref. [32] show just that $|M|_q \in \mathbb{C}$ does not depend in fact on the choice of the triangulation $\mathcal{X}$, namely it is a bona fide topological invariant.

Moreover, Turaev and Viro show that

$$|M|_q = |I_k(M)|^2 \ , \tag{4.3}$$

where $I_k(M)$ is the Dehn surgery invariant for M discussed in Sect. 2, and $q = \exp\left(\dfrac{2\pi i}{k + 2} \right)$.

Acknowledgements

The authors gratefully acknowledge inspiring discussions with M. Carfora, S. Carlip, J. Fröhlich, G. Jona-Lasinio, T. Kohno, L. Kauffman, P. Menotti and T. Regge. The substantial help of G. Bonacina in preparing the manuscript is also warmly acknowledged.

references

1. E. Witten, Nucl. Phys. **311B** (1988/89) 46.

2. See *e.g.* D. Rolfsen, *Knots and Links*, Publish or Perish, Inc., (Berkeley, 1976).

3. A. Sedrakyan, Phys. Lett. **137B** (1984) 397;
   A. Kavalov and A. Sedrakyan, Phys. Lett. **173B** (1986) 449, Nucl. Phys. **285B** [FS19] (1987) 264;
   V. Dotsenko, Nucl. Phys. **285B** [FS19] (1987);
   V. Dotsenko and A. Polyakov, Adv. Studies in Pure Math. **16** (1988) ;
   A. Polyakov, *Gauge Fields and Strings*, Harwood Academic Publishers (1987).

4. D. J. Gross and A. A. Migdal, Phys. Rev. Lett. **64** (1990) 127;
   Nucl. Phys. **340B** (1990) 333;
   E. Brézin and V. Kazakov, Phys. Lett. **236B** (1990) 144;
   M. Douglas and S. Shenker, Nucl. Phys. **235B** (1990) 635.

5. M. Rasetti, *Ising Model on Finitely Presented Groups*, in *Group Theoretical Methods in Physics*, M. Serdaroglu and E. Inönü eds.; Springer Verlag, Lecture Notes in Physics **180**; Berlin, 1983
   plus references therein and in
   G. Jacucci and M. Rasetti, J. Phys. Chem. **91**, 4970 (1987).

6. M. Dehn, Math. Ann. **72**, 413 (1912)

7. E. Witten, Nucl. Phys. **323B** (1989) 113

8. R. Seeley, Proc. Symp. Pure Math. **10** (1966) 288.

9. D. B. Ray and I. Singer, Adv. Math. **7** (1971) 145;
   Ann. Math. **98** (1973) 154.

10. J. Cheeger, Ann. Math. **109** (1979) 259;
W. Müller, Adv. Math. **28** (1978) 233

11. J. Milnor, Bull. Ann. Math. Soc. **72** (1966) 358

12. D. Fried, Invent. Math. **84** (1986) 523

13. W. Alexander, Trans. A. M. S. **30** (1928) 275.

14. W. B. R. Lickorish, Ann. of Math. **76** (1962) 531.

15. W. Burau, Abh. Math. Sem. Hanischen Univ. **11** (1936) 171.

16. J. Milnor, Ann. Math. **76** (1962) 137

17. W. Franz, J. Reine Angew. Math. **173** (1935) 245

18. V. G. Turaev, Uspekhi Mat.Nauk. **41** (1986) 97

19. M. Carfora, and A. Marzuoli, *Reidemeister Torsion and Simplicial Quantum Gravity*, Intl. J. Mod. Phys. **B**, 1993, in press

20. W. P. Thurston, Bull. Am. Math. Soc. **6** (1982) 357;
The Geometry and Topology of Three-manifolds, Princeton Lecture Notes (1979).

21. M. Rasetti, *The Mapping Class Group in Statistical Mechanics: a concise review*, in *Symmetries in Science 111*, Edited by B.Gruber and F.Iachello, Plenum Publ.Co., (1989) and references therein;

22. H. Zieschang, Asterisque **163-164** (1988) 247

23. T. Kohno, Topology **31** (1992) 203, Nagoya preprint **6** (1990).

24. N. Reshetikhin and V. Turaev, Invent. Math. **103** (1991) 547;
W. B. R. Lickorish, Math. Ann. **290** (1991) 657.

25. T. Kohno, private communication.

26. J. S. Birman, *Braids, Links and Mapping Class Group* Ann. Math. Studies **82**, PUP (1975).

27. S. E. Cappell, R. Lee and E. Y. Miller, *Invariants of 3-Manifolds from Conformal Field Theory*, Courant Institute preprint (1990).

28. M. Cohen, *A Course in Simple Homotopy Theory*, Springer-Verlag (New York, 1973).

29. B.A. Dubrovin, A.T. Fomenko, and S.P. Novikov, *Modern Geometry – Methods and Applications*, Springer-Verlag (New York, 1990).

30. J.H.C. Whitehead, Amer. J. Math. **72**, 1 (1952)

31. M. Carfora and A. Marzuoli, *Finiteness Theorems in Riemann Geometry and Lattice Quantum Gravity*, Contemporary Math., 1993, in press

32. V.G. Turaev, and O.Y. Viro, Topology **31** (1992) 865; L. H. Kauffman and S. Lins, Manuscripta Math. **22** (1991) 81

33. D. Fried, *Counting Circles*, in *Dynamical Systems*, Springer-Verlag Lect. Notes Math. **1342**, 196 (1988)

34. H. Moscovici and R.J. Stanton, Invent. Math. **105**, 185 (1991)

35. J. Milnor, *Infinite Cyclic Coverings*, in *Topology of Manifolds*, Prindle, Weber & Schmidt, Boston, 1968

36. R.J. Baxter, *Exactly Solved Models in Statistical Mechanics*, Academic Press, London, 1982

37. A. Hatcher and W. Thurston, Topology **19**, 221 (1980)

38. J. Harer, Inventiones Math. **72**, 221 (1983)

39. B. Wajnryb, Israel J. Math. **45** (1983) 157.

40. A. Montorsi and M. Rasetti, *The mapping Class Group : Homology and Linearity*, in *Group Theoretical Methods in Physics*, H. D. Döbner and T. D. Palev, eds.; World Scientific Publ. Co.; Singapore, 1988.

41. N. V. Ivanov, Uspekhi Mat. Nauk **42**:3, 49 (1987).

42. A. Fathi, F. Laudenbach and V. Poenaru, Astérisque, **66- 67**, 33 (1979)

43. R. Mandelbaum and J. R. Harper, Can. Math. Soc. Conf. Proc. **2**, 35 (1982).

44. M. E. Fisher, J. Math. Phys. **7**, 1776 (1966).

45. H. Zieschang, *Finite Groups of Mapping Classes*; Springer Verlag, Lecture Notes in Mathematics **875**; Berlin, 1981.

46. C.P. Rourke, and B.J. Sanderson, Ann. Math. **87**, 1, 256, and 431 (1968).

47. A. Kerber, *Representations of Permutation Groups*; Springer Verlag, Lecture Notes in Mathematics **240**; Berlin, 1971.

48. C. Itzykson, Nucl. Phys. **B 210** [**SF 6**], 448, aND 477 (1982) and S. Samuel, J. Math. Phys. **21**, 2806, 2815, and 2820 (1980).

49. E. Witten, Comm. Math. Phys. **121** (1989) 351.

50. J. Murakami, *The Parallel Version of Link Invariants*, Osaka Preprints (1987);
    H. R. Morton and P. M. Strickland, "Jones Polynomials Invariants for Knots and Satellites", to appear on Math. Proc. Camb. Phil. Soc. (1991).

51. L. H. Kauffman and H. Saleur, Comm. Math. Phys. **141**, 293 (1991)

52. L. H. Kauffman, Topology **20** (1981) 101.

53. V.I. Arnol'd, private communication, to appear in the Proceedings of the Stony Brook Symposium in honour of J. Milnor, A. Phillips, ed.; Publish or Perish Press, Boston

54. A. Alekseev, L. Faddeev, and A. Volkov, *The unravelling of the quantum group structure in the WZNW theory*, CERN preprint TH-5981/91

55. S.G.Naculich, and H.J.Schnitzer, Phys.Lett. **244**, 235 (1990).

56. M. Saveliev and A. Vershik, Comm. Math. Phys. **126** (1989) 367;
    S. Levendorskii and Yan Soibelman, Comm. Math. Phys. **140** (1991) 399.

57. K. Ueno, and K. Takasaki, Adv. Studies in Pure Math., **4**, 1 (1984).

58. G. Ponzano and T. Regge, in *Spectroscopy and Group Theoretical Methods in Physics*, North-Holland Publ.Co. (Amsterdam 1968).

59. T. Regge, Nuovo Cimento **19** (1961) 551.

60. H. Ooguri and N. Sasakura, Mod. Phys. Lett. **A6** (1991) 3591

61. S. Mizoguchi and T. Tada (Phys. Rev. Lett. **68** (1992) 1795) argued that the Turaev-Viro model provides a q-analogue lattice regularization of the Ponzano-Regge model, where the cut-off is given by $\dfrac{1}{\ln q}$. In their approach the equivalence with the 3D Euclidean quantum gravity in the form (1.1) follows as continuum limit (equivalent to our semiclassical limit $q \to 1$) of the Turaev-Viro piecewise-linear model.

62. V. G. Turaev, Inventiones Math. **92** (1987) 527;
    A. N. Kirillov and N. Yu. Reshetikhin, *Representations of the $U_q(sl(2))$ Algebra, q-Orthogonal Polynomials and Invariants of Links*, LOMI preprint E-9-88.

63. See for example W. B. R. Lickorish and K. C. Millett, Topology **26** (1987) 107.

VASSILIEV KNOT INVARIANTS AND THE STRUCTURE OF RNA FOLDING.

Louis H. Kauffman [*]   and Yuri B. Magarshak[†]

September 10, 1993

1   Introduction.

It is the purpose of this paper to introduce certain combinatorial structures into the study of RNA folding. These structures are useful for the classification of foldings and for the topological classification of the embeddings of these foldings into three-dimensional space. Both the abstract classification and the topological classification are highly relevant to problems in molecular biology - where these folded structures are instantiated as molecules in a three - dimensional ambient physical space.

The paper is organized as follows. In section 2 we discuss the basic idea of a folding (folded molecule) and graphical models for such foldings. We introduce the use of the Brauer monoid for the classification of non- embedded foldings. This introduces a multiplicative structure into the set of foldings and we discuss the structure of the resulting algebra. Section 3 discusses the relationship of foldings and topological invariants of embedded rigid - vertex graphs. Vertices arise in foldings as loci of a linear sequence of base pairs. We translate these folding vertices into standard 4-valent vertices and thereby obtain a translation of rigid vertex invariants to the category of folded molecular structures. This section discusses both Vassiliev invariants and a more general scheme that produces invariants of embedded foldings from any topological invariant of knots and links. Section 4 gives specific information about the Vassiliev invariants. In particular, we show how to construct a Vassiliev invariant of type 3, and we illustrate how Lie algebras give rise to Vassiliev invariants. The appendix discusses this last point in more detail. This key relationship between Lie algebras and Vassiliev invariants provides an interconnection among topological invariants, Lie algebras, Feynman diagrams and significant indices for protein

[*]Department of Mathematics, Statistics and Computer Science, The University of Illinois at Chicago, Chicago, Illinois, 60680. [†] Biomathematical Sciences Department, Mount Sinai School of Medicine of City University of NY, NY 10029.

foldings. These connections are just beginning. We conjecture that these relationships occur at biological as well as topological levels of natural structure.

Acknowledgments. Research for this paper on the part of the first author was partially supported by the Program for Mathematics and Molecular Biology, University of California at Berkeley, CA, and by NSF Grant # DMS 9205277. On the part of the second author, research was partially supported by NSF Grant No. CHE 9123802.

2   Foldings and the Brauer Monoid.

The purpose of this section is to introduce our abstraction of an RNA folding, and to give a method of enumerating such foldings in terms of the Brauer monoid [BR], an algebraic structure that generalizes the symmetric group on n letters. The Brauer monoid is of independent interest via its relationship with the theory of group representations [BR] and with the theory of invariants of knots and links ([BW], [K3], [KV]).

In order to begin, we need an appropriate mathematical abstraction for RNA. To this end, let us discuss some of the properties of the RNA molecule. The molecule is a long chain consisting in a sequence of the bases A (adenine), C (cytosine), U (uracil) and G (guanine). The pairs [A and U] and [C and G] are capable of bounding with each other. It is characteristic of RNA that the molecule can bond with itself. We say that two bases are paired if they are so bonded.

Thus, an abstract RNA molecule is just a linear sequence of the letters A, C, G and U. A *folding* of the molecule is a possible pairing structure with respect to the given sequence of bases. For example, we could have the chain $\cdots CCCAAAACCCCCCUUUUCCC\ldots\cdots$ and the corresponding folding

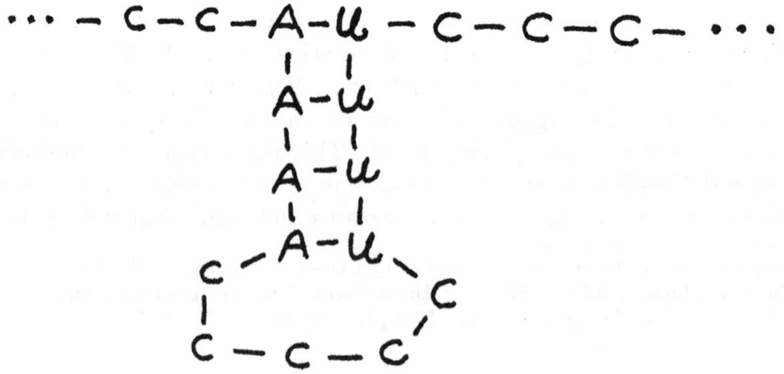

This folding can be indicated on the chain itself by a diagram (see also [M], [KMM]) with arcs connecting the paired bases:

We may abstract this to a diagram that simply indicates the form of the pairing:

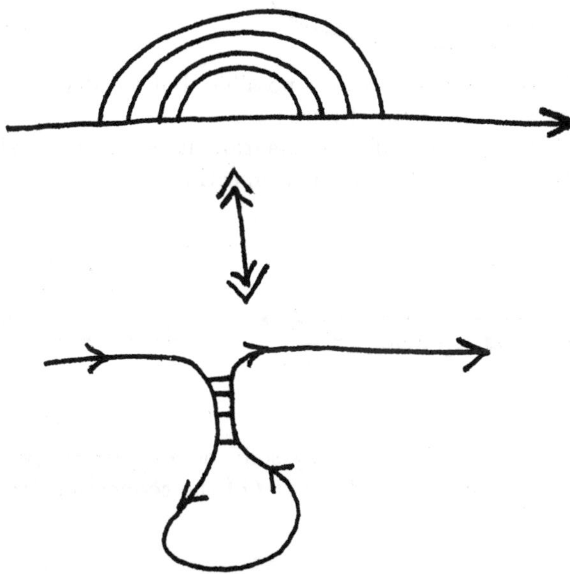

It is often the case that a sequence of repeated bases pairs with another sequence of repeated bases. This results in a *basic pairing node* of the form

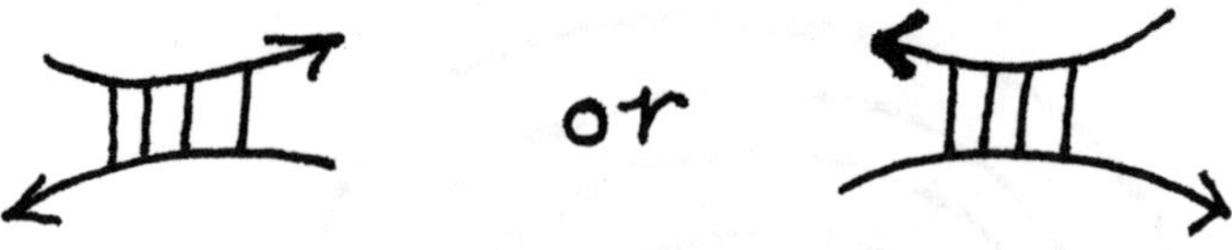

Note the directions on the arcs of this node that correspond to the sequence of bases. The arcs are oppositely oriented just as in our example:

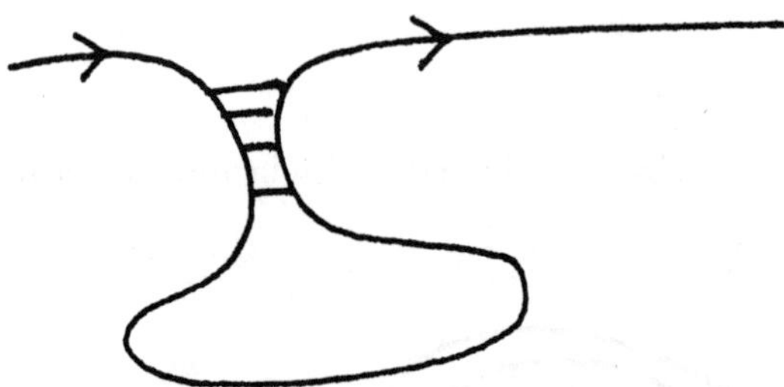

Any multiplicity of connecting arcs is possible, but we shall adopt the convention of four such arcs for pictorial purposes.

On the other hand, in an unfolded diagram it is useful to abbreviate a multiplicity of connecting arcs to a single connecting arc as in

Abbreviated arcs will be indicated by solid nodes ⟶ as shown above. The solid nodes will be called the *feet* of the connecting arcs. Thus, we have the correspondence

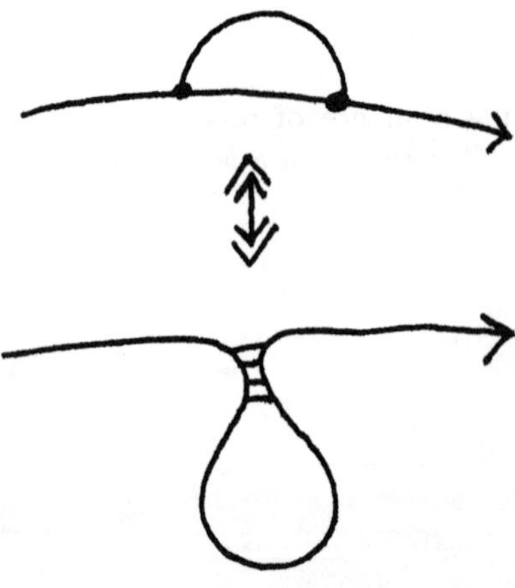

With these conventions, we can indicate the form of a great multiplicity of foldings (see Fig.1).

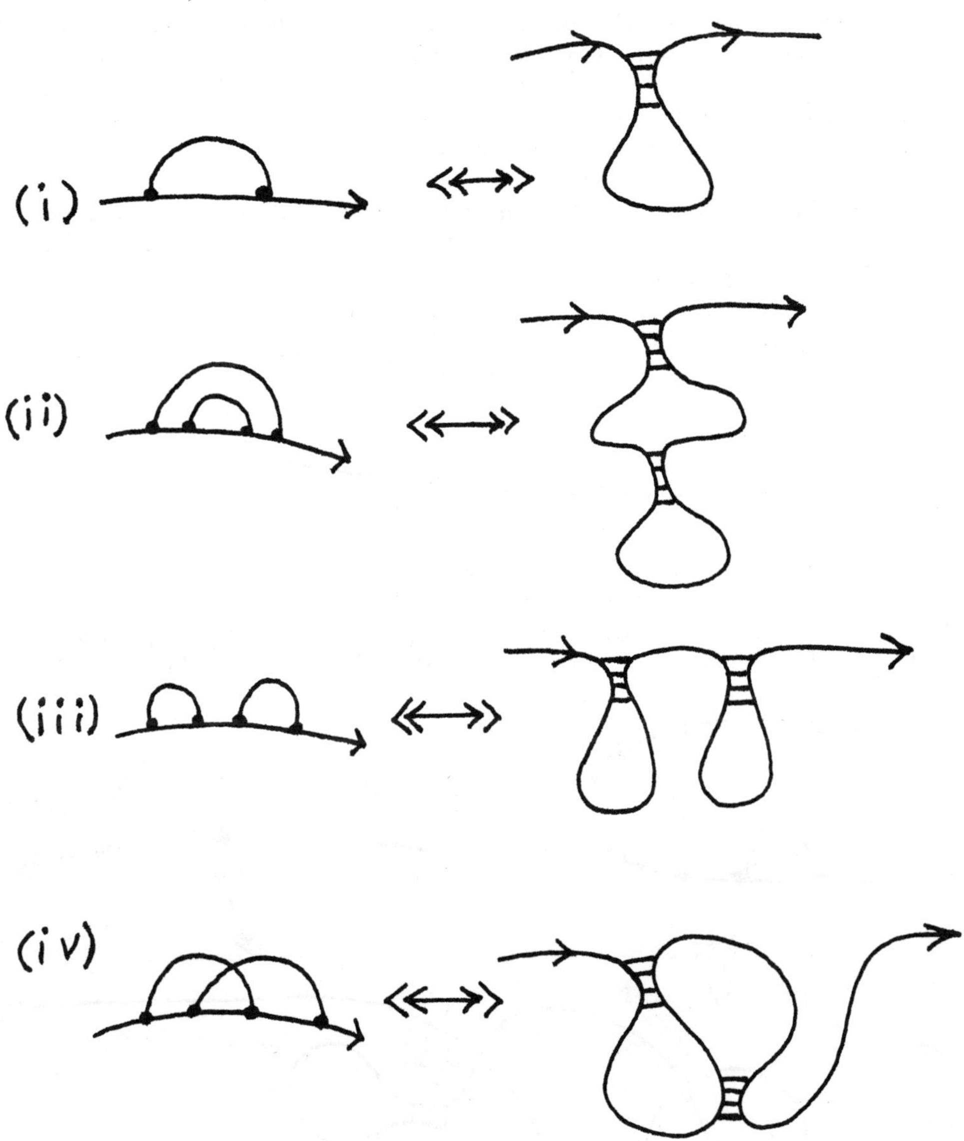

Figure 1

It is these forms that we are interested in classifying, first abstractly and then with respect to their possible embeddings in three dimensional space. The remainder of this section is devoted to the abstract classification.

First, note that the collection of arc diagrams bifurcates into those diagrams with non-intersecting arcs (examples (i), (ii) and (iii) in Fig.1) and those with arcs that necessarily intersect (example iv on Fig.1). We shall refer to foldings that correspond to arc diagrams free of intersections as secondary structures. The rest are *tertiary structures*. The simplest tertiary structure is the "pseudo-knot" illustrated as example (iv) in Fig.1.

We now discuss the following simple strategy for enumerating all arc diagrams (for secondary and tertiary structures). First bend the backbone (i.e. the line representing the linear sequence) into a "finger" as shown below.

A given arc diagram involves the pairing of 2N points for some positive integer N. Array the first N of these points on the bottom arc of the finger, and the second N on the top arc. Draw the connecting arcs in the bounded space of the finger. The example below shows this correspondence in a special case.

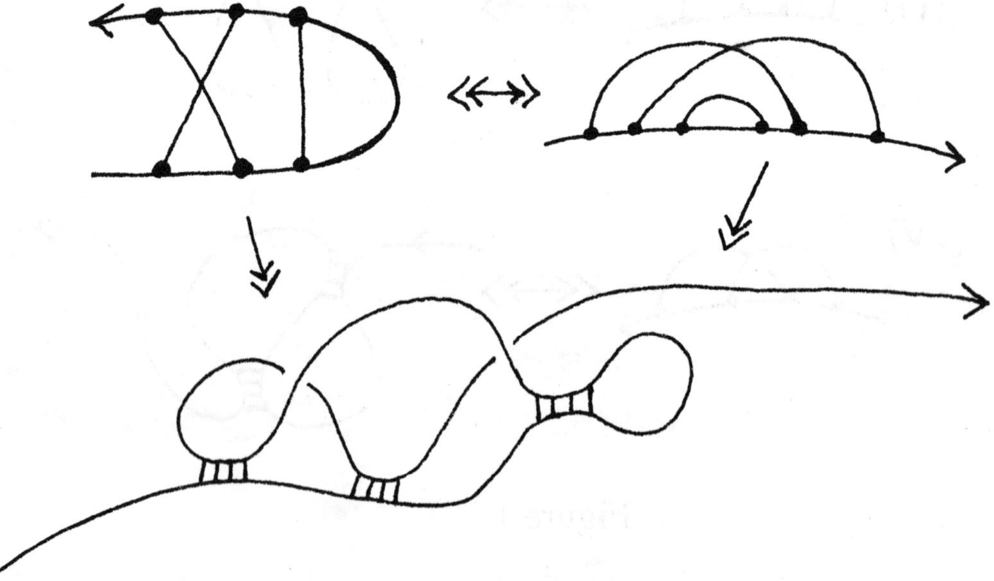

It is now a small step from the finger diagram to the *tangle diagram* consisting in two rows of N-points with arcs connecting the total 2N points in pairs. The arcs are restricted to the space between the two rows of points as shown below (Figure 2).

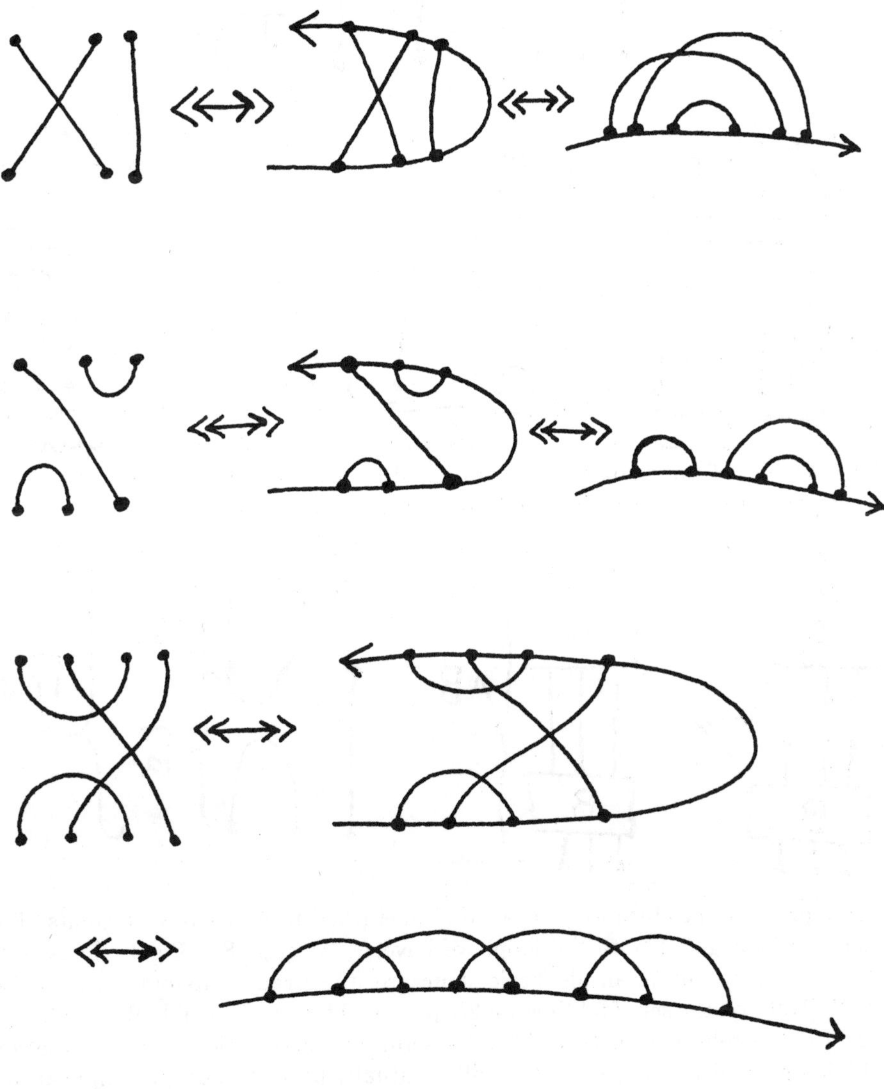

Figure 2

Thus we have shown that **foldings on 2N points are in one-to-one correspondence with tangle diagrams with two rows of N points.**

Let T_N denote the set of tangle diagrams with two rows of N points. Note that two such diagrams are equivalent if and only if they denote the same pattern of point connections. The pattern of intersections of arcs in the tangle is, however, relevant to the structure of the entire set of tangle diagrams. Every tangle diagram can be decomposed as a product of elementary diagrams of the following forms:

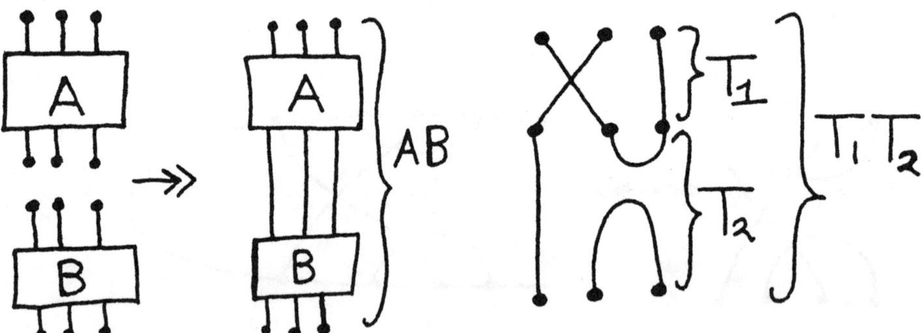

The product of tangle diagrams is obtained by attaching the bottom row of one diagram to the top row of the next diagram as indicated below.

Some products produce loops that are unattached to either row of points. For example letting δ denote the loop, we have $U_i^2 = \delta U_i$. See Fig.3. Thus, setting $\delta = 1$, we obtain an algebraic structure, the Brauer Monoid [B] (see also [BW],[K3]), on the set of foldings on 2N-points. The set of possible RNA-foldings has a rich algebraic structure. This is useful for classification and enumeration of foldings, and we suggest that it will eventually have even deeper implications for molecular biology. It is worth stating some of the algebraic structure of the Brauer monoid explicitly for arbitrary loop value δ. Here is a summary:

    i. δ commutes with every element of T_N.

    ii. I_N is an identity element.

    iii. $T_i^2 = I_n$, $T_iT_{i+1}T_i = T_{i+1}T_iT_{i+1}$

    iv. $U_i^2 = \delta U_i$

    v. $T_iU_i = U_iT_i = U_i$

    vi. $T_iU_{i+1} = T_{i+1}U_iU_{i+1}$, $U_iT_{i+1} = U_iU_{i+1}T_i$.

These relations abstractly specify the complete structure of (T_N, δ), the Brauer monoid of N-tangle diagrams with loop value δ. In Fig.3 we have indicated the diagrammatic picture that accompanies each of these relations. In Fig.4 we have listed all elements of T_3 as tangle diagrams, arc diagrams, and some corresponding foldings.

Fig.4 illustrates the sort of taxonomy provided by the Brauer Monoid. There are 15 structures in all, 15 being the number of ordered foldings on six sites. We have labelled these structures $(1), (2),...,(15)$. Note that $(1) \rightarrow (5)$ are secondary structures. These form the subalgebra of the Brauer Monoid that is generated by $\{I_n, U_1, U_2, ..., U_{N-1}\}$. Call this subalgebra $T\mathcal{L}_N$. it is the multiplicative structure of the *Temperley-Lieb algebra* [BA] (see also [K3],[K1],[KV].) This algebra occurs in statistical mechanics, and it is the basis of the construction of the Jones polynonial in the theory of knots. The structures $(6) \rightarrow (10)$ are tertiary structures occurring as products of $T_1, T_2, ..., T_{N-1}$. The algebra generated by $\{I_N, T_1, T_2, ..., T_{N-1}\}$ has $N!$ elements and is isomorphic with the symmetric group on N letters. In other words, these foldings are in one-to-one correspondence with all permutations of N objects. Finally, the foldings $(11) \rightarrow (15)$ are mixed structures- products of U_i's and T_j's. The first pseudo-knot occurs at (7). We have illustrated particular embeddings in three-space associated with the structures (10) and (15). It will be interesting to compare this approach to the combinatorics of foldings with other methods. In particular, it appears to us that there is a fruitful interaction of the Brauer monoid technique with the methods of Magarshak and Benham in [MB]. This will be the subject of another paper. It is also of interest to compare our approach to that of Penner and Waterman [PW]. Penner uses the secondary structures, with some extra conditions, to create a cell structure for a moduli space for hyperbolic structures on Riemann surfaces. This gives a point of view on the topology of the space of all (unembedded) secondary structures.

We now turn to the matter of embeddings.

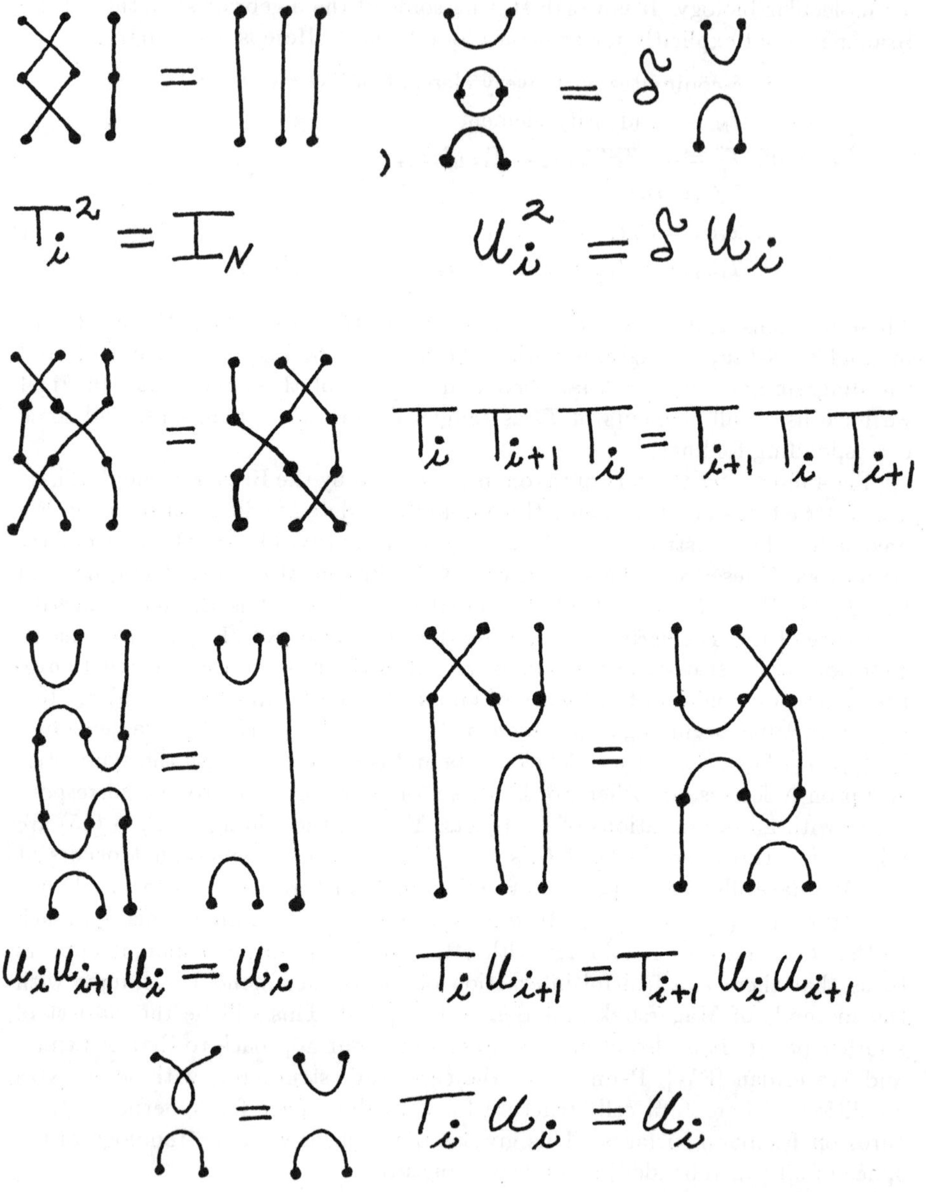

$$T_i^2 = I_N \qquad U_i^2 = \delta\, U_i$$

$$T_i\, T_{i+1}\, T_i = T_{i+1}\, T_i\, T_{i+1}$$

$$U_i\, U_{i+1}\, U_i = U_i \qquad T_i\, U_{i+1} = T_{i+1}\, U_i\, U_{i+1}$$

$$T_i\, U_i = U_i$$

Figure 3

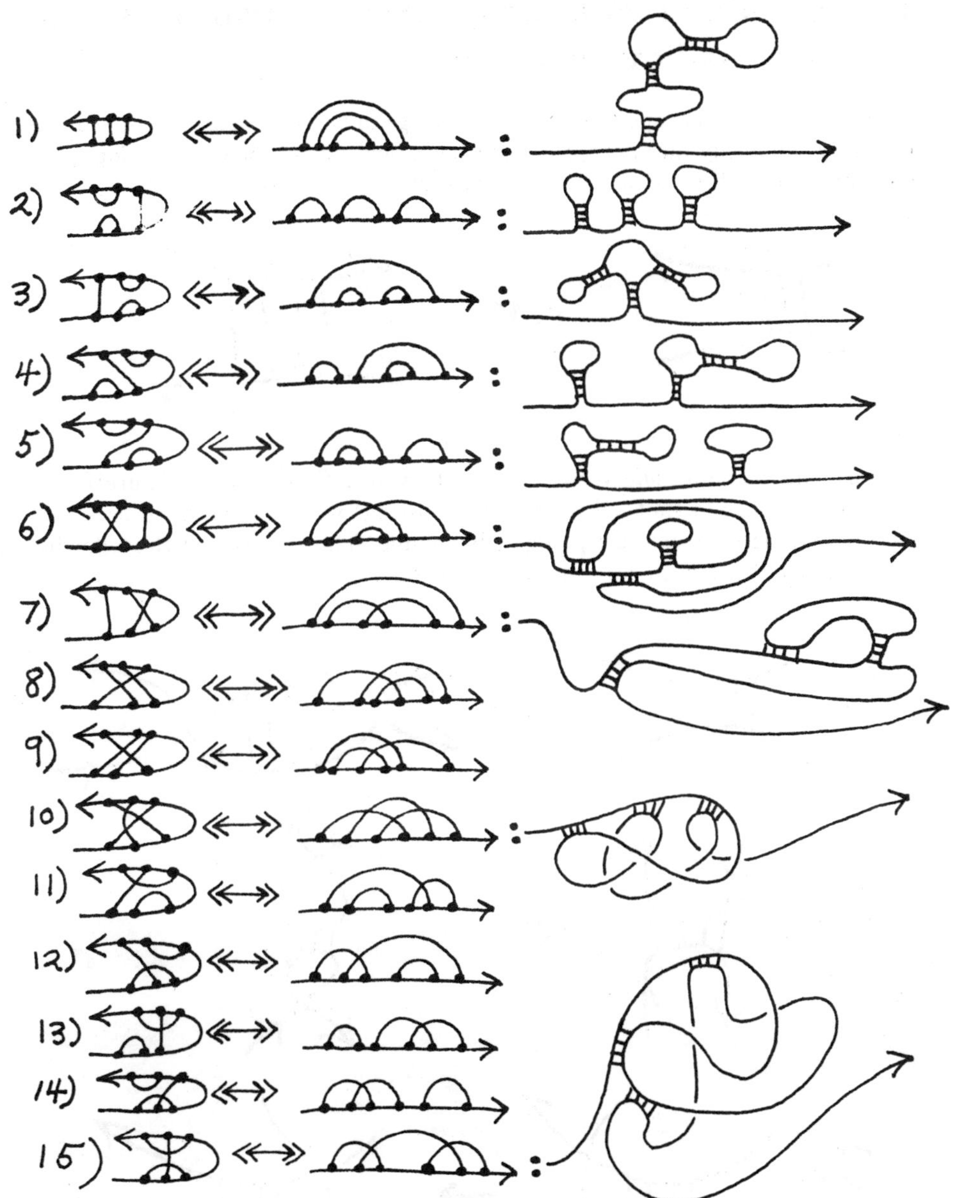

Figure 4

3 Embedded Foldings, Graph Invariants and the Vassiliev Invariants.

In order to study the topology of RNA foldings in three dimensional space it is necessary to specify an appropriate mathematical model for this topology. We take the lead for this model from the form of our basic bond vertex:

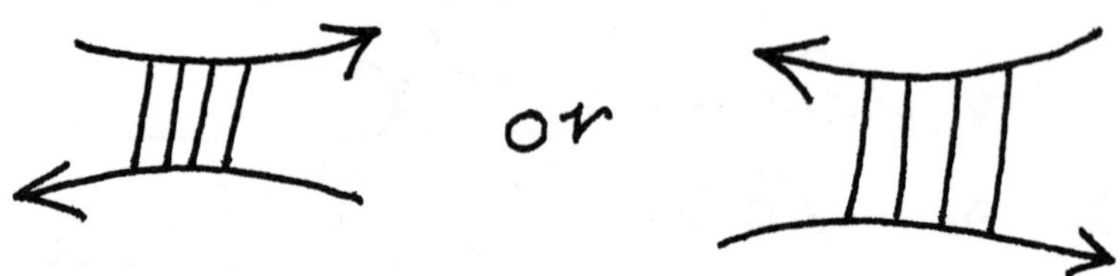

We take this to be a *rigid* vertex by which is meant that the configuration of bonding arcs is rigid (not subject to any twisting) while the oriented arcs that enter or leave the vertex are topologically flexible. This means that the following moves are available with respect to the vertex:

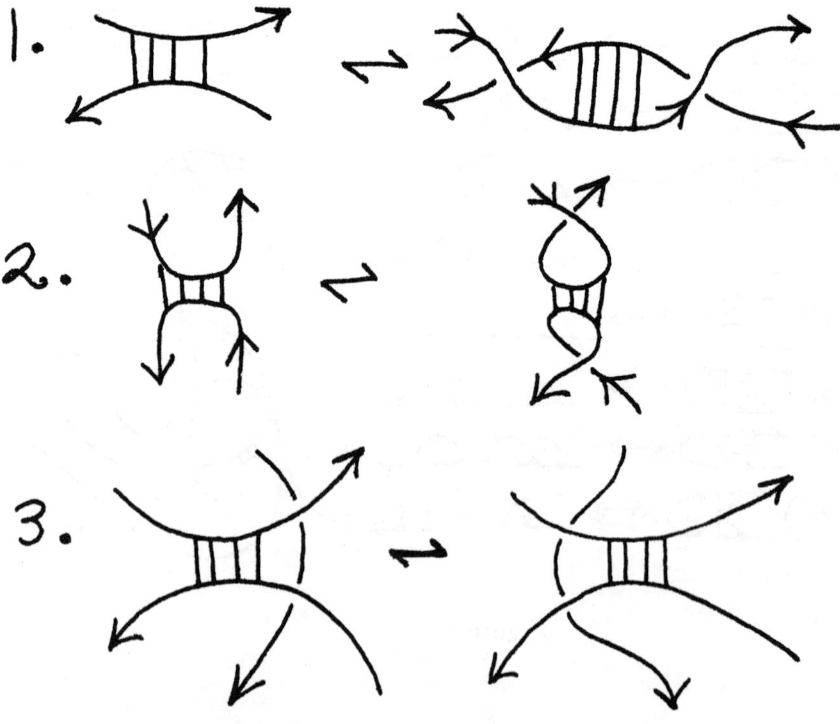

These moves (and obvious symmetries obtained by mirror imaging) plus the usual Reidemeister moves [K1] away from the bonds constitute our topological model for rigid vertex isotopy [K4]. For the record, the basic Reidemeister moves are shown below:

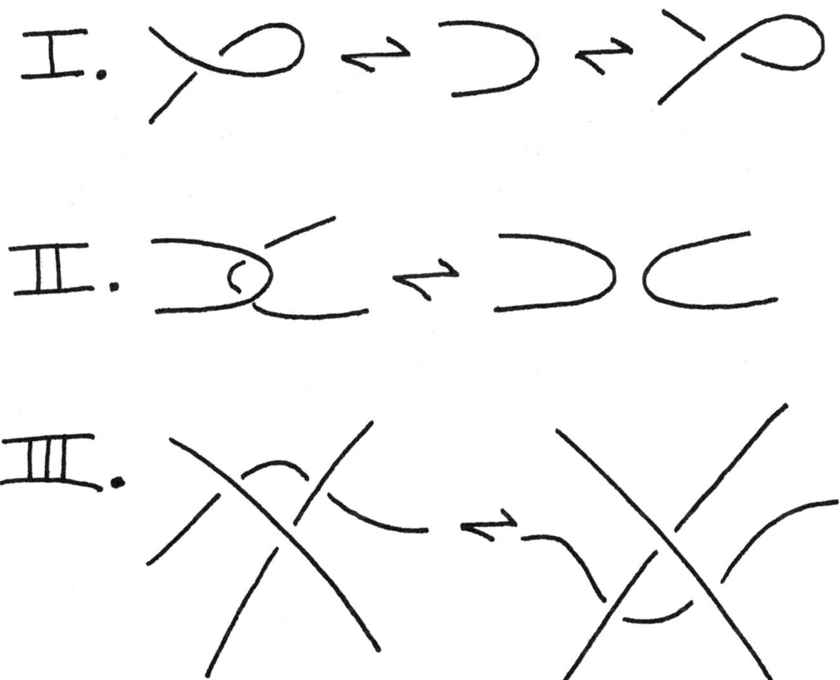

The invariants that we are about to discuss are indeed invariants of rigid-vertex graph embeddings in three dimensional space. However, they are formulated in the mathematical literature with respect to a rigid vertex with a different structure. This structure is as shown below.

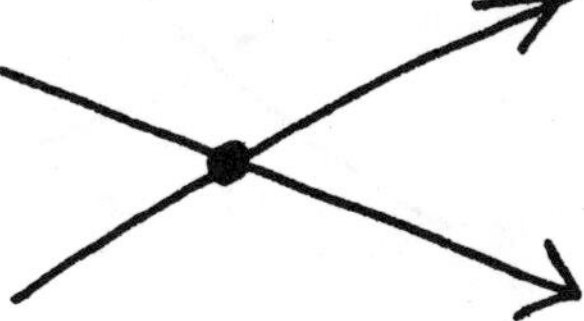

In this vertex the strands that bond go cross-wise to one another, forming a vertex with two in-going and two out-going lines. Mathematical formulations with respect to this *transverse vertex* are particularly convenient and symmetrical. Consequently, we shall define a conventional relationship between the transverse vertex and the bond vertex so that they can (up to a translation) be used interchangeably.

Definition. By convention, we take the following relationship between bond vertex and transverse vertex.

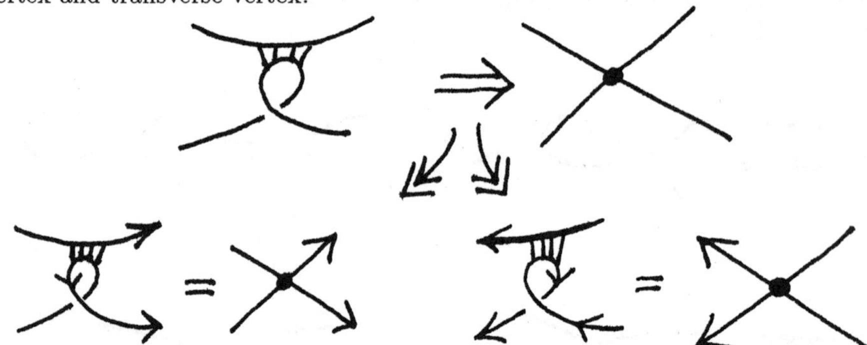

From the point of view of RNA-folding, this serves as the definition of the transverse vertex. An invariant of rigid isotopy for transverse vertices will automatically be an invariant of rigid isotopy for bond vertices (and conversely). Thus we shall discuss invariants in the context of the transverse vertices. The translation between bond vertices and transverse vertices is simple, but it does involve a definite shift of context. For example (and this is quite important), we can "resolve" a transverse vertex into a crossing of two lines that do not touch in two possible ways:

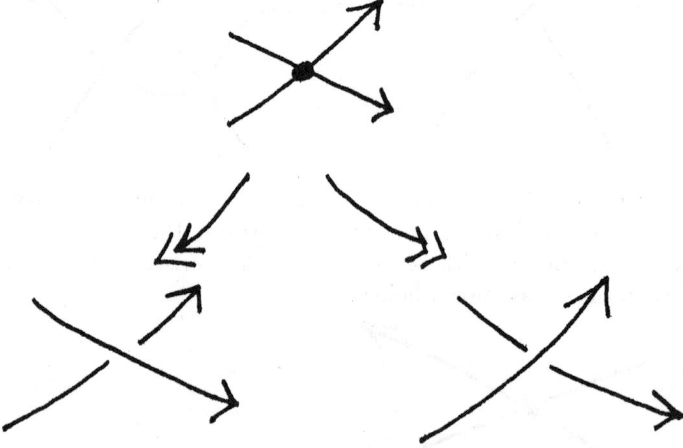

These resolutions are most natural in the transverse context, where *we visualize the vertex as a stage in the act of passing one line through the other.*

Translating this scenario to the bond vertex, we find

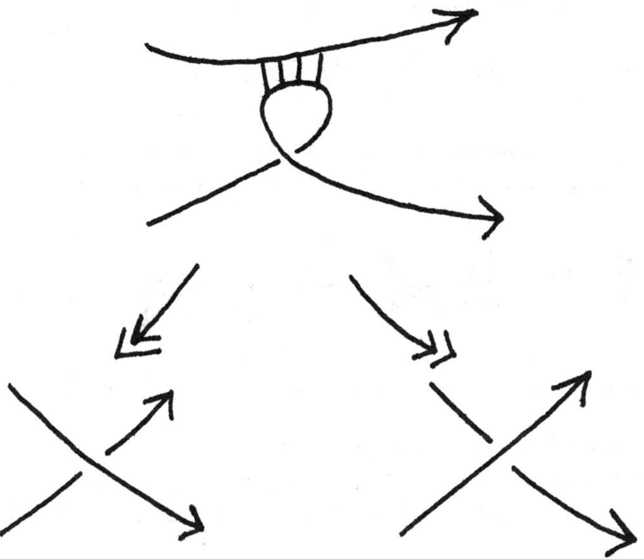

Thus the resolution of the bond vertex involves a recombination that may have no biological significance in the context of RNA. It is interesting to speculate about the possible meaning of mathematical operations in terms of biology. In this case, we justify including such recombinations because they allow us to calculate topological invariants.

Now, let us begin the topology. First of all, suppose that $\mathcal{I}_K$ is an invariant that assigns a number to an oriented knot or link K so that if K and K' are related by a sequence of Reidemeister moves, then $\mathcal{I}_K = \mathcal{I}'_K$. Given such an $\mathcal{I}_K$, we can define an extension of $\mathcal{I}$ to include rigid vertex graphs. We do this by the equation

$$\mathcal{I}\,\mathord{\times\!\!\!\times} = \mathcal{I}\,\mathord{\nearrow\!\!\!\nwarrow} - \mathcal{I}\,\mathord{\times\!\!\!\times}$$

(this is part of a more general scheme, see [KV]). More specifically, let G be a graph embedding with rigid vertices $V_1, V_2, ..., V_n$. Let $\vec{\epsilon} = (\epsilon_1, \epsilon_2, ..., \epsilon_n)$ be a vector with $\epsilon_i = \pm 1$ for each i. Let $[2^n]$ denote the set of these vectors. Let $G(\vec{\epsilon})$ denote the link or knot that is obtained from G by replacing V_i by a crossing of type ϵ_i. Here $\mathord{\times\!\!\!\nearrow}$ has type $+1$ and $\mathord{\times\!\!\!\searrow}$ has type -1. Let $|\vec{\epsilon}|$ denote the number of $\epsilon_i = -1$ in $\vec{\epsilon}$. Now define $\mathcal{I}_G$ by the formula

$$\mathcal{I}_G = \sum_{\epsilon \in [2^n]} (-1)^{|\vec{\epsilon}|} \mathcal{I}_{G(\vec{\epsilon})} \qquad (1)$$

This formula gives a well-defined value to $\mathcal{I}_G$ in terms of the values of $\mathcal{I}$ on knots and links and it is obvious that $\mathcal{I}$ satisfies the formula

where the small diagrams are regarded as parts of an otherwise unchanged larger diagram.

We now have the following basic lemma [KV].

Lemma. If $\mathcal{I}$ is an ambient isotopy invariant of knots an links, then its extension to rigid vertex graphs, as defined above, is an invariant of rigid vertex isotopy.

Proof. We just check one case of rigid vertex isotopy, leaving the rest as an exercise for the reader.

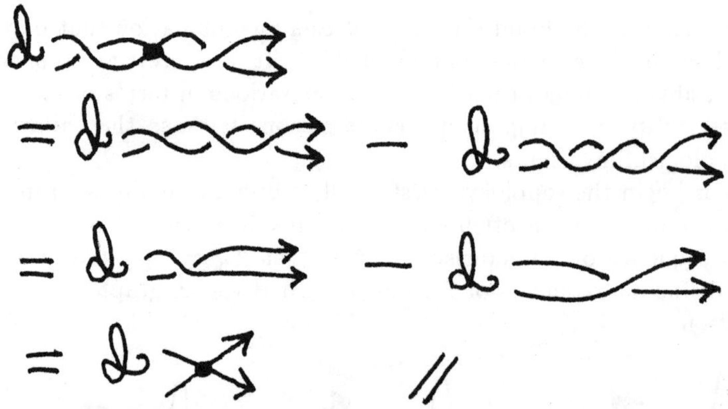

Example. Let $\nabla_K(z)$ denote the Conway (Alexander) polynomial of K, for K a knot or link. Then ∇_K is determined by the axioms [K1]

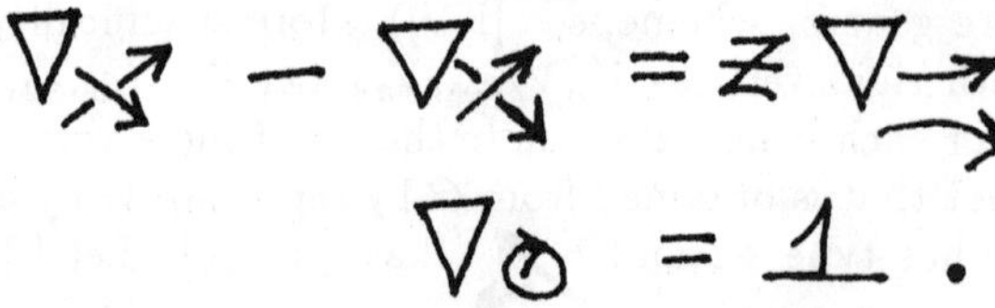

Thus, in the graph extension, we have

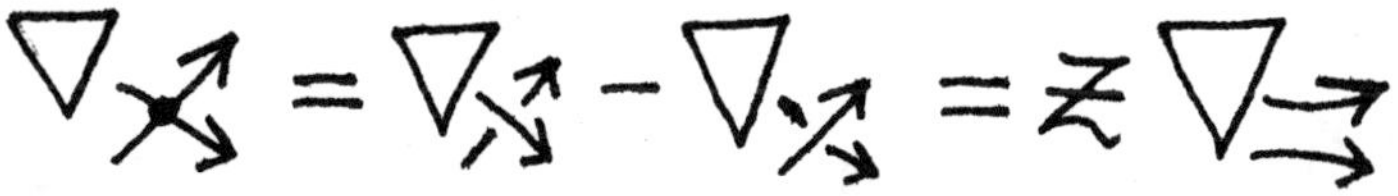

$$\nabla_{\!\! \times} = \nabla_{\!\! \times} - \nabla_{\!\! \times} = z\, \nabla_{\!\!\longrightarrow}$$

From this we see that if G has n vertices, then ∇_G is divisible by z^n. If we think of ∇_K as a set of numerical invariants (the polynomial coefficients), then we have

$$\nabla_K = \sum_{i=0}^{\infty} c_i(K) z^i$$

(c_i eventually zero for any given knot or link K). The statement that G with n vertices $\Rightarrow z^n | \nabla_G$ then becomes: $c_i(G) = 0$ if G has $> i$ vertices. We say that the invariants c_i are of finite type. Note that they also satisfy the identity

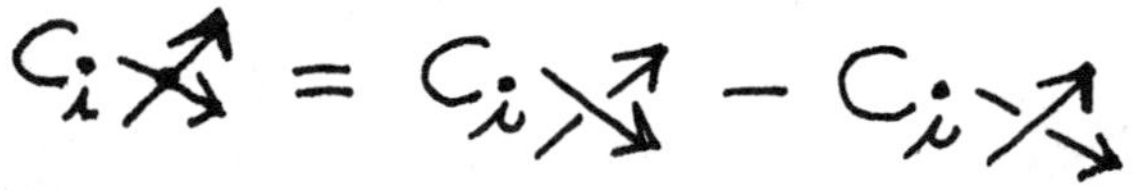

$$C_i \,{\times} = C_i \,{\times} - C_i \,{\times}$$

This leads to the following

Definition. An invariant $\mathcal{I}$, of rigid-vertex graphs, is said to be a Vassiliev invariant of finite type i, if it satisfies the following rules [V], [BL]:

1. 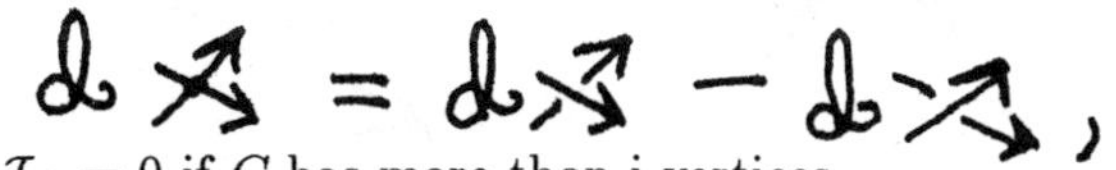 $$\partial_{\times} = \partial_{\times} - \partial_{\times},$$
2. $\mathcal{I}_G = 0$ if G has more than i vertices.

Thus we have shown that the rigid- vertex graph extensions of the coefficients of the Conway polynomial, $c_i(G)$, are Vassiliev invariants of type i.

The virtue of the Vassiliev invariants is that, being of finite type, they are determined by their behavior on a finite collection of graphs. These graphs can be interpreted as RNA foldings! Thus the Vassiliev invariants can give us information about the structure of embeddings of RNA foldings *and* they are also a way to look at the abstract structure of these foldings.

We shell give specific examples shortly. But first, it is necessary to look more closely at the translation from foldings to graphical nodes: in a Vassiliev invariant we have

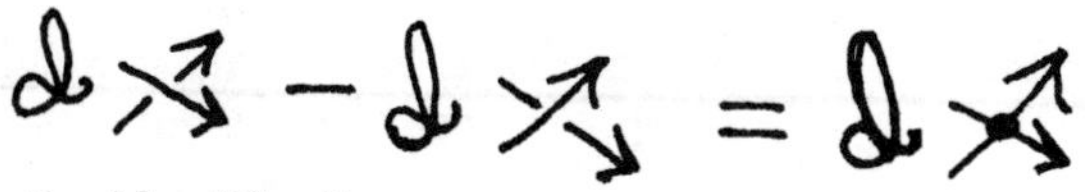

$$\partial_{\times} - \partial_{\times} = \partial_{\times}$$

and we have made the identification

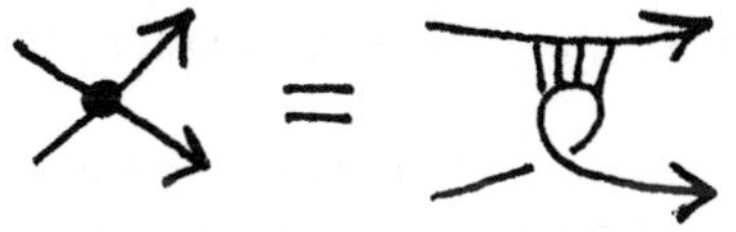

$$\times = \bigcup$$

Therefore

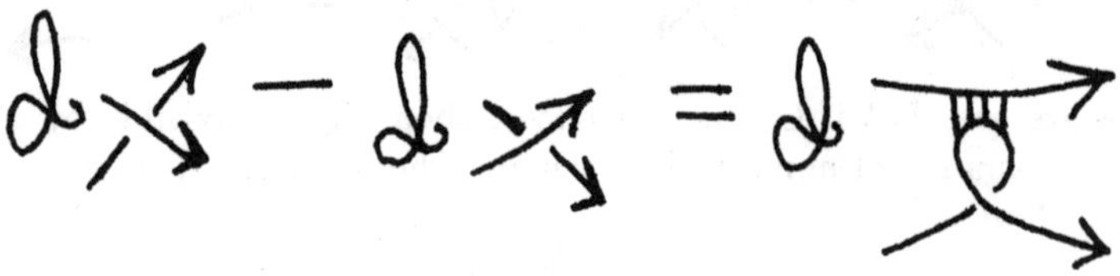

$$\mathcal{d} \nearrow - \mathcal{d} \searrow = \mathcal{d} \,\, \asymp$$

whence,

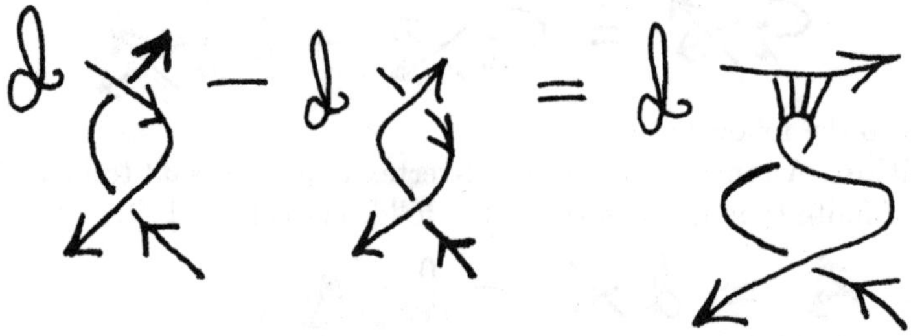

$$\mathcal{d} \nearrow - \mathcal{d} \nearrow = \mathcal{d} \,\, \asymp$$

Hence

$$\boxed{\mathcal{d} \,\uparrow\uparrow - \mathcal{d} \,\nearrow = \mathcal{d} \,\, \mathrm{III}}$$

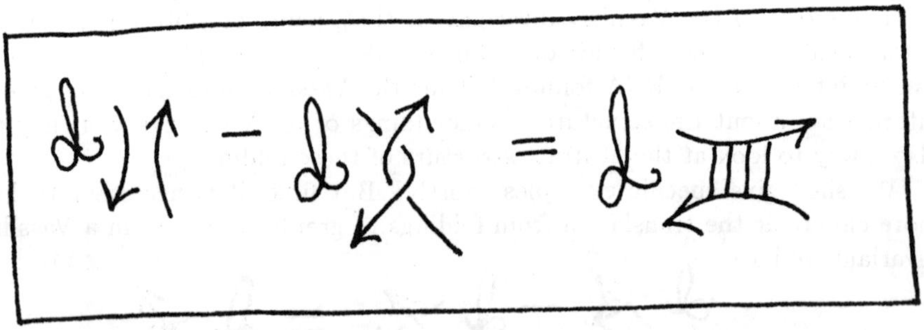

This is the basic equation for computing the Vassiliev invariant in the language of foldings.

It is this equation that may be of direct use to the microbiologist interested in the topology of foldings. Note that we can prove the basic invariance lemma directly:

Lemma. Let $\mathcal{I}_K$ be an ambient isotopy invariant of oriented knots and links K. Define a function on foldings via

(This is formalized just as in our discussion of $\mathcal{I}$).

Then, for a folding F, $\mathcal{I}_F$ is an invariant of rigid-vertex isotopy of the folding.

Proof. Again we just check one of the cases of rigid vertex twist:

As an example, consider the embedding of the tertiary structure T shown below:

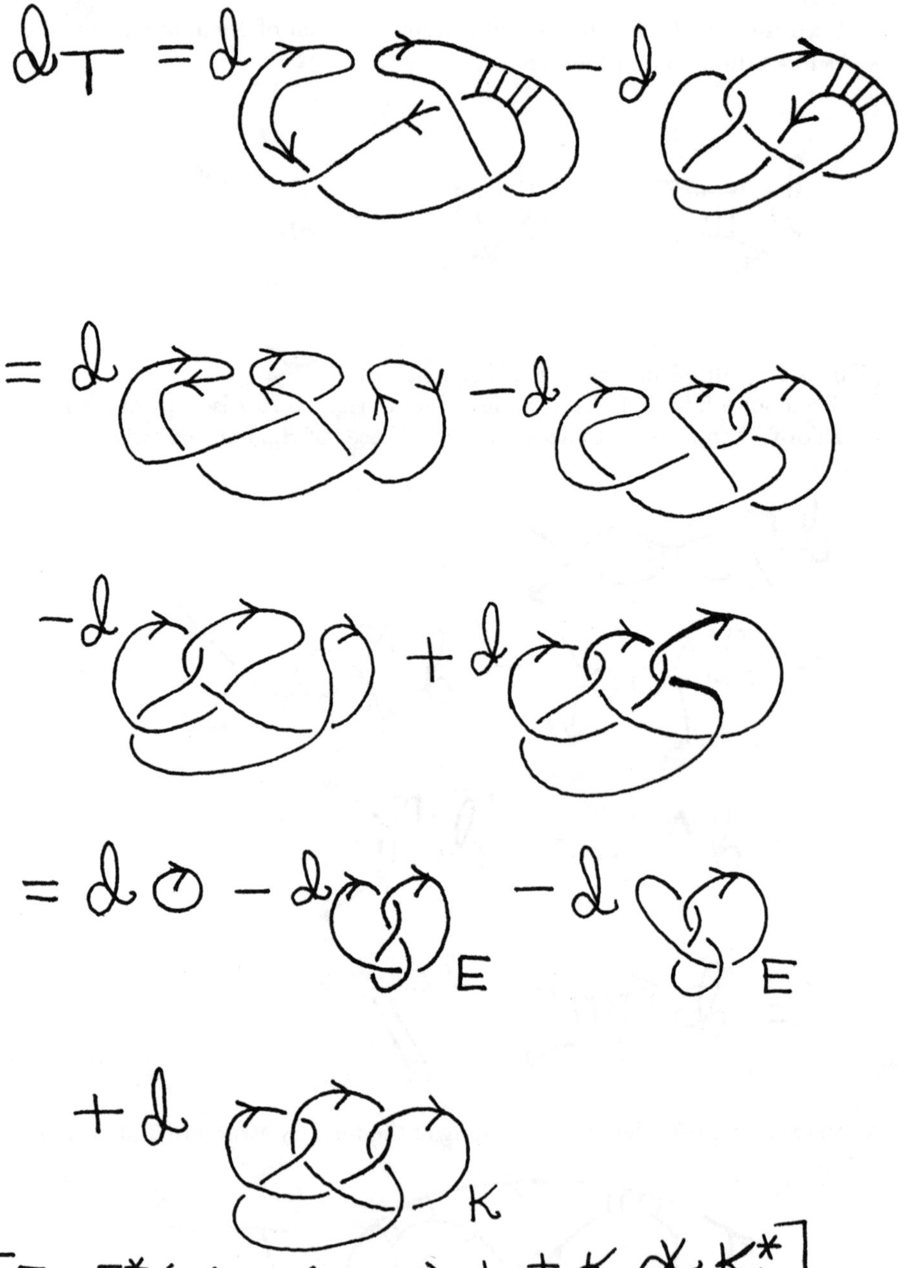

$$\partial_T = \partial \quad - \partial$$

$$= \partial \quad - \partial$$

$$- \partial \quad + \partial$$

$$= \partial \bigcirc - \partial \Big)_E - \partial \Big)_E$$

$$+ \partial \Big)_K$$

$$[E \sim E^* \ (\text{mirror image}), \ \text{but} \ K \not\sim K^*]$$

Then by using the facts that the two knots shown in the above expansion are actually knotted and inequivalent to their mirror images, we conclude that T is not rigid-vertex isotopic to T',

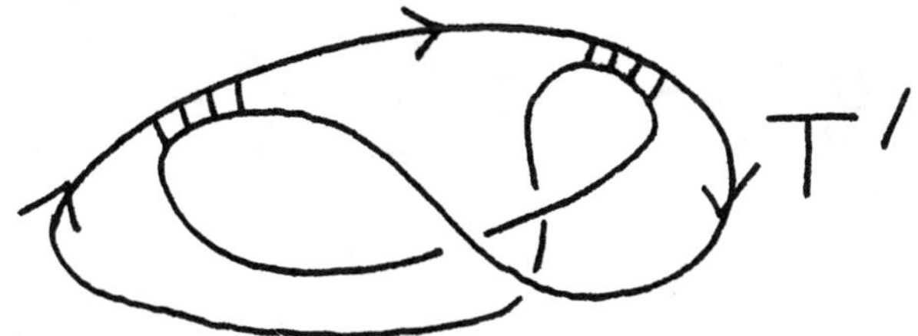

nor is T rigid vertex isotopic to its mirror image. In this case, we have not assumed that our invariants were of finite type. Nevertheless, the formulation

$$\mathcal{d} \overline{\underline{\mathbb{I}}} \; = \; \mathcal{d} \, \nearrow \; - \; \mathcal{d} \, \nwarrow$$

corresponds directly, via our conventions, to the basic identity for Vassiliev invariants

$$\mathcal{V} \times \; = \; \mathcal{V} \times \; - \; \mathcal{V} \times$$

and hence, computations of $\mathcal{I}_K$ can be interfaced with computations of Vassiliev invariants.

In the next section we shall supply more information about Vassiliev invariants. For the reminder of this section, we show how to reformulate our more general rigid vertex graph invariants [K4] for the case of protein folding.

4 Generalizing The Invariant $\mathcal{I}$.

Let

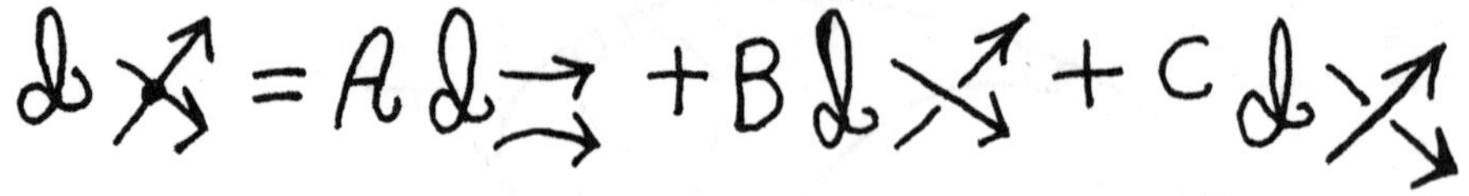

define a graph invariant corresponding to a given oriented link invariant (as explained in [K4]). Then we can reformulate $\mathcal{I}$ in folding terms via:

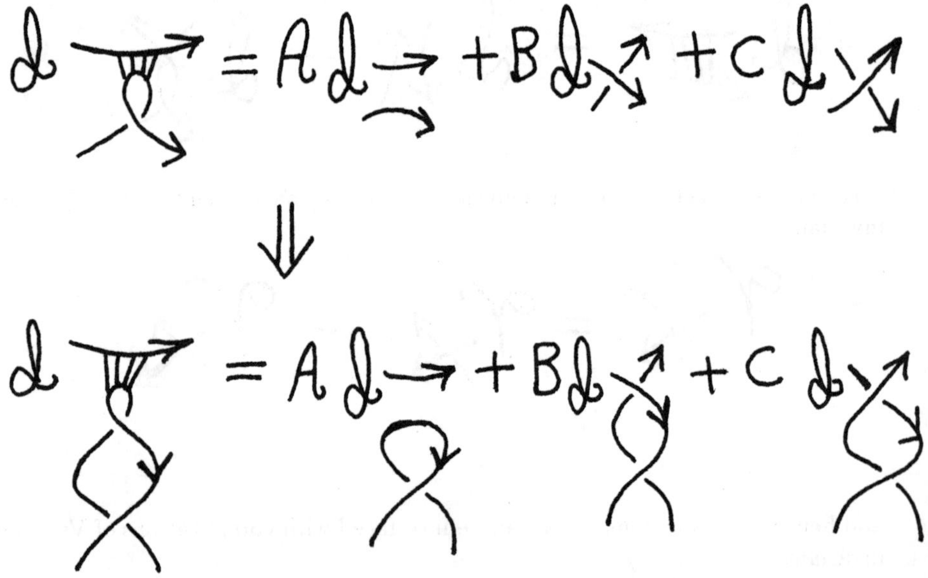

Hence (assuming $\mathcal{I}$ on links is an ambient isotopy invariant)

$$\mathcal{I}\left(\vcenter{\hbox{⧓}}\right) = A\,\mathcal{I}\left(\vcenter{\hbox{⇄}}\right) + B\,\mathcal{I}\left(\vcenter{\hbox{)(}}\right) + C\,\mathcal{I}\left(\vcenter{\hbox{✕}}\right).$$

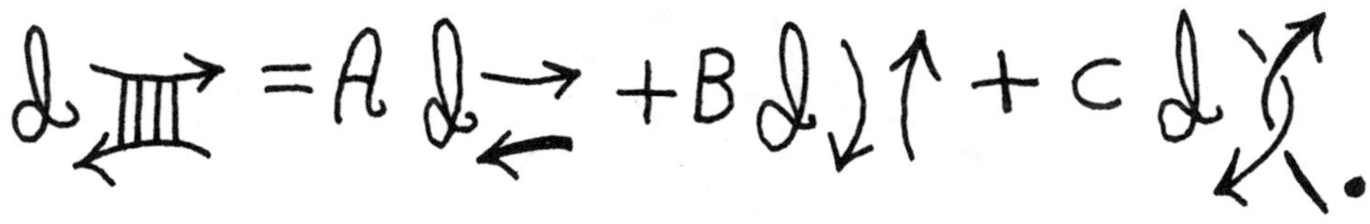

This completes our description of the *generalized invariant of foldings*. It is quite useful for studying the topology of foldings via direct unfolding, recombination and linked recombination. If one knows either an ambient isotopy invariant $\mathcal{I}$ or the specific isotopy classes of the knots and links that occur through this process of resolution, then a great deal of information about the topology of the folded molecule is obtained as a consequence.

5 More about Vassiliev invariants.

So far, except in the case of the Conway polynomial, we have not discussed the crucial matter of finite type in relation to graph invariants satisfying

$$V\!\left(\vcenter{\hbox{✕}}\right) = V\!\left(\vcenter{\hbox{over}}\right) - V\!\left(\vcenter{\hbox{under}}\right)$$

Therefore, suppose that V is of type i. Let $\#\,G$ denote the number of 4-valent nodes in a graph G. Then we have the important
Fact. If V is of type i, then for $\#\,G=i$, V is independent of the embedding type of G in R^3.

   Proof. Suppose G is embedded with a crossing of the form $\times$ in its list of crossings. Then let $G(\times)$ denote this embedding and $G(\times)$ the embedding obtained by switching this given crossing. Then we have:

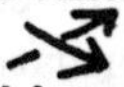

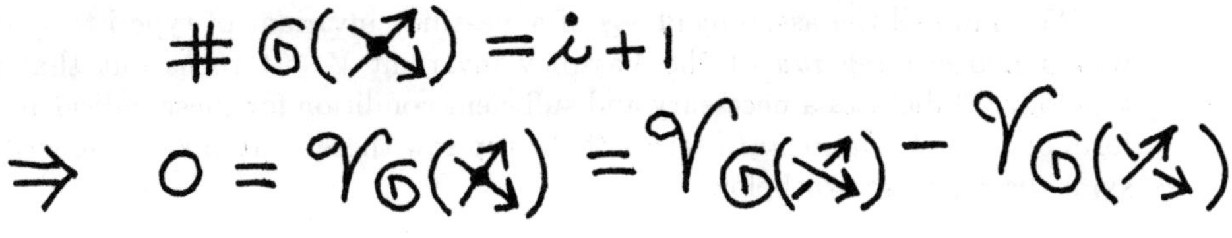

$$\#\,G(\times) = i+1$$

$$\Rightarrow\quad 0 = V_{G(\times)} = V_{G(\times)} - V_{G(\times)}$$

Thus $V_{G(\times)} = V_{G(\times)}$. From this independence of crossings, it follows

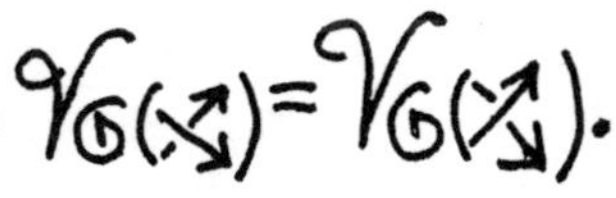

that $\mathcal{V}_G$ depends only upon the abstract graph G. This completes the proof. //
In this section we shall work entirely in the language of 4-valent nodes. Thus a diagram

is an embedded graph with nodes labelled 1 and 2. The abstract structure of G is represented by the pairing diagram

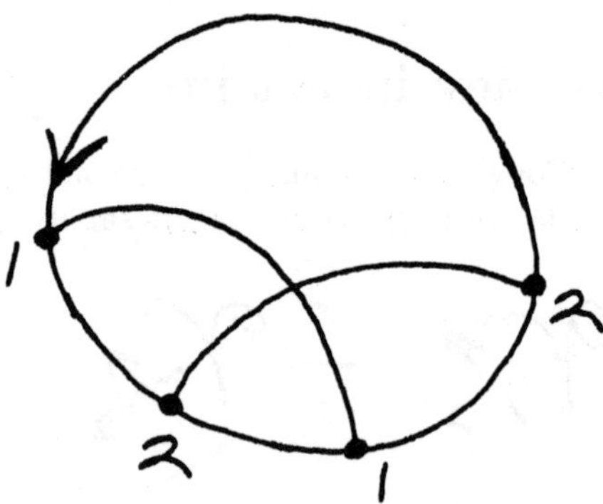

This same diagram represents a folding structure. We see that *Vassiliev invariants of type i assign (topological) indices to abstract folding structures with i pairings.* These indices do not depend upon the embedding type of foldings with i pairings *and* they can be used to obtain information about embeddings of foldings with fewer than i pairings (as we did in the last section).

We shall call the assignment $\mathcal{V}_G$ of a Vassiliev invariant of type i to graphs with i nodes a *top row* of the Vassiliev invariant V. It turns out that the topology [S] dictates a necessary and sufficient condition for these indices in the form of a set of relations. In terms of the top row these relations can be written symbolically as shown below:

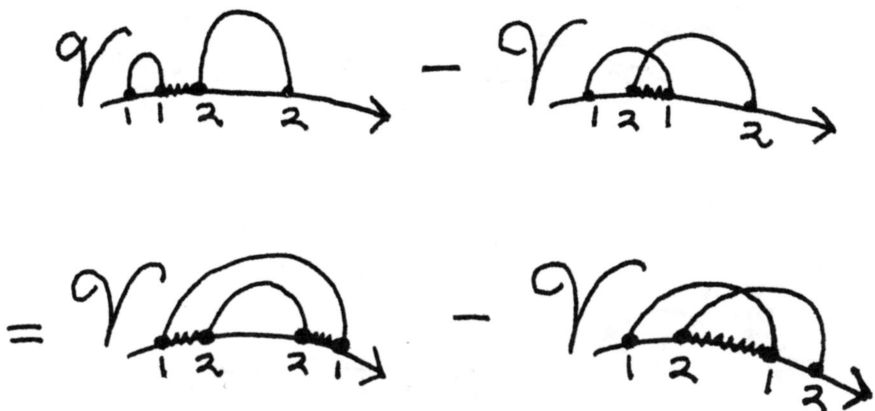

Here the wiggly line indicates that the indices 1 and 2 on either end of it are actually neighbors with no other intervening connections.

For example, at i=3 we have:

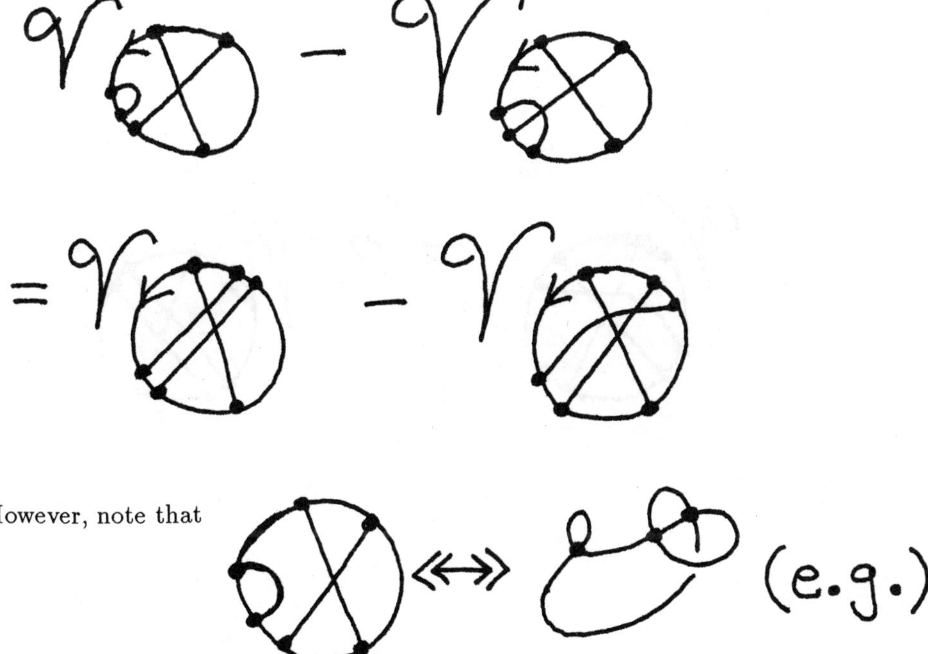

However, note that $\quad\longleftrightarrow\quad$ (e.g.)

$$\mathcal{V}_\sigma^{\rightarrow} = \mathcal{V}_\partial^{\rightarrow} - \mathcal{V}_{-\sigma}^{\rightarrow}$$

$$= 0.$$

$$\Rightarrow \mathcal{V}\,\bigcirc = 0 \quad \text{at } i = 3.$$

Thus

$$-\mathcal{V}\,\bigcirc = \mathcal{V}\,\bigcirc - \mathcal{V}\,\bigcirc$$

Hence

$$\mathcal{V}\,\bigcirc = 2\,\mathcal{V}\,\bigcirc$$

Whence

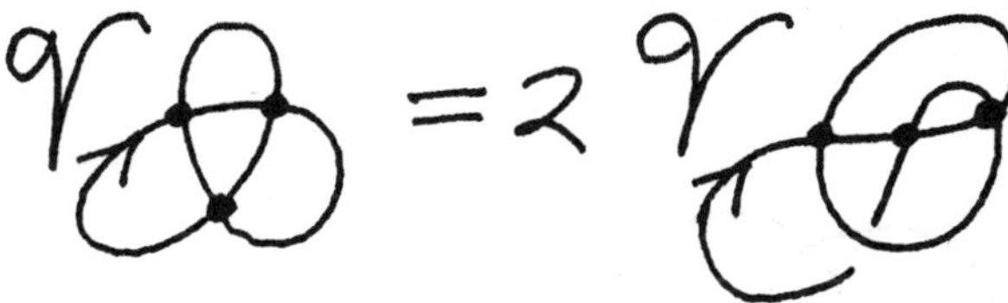

if 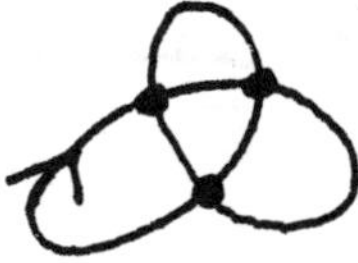 has *type 3*.

In terms of constructing Vassiliev invariants of type 3, this means that one only has to consider the 3-noded graph

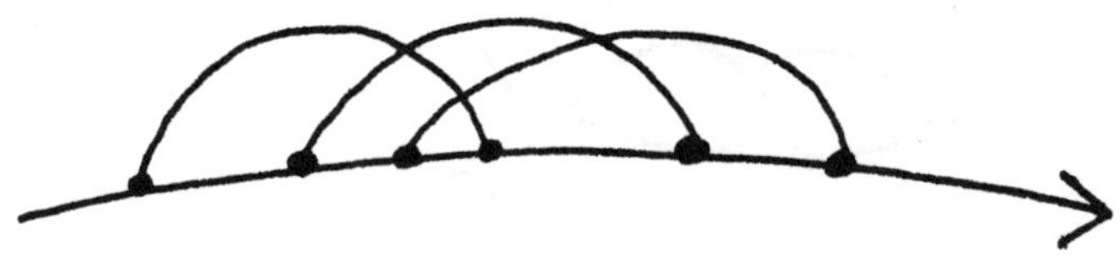

or its corresponding folding

This makes Vassiliev invariants of type 3 rather easy to compute. We assign

Now go back to 2-noded graphs such as

In fact, we see that abstractly this is the only 2-noded graph of relevance. Assign it the abstract value 1, and *define*

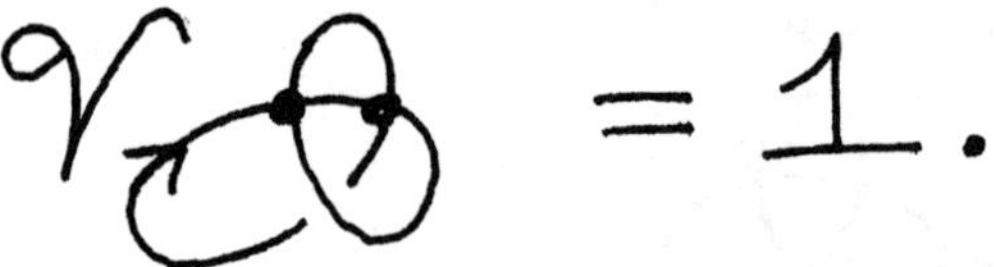
$$= 1.$$

Then, any other embedding of

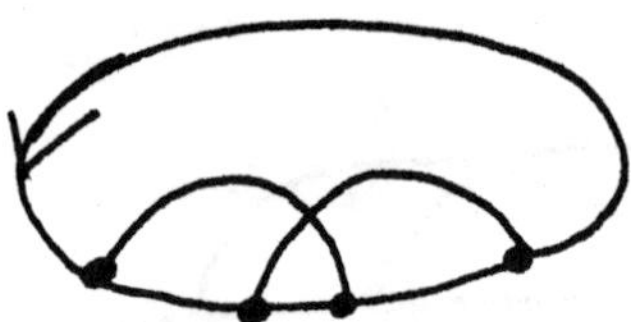

is determined by the values at level 3 and the switching relations. For example,

$$-\ \bigg/\!\!\!\bigcirc\!\!\!\bigcirc\ +\ \bigg/\!\!\!\bigcirc\!\!\!\bigcirc\ =\ \bigg/\!\!\!\bigcirc\!\!\!\bigcirc\ = 2,$$

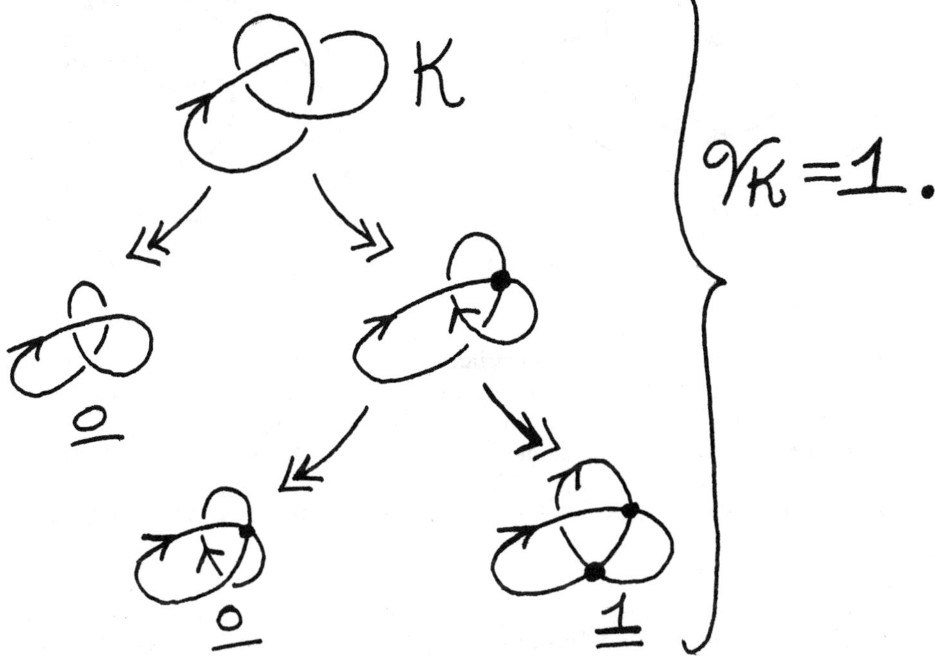

$$\Rightarrow \quad V = -2 + 1 = -1.$$

Similarly

$$V_K = 1.$$

Since it is easy to check that

$$V = V_K - V_{K^*}.$$

This shows that this Vassiliev invariant of type 3 detects the topological difference between the trefoil knot K and its mirror image K^*.

By the same token, we have shown that the graph embeddings

and

are not rigid vertex isotopic, and hence neither are the folded embeddings shown below isotopic.

There are non-trivial Vassiliev invariants of all orders. These can be obtained from the well-known skein polynomials via truncation of power-series substitutions. For example, we have the theorem of Birman and Lin.

Theorem.[BL] Let $V_K(t)$ denote the original Jones polynomial [J] as a Laurent polynomial in K. Let $V_K(e^x) = \sum_{n=0}^{\infty} v_n(K)x^n$ be the power series resulting from substituting e^x for t. Then the coefficients $v_n(K)$ are Vassiliev invariants of type n.

Proof. See [BL] or [K5].

Another very striking result is the construction of Top Rows by Bar-Natan [BAR] via Lie algebra and "Feynman diagrams". In Bar-Natan's construction, the "chord diagrams" (our folding diagrams) are extended to allow a 3-valent interior vertex such as

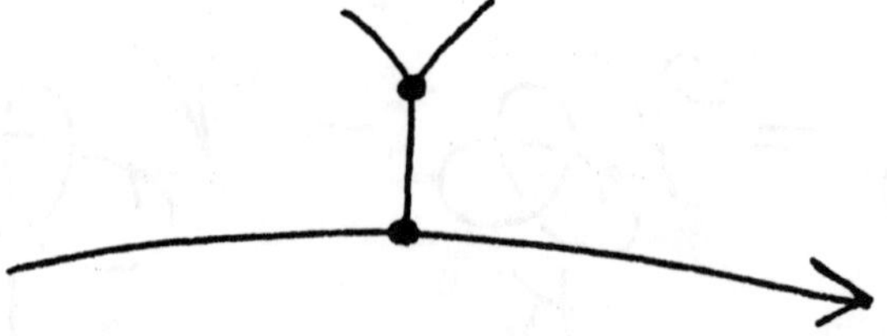

or

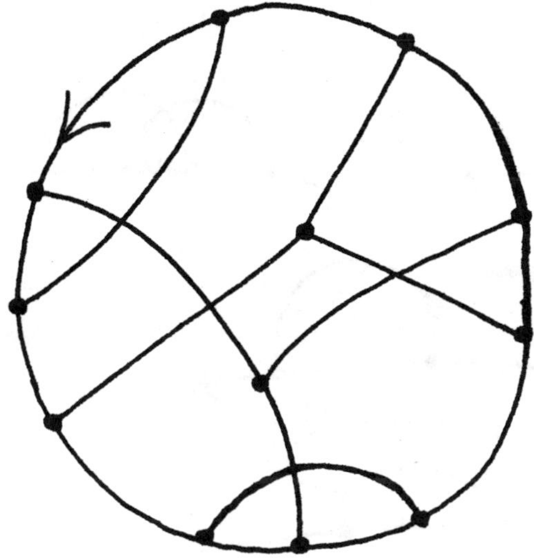

Bar-Natan takes as axiomatic the relation

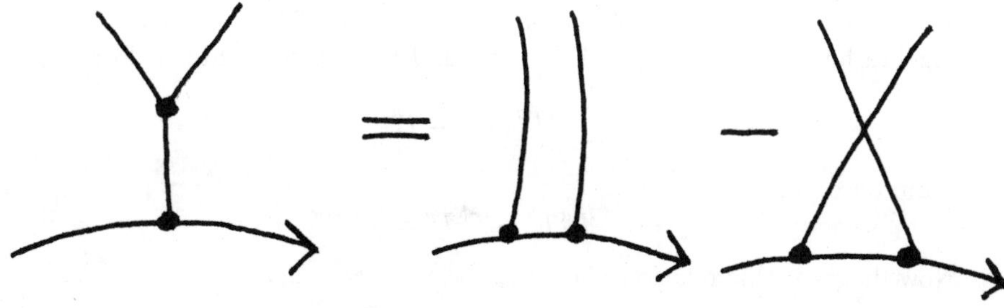

Call this relation the *STU* relation.

Proposition. The STU relation implies the topological 4-term relation on top row diagrams.

Proof.

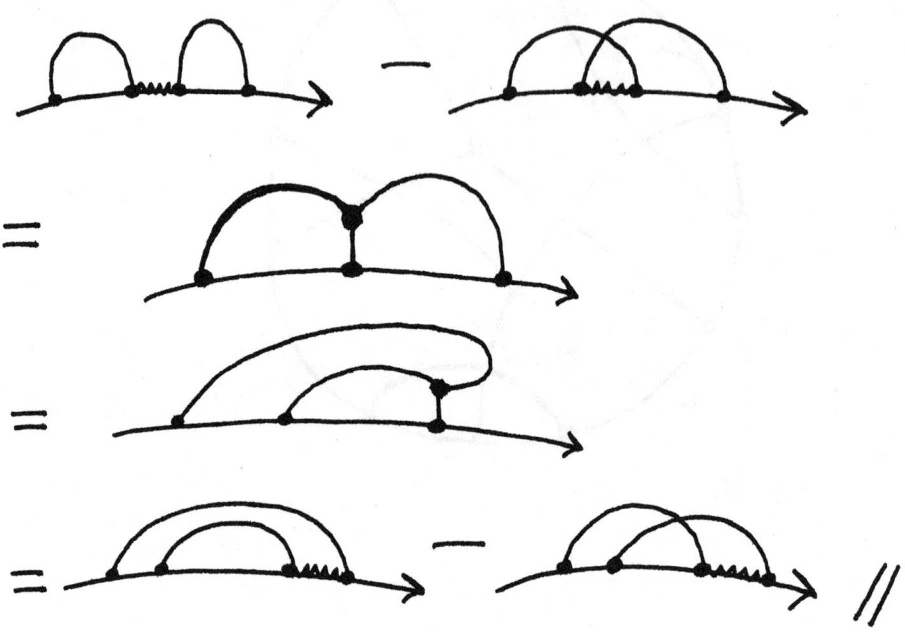

The fantastic thing about this observation is that (in the context provided by the Vassiliev invariants) it provides the core explanation why Lie algebras implicate topological invariants of knots, links and graphs! For the STU relation is actually an abstract way to state that in representing a Lie Algebra, *the representation of a commutator is the commutator of the representations.* A Lie algebra has a basis $\{T^a | a = 1, 2, ..., n\}$ and a basic commutator formula

$$[T^a, T^b] = f_c^{ab} T^c$$

(sum on c), or

$$T^a T^b - T^b T^a = f_c^{ab} T^c$$

Now diagram this relation via

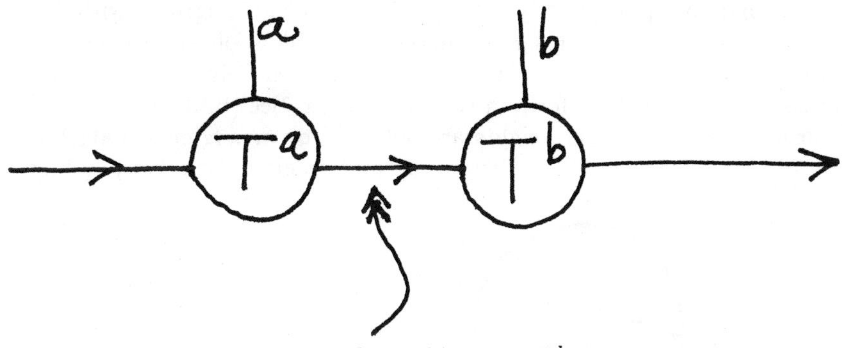

product of T^a with T^b.

$$\overset{a\qquad b}{\underset{c}{\mathbf{Y}}} = f^{ab}_{\;c}$$

Then

$$T^a T^b - T^b T^a = f^{ab}_{\;c}\, T^c$$

$$\Updownarrow$$

Thus, we see that by appropriately labelling the trivalent diagrams with Lie algebra generators (or their representing matrices) we shall obtain weight systems that give rise to Vassiliev invariants (We have deliberately left out certain technicalities about the Killing form in the Lie algebra. See [BAR].)

For pairings themselves, the weights are obtained by arranging Lie algebra generators at the paired points and summing and tracing the corresponding matrices. Thus

$$\langle\!\langle \leftrightarrow \rangle\!\rangle \;\leftrightarrow\; tr\left[\sum_{a,b} T^a T^b T^a T^b\right]$$

$$(tr \text{ denotes matrix trace.})$$

(again we have deliberately left out the Killing form). The approach sketched here is valid when f^{ab}_c is totally anti-symmetric in the three indices- such bases are available for the classical semi-simple Lie algebras.

It is fascinating to speculate on deeper relationships between molecular biology and this pattern of assigning Lie algebra elements to the pattern of paired bases in a protein folding.

6 Discussion.

1. **4D-Interval and nucleotide algebra.** Self-splicing and RNA enzymatic activity suggest that not only secondary structure of RNA is biologically important, and under control, but secondary structure transitions are important as well ([PW], [MK]). In this paper secondary structures of RNA have been analyzed without particular analysis of nucleotide sequence. To study secondary structure transitions, it is necessary to find mathematical formulation of complementarity of nucleotides, as well as basepairing.

As has been demonstrated recently (see [M] and [M2]), from the formal definition of nucleotides as two hierarchical negations, follows the representation of nucleotides by four 4×4 matrices of the form:

$$[a, b|c, d] = \begin{pmatrix} a & d & b & c \\ c & a & d & b \\ b & c & a & d \\ d & b & c & a \end{pmatrix} \tag{2}$$

Namely, one pair of complementary nucleotides is represented by unit matrices $[1, 0|0, 0]$, $[0, 1|0, 0]$. Another pair is represented by matrices $[0, 0|1, 0]$ and $[0, 0|0, 1]$. One can check that **multiplication of four matrices**

$$\breve{1} = [1, 0|0, 0], \quad -\breve{1} = [0, 1|0, 0], \quad \breve{i} = [0, 0|1, 0], \quad -\breve{i} = [0, 0|0, 1] \tag{3}$$

coincides with that of unit complex numbers 1, -1, $\imath$ and $-\imath$ **correspondingly** (see [M2]).

From the postulate that *complementary nucleotide matrices sum to zero* it follows, that

$$0_W = [1, 1|0, 0] \leftrightarrow 0 \qquad and \qquad 0_C = [0, 0|1, 1] \leftrightarrow 0 \tag{4}$$

This condition is satisfied if one determines *four-dimensional complementary interval*, generated by linear combinations

$[a, b|c, d] = a[1, 0|0, 0]+b[0, 1|0, 0]+c[0, 0|1, 0]+d[0, 0|0, 1]$ with non-negative coefficients a, b, c, d of unit matrices.

Let the 4D-complementary interval be defined as follows:

$$\mathcal{I}(a, b|c, d) = \begin{cases} (a - b, 0|c - d, 0) = (a - b)\breve{1} + (c - d)\breve{i} & if \quad a \geq b \text{ and } c \geq d \\ (a - b, 0|0, d - c) = (a - b)\breve{1} + (d - c)(-\breve{i}) & if \quad a \geq b \text{ and } d \geq c \\ (0, b - a|c - d, 0) = (b - a)(-\breve{1}) + (c - d)\breve{i} & if \quad b > a \text{ and } c \geq d \\ (0, b - a|0, d - c) = (b - a)(-\breve{1}) + (d - c)(-\breve{i}) & if \quad b > a \text{ and } d > c \end{cases} \tag{5}$$

So defined, the 4D-interval is a 4×4 matrix. Multiplication and addition of intervals 5 is defined as usual matrix operation. It is rather straightforward to prove that the set of intervals 5 is a field. Moreover, 4D-intervals 5 is a commutative division algebra of the order 2 over the algebra of 2D-intervals

$$\mathcal{I}[a,b] = \begin{cases} (a-b)\begin{pmatrix} 1 & 0 \\ 0 & 1 \end{pmatrix} & if\, a \geq b \\ (b-a)\begin{pmatrix} 0 & 1 \\ 1 & 0 \end{pmatrix} & otherwise \end{cases} \tag{6}$$

The set of all 2D-intervals, with multiplication and addition defined as corresponding matrix operations, is isomorphic to the algebra of real numbers [M2].

The set of all 4D-intervals, with multiplication and addition defined as corresponding matrix operations, is isomorphic to the algebra of complex numbers.

One can interpret the 4D-complementary interval geometrically (see details is [M2]). To any matrix $[a,b|c,d]$ there corresponds another matrix $P = I[a,b|c,d]$. The matrix P can be considered as a projection of matrix $[a,b|c,d]$. In this interpretation, the 4D-interval I is a projection operator. The set of matrices $P = I[a,b|c,d]$ with $a,b,c,d \geq 0$ constitutes a 4D-octant. Interval (projective) matrix P belongs to the first quadrant of one of four planes $[X,0|Y,0]$, $[X,0|0,Y^c]$, $[0,X^c|Y,0]$, $[0,X^c|0,Y^c]$. These **four mutually orthogonal quadrants with the projective operation 5 make a complex plane over the set of matices $[a,b|c,d]$** (with $a,b,c,d \geq 0$) ! This surprising result proves the validity of complex number representation of nucleotides, proposed recently by Magarshak and Benham [MB].

2. **Complex numbers representation of nucleotides.** Now we sketch the approach of Magarshak and Benham [MB] for the representation of secondary structures. In this approach complementary base pairs G and C receive the labels 1 and -1, while A and T receive the labels i and $-i$ ($i^2 = -1$). In a diagram for the folding the bases are labelled 1,2...,n and an $n \times n$ matrix M is formed so that

$$M_{ij} = \begin{cases} -1 & i \text{ is paired with j } (i \neq j) \\ 0 & otherwise \end{cases} \tag{7}$$

The sequence of bases is mapped to a vector $\vec{v}$ with entries $1, -1, i$ and $-i$ as explained above. It is then easy to see that $M\vec{v} = \vec{v}$. Thus, *for a given nucleotide vector $\vec{v}$ the set of possible foldings is in 1-1 correspondence with the matrices M such that $M\vec{v} = \vec{v}$.*

Matrix M is a solution with eigenvalue 1 of the equation $M\vec{v} = \vec{v}$, where (what is unusual), vector $\vec{v}$ is given, and matrix M is unknown [MB].

The matrix M can be interpreted either as a structure matrix, or as a transition operation, which transforms free (i.e.totally dispaired) state of RNA molecule into the final secondary structure. More generally, one can define transition matrix $T = M_2 M_1 = M_2 M_1^{-1}$ which transforms initial secondary structure M_1 into the final secondary structure M_2. A well formulated account of this method is given in [MB].

Vassiliev knot invariants, analyzed in this paper, can be interpreted as these of solutions of the equation $M\vec{v} = \vec{v}$. In a sequel paper the topological invariants of transition matrices and secondary structure dynamics will be studied.

3. <u>Non-Watson-Crick basepairs and Hermitian forms</u>. The theory presented in the present paper is developed specifically for the analysis of the RNA and DNA structures with Watson-Crick basepairing. But it can be generalized. If non-Watson-Crick nucleotides U and G (number j and k) are paired, the structure matrix S becomes Hermitian. Let the transition from the secondary structure S_1 to the secondary structure $S_2 = T_{21} S_1$ be performed. The transition matrix of the inverse process T_{21} is complex conjugate to the matrix T_{12}^*, so any transition matrix $T_{21} = S_2 S_1^{-1} = S_2 S_1$ is unitary.

Let analyze Hermitian form

$$\mathcal{H}(S) = \mathbf{g}^\dagger S \mathbf{g} = \sum_{i=1}^{n} \sum_{k=1}^{n} S_{ik} g_i^* g_k \tag{8}$$

If complementary nucleotides g_i and $g_k = g_i^c$ are *real*, i.e. equal to ± 1, then element (i, k) of the Hermitian form 8 $\mathcal{H}(S)_{ik} = \mathcal{H}(S)_{ki} = S_{ik} g_i^* g_k = S_{ki} g_k^* g_i = (-1)(-1)1 = 1$. If complementary nucleotides g_i and $g_k = g_i^c$ are *imaginary*, i.e. equal to $\pm i$, then any nonzero element (i, k) of the Hermitian form 8 $\mathcal{H}(S)_{ik} = \mathcal{H}(S)_{ki} = 1$. We see that in any case the Hermitian $\mathcal{H}(S)$ is positive. Moreover, for arbitrary nucleotide sequence $\mathbf{g}$ and any secondary structure S, which this sequence can form, $\mathcal{H}(S) = const = n$, i.e. is just equal to the number of nucleotides in the chain. Sure, this trivial model can be modified. We begin with definition of the matrix $\mathcal{E}$ such that:

    i. element (k, k) is equal to the energy E_k, associated with non-paired nucleotide number k,

    ii. element (j, k) is equal to $-(E_j + \Delta E_{jk}/2)$, if and only if nucleotides j and k are paired, and

    iii. all other elements of matrix $\mathcal{E}$ are equal to zero.

The Hermitian form

$$\mathcal{H}(\mathcal{E}) = \mathbf{g}^\dagger \mathcal{E} \mathbf{g} = \sum_{i=1}^{n} \sum_{k=1}^{n} \mathcal{E}_{ik} g_i^* g_k \tag{9}$$

is increasing by the ΔE_{jk} if the bond (jk) is making. This property can be used for calculation of a *partition function* of RNA secondary structure formation.

4. Partition function of complementary structures (Co-partition).

Finally, we wish to point out here that the Magarshak-Benham approach to foldings gives rise to a natural partition function of the form

$$Z_{\vec{v}} = \sum_{M:M\vec{v}} = \vec{v}e^{-\frac{1}{kT}E(M)} \tag{10}$$

where $E(M)$ is appropriately chosen energy functional for the foldings, and T is temperature, k Bostzmann's constant. For example, one may take $E(M) = 3[G - C] + 2[A - U]$ where $[G - C]$ denotes the number of $G - C$ pairs while $[A - U]$ denotes the number of $[A - U]$ pairs. (The 3 and the 2 are the number of hydrogen bonds needed in each case)[1].

Analysis of the algebra, generated by nonconventional basepairs, goes out of the frame of this contribution and will be discussed elsewhere. Partition function analysis also will be the subject of a sequel to this paper.

5. In this paper the 3-dimensional disposition of atoms was not introduced directly. But the model can be modified. Namely, one can introduce a complementary field, which acts on complementary nucleotides only [M2]. This is a short-range force field, responsible for basepairing. In complex-numbers designations, complementary force is proportional to $\delta_{g_i,-g_k}$, where g_i and g_k are complex values of nucleotides number i and number k. In other words, compelmentary force attracts nucleotides, if and only if they satisfy the trivial equation:

$$g_i + g_k = 0 \tag{11}$$

and orients them in accord with their disposition in DNA duplex. One must include the complementary force into the model in addition to the force fields, typical for double-stranded DNA (see, for instance, [SO]). So there is a hope to compute 3-dimensional structures of one-stranded RNA, as well as RNA secondary structure formation.

6. Topological code. Topological properties of equations, which describe some key biological processes, often determine the solutions of these equations. For instance, it is known that solution of equations of stationary enzyme kinetics is determined by the graph of corresponding enzyme reaction (see, for instance [VG] and [VM]). The same statement is true for electron transfer rate in proteins and some other biologically important macromolecules

---

[1] We thank Nancy Wood for suggesting this particular energy function.

(see [OBG] and [MMJ]). In this paper, a relationship between RNA secondary structure and Vassiliev polynomials has been found. So the hypothesis, that **one of the languages, which nature uses in vivo, is topological language**, seems to be reasonable.

Appendix 1. It is the purpose of this appendix to give a quick derivation of the necessity of the 4-term relation (see [S])

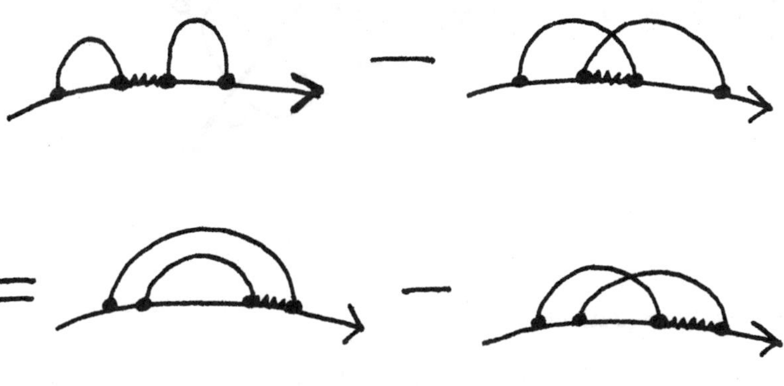

in a Vassiliev invariant. To this end, consider the following embedded isotopy:

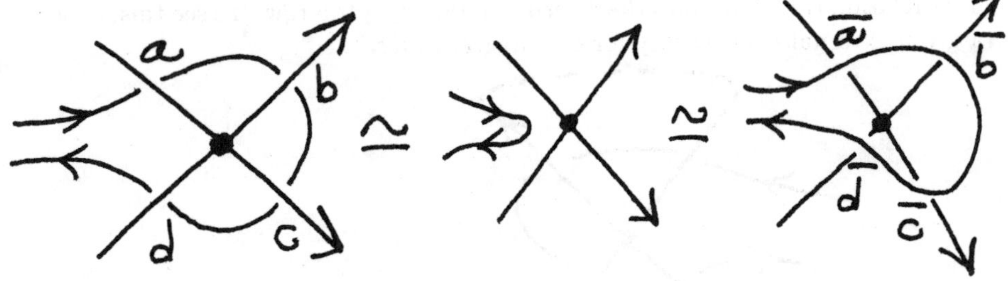

Let us write the symbol $\langle abcd \rangle$ for the Vassiliev invariant of the left-hand configuration, and $\bar{x}$ for the switch of a given crossing labelled x. Then we can write

$$\langle abcd \rangle - \langle \bar{a}bcd \rangle = \langle \bullet bcd \rangle$$

where $\langle \bullet bcd \rangle$ denotes the replacement of the crossing a by a 4-valent node ($\bullet$).

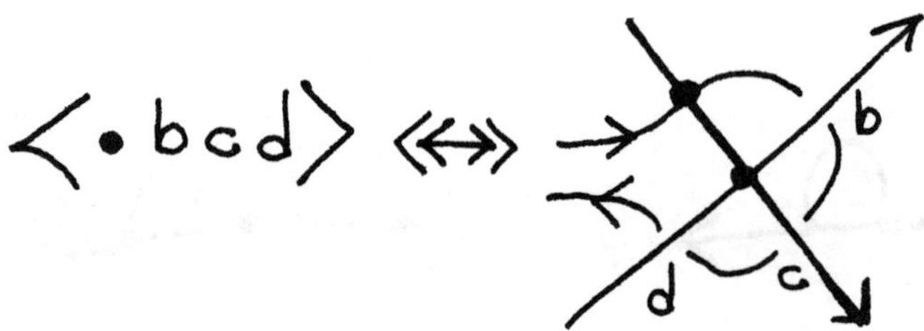

Then we have the equations

$$\begin{cases} \langle abcd \rangle - \langle \bar{a}bcd \rangle = \langle \bullet bcd \rangle \\ \langle \bar{a}bcd \rangle - \langle \bar{a}\bar{b}cd \rangle = -\langle \bar{a} \bullet bcd \rangle \\ \langle \bar{a}\bar{b}cd \rangle - \langle \bar{a}\bar{b}\bar{c}d \rangle = -\langle \bar{a}\bar{b} \bullet d \rangle \\ \langle \bar{a}\bar{b}\bar{c}d \rangle - \langle \bar{a}\bar{b}\bar{c}\bar{d} \rangle = \langle \bar{a}\bar{b}\bar{c}\bullet \rangle \end{cases} \qquad (12)$$

Since the isotopy shows that $\langle abcd \rangle = \langle \bar{a}\bar{b}\bar{c}d \rangle$, we conclude that

$$\langle \bullet bcd \rangle - \langle \bar{a} \bullet bcd \rangle - \langle \bar{a}\bar{b} \bullet d \rangle + \langle \bar{a}\bar{b}\bar{c}\bullet \rangle = 0$$

This relation translates into the 4-term relation on a top row. To see this, choose a given external connectivity for the diagram $\langle abcd \rangle$:

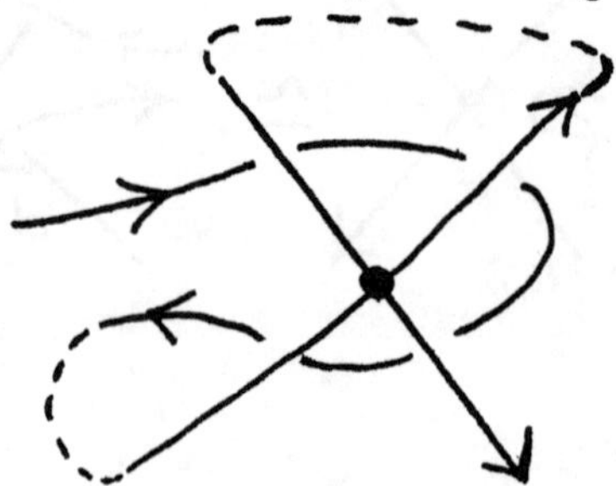

and translate into chord diagram language:

$$\langle \bullet bcd \rangle - \langle \bar{a} \bullet cd \rangle - \langle \bar{a}\bar{b} \bullet d \rangle + \langle \bar{a}\bar{b}\bar{c} \bullet \rangle = 0$$

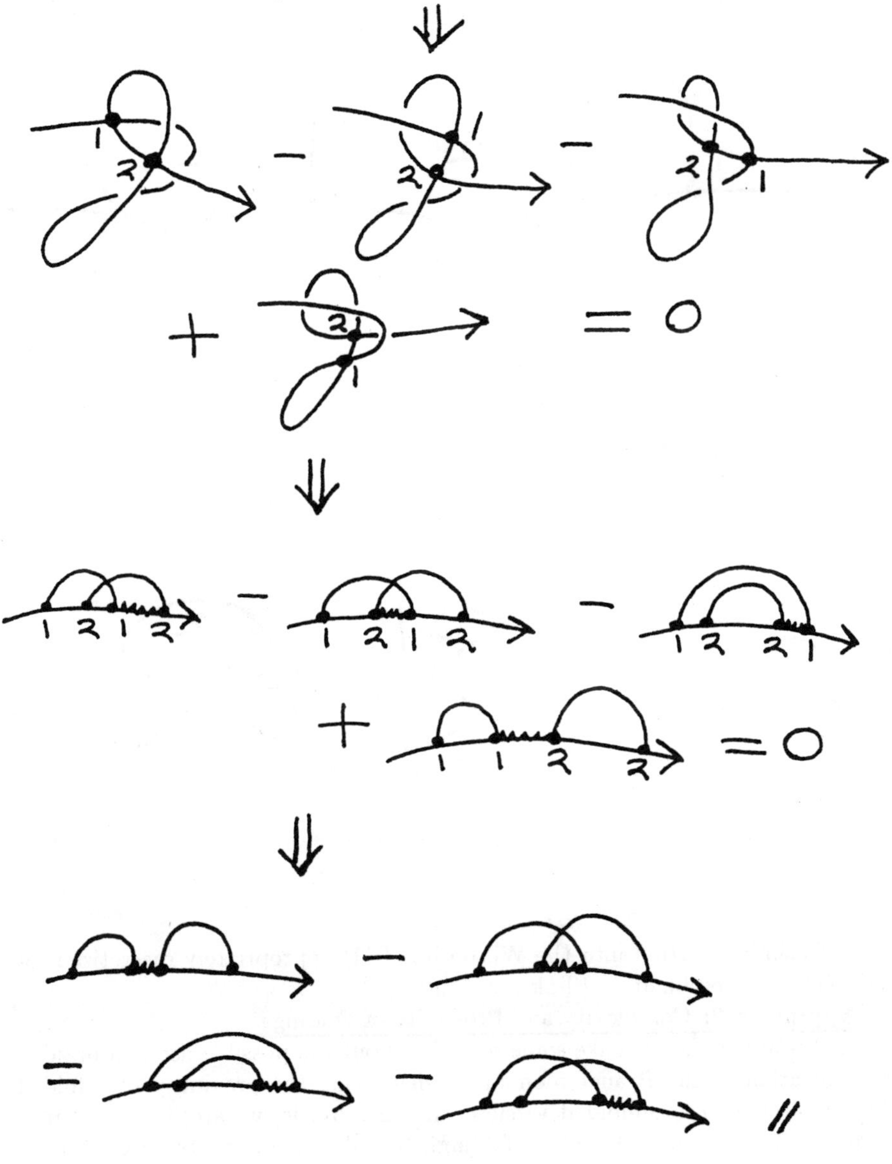

$$\Downarrow$$

$$\Downarrow$$

$$\Downarrow$$

We must emphasize once again that it is quite extraordinary how this simple topological requirement is tied so directly to the Lie algebra patterns (see end of the last section of this paper) via the intermediate STU identity

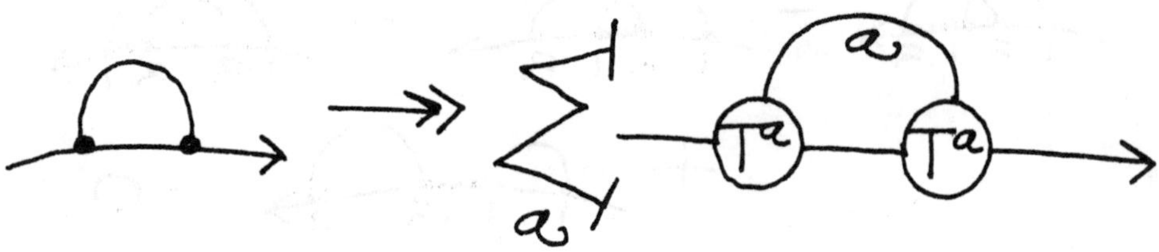

It should also be mentioned that

is the Casimir insertion into the Wilson line [K1] appropriately discretized for the content of combinatorial link invariants.

<u>**Appendix 2:** Complexity and Proximity of Foldings.</u>

In this appendix we make some remarks about the possible use of a braided generalization of the Brauer monoid for measuring the proximity of a folded RNA molecule to its unfolded version. In other words, we are interested in a measure of the complexity of the folding. Recall that in section two we have represented abstract foldings via patterns such as

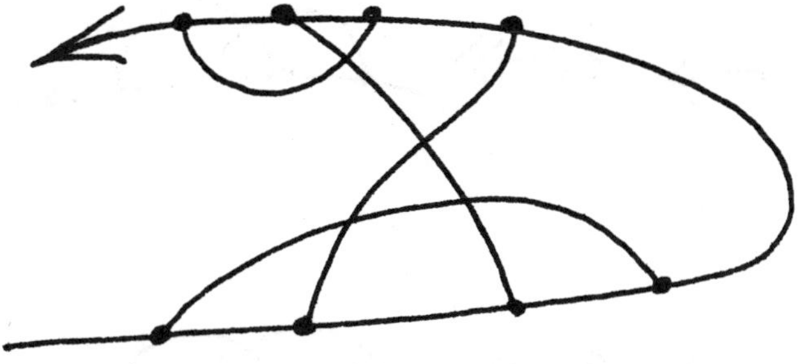

giving a correspondence with elements in the Brauer monoid.

An embedded folding can be (if the chain itself is unknotted) "pulled" into Brauer monoidal form, but the monoid strands may be woven about one another as in the simple pull of a pseudo-knot shown below.

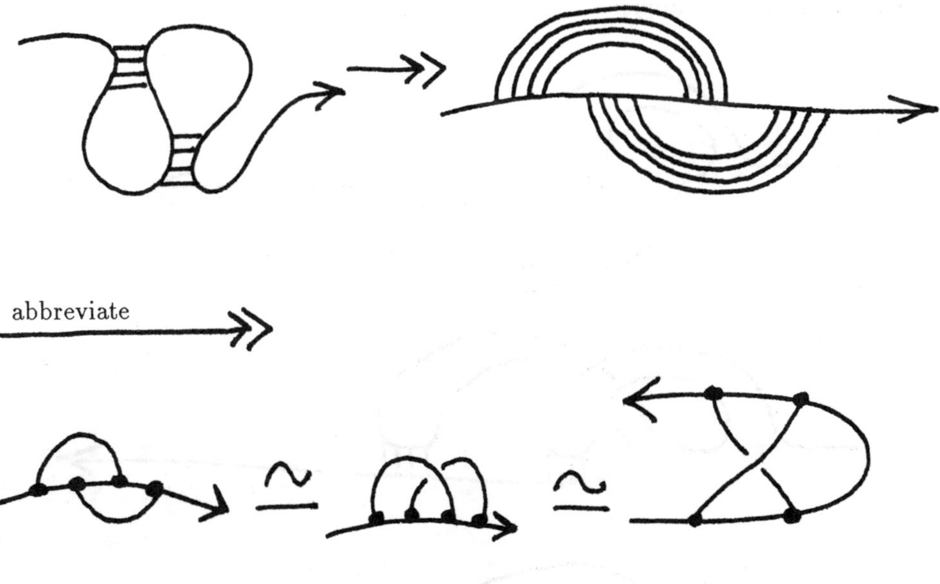

It is clear that it makes sense to measure the complexity of the pseudoknot by including the braiding structure of the attaching arcs. However, in some cases, more than one braiding structure will correspond to the same structure of attachment. For example, in the case above we have

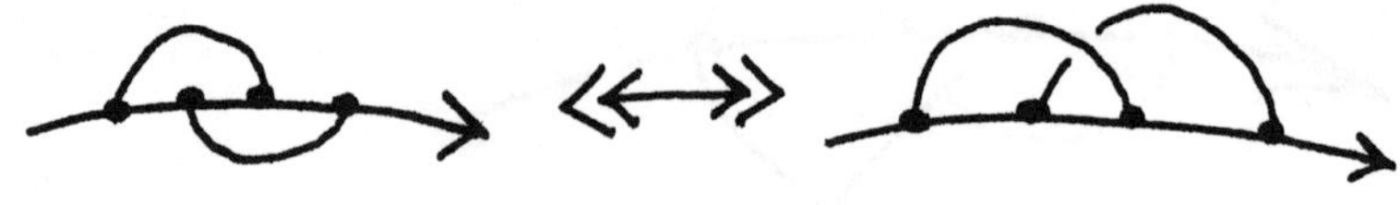

and also

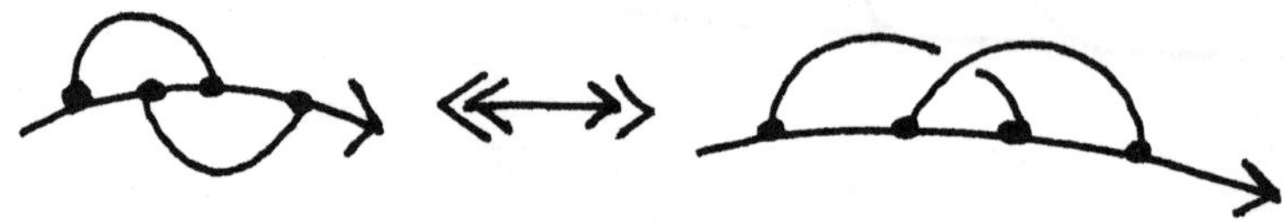

(by swinging the given arc in one of the two possible directions around the oriented axis). This is a reflection of the fact that the following three pseudo-knots are almost isotopic:

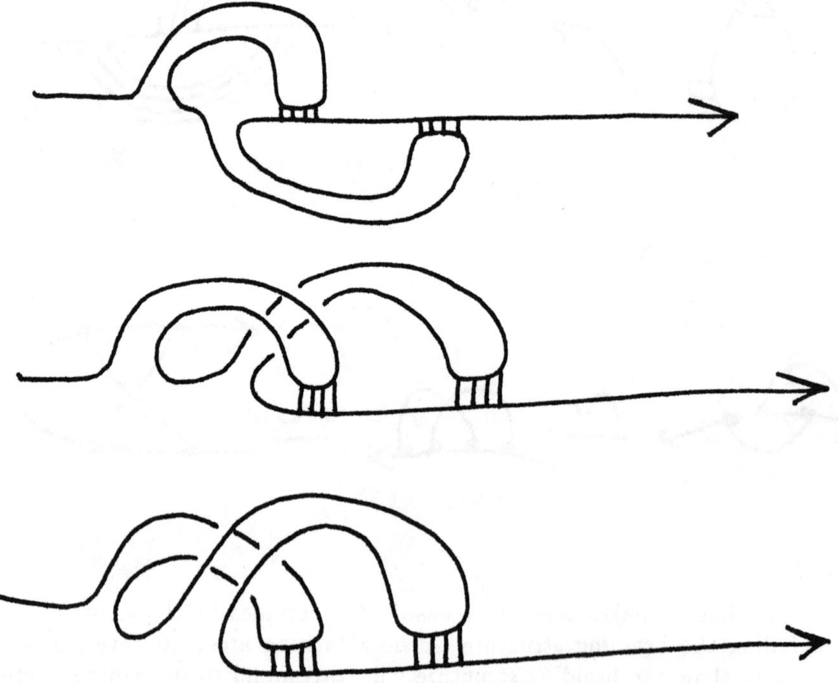

We say <u>almost isotopic</u>, because the three pictured pseudo-knots are isotopic up to twists of the sort shown below in a regular isotopy:

There are some interesting subtleties in even the simplest examples. For example, call a pseudoknot <u>simple</u>, if it gives rise to an unknotted embedding of its axis when all the pairing nodes are eliminated:

The standard pseudo-knot is simple:

More complex examples of simple pseudo-knots are easily manufactured. For example,

$\mathcal{P}$ is a simple pseudoknot with only one (multiple) pairing node. Translating into the language of attaching arcs, we have

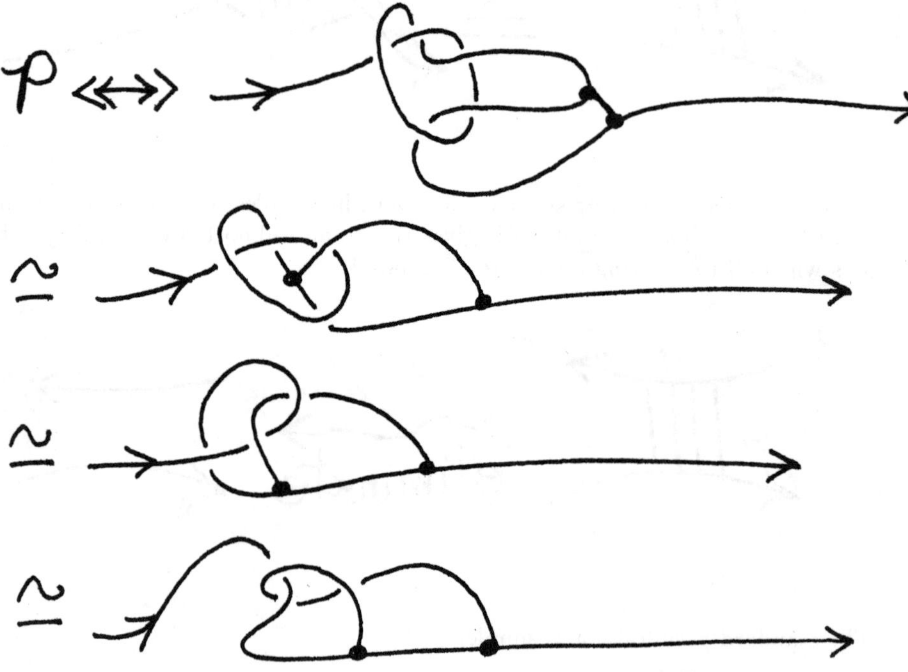

In this last representation we have an unknot · axis, but the attaching arc is entangled with the axis. This weaving reflects the complexity of the pseudoknot $\mathcal{P}$.

One way to analyze the complexity of $\mathcal{P}$ is to compute the invariant $\mathcal{I}_{\mathcal{P}}$ defined via (see section 3 of this paper). This is the analogue of the Vassiliev invariant for folding. We obtain:

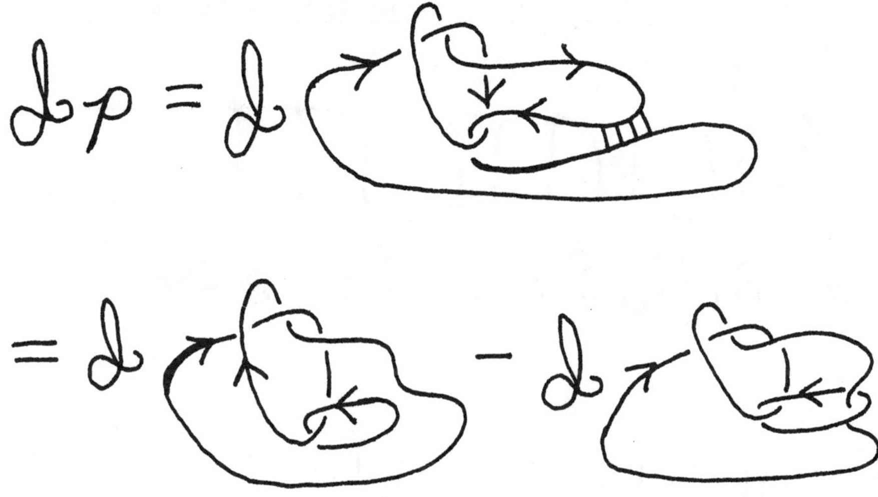

The two links obtained in this resolution are non-trivial and so we see that this pseudoknot is quite distinct from

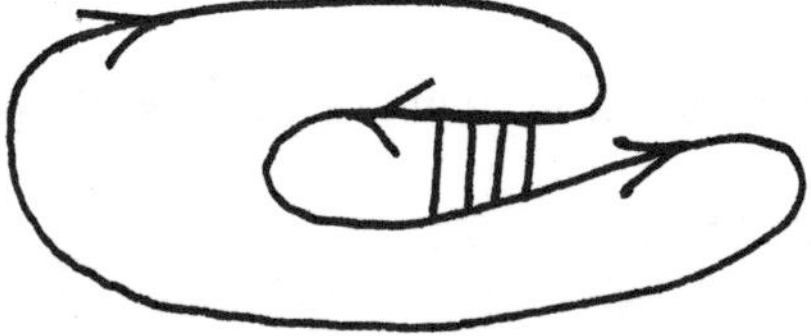

and in fact it is quite complex. For example, it is topologically distinct from its mirror image.

If we try analyzing $\mathcal{P}$ via the braid monoid (generalized Brauer monoid) picture, then another complexity arises:

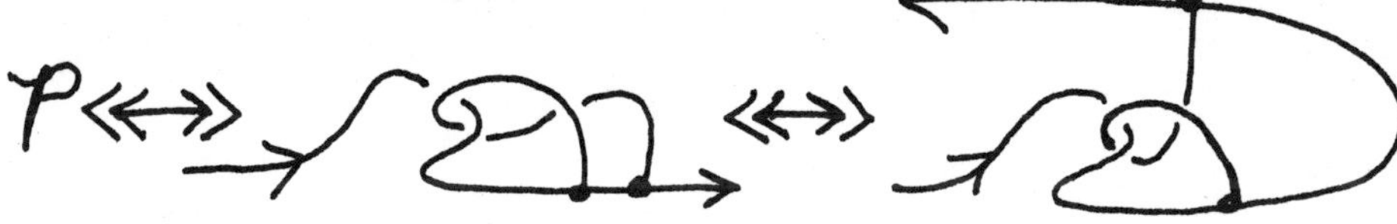

The entwinement of the axis with the attaching arc yields a new phenomenon
for articulation. We can capture this phenomenon by adding to the vocabulary
of the braid monoid <u>top and bottom fixture elements</u> of the form:

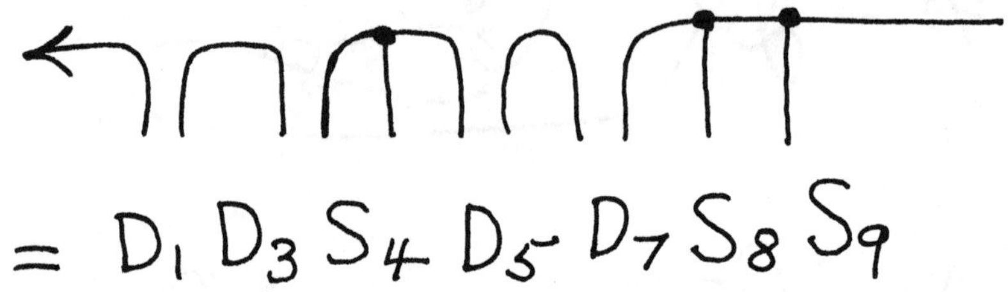

$$= D_1 \, D_3 \, S_4 \, D_5 \, D_7 \, S_8 \, S_9$$

or

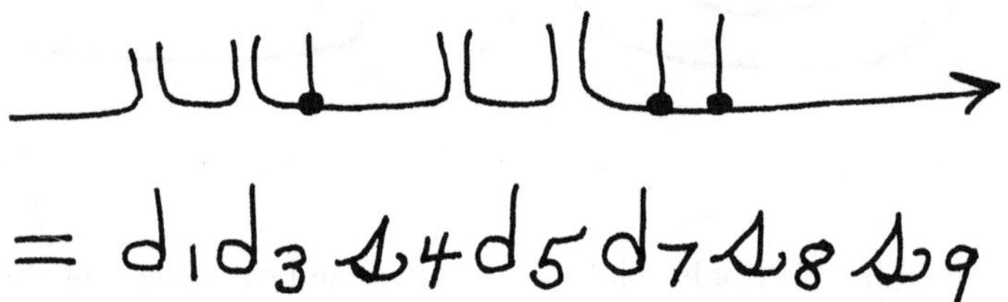

$$= d_1 \, d_3 \, \not{d}_4 \, d_5 \, d_7 \, \not{d}_8 \, \not{d}_9$$

The basic parts of a top fixture consist in double strands such as

and an attachment as in

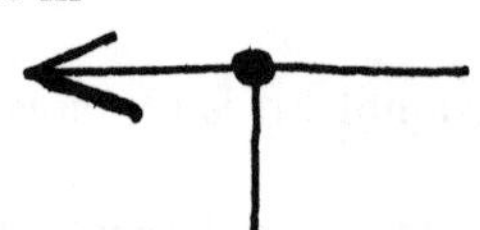

We use lower case letters to describe bottom fictures.

Then, any simple pseudo-knot can be expressed in the form of a product:
(Top Fixture)(Word in Braid Monoid)(Bottom Fixture). For example,

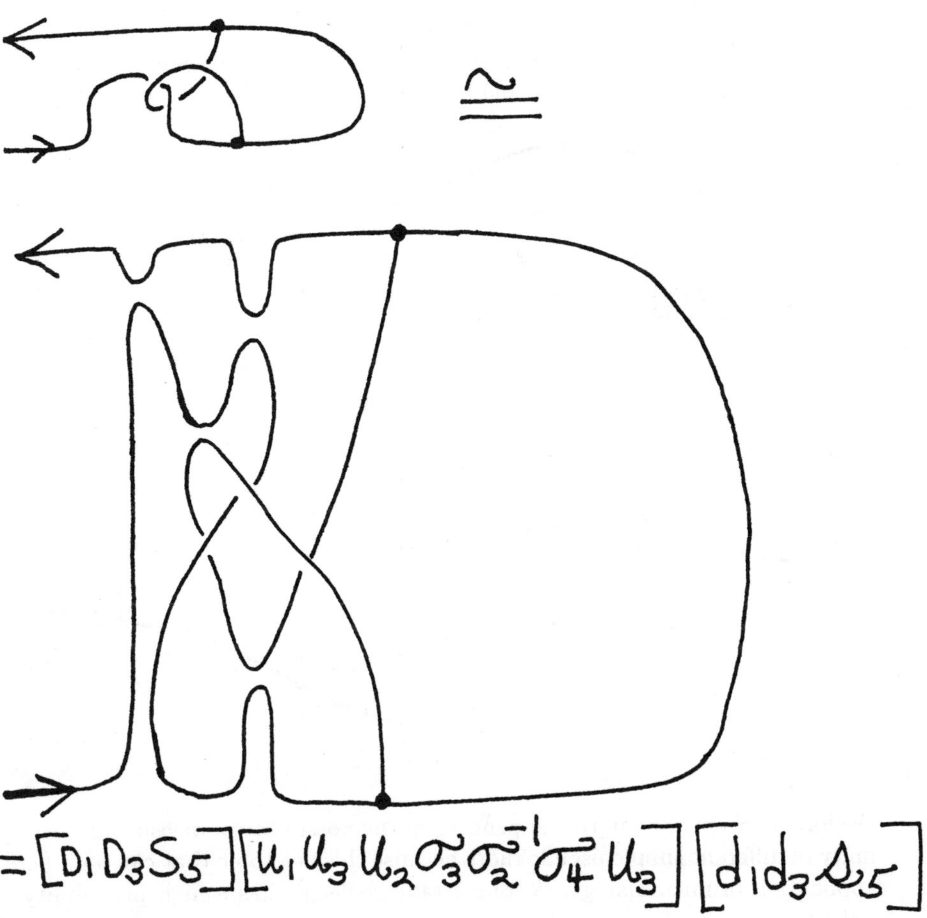

$$= [D_1 D_3 S_5] [U_1 U_3 U_2 \sigma_3^2 \sigma_2^{-1} \sigma_4^2 U_3] [d_1 d_3 s_5]$$

A simpler example is

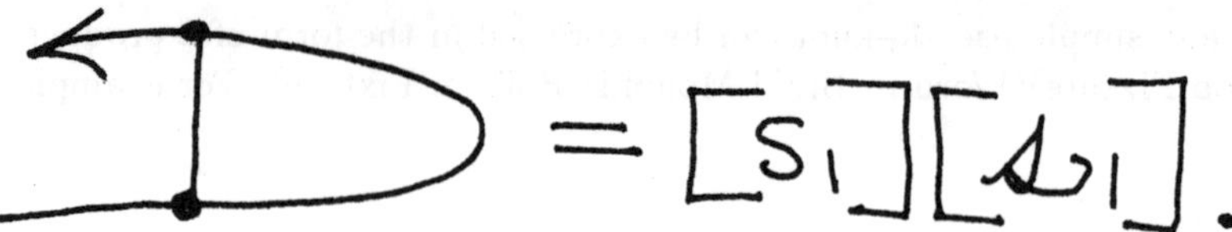

The difference in complexity of $[S_1][s_1]$ and
$[D_1 D_3 D_5][U_1 U_3 U_2 \sigma_3 \sigma_2^{-1} \sigma_4 U_3][d_1 d_3 s_5]$ can be regarded as a measure of the
"topological proximity" of the foldings

and

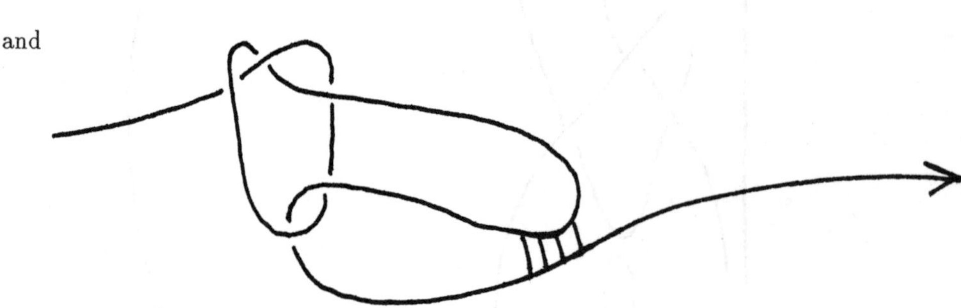

We have concentrated in this appendix, on the possibility of measuring the proximity of different simple pseudoknots because this is a topic that can have actual
application in microbiology. A long strand molecule has a high probability for
unknotted self- entanglement. If this molecule can undergo self-binding (as in
RNA) then some of these unknotted states of the strand can become simple
pseudoknots of varying degrees of complexity.

REFERENCES.

[BAR] Dror Bar-Natan.*On the Vassiliev Knot Invariants*. (preprint 1992).

[BAX] R.J.Baxter. *Exactly Solved Models in Statistical Mechanics*. Academic Press (1982).

[BL] J.S.Birman and X.Lin.*Knot polynomials and Vassiliev's invariants. Columbia Univ.preprint* (1991).

[OBG] J.N.Onuchic, D.N. Beratan and H.B.Gray. *Pathway Analysis of protein electron-transfer reactions*. Ann. Rev. Biophys. Biomol.Struct. 21, (1992) pp. 349-377.

[BR] R.Brauer. On algebras which are connected with the semisimple continuous groups. Annals of Math. Vol.58, N^0 4, October (1937).

[J] V.F.R.Jones. *Hecke algebra representations of braid groups and link polinomials*. Ann.of Math. 126 (1987) pp.335-388.

[K1] L.H.Kauffman.*Knots and Physics*. World Sci.Pub. (1991).

[K2] L.H.Kauffman. *State models for link polynomials*. L'Enseignment Math.36 (1991) pp.1-37.

[K3] L.H.Kauffman. *An invariant of regular isotopy*. Trans. Amer.Math. Soc., Vol. 318, N° 2, (1991), pp.417-471.

[K4] L.H.Kauffman. *Invariants of graphs in 3-space*. Trans. Amer. Math.Soc. Vol. 311, 2 (Feb.1989) pp.697-710.

[K5] L.H.Kauffman. *Vassiliev Invariants and the Skein Polynonials*, (to appear).

[KV] L.H.Kauffman and P.Vogel. *Link polynomials and a graphical calculus*. Journal of Knot Theory and its Ramifications. Vol.1 N° 1 (March 1992). pp.59-104.

[MK] Mironov,A.A and Kister,A.E. *RNA secondary Structure Formation during Transcription*J. *Biomol.Str. Dyns*. 1986,v. 4, pp. 1-9

[KMM] A.Kister, Y.Magarshak, J.Malinsky, *The Theoretical Analysis of the process of RNA molecule self-assembly, in book Computer Genetics*, Elsevier Science Publishers (**1993**).

[M] Y.Magarshak, *Quaternion Representation of RNA Sequences and Tertiary Structures*. in book *Computer Genetics*, editors P.Pevsner and M.Gelfand, Elsevier Science Publishers, (**1993**).

[M2] Y.Magarshak. *Nucleotide algebra and nucleic relativity*. To appear.

[MB] Y.Magarshak and C.J.Behnam. An algebraic representation of RNA secondary structures. J of Biom.Str. & Dyn. ISSN 0739-1102, Vol.10, N° 3 (1992) pp. 465-488.

[MMJ] Y.Magarshak, J.Malinsky & A. Joran. Diagrammatic techniques for solving Schwinger-Dyson equation: Electron transfer pathways in biological molecules, J.Chem Phys., 95,1 (1991), pp.418-432.

[PW] R.C.Penner and M.S.Waterman. Spaces of RNA Secondary Structures (to appear in Advances in Mathematics).

[PI] S.Piunikhin. Weights of Feynman diagrams, link polynomials and Vassiliev knot invariants (to appear in Journal of Knot Theory and its Ramifications).

[S] T.Stanford. Finite-Type invariants of knots, links and graphs (preprint 1992).

[SO] T.Schlick and V.K.Olson. *Supercoiled DNA energetics and Dynamics by Computer Simulation.* J.Mol.Biol. 223, (1992), pp.1089-1119.

[V] V.A.Vassiliev. *Cohomology of knot spaces in Theory of Singularities and Its Applications* (ed.by V.I.Arnold), Adv. in Soviet Math., Vol.1, AMS (1990).

[VG] M.V.Volkenstein and B.N. Goldstein.*A new method for solving the problems of the stationary kinetics of enzymological reactions.* Biochimica et Biophysica acta, 115, (1966) pp.471-477.

[VM] M.V.Volkenstein, Yu.B.Magarshak.*Application of the diagrammatic method of solving kinetic equations of cyclic enzymatic reactions.* Biophysics, 15, 5 (1970), , pp.805-813.

The Entanglement Structures of Polymers

Alison MacArthur
Department of Polymer Science, The University of Akron, Akron, OH 44325

1. Introduction

1.1 Tangled Strands

The simplest view of a flexible polymer chain is as a single very long strand of cooked spaghetti sitting in a large serving bowl with many other spaghetti strands. Entanglement becomes obvious when one attempts to remove a small portion of spaghetti from the bowl of many strands called the bulk state of the polymer. An entanglement is less obvious but is still present when the same amount of spaghetti is still in a very large pot of the cooking water. Such a diluted state could be called a "solution". In this simplest discussion of polymers all chains will be considered to be in a liquid-like (i.e., non-crystalline) state. Also, like spaghetti (uncooked or cooked) the chains have no branches and are therefore called linear. In all cases the physical properties depend strongly on the length and fineness of the strands, and vary from fluid to solid. In fact, there is always a critical length below which entanglement is not observed in the physical properties. Consider how easily chopped spaghetti is eaten by little children. Remember however, that the spaghetti strands have an inherent stickiness to each other compared to the very light molecular friction between molecular chains.

1.2 Characteristic Familiar Behavior of Bulk Polymers

A reader who is technically unfamiliar with polymer molecular behavior might have observed the following types of behavior that are typical of linear polymers. A child's toy sold under various brand names (which generally include the word "putty") shows the three regions characteristic of polymer behavior on a convenient time scale at room temperature. If the material is put on the table and allowed to rest, it will flow under gravity as a liquid. It can easily be hand molded into a ball. If the ball is thrown against the table, it will bounce like a typical rubber. When it is put on a hard surface and struck sharply with a hammer, it will shatter, exhibiting glassy behavior.

While many materials exhibit a glass transition, the rubbery region is characteristic of high molecular weight flexible polymers and is generally attributed to entanglement. A number of other polymeric materials (sold as toys) show similarly interesting behaviors for the curious and serious investigator. In one case enough orientation was achieved after much manipulation to allow a quite long-lived fiber to be spun from the gel. Unfortunately the production and marketing of these materials is quite haphazard, the names change so frequently as to make referring to specific ones useless.

2. Background of Polymer Molecular Entanglement

2.1 Manifestations of Polymer Entanglement

Several excellent reviews[1,2,3] describe the experimental evidence for chain entanglement and the various parameters and models used to describe it. The major entanglement effects cited by Porter and Johnson[4] for uncrosslinked polymers include: (i) the discontinuity in the power law dependence of viscosity on the molecular weight at low shear, (ii) the non-Newtonian behavior of very concentrated polymer solutions at high shear, (iii) changes in the relaxation times measured by NMR, and (iv) various visco- elastic properties of bulk polymers such as the plateau modulus.

The material properties and time-dependent behavior of polymers discussed above stem in large part from the entanglement of the polymer chains. However the processes by which entanglement takes place and the interactions that give tangled systems their stability are not well understood using stochastic processes with topological attention to the chain connectedness. Although the connectedness of single chains is often mentioned there has been little study of possible entanglement structures, their properties, and their formation.

Polymer systems can be formed in many different ways, and can be modified after formation in ways that affect the entanglement process. While the strands in many macroscopic strand systems are static after they have been formed, polymer molecules above the glass transition are constantly in motion. Polymer chain entanglement is a dynamic equilibrium process that responds continuously to changing conditions of stress and temperature. As a result in some polymer systems a knowledge of the precise conditions of formation and post-formation history may be critical to understanding the behavior of the material in its manufacture and use. Only a careful study of the effect of detailed long term history on properties will separate these effects.

It is necessary to formulate a more realistic topological formalism that will describe the entanglement structure by explicitly taking account of the chain continuity in such a way that the mechanisms of entangling and disentangling can be reduced to a series of simple kinetic motions that can, in principle, have an equilibrium or steady state value. When all possible entanglement mechanisms are considered, the consequences of the resulting global structures might lead to better ways of analytically describing multistrand interactions seen in other fields as well.

2.2 Early Entanglement Theory

Chain entanglement has been used for over sixty years to explain deviations from the expected behavior of phantom flexible polymer chains.[5] The tensile mechanical behavior of stretched uncrosslinked rubber was attributed to a physical interlocking of chains by Busse in 1932.[6] Treloar in 1940[7] considered entanglements to be isolated regions of high viscosity. Flory in 1944[8] attributed experimental deviations from the calculated behavior of crosslinked rubber partly to trapped entanglements. Edwards[9], Edwards and Alexander[10], and Vologodskii, Lubashin,

Frank-Kamenetskii, and Anshclevich[11] considered the entropy loss due to simple knots in the chain. In all of the above models the topological constraints are either unspecified or assumed to be knotted, and no explicit process is mentioned for the formation or rearrangement of the entanglements. The entanglements in most of these models are therefore assumed to be localized and are frequently treated like crosslinks. Flory[12] in 1985 stated that entanglements are diffuse and can not be equated to crosslinks, but he did not suggest the specific nature of those interactions and gave no mechanism for their formation or rearrangement.

The reptation model proposed by deGennes[13] is one process by which entanglements can be formed, undone, and rearranged. This model assumes that a single chain in the bulk polymer (or concentrated solution) is so constrained by other chains in the neighborhood that it may be considered to be trapped in a tube. As a result it can be assumed to move only along its axis. A section of the tube vanishes when the chain moves out of it at one end and a new section is extended at the other end as the chain advances. In time as the chain moves randomly back and forth the tube is entirely re-created. Recent modifications[14] of the model consider some rearrangements of the tube caused by the motion of other strands; however, the basic concept remains the same. The reptation model is intrinsically knotted since the single chain explicitly moves its ends through a matrix formed by strands of other chains or even other parts of the same very long chain.

3. Introduction to the Topology of Entanglement

The structure and properties of entanglements are also important in many other fields. It may be desirable to control or understand entanglements in diverse molecular or macroscopic systems. Systems that tangle by different processes do have different final structures and thereby exhibit different physical properties. only with a detailed understanding of all of the processes of entanglement undergone by the material in its history, can the relationship between entanglements and the properties of the system be made clear. A full understanding of entanglement will aid in the development of techniques for controlling and modifying the entanglement to produce the desired final properties.

The first step to a broad understanding of entanglement requires an awareness of the different basic types of possible interactions. Much of the material in this chapter applies to any tangled system. First a discussion of global structure is given that will be helpful for choosing a model structure for tangled systems in any field. In this section I assume little explicit background either in knot theory or in polymers. I cite polymer examples to emphasize the interrelationships between formation and equilibration in a complex system of mobile strands. In these examples it is necessary to distinguish the entanglement structure that was "built in" during strand formation from the superimposed equilibrium structure that could be expected from unrestricted from strand motions. To keep track of such superimposed topologies a new type of analysis and classification of strand interactions is introduced.

Various textile structures formed from linear strands are considered to study the intrinsic properties of the various types and to serve as starting points for possible models of entanglement. The method of formation, global structure, and physical properties of several are compared. Reasonable extensions to a random three-dimensional system of polymer chains are considered for a few. Observations of tangling in macroscopic random systems are also discussed. A loop model of unknotted entanglement is proposed and described briefly. This model does provide an insight to a qualitative understanding of the dynamics of long chain flexible polymers. Hopefully it can be made functionally predictive as well.

3.1 Background in Knot Theory

The extensive past work in knot theory has been chiefly focused on invariants of the knot. It has also only considered systems with no ends and no branch points. As will be shown, interactions between the middles of strands, that do not change the knotting, can be very important in some types of entanglement. The half-cycle analysis introduced in Section 5.3 was developed in order to enable a study of all types of interactions.

3.2 Commonplace Problems and Products in Which The Understanding of Entanglement is Important

Three examples will illustrate the importance of the control of molecular entanglement in various processes: (i) Our current understanding of entanglement suggests that much of the shop–floor "know-how" of rubber compounding improve the properties or the processibility of rubber products by acting on the entanglement of the rubber molecules. (ii) The role of chaperon molecules that control large loops of a protein[15] molecule to prevent their entanglement before they can assume their correctly folded form has been recently better understood. (iii) The technology for the production of ultra high modulus fibers includes a spinning and drawing process that allows the molecules to untangle and align completely in a controlled fashion so that the integrity of the fiber is maintained by the entanglements until the alignment is sufficient to give the necessary strength.

In less high-tech, more everyday processes much effort is devoted in the cosmetics industry to controlling the entanglement of hair. The non-woven fabric industry makes extensive use of several entanglement methods. On a still larger scale, construction workers make a large chain braid of their extension cords to keep them from tangling when not in use.

3.3 Methods of Studying Tangled Systems

The investigation of a tangled system, either macroscopic or molecular, must begin with a careful analysis of all possible mechanisms that could be responsible for the formation and rearrangement of the entanglement and the relative rates and probabilities of those mechanisms. Based on these considerations an appropriate model global structure for the entanglement may be chosen.

Natural structures such as bog mats and rain forest canopies that are formed by growth at the ends of relatively non-mobile branching strands will derive their physical properties from different interactions than those formed by the random motion of mobile strands with few, if any, branch points.

Experimentally, it is sometimes possible to observe entanglement in macroscopic systems by a low power microscope or a loupe. Even in dense systems it may be possible to find a way to make most strands invisible and observe only a few marked strands, assuming that interactions between those strands are representative of the whole. This enhancement of detail can be done by using a few strands of different refractive index, then immersing the tangle in a medium whose refractive index matches the unmarked strands. In molecular systems the scattering or fluorescence of radiation by the marked strands may be examined. These marked strands can be representative only if they are identical in flexibility, mobility, and length to the other strands since differences in mobility can markedly affect the entanglement.

More recently the atomic force field microscope can see spatial relations at 0.1 nm levels. Unfortunately the field is too small to allow the strands to be followed far enough to determine much about the type of entanglement.

3.4   Observations of Macroscopic Tangling

Many common linear objects resemble linear polymer chains in continuity and flexibility. They can and frequently do tangle and these tangles can not be easily shaken out. Untangling these accidental tangles is, however, very different from untying a large knot. Observations of various accidentally and intentionally tangled systems of long macroscopic strands[16] have shown that the stability of the tangle is not dependent on knotting. Random tangles are seldom knotted when they have been formed accidentally in macroscopic flexible strands by processes that do not act specifically on the ends. Notwithstanding, these tangles can be quite tenacious, and if the strand is long and the tangle complex, pulling on an end will not free it. In such systems as kite strings, fishing lines, and sewing threads the ends are often constrained in such a way that knots can not be formed, yet the tangles that occur can have a frustrating degree of stability. The knots that are found in examining these random tangles are infrequent, small, usually near an end, and therefore insufficient to explain the apparent stability of these systems. These systems are generally most efficiently untangled by manipulating loops along the strand. (In contrast, knots must be untied by threading the ends through and between other parts of the strand.) Sections of the tangled strand apparently interact in some manner to form a structure that decays on long time scales but resists deformation on short time scales. All parts of the strand appear to be actively involved in the entanglement, and as a consequence these systems also seem to have a critical length for stable entanglement. Fig. 1 shows a large unknotted tangle.

It will be seen later that the interactions that are found in systems of long strands that have become tangled through the mobility of the entire strand do not change the knotting of the system. Methods of knot theory, such as the recently developed knot polynomials[17,18] that seek invariants of the knot, thus can not be used in the

Fig. 1.   An unknotted random tangle in which many loop interactions may be seen.

study of these systems. The systems are also frequently too large to make complete strand analysis possible.

4. What is a Polymer

A polymer is a molecule that is made up of many sub-units. These sub-units, called monomers may be all alike or of several types. Space does not permit a thorough description of polymer structure and properties but many good references may be consulted.[19,20] The monomers may be connected in various ways. We will not be concerned here with the detailed chemistry of how these forms are achieved since our interest is in the properties of their connectedness. However, the method of polymerization can influence the type of entanglement that is "built in" during the polymerization process. I have attempted here to separate structures caused by the chain formation from those due to the motion of the chains. Several typical classes of polymerization will be discussed briefly. A few basic forms of polymer architecture are also described below because some of these variations may be expected to severely limit the mobility of some sections of the chain and thereby affect their entanglement behavior.

The type of polymerization can also affect the architecture of the polymer chain. The entanglements found in a network of short chains with many multi-functional branch points can be expected to be very different from that found in a linear chain of several hundred thousand monomer units. Their properties are also very different.

4.1   Polymerization by Attachment of Monomers at Only One End

Many methods of polymerization add individual monomer units to the end of the polymer chain. This reaction may take place in the bulk or it may occur at a specific site such as an enzyme or a solid catalyst surface. In these diagrams of polymerization reactions ♦ represents one of the repeating units in the chain.

$$\blacklozenge \blacklozenge \blacklozenge + M \longrightarrow \blacklozenge \blacklozenge \blacklozenge \blacklozenge$$

The entanglement structure of a chain polymerized (elongated) *in situ* from monomer units that can and do swell an existing polymer network can differ significantly from that of the chain elongated while held in position at an enzyme site in a plant. When a polymerization is carried out in pure monomer, the monomer acts both as a solvent for the growing chain and as the extending unit for the polymerization If this type of polymerization is continued until all monomer is exhausted with few growing chains, the resulting structure will certainly be knotted.. Depending on the precise chemistry, a new end may occasionally begin in the middle of the chain, causing a branch to form.

When natural rubber (polyisoprene) is formed enzymatically in a plant, the enzyme is large, compact, and water soluble, the unit to be added is water soluble, and the long chain being zipped together is not itself soluble in water. The addition takes place at an active site on the enzyme surface and the newly formed polymer reels off into a small particle. In this case the newly formed polymer must be unknotted. The same would be true of a polymerization that takes place continuously, while being attached at a solid catalyst surface or at a liquid-liquid interface.

4.2 Polymerization by Linking Addition

Another polymerization scheme involves the successive linking of growing units because each unit has two reactive ends. At the beginning two monomer units react to form a dimer, (perhaps with the release of some small molecule). The resulting dimers have two reactive ends and can react at either end with a monomer, another dimer, or a larger fragment. In this way larger and larger chains are formed.

$$A \blacklozenge B + A \blacklozenge B \longrightarrow A \blacklozenge \blacklozenge B + X$$

$$A \blacklozenge \cdots \blacklozenge B + A \blacklozenge B \longrightarrow A \blacklozenge \cdots \blacklozenge \blacklozenge B + X, \text{ or}$$

$$A \blacklozenge \cdots \blacklozenge B + A \blacklozenge \cdots \blacklozenge B \longrightarrow A \blacklozenge \cdots \blacklozenge \blacklozenge \cdots \blacklozenge B + X$$

If some of the monomer units are purposely made to have more than two active groups, then branched and crosslinked structures may be formed.

$$\begin{matrix} A \blacklozenge B + A \blacklozenge B & \longrightarrow & A \blacklozenge \blacklozenge B + X, & \text{or} \\ B & & B & \end{matrix}$$

$$\begin{matrix} A \blacklozenge B + A \blacklozenge B & \longrightarrow & A \blacklozenge B + X \\ B & & \blacklozenge \\ & & B \end{matrix}$$

$$\begin{matrix} A \blacklozenge B + A \blacklozenge \blacklozenge B + A \blacklozenge B & \longrightarrow & A \blacklozenge \blacklozenge \blacklozenge B + X \\ \blacklozenge & & \blacklozenge \\ B & & B \end{matrix}$$

4.3 Polymer Backbone Architectures

A linear chain (Fig. 2) is one in which all monomers except the ends are joined to exactly two other monomers. (Note that the term linear describes the topology of the chain, not its path which is anything but straight.) In a cyclic chain (Fig. 3) every monomer is connected to two other monomer units and the chain has no ends. Chains of this type have been studied to determine the effect of ends on the entanglement[21]. In a branched chain one or more monomers are connected to more than two other monomers. These are called branch points. A branched polymer may have short chain branches (Fig. 4) in which shorter chains are linked to a longer backbone of the same or different composition or it may have a few long branches (Fig. 5).

Fig. 2.　A linear chain　　　　Fig. 3.　A cyclic polymer　　　　Fig. 4.　Short chain branching

An extreme type of long chain branching is the star branched polymer in which several long and approximately equal arms are attached to a central hub. The effect of the various types of branching on the entanglement are very different. The polymer may also form a network with many closed cycles and dangling ends.

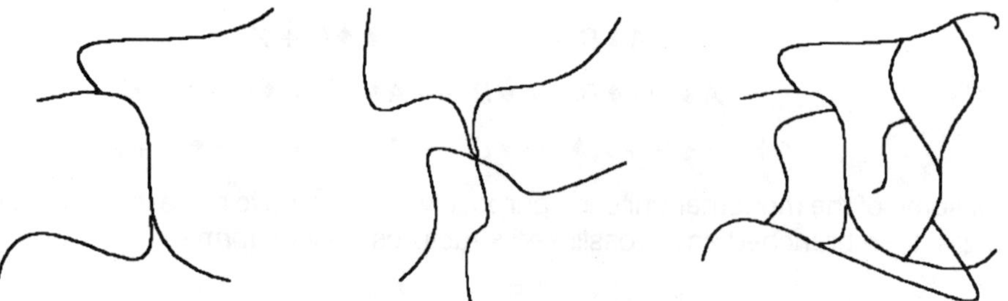

Fig. 5.　Long chain branching　Fig. 6.　A star polymer　　　　Fig. 7.　A network

This network formation may be achieved (i) during the polymerization process, by joining chains with active ends using a chemical linking agent, or (ii) after polymerization by using a crosslinking agent appropriate to the particular polymer. The interaction between the competing steps in network formation and entanglement rearrangement together with the length of the resulting strands will determine its effect on the final properties of the system.

4.4　Molecular Weight

Unlike the case of small molecules where all molecules of a given compound are exactly alike, the molecular weight of polymers can vary over a wide range. The shape and breadth of the molecular weight distribution depends on the chemistry of the polymerization method and also on any subsequent chain degradation. Several moments of the distribution and ratios of moments are used to characterize a polymer sample. The most frequently used averages are $M_n = \dfrac{\sum_{i=1}^{N} n_i M_i}{N}$, (the number-average) and $M_w = \dfrac{\sum_{i=1}^{N} w_i M_i}{\sum_{i=1}^{N} w_i}$, (the weight-average). Other moments are sometimes used. The molecular weight distribution is often characterized by $\left(\dfrac{M_w}{M_n}\right)$ or the dispersion, $1 + \sigma^2$. The dispersion may be altered by various degradation processes. Fractionation methods are used to obtain samples with a narrow molecular weight distribution for experiments in which the molecular weight dependence is critical. The entanglement of polymers as measured by their ability to sustain stress with little deformation typically depends upon the molecular weight to the 3.4 power. The overall mobility is also strongly dependent on the chain length and the shorter chains can move more easily than the longer ones. For these reasons high molecular weight polymers require long times to reach equilibrium following a change of conditions and show many interesting time and frequency dependent behaviors. A very small fraction of high molecular weight polymer can greatly affect the flow properties.

4.5 Arrangements of Monomers

A polymer may contain one or more different monomers. A polymer chain formed from one type of monomer is called a homopolymer. One which contains more than one type of monomer (a heteropolymer) is called a copolymer. The way the differing monomers are distributed along the chain greatly affects the properties. Similar units may be grouped together in a block or they may be distributed randomly through the chain. True molecular mixing of differing homopolymers is a complicated subject but for our purposes it is enough to know that unlike segments are usually not compatible and will try to form separate domains if the blocks are large enough. Typical examples shown graphically below include a homopolymer, (Fig. 8); a random co-polymer, (Fig. 9); a block co-polymer, (Fig. 10); a triblock co-polymer, (Fig. 11); a graft co-polymer, (Fig. 12), where branches of one polymer have been attached to a backbone of another; and a star polymer, (Fig. 13), where each arm is a di-block. In these diagrams the different monomers are represented by different lines.

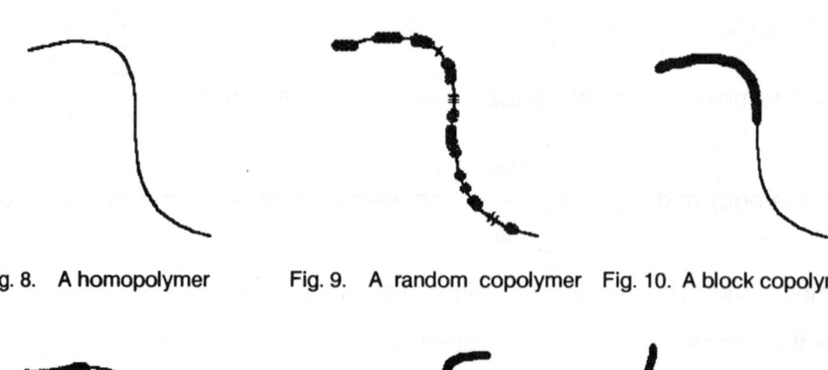

Fig. 8. A homopolymer   Fig. 9. A random copolymer   Fig. 10. A block copolymer

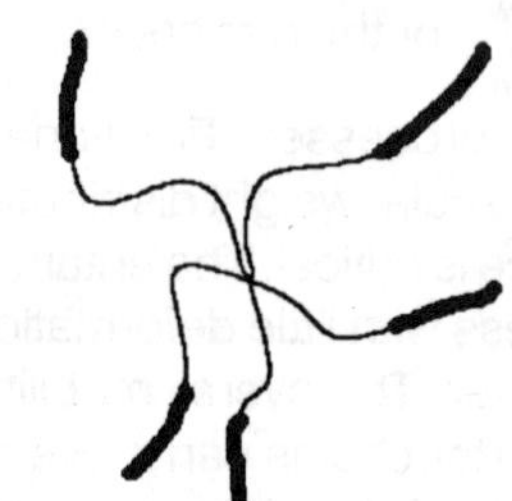

Fig. 11. A triblock copolymer   Fig. 12. A graft copolymer   Fig. 13. A star-branched
                                                                polymer with grafted
                                                                arms

5. Analyzing Tangled Systems

5.1 Tangle Projections and Strand Crossings

In physical systems entanglement exists in three dimensions but the topology of a finite tangle can be studied without loss of generality by projection of the strands onto a plane in such a way that all multiple points are double points at which the two strands cross[22]. Fig. 14 and Fig. 15 show the types of crossings that are disallowed and how each may be changed into acceptable crossings by a small change in the angle of projection.

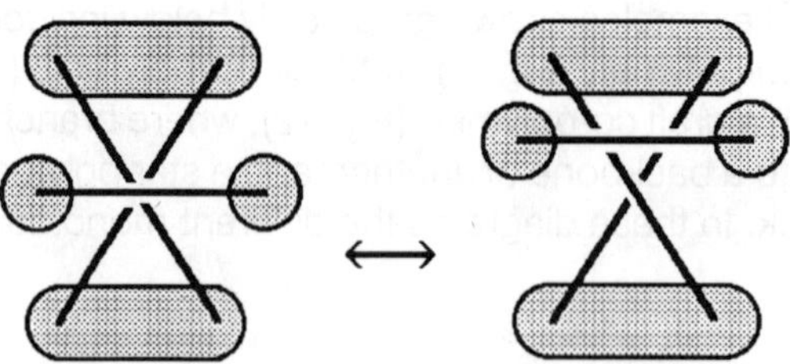

Fig. 14. If more than two strands cross at a point as shown on the left, they may be resolved into several regular crossings by a small change in the angle of projection as shown on the right.

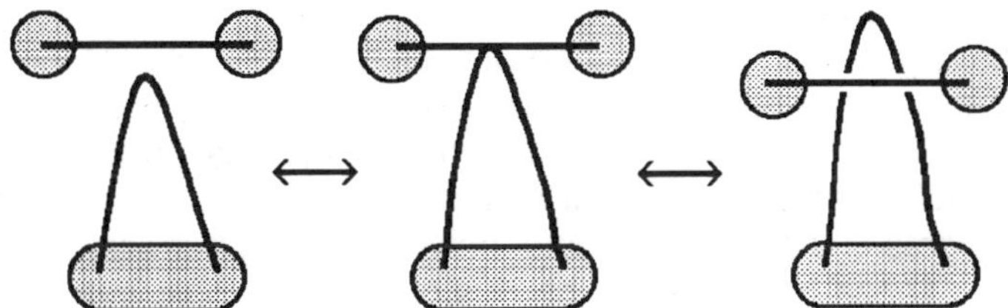

Fig. 15. If two strands touch at a point without crossing as in the center, it may be resolved into two crossings (right), or none (left).

A convention must be used that allows identification of the upper and lower strands at each crossing. The convention used in this chapter shows the upper strand as continuous while the lower strand has a short gap across the intersection. The crossings are drawn far enough apart so that no ambiguity is introduced by this convention. In making such a projection all information about spatial distances perpendicular to the plane of the projection have been sacrificed except for identification of the strand that is nearer to the observer. Since the two strands at a crossing may in fact be very far apart, no conclusions about strand interactions may be drawn from <u>single</u> crossings, but only from the topological interactions of <u>groups</u> of crossings.

5.2   Definition of Knotting

The concepts of knotted and unknotted structures are needed to discuss the difference between these processes. Because the word "knot" can have several meanings in various contexts, and because the concept of an unknotted tangle is central to this discussion, knotted and unknotted tangles are defined briefly here. Extend the ends of the strand to infinity or join them outside the tangle. The tangle is unknotted if it can be untangled without cutting the cord. A knotted tangle can not be untangled in this way. The tangle shown on the right in Fig. 16 is unknotted and that on the left is an overhand knot. The term knot will be reserved for interactions that are knotted as defined above. The term tangle will be used for strand interactions that are unknotted, or that contain both knotted and unknotted or unknown interactions,.

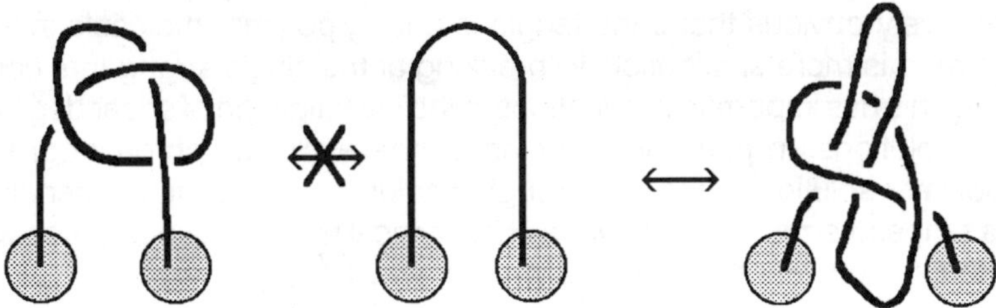

Fig. 16. The overhand knot on the left can not be untied without freeing an end. The unknotted tangle on the right can be undone.

5.3 Half-cycle Analysis

Each crossing in a tangle projection defines two half-cycles of the tangle. The interactions between the half cycles, and the interactions within each half cycle may then be distinguished. This approach retains the information in all crossings. It allows the identification of sections of a compound tangle. This analysis leads to interesting insights in the study of knots; however space permits only the briefest of summaries. These concepts will be illustrated with single strand tangles. An extension to multiple strand systems is in progress.

The crossing list is formed in the following way. A starting point and direction are chosen arbitrarily, then the crossings are listed in the order in which they are encountered from the start. The sign of each of the crossing list entries is determined by the position (top or bottom) of the strand being traced at that point. Each crossing is, of course, visited twice to complete the cycle. A half-cycle of the crossing list is the section of the list between one occurrence of a crossing and the return to that crossing. The crossings at which the two half-cycles intersect are those that occur once in each half-cycle. These crossings make up the boundary set of that crossing. An interaction matrix derived from the boundary sets of all crossings gives additional insight into the structure of the tangle. The size of the boundary set is an indication of the degree of localization of the interaction.

The interaction class of a crossing can be defined and consists of the crossing, its boundary set, the boundary sets of those crossings, etc. until no new crossings are added. The crossings that occur twice in the same half-cycle are of two types. They represent interactions within the half-cycle that may belong to the same interaction class or to other interaction classes. Those that belong to the other interaction classes are paired and form entire interaction classes. Those that belong to the same interaction class make up the internal set of that crossing. With a few trials you can convince yourself that neither the boundary sets, nor internal sets, nor the interaction classes depend on the starting point or direction of travel.

There are no interactions between different interaction classes. Crossings belonging to one interaction class are separated in the crossing list only by one or more entire interaction classes. These interaction classes are equivalence classes since membership in an interaction class is symmetric, reflexive, and transitive.

It is intuitively obvious that if the tangle is a long polymer molecule in a shear field, the strand is more susceptible to breaking at the single strand that connects the classes. This has important implications in the investigation of shear degradation of polymer solutions. In particular, studies of size exclusion chromatography, in which a polymer solution is passed through a column packed with material having pores of a range of sizes, have shown a chain critical length that correlates with the critical entanglement length measured in other ways[23].

5.4 Subtangles

A simple closed curve in the plane must cut the tangle an even number of times if no end of the tangle lies inside the curve. A subtangle of order n may be defined to be a section of the tangle with 2*n* ends. A section of the tangle with *n* ends thus represents the interaction of *n*/2 strands (or parts of the same strand). A strand passing through the subtangle may interact only with other strands or it may interact with itself at one or more crossings. The subtangles are the mechanism by which stress is communicated between separate parts of the structure and it appears that the nature of the subtangles in a fabric structure is important in determining the way the structure responds to stress.

5.5 *Types of Subtangle*

A subtangle of order 2 is an interaction of 2 strands or (independent parts of the same strand) and is connected to the rest of its interaction tangle by four strands. Three types of subtangle may be distinguished. Type A, not usually considered a subtangle, is a single crossing. It is useful in the context of woven structures that have no other subtangles and in which no localized knotting occurs. Type B contains one or more twists but neither of the strands intersects itself. In type C subtangles, at least one strand crosses itself within the subtangle. It will be instructive to consider the way these three types allow equilibration within the subtangle and transmit stress to other parts of the fabric.

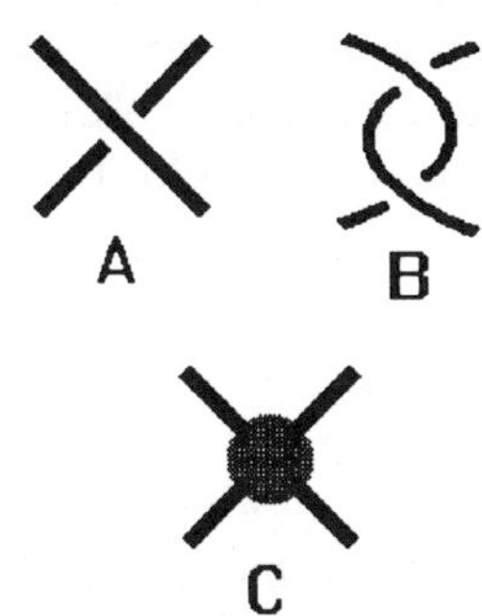

Fig. 17. Three types of subclass are useful in interpreting fabric structures and tangles.

A type A subtangle can only communicate stress through friction, and then only if the strands are in contact. A type B subtangle allows both strands to equilibrate by overcoming only frictional forces so that the stress on all four strands may equalize. It will transmit stress unless the motion is very small. A type C subtangle tends to tighten under stress especially if the two parts of the same strand are pulled in opposite directions. This type transmits stress to the other strand but because of the possibility of "jamming" the four strands may have different stress.

The range of the interaction can be investigated through the distribution of sizes of the interaction classes, subtangles, and boundary sets. Ongoing investigations are exploring the relative effect of these various parameters in determining the properties of the different structures.

6. Selection of Model Structures

A study of many types of traditional and modern fabrics and other stable macroscopic structures was made to establish a data base to relate model structures to known physical properties. The structures investigated were those that determine the cohesiveness of the underlying fabric, as opposed to elements such as knotted pile or embroidery that are added to the basic structure.

This attempt was made to develop analytical ways of describing the topology of these structures and to determine what topological aspects determine the properties. It seemed that the previously discussed crossing list interactions could be a useful set of parameters to explore the relationship between the structure of a fabric and its properties. That is, the fabric's global structure can be related to parameters that may be derived from the strand projection.

In addition the processes by which these structures form were later studied in order to extend this parameter description to other possible random three–dimensional structures formed from a system of long strands entangling in random motion. Such systems could involve molecular strands that are too large, too complex or too irregular to allow a complete analysis or description of all crossings. It would be extremely useful, however, just to describe the types of interactions that occur.

6.1   Fabric Structures

Throughout history and prehistory considerable effort and ingenuity has been used to produce stable two-dimensional fabrics from threads and fibers of various sorts. Since most types of cord or thread are not readily extensible, with a few modern exceptions, the type of thread can be ignored and any extension the object may have may be attributed to the fabric structure rather than the strand. The ingenuity and variety of techniques used in the construction of various utilitarian objects can be apreciated by a study of Peter Collingwood's book, *The Makers Hand*[25]. The author describes the structure and possible methods of construction of a number of artifacts from his personal collection and varied experience. While it is not intended to be an exhaustive study of possible structures, Collingwood provides an efficient and most enjoyable overview of the possible interactions of various types of linear strands and gives a good feeling for the interrelation between the nature of the strands, the structure of the fabric, and the ultimate properties and use of the fabric.

Be aware that handiwork from very different cultures that fulfill related purposes frequently have similar structures even when the techniques of construction may be quite different. There is an intrinsic relationship between structure and properties that to some extent transcends the nature of the strand itself.

6.2   Classification of Fabric Structures

Emery in *The Primary Structure of Fabrics*[26], classifies the structures according to the number of elements and sets of more or less parallel elements and the type of interactions between the sets of elements. The sets-of-elements (SOE) classification is not easily extended to random systems of long strands where the identity of strands interacting at a given neighborhood may not be known. In some fabric structures such as bobbin lace, one strand may alternate between intertwining with other strands in a 1SOE type of structure and at another place acting as a warp or weft thread in a 2SOE woven structure. Since our need is to describe random systems, we will use a different approach. We will attempt to classify the structures in terms of the fundamental type of motion (loop or end) involved in the formation, by the length of the strands, and by the localization of the interactions.

Fabric construction techniques divide naturally into two major classes. The first class contains techniques such as weaving, netting, and tatting that use a bobbin or shuttle to shift the end of the strand through other parts of the structure. The second class includes techniques that make use of the standing part of the strand such as knitting, sprang, and crochet. A few techniques use both types of interaction. Several representative structures are described below.

6.3   Structures Made Using a Moving End

6.31   Netting

In ordinary fish net the two strands are knotted together at evenly spaced intervals. On the next round the knots are placed in the middle of the interval in the previous round. By locating more than one knot on a given segment, the total number of knots may be increased to change the dimension or shape of the finished net. In decorative netting the spacing may be varied. The knots may be seen to be localized, and the crossings of one knot do not interact with the crossings of another. The knots in netting are quite localized and some boundary sets are small. The boundary set of one crossing (shaded) may be seen to consist of only four crossings.

Fig. 18. Netting

Netting has little elasticity. It can be deformed to the extent allowed by the strands between the knots, but it has little tendency to resume its original shape. It does not feel "stretchy". The spacings do not change in response to stress because the knots tighten and hold more firmly thus only the angles may vary. A three dimensional random netted structure could be made.

6.32   Weaving

In weaving one strand (or group of strands) goes regularly under and over another strand (or group of strands). Weaving may be thought of as a generalized knot and has an infinite boundary sets since no strand crosses itself. The boundary set of a crossing in this case consists of all other crossings. Weaving also has little elasticity. It is unstable if loosely done. If it has not been stabilized by post-weaving treatments such as calendering or waulking (mild felting) raw edges unravel easily.

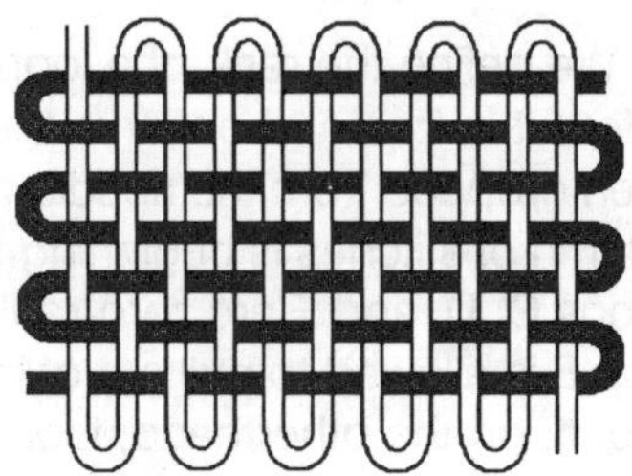

Fig. 19. Tabby weave

Stress can be communicated to mainly perpendicular strands and only by frictional forces. Because stability arises from the regularity and closeness of the under and over crossings it is hard to imagine a random three dimensional structure that would show the observed creep and recovery behavior. If a single strand is pulled through the perpendicular strands a bit, random agitation of the entire structure does not cause it to return to its previous position.

6.4   Structures Made Using Loops

6.41   Knitting

The underlying structure of the entanglement of long mobile strands proposed in this work is similar to that which holds a piece of knitting together. For that reason the process by which the interactions are formed and the effect of a flaw in the ordered structure will be described in detail. Knitting is formed by an ordered system of loop interactions in a single strand. During the knitting process the loops are prevented from unraveling by the knitting needles, which pass through them. Fig. 20 shows a piece of hand knitting in progress. A new stitch is being formed on the right hand needle. The needle has been passed through the next stitch in the previous row, stitch A, at the tip of the left hand needle. The standing part of the yarn is wrapped around the right hand needle and will be pulled through loop A to form the next stitch in the upper row. The left hand needle will then be withdrawn from loop A, which is now locked by the newly formed loop. The process is repeated with the next stitch on the left hand needle and continues to the end of the row when all the stitches are on the right hand needle. The work is turned around so that the needle holding the newly formed loops is held in the left hand. The empty needle is held in the right hand, the tip is inserted into the first stitch of the new row and the process continues as before. Note that the ends play no part in this process. One end of the strand is inside the ball of yarn and the other is at the beginning of the piece of knitting. The knitting process is done entirely with the interior of the strand. When a new color is desired the old strand is allowed to hang at the back of the work and the process continues with the new color. Several strands may be used to produce an intricate design. The ends are fastened later to ensure durability but this does not affect the interactions of the loops. Except near an unfastened end, the integrity of the piece does not depend upon the number of different strands used in its construction.

We define the rank of a loop to be the number of loops that must be removed before it is free and observe the effect of a flaw in the knitting. One stitch, D, has been dropped from the needle without being locked by loop C in the new row. The column of stitches is beginning to unravel and loop D has already been pulled out. Loops C, D, and E are zero-rank, loop F is first-rank, and loop G is second-rank. If loop E is allowed to pull out of loop F, loops F and G will each decrease in rank by one. If, on the other hand, loop D is reinserted through loop E; loops E, F, and G will each increase in rank by one. Note that this sequence can only move up and down the column and does not affect the rank or stability of the loops in other columns.

The process may be modified to change the number of loops locked by each loop or the number of loops locking each loop or by twisting the loops. A series of

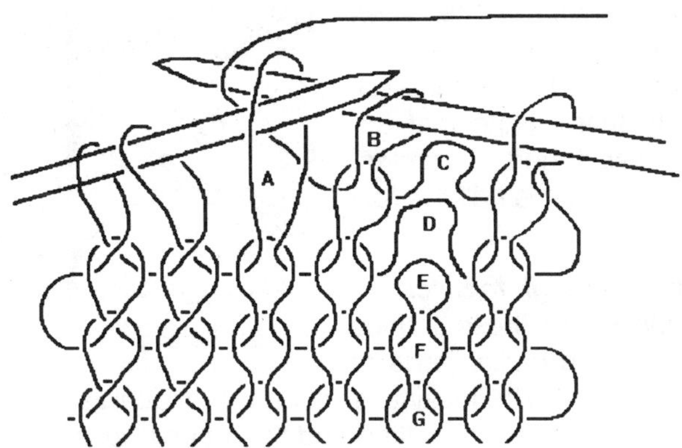

Fig. 20. Stockinette stitch. A new stitch is being formed and the effect of a dropped stitch is shown. The stitches in the left two columns have been twisted. Note the change in the subclass structure.

samples using the same yarn and needles but knitting stitches that differ in the number of loops that are locked by each stitch have been prepared. They show considerable variation in feel.

There are many collections of knitting stitches. Hiatt[27] gives a clear description of many aspects of knitting.

The location and orientation of the subclasses in knitting may aid in the equilibration of the structure in which local variations in gauge decay with use. Each subclass consists of a simple twist and is unknotted allowing equilibration of the strands connecting the subclass to the rest of the structure. All boundary sets in single knitting consist of an entire row. Novice knitters often become discouraged because the newly formed stitches do not appear as even as those in garments they have seen but the stitches in hand knitted garments become more even with wear and washing. If one loop is pulled out in a snag, it will tend to return to its previous position if the fabric is randomly agitated.

The stitches may also be twisted as they are knit as shown in the left hand column of stitches in Fig. 20. A fabric using this stitch exclusively is less elastic than uncrossed knitting. A similar structure (needle netting) is very ancient and is made by another method.

6.42 *Sprang*

Sprang is an ancient technique that is not well-known today. It is thought to be older than knitting and was used for garments where stretchiness was desirable[28]. Its formation begins with a series of parallel strands. The fabric is constructed by twisting adjacent strands. Note that in the example shown, if the black rod were removed, the entire structure could unravel. A similar structure, although frequently made by another technique is often seen in hammocks and carrying bags. The bags stretch to conform to the shape of whatever is put in them. Note that the subclasses are those found in knitting. The average boundary set is half the row length.

6.43   *Crochet and Afghan Stitch*

In the construction of these fabrics loops are pulled through earlier parts of

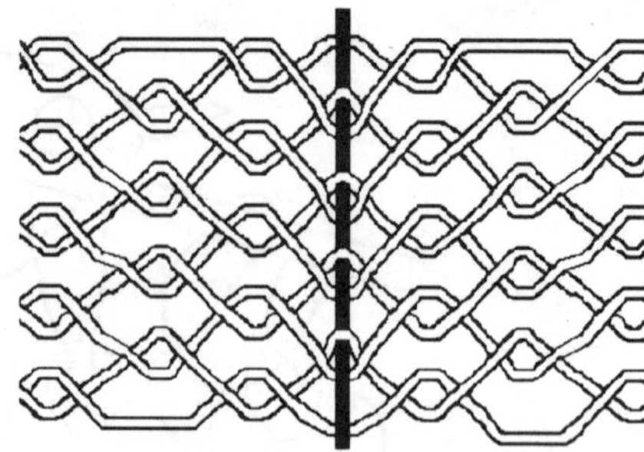

Fig. 21. Sprang

the structure with a hook. The locking is much more localized as reflected in the size of the boundary sets than in knitting and the fabric is less elastic.

6.5   *Mixed Constructions*

6.51   *Gauze Weaves*

Gauze weaves combine center-of-strand actions between warp strands with end actions by the weft. The interactions between adjacent warp strands stabilize the weft strands and maintain the spaces in the loose weave. An equally loose plain weave would distort easily and would not wear well. The sub-class type interactions between adjacent warp threads in gauze weaves appears to be an important factor in their increased stability compared to loose plain weaves. Many very intricate gauze weaves were used in Central and south America.[29,30]

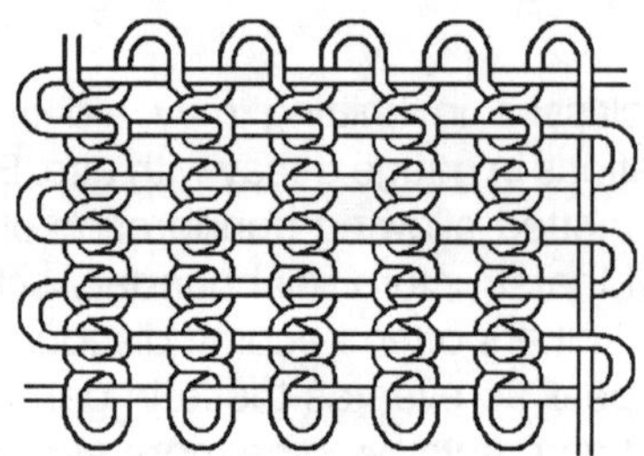

Fig. 22. Simple Gauze weave

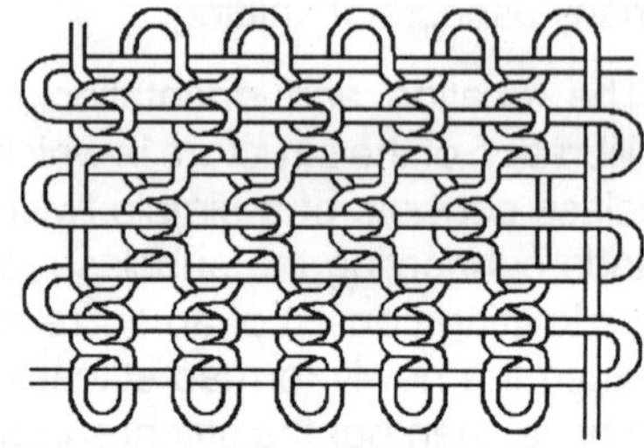

Fig. 23. Alternating Gauze weave

6.52   *Felt*

Felt is formed by the wet agitation of wool fibers. The process is used either as the primary structure by felting a mass of loose wool fibers or as a post-formation treatment of a wool fabric formed by another technique (usually weaving or knitting). It can be done without sophisticated equipment. The scales on the surface of the fibers open in the presence of water and heat and the fibers can quickly become entangled. Several mechanisms have been proposed for this entanglement[31] but none are considered entirely satisfactory. Loop formation has been observed in individual fibers. This raises the possibility that felting occurs by a loop mechanism similar to that described in Section 8.. However because of the directional nature of the scale structure and the relatively short length of the fibers, the possibility of motion along the strand can not be ruled out. Shrinkage of wool can be reduced by coating the scales or by chemically altering them. It is known that the scale structure is important in felting and that the curliness of the fiber also plays a role.

Synthetic fibers are felted by forcing needles or jets of fluid through the mass of randomly oriented fibers at several angles.

6.6   Comparison of Common Fabric Structures Using Half-cycle Analysis

Formation	Length of Strands	Range of Interactions	Technique	Subclasses	Boundary Set
End	Long	Localized	Netting	Knot	Small
			Tatting	Knot	Small
			Macrame	Knot	Small
		Generalized	Weaving	Only at selvage	Infinite
			Braiding	Frequently none	Infinite
		Mixed	Bobbin lace	Variable	Variable
			Needle lace	Variable	Variable
Loop	Long	Localized	Crochet	Variable	Variable, some small
			Afghan stitch	Variable	Variable, some small
		Generalized	Knitting	Type B	Row
		Mixed	Crossed	None	Row and stitch
			Twice-knit		
		Generalized	Knit below		
			Sprang		
			Interlinked	Type B	Column
			Interwoven	Type A	Column
			Needle felt		Unknown
			Fluid jet non-woven fabrics		Unknown
		Mixed	Crossed loop needle netting	None	
Mixed	Short	Generalized	Wool felt		
	Long	Generalized	Gauze weaves		Infinite for weft strands, length of warp for warp strands
			Card weaving		

7. Entanglement Models

From the previous examples it may be seen that: (i) the type of interaction between strands influences the global structure, and (ii) the global structure is very important in determining the physical properties of the system. Therefore any model of entanglement must address implicitly or explicitly the type of global structure desired and state the strand interactions that are allowed for its formation and equilibration.

7.1 Network Models

The structure envisioned by the early elastic network model used in polymer science may be compared to three-dimensional netting where the length between knots is random or uniform depending on the method of preparation. In end-linked systems the number of strands entering each knot may be varied by changing the functionality of the linking agent. Entanglements are viewed as transient knots, some of which may be trapped when the system is linked. In a model of this type the distribution of lengths between knots is an important parameter to describe the system. It will be seen later in this chapter that the interaction between the crosslinking steps and the global entanglement is much more complex so that new physical properties may be possible by controlling and optimizing the two physical processes.

7.2 Tube Models

In pure reptation and other related tube models the entanglement is changed only by action of the ends. Because the ends move through the matrix formed by the other chains and by other parts of the same chain, the global structure is knotted. Conversely, the same sort of end rearrangements are needed to remove or modify the entanglement.

Polymer entanglement experimentally continually increases with molecular weight, in fact as a power law! Thus the entire length of the chain must be involved in the entanglement. In a knotted entanglement the ends are the only active part, and the rest of the chain is passive. As a result, a knotted entanglement model requires long segments of the chain to be shifted over relatively large distances in order to involve parts of the chain far from the ends. Up to one-half of the length of the chain must be moved across another strand in order to remove a knotted entanglement.

Because the entanglement rises with the 3.4 power of the molecular weight (M) and also the concentration of ends is proportional to M^{-1}, the time to reach equilibrium should rise as $M^{4.4}$. A knotted entanglement model would thus require unrealistically long annealing times to reach an equilibrium value. Conversely, there would be limited entanglement structure at short times for high molecular weights.

7.3 The Loop Entanglement Model

The global structure resulting from the entanglement model described here using loop interactions would be analogous to a random three-dimensional knitting with many dropped stitches. Both the formation and structure of knitting and also the effect of flaws in the regularity of the linking were discussed above. Although the parameters necessary to describe a system of this type are not immediately obvious, later in this chapter they are shown to include the distribution of loop ranks at equilibrium, a measure of mesh size that is characteristic of the instantaneous system, and a characteristic time for loop rearrangement. These quantities are related to the flexibility and mobility of the chain. A fractal dimension would also be necessary to describe the structure formed by molecular chains. The nature of this fractal behavior is under further investigation.

Whether the tangles be macroscopic or molecular, the entanglement is a property of the entire system. The constraints are not localized at "points of entanglement" and can not be described in terms of a "distance between entanglements" as used in the description of the rubber network model. Also keep in mind that the ends of a high molecular weight polymer are present in very low concentration. Consequently, at short times the rearrangements of the middle of the strand are much more likely to occur than rearrangements of an end of sufficient magnitude to measurably affect the overall properties.

8. Entanglement by Interlocking Loops

A notation is developed for describing the interactions of discrete loops. This notation leads to a concept of loop rank that is useful for understanding large single or multi-strand systems. An extension of this idea to a system of mobile loops has been developed and extended to the dynamic tangling behavior of moving strands. This loop analysis appears to offer insight into many aspects of the behavior of long chain flexible polymers. The concepts developed can be applied to a number of polymer problems and are found to be useful in understanding diverse polymer behavior both in bulk and in solution. The loop model requires no ends and is thus equally applicable to branched, cyclic, and lightly crosslinked polymers. The stress is resisted in all cases by the entanglement. In particular, crosslinks interact with the entanglement in such a way as to hinder untangling more than tangling. Thereby crosslinks not only maintain, but in fact increase the level of entanglement as discussed below.

The above model and its analytical formulation developed in the course of this study is consistent both in the approach to a final structure and in the final structure with observations and analysis of entanglement by random tangles formed spontaneously in long flexible macroscopic strands under agitation.

8.1 *Static Interactions of Separate Loops - The Linking Sequence Digraph*

The linking sequence describes interactions between distinct loops in a tangle and is useful for understanding the behavior of tangled systems. For simplicity in depicting the tangle projections in this section the loops in Figs 24 through 24 are sketched as short, regularly shaped and rather narrow. The ends are assumed to be fixed in the matrix and to be unavailable. Neighboring non-interacting strands are omitted. Only one sketch will be shown for each linking sequence discussed.

Fig. 24 is the simplest example of a loop interaction and illustrates the way in which such systems resist deformation. Loop B cannot be removed by pulling on the "standing parts" (i.e., the strands that enter the matrix) unless loop A is first withdrawn. Note that this corresponds to two applications of the second Reidemeister[32] move and does not change the knotting. If a third loop, C, passes through loop A as shown in Fig. 25, loop B is not free until loops C and A are removed in that order.

These interactions may be represented by the linking sequence digraph, a directed graph in which: (i) a vertex (V_i) represents a loops and an edge, $(V_i \rightarrow V_j)$, signifies that loop V_i passes through (or locks) the loop V_j. Thus the linking sequence digraph for the tangle shown in Fig. 24 is A → B and is read as A locks B. The linking sequence digraph for Fig. 25 is C → A → B. The loop that passes entirely over or entirely under another loop does not interact with that loop and does not affect the linking sequence. The letters associated with the loops are arbitrary. Tangles with linking sequence digraphs containing cycles are knotted and cannot be formed by loop interactions.

The adjacency matrix, **J**, for a linking sequence diagram of n vertices is an [n x n] matrix whose elements, $\mathbf{J}(i,j)$ are 1 if an arrow goes from V_i to V_j and are zero otherwise. The element, $\mathbf{J}^n(i,j)$, of the n power of an adjacency matrix is the number of paths of n steps or n + 1 vertices going from V_i to V_j. If the digraph is finite and contains no cycles, each path must contain at most n-1 steps and all powers of the adjacency matrix greater than n-1 must be zero. [Note: In addition to a projection of the tangle, the linking sequence digraph, the adjacency matrix, **J**, and its non-zero powers are given for the simple tangles (Fig. 24, 25) discussed in this section and for the more complex tangles discussed below (Fig. 26, 27).]

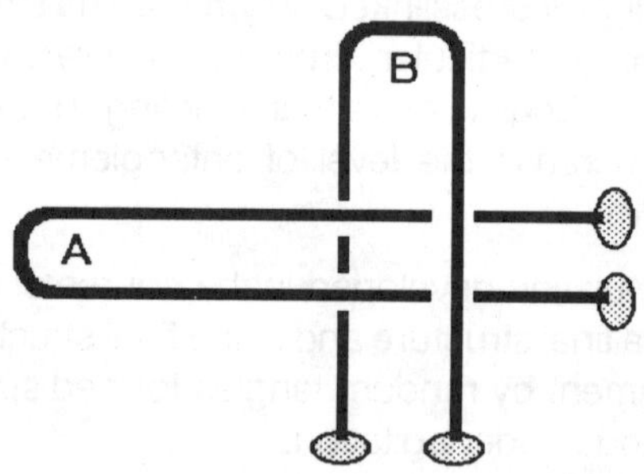

Linking sequence digraph

$$A \rightarrow B$$

Powers of adjacency matrix

$$\mathbf{J} = \begin{bmatrix} 0 & 1 \\ 0 & 0 \end{bmatrix}, \mathbf{J}^2 = 0$$

Fig. 24. Loop A locks loop B and prevents it from moving downwards.

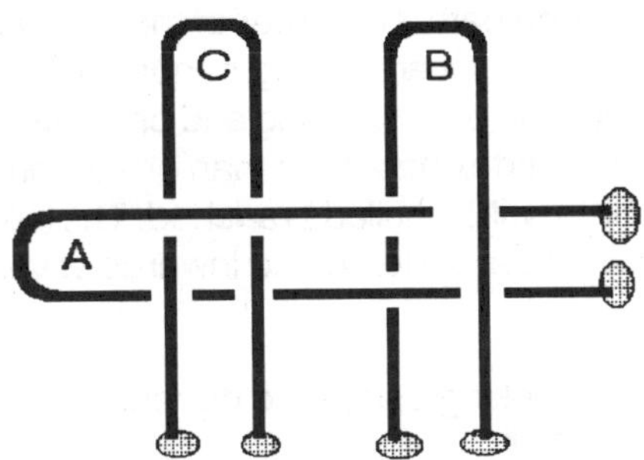

Linking sequence digraph

$$C \to A \to B$$

Powers of adjacency matrix

$$\mathbf{J} = \begin{bmatrix} 0 & 1 & 0 \\ 0 & 0 & 0 \\ 1 & 0 & 0 \end{bmatrix}, \mathbf{J}^2 = \begin{bmatrix} 0 & 0 & 0 \\ 0 & 0 & 0 \\ 0 & 1 & 0 \end{bmatrix},$$

$$\mathbf{J}^3 = 0$$

Fig. 25. Loop C now locks loop A, further restricting the motion of loop B. Both loop C and loop a must be removed (in that order) before loop B is free.

8.2 The Concept of Loop Rank

Define the rank of a loop to be the number of loops which must be successively withdrawn before the loop is free. The rank of a tangle (generated by loops) is that of the highest rank loop in the tangle. The rank of the tangle is therefore the highest power of the adjacency matrix having one or more non-zero elements. Fig. 24 shows a first-rank tangle. Loop A is zero- rank and loop B is first-rank. Fig. 25 is a second-rank tangle. Loop C is zero-rank and the ranks of loops A and B have increased by one. The rank is independent of the number of separate strands involved. The loops participating in a given tangle may belong to the same strand or to different strands. Furthermore, twisting of the loops does not affect the rank of the entanglement. The strands continue on to another loop of the same tangle, another tangle, or other random behavior.

Untangling must begin with a zero-rank loop, and no loop can be withdrawn until it has become zero-rank by the removal of all loops passing through it. The addition of another zero-rank loop increases the rank of the tangle only if it locks one of the zero-rank loops that determine the rank of the tangle. The linking sequence of an unknotted tangle must have one or more zero-rank loops. The column representing a zero-rank loop in the adjacency matrix has no non-zero elements.

8.3 Types of Linking Sequence Digraphs

The tangle shown below in Fig. 26 must be diagrammed as two separate tangles since loop A can be withdrawn far enough to free loop B without the prior removal of loop C. In contrast, loop C in Fig. 25 must be withdrawn from loop A before loop A can be withdrawn from loop B. The digraph for Fig. 25 will be said to be connected.

In the linking sequence digraph for the tangle shown in Fig. 27, vertex D can be reached from vertex A by two paths of two steps each. A "simple" linking sequence will be defined to be one which is both connected, and also in which no vertex can be reached from another vertex by more than one path. No power of the adjacency matrix contains any element other than 0 or 1. A "linear" linking sequence will be defined to be one for which no vertex has more than one incoming and one outgoing arrow. A linking sequence in which at least one vertex has more than one incoming arrow, or more than one outgoing arrow, or both, will be called branched. The linking sequence shown in Fig. 27 is branched outwards at vertex A and inwards at vertex D.

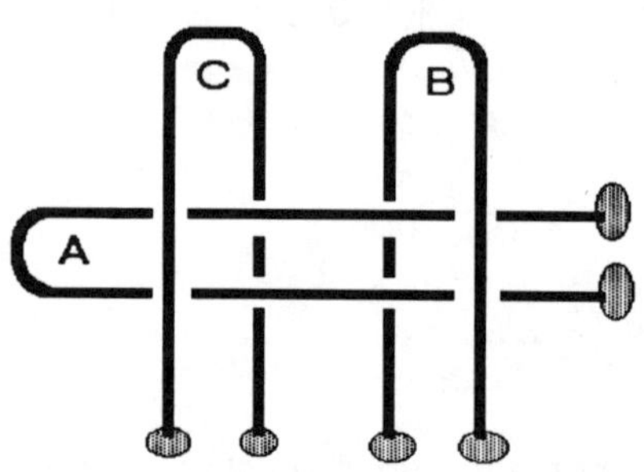

Fig. 26. The addition of loop C in this tangle does not affect the constraint on loop B. Compare the position of loop C and its effect on loop B in this tangle and in the previous figure. .

Linking sequence digraph

$$A \to B, \qquad A \to C$$

Powers of adjacency matrix

$$\mathbf{J_1} = \begin{bmatrix} 0 & 1 & 0 \\ 0 & 0 & 0 \\ 0 & 0 & 0 \end{bmatrix} \quad \mathbf{J_1^2} = 0$$

$$\mathbf{J_2} = \begin{bmatrix} 0 & 0 & 0 \\ 0 & 0 & 0 \\ 1 & 0 & 0 \end{bmatrix} \quad \mathbf{J_2^2} = 0$$

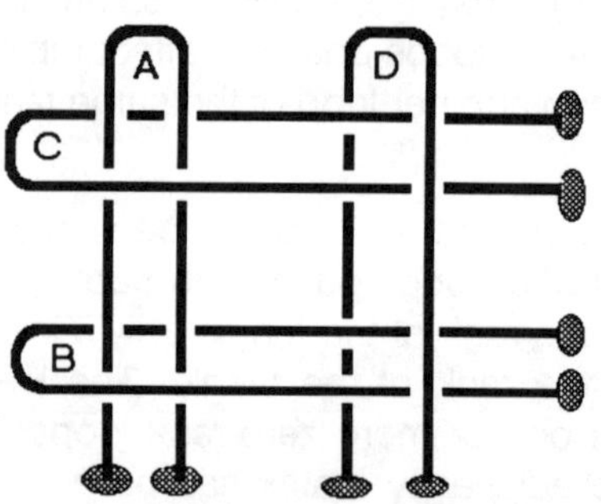

Fig. 27. Loop A locks both loops B and C. Loops C and B both lock loop D.

Linking sequence digraph

$$\begin{array}{ccc} A & \to & B \\ \downarrow & & \downarrow \\ C & \to & D \end{array}$$

Powers of adjacency matrix

$$\mathbf{J} = \begin{bmatrix} 0 & 1 & 1 & 0 \\ 0 & 0 & 0 & 1 \\ 0 & 0 & 0 & 1 \\ 0 & 0 & 0 & 0 \end{bmatrix}, \quad \mathbf{J^2} = \begin{bmatrix} 0 & 0 & 0 & 2 \\ 0 & 0 & 0 & 0 \\ 0 & 0 & 0 & 0 \\ 0 & 0 & 0 & 0 \end{bmatrix}$$

$$\mathbf{J^3} = 0$$

8.4 *Dependence of Physical Properties on the Distribution of Loop Ranks*

The distribution of loop ranks describes the interlocking of the system and may be expected to determine many properties of an elastomeric system above its glass transition and may be a sensitive indicator for the onset of glassy behavior. Loop rank can not be measured directly. Unfortunately the analytic dependence of the physical properties like flow on the loop rank can not presently be stated. However, the improvement of some viscoelastic properties can be expected to parallel an increase in the interlocking of the system, at least up to some optimum level. Consider a macroscopic tangle like felt. The strength of needled felt shows a strong dependence on the length of the strand and increases with needling time up to a maximum beyond which it decreases due to the breakage of the strands.[32] A further program of both analysis and coordinated experiments[33] is in progress to explore the relevant experimental variables and their interdependence. Therefore, only the general arguments are presented in this paper.

9. An Unhindered System of Mobile Loops

Several applications of this model to polymer systems will be sketched. The detailed calculations are too long to include here and will be published elsewhere.

9.1 Random Loop Interactions

This calculation considers one way of generating a distribution of loop ranks. It does not address the number of loops locked by a given loop or the effect of branching in the linking sequence digraphs. A polymer above its glass transition (i.e., capable of slow deformation) may be thought of a system of flexible strands in constant random motion. Loops are constantly being thrust out and withdrawn. If the strands are very long, there are few ends and loop interactions will predominate. In the present discussion the effect of the ends is ignored so that the conclusions apply only to long chains. Above the level of entanglement needed to form a network, it is not necessary to distinguish between inter-molecular and intra-molecular interactions. As the loops move, the rank of a given loop will change on a time scale that depends on the mobility of the polymer. In one single step the rank of the loop may increase by one, decrease by one, or remain the same. Only the zero-rank loops of each sequence are active in the change in rank, and thus the probability of change can not depend on the rank of the sequence. The probability of a loop increasing, decreasing, or remaining the same in rank is independent of the rank of the loop except at zero-rank and possibly at first-rank. Therefore if the system is not under stress, the rank of an individual loop forms a Markov chain with possible rank states 0, 1, 2, ..., with a retaining barrier at 0 rank. If the probability of the zero-rank to first-rank transition is assumed to be equal to that of the transition from nth-rank to $n + 1$st-rank, the matrix of one-step transition probabilities (P) is given by Eq. 1 in which: i) $p + q + r = 1$, and ii) the entry in column j of row i represents the probability that a loop of rank i will move to rank j at the next step.

$$P = \begin{bmatrix} q+r & p & 0 & 0 & \cdots \\ q & r & p & 0 & \cdots \\ 0 & q & r & p & \cdots \\ 0 & 0 & q & r & \cdots \\ \cdots & \cdots & \cdots & \cdots & \cdots \end{bmatrix} \qquad (1)$$

In this matrix p is the probability that a loop will increase in rank by one, q is the probability that the rank of the loop will decrease by one, and r is the probability that the rank of the loop will remain unchanged. If $0 < p < q$, an equilibrium distribution exists for the system and the probability, W(n), that a loop is at the rank n at equilibrium is given by Eq. 2.

$$W_{(n)} = \left(\frac{p}{q}\right)^{n-1} \left(1 - \frac{p}{q}\right) \tag{2}$$

(Note that this equilibrium distribution depends only on the ratio of p to q and is independent of r.)

The number average rank, $<N>$, at equilibrium in this system is

$$<N> = \sum_{n=0}^{\infty} n\rho^n (1 - \rho) = \frac{\rho}{(1 - \rho)} \tag{3}$$

9.2   *The Effect of Stress on a System of Mobile Loops*

Stress can be transmitted to a loop only along the strand. When the stress is resolved into components acting parallel and perpendicular to the loop axis, only the cases shown in Fig.s 28 and 29 are effective in untangling. If a loop is zero-rank, it can respond to a perpendicular stress, as in Fig. 28, by straightening out, and then it is not available for further tangling. Its rank in the latter state is defined as 0^*. When a zero-rank loop responds to a parallel stress, as in Fig. 29, it lowers the rank of any tangle it is removed from, but, unless a perpendicular stress is also imposed, it is available for further tangling at its new location. Its rank is unchanged. If either type of stress is applied to a loop that is not zero-rank the stress is transmitted to the loops trapped inside. However, the rank of the tangle is not changed by the motion of the first loop.

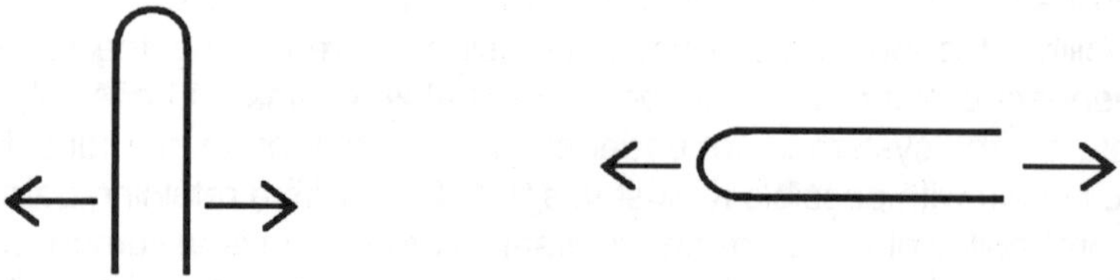

Fig. 28. A zero-rank lop can respond to this stress by straightening out.

Fig. 29. A zero-rank loop subjected to this type of stress may move but will not straighten out.

The calculation above can be extended to include the effect of stress. When a loop is placed under a stress perpendicular to the loop axis, the rank 0^* is an absorbing state. If the probability of a loop coming under a suitable stress is κ, the matrix of transition probabilities is then given by Eq. 4 in which the first row and column now refer to the state 0^*.

$$
P = \begin{bmatrix}
1 & 0 & 0 & 0 & 0 & \dots \\
\kappa & (q+r)(1+\kappa) & p(1-\kappa) & 0 & 0 & \dots \\
0 & q & r & p & 0 & \dots \\
0 & 0 & q & r & p & \dots \\
0 & 0 & 0 & q & r & \dots \\
\dots & \dots & & \dots & \dots & \dots
\end{bmatrix}
\tag{4}
$$

The probability κ depends on the amount of stress, confinement, and mobility. In this limited presentation the stress is assumed to be constant and uniform, and therefore there is no need to consider further the dependence of κ on stress.

In an uncrosslinked system the actual untangling proceeds in this step by step manner until the time when the entanglement is insufficient to form a network under the applied stress. After that time the system consists of clusters of individual chains that are only tangled with members of the same cluster but are not tangled with chains of other clusters. The system can begin to undergo elongational flow at this time. The extension is then due both to untangling and to viscous flow in a system of decreasing "cluster size". The cluster size is infinite before the time when viscous flow occurs and is described as t_f. After t_f the cluster size decreases to the limit of the single molecule size.

9.3 Creep in an Uncrosslinked Elastomer

When the calculation of creep is carried out, the curves are similar to those obtained by measuring creep in an uncrosslinked elastomer under constant load. That is, a system of short chains flows immediately; but as the number of loops is increased, a plateau in time of arrested creep develops that lengthens as the number of loops is increased further. If a few of the chains are much longer than the average, the flow is altered dramatically with important consequences for processing polymer melts. When the cluster size is much larger than any of the chains the flow proceeds normally. However, when the cluster size approaches the size of the largest chains, the cluster size distribution is skewed by the presence of these chains and the flow is slowed.

9.4 Stress Softening

When any tangled network is deformed, some of the zero-rank loops are removed from the equilibrium by extension. Over the time that the stress is applied more loops shift to zero-rank and are extended. When the stress is removed, all the extended loops return to zero-rank. If the stress is re-applied before the entanglement as been allowed to return to equilibrium, the number of zero-rank loops available for immediate extension will be greater than the original equilibrium number. This increase in zero-rank loops allows the network to respond more readily up to the previous extension, where it will elongate as though it had not been previously stressed. This behavior corresponds to the well-known phenomenon of "stress softening".

9.5   Entanglement from Newly Melted Single Crystals

When high molecular weight polyethylene is crystallized from dilute solution it is known to form folded chain crystals (in the form of platelets) in which the chain axis is perpendicular to the broad surface (plate) of the crystal except at the ends where it folds. The folds of the molecule are regular with rather tight bends. That is, the polymer chains are folded back and forth like a newly bought shoe lace. If the crystals are collected from solution and kept below the crystalline melting temperature while being dried, a mat of crystals is obtained from which fibers of fully extended chains can be drawn. If the temperature is allowed to rise above the melting point, even briefly, the chains become entangled and the fiber can not be drawn out to the same degree. However, the time is much too short to allow entanglement using only the ends of the chain. When the crystal melts, however, the tightly folded loops have the energy to become random in motion, undulate in all directions, and tangle with adjacent loops.

10. Application to Systems with Selective Immobilization

10.1 Introduction

This discussion applies to any type of selective immobilization of some loops of a tangled system. This immobilization can occur, for example, through chemical crosslinking, by attachment to a surface of an inert filler particle, by irregular or incomplete crystallization, or by the locking of loops into an adjacent glassy domain. The range of factors affecting the equilibration of entanglement include overall strand mobility, local variations in mobility (whether temporary or permanent) and flexibility.In macroscopic systems the addition of twigs or similar material into a tangle appears to increase the complexity of the tangle and greatly increases the difficulty of untangling the snarl. The precise effect will depend on the location of the immobilized loops in the matrix, on the number of loops involved and also on the rate at which the immobilization occurs with respect to the motion of the strand.

Unless it is specified otherwise, the distribution of the immobilized loops is assumed to be random. Although the immobilization need not be permanent, it must change on a time scale that is much slower than the step frequency of the Markov chain controlling the entanglement. In the calculations that follow it is assumed that the system is at equilibrium before the immobilization. In the first discussion of crosslinking, the chemical reactions are assumed to be instantaneous and permanent. Later the effect of crosslink mobility and the rate of crosslinking will be considered.

10.2 Notation for Crosslinks in the Linking Sequence Digraph

Crosslink sites, shown as a thick bar connection between the strands, may be indicated on the linking sequence digraph by writing the hindered loops in bold type. A crosslink at the site on loop B in Fig. 30 does not prevent the disentanglement of loop A. This may be written symbolically as A → **B**. A crosslink at the site indicated in Fig. 31 will prevent the removal of loop A and will thus prevent loop B (now of first-rank) from reaching zero-rank. This interaction is written **A** → B. Loop A can then move only in cooperation with the strand to which it is crosslinked. This motion is shown in Fig. 32 where loop B remains first-rank. Loop B can, however, increase in rank by the passage of a loop, such as loop D, through loop A as shown in Fig. 33. This can be written D → **A** → B. The probability that this will occur is assumed to be unaffected by the crosslink at A.

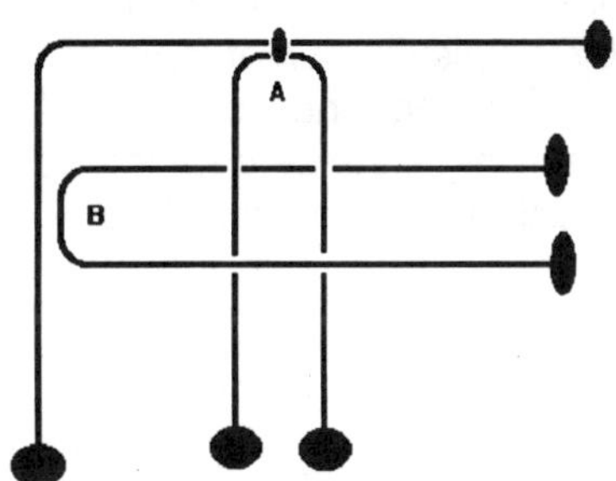

Fig. 31.

Fig. 30.

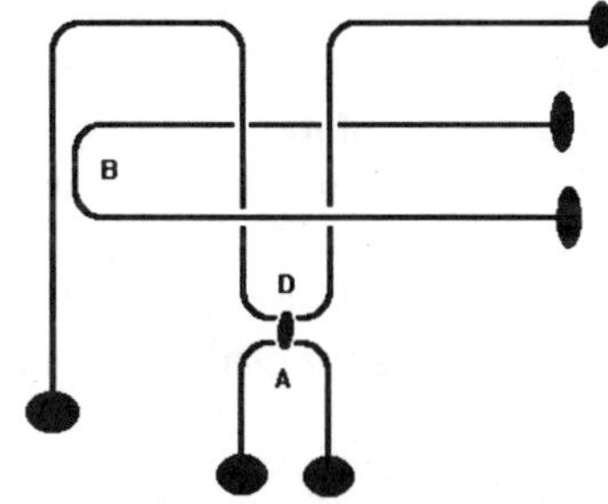

Fig. 32.

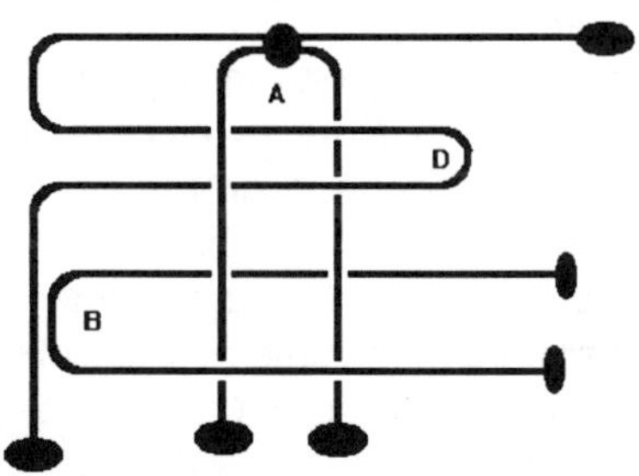

Fig. 33.

10.3 Loop Dynamics in a Hindered System

The probability that loop B in Fig. 30 will increase in rank depends on the number of loops in the neighborhood of loop A that are zero-rank and unhindered. In the unstressed system loop B is then governed by a one-step transition matrix with a retaining barrier at first-rank as shown in Eq. 5, where p', q', and r' refer to the crosslinked state.

$$P = \begin{bmatrix} 0 & 0 & 0 & 0 & 0 & \dots \\ 0 & q'+r' & p' & 0 & 0 & \dots \\ 0 & q' & r' & p' & 0 & \dots \\ 0 & 0 & q' & r' & p' & \dots \\ 0 & 0 & 0 & q' & r' & \dots \\ \dots & \dots & \dots & \dots & \dots & \dots \end{bmatrix} \qquad \text{Eq. 5.}$$

The matrix of transition probabilities for a hindered system must therefore be decomposed into the weighted sum of a series of matrices with barriers at each rank. The weights of the matrices are determined by the distribution of ranks at the instant of crosslinking. If the zero-rank loop controlling a sequence is hindered, a loop of rank n in the sequence cannot go below that rank. If the zero-rank loop is unhindered and the first-rank loop is hindered, the rank of the loop can go only as low as rank n-1.

Space does not permit the calculation to be followed further but it is obvious that when only a few crosslinks are added, untangling will be hindered more than tangling. Thus the equilibrium average rank will increase. Finally if all loops are hindered, no loop can change rank. Thus the average rank for a totally crosslinked system will be that which it had before crosslinking.

10.4 Distribution of Loop Ranks in the Crosslinked System: The Effect of Temperature and Stress

Note that this discussion implies that the average rank of the hindered system depends not only on the number of crosslinks, but also on the average rank of the entanglement before it was crosslinked. The average rank of the hindered system will therefore depend on the stress history of the material before crosslinking. In addition, since the equilibrium average rank depends on the mobility and therefore on the temperature, the average rank after crosslinking will also depend on the temperature at the time of crosslinking. Therefore the rank after crosslinking will be higher when carried out at lower temperatures.

The distribution of ranks is lowered by mixing or milling due to the continued extension of loops. An increase of equilibrium average rank can be obtained after any milling or mixing processes if the compounded material is allowed to rest before vulcanization. A time delay allows the entanglement to come to equilibrium before the crosslinking occurs.

In some situations the hindrances to the loop motion are not permanent, but break and reform as a function of time. The shift in average rank will be dependend on time in this situation and will approach a limit that is higher than that obtained for the same number of permanent crosslinks formed simultaneously. This increase in rank occurs because a crosslink that breaks and reforms after the entanglement has already begun to move towards its post-crosslinking equilibrium, interacts with a new distribution of higher average loop rank. It is clear from this section's argument that the effect of a given number of permanent crosslinks will also depend on the rate at which they are introduced relative to the rate of equilibration of the entanglement.

10.5 Comparison of Crosslinking Calculation to Observed Behavior

The discussion above describes the effect of crosslinks on the average rank of the entanglement. It is seen that crosslinks alter the probability of untangling some of the loops so that the equilibrium entanglement increases. The crosslinks do not enter directly into the load-bearing process but only need to withstand the force of the random chain motions. That is, the stress is carried at all times by the knitted structure of the entanglement. The creep behavior of crosslinked and uncrosslinked elastomers is similar through the plateau zone until the point at which the un-crosslinked system begins to flow. The crosslinked system shows no sharp change in creep behavior but it simply does not flow. This similarity in behavior suggests strongly that the extension is controlled in the same manner both in crosslinked and uncrosslinked systems.

11. References

1. R. S. Porter and J. F. Johnson, The entanglement concept in polymer systems, *Chem. Rev.*, **66**(1), Jan. 25, 1966, 1- 27.

2. W. W. Graessley, *Advances in Polymer Science*, **16**, (1974).

3. P. G. DeGennes, *Physics Today*, June 1980, 33-39.

4. R. S. Porter and J. F. Johnson, op. cit.

5. W. W. Graessley, op. cit.

6. W. F. Busse, *J. Phys. Chem.* **36**, 2862-2879 (1932).

7. L. R. G. Treloar, *Trans. Faraday Soc.*, **36**, 538-549 (1940).

8. P. J. Flory, *Chem. Rev.*, **35**, 51-75 (1944).

9. S. F. Edwards, *Proc. Phys. Soc.*, **91**(3), 513-517 (1967).

10. R. Alexander-Katz and S. F. Edwards, *J. Phys .A*, **5**(5) 674-81, (1972).

11. A. V. Vologodskii, A. V. Lubashin, M. D. Frank-Kamenetskii, and V. V. Anshclevich, *Sov. Phys. JETP*, **39**(6), 1059-1063, (1974).

12. P. J. Flory, *Polym. J.*, **17**(1), (1985).

13. P. G. deGennes, *Scaling Concepts in Polymer Science*, Cornell University Press, Ithaca, NY, 219-241, (1979).

14. M. Doi and S. F. Edwards, *The Theory of Polymer Dynamics*, Clarendon, Oxford, (1986)

15. N. Angier, When Proteins Come to Life, 'Chaperones' Show the Way, *The New York Times*, Science Section, (2/11/92).

16. A. MacArthur, University of Akron Ph.D. Dissertation, (1984).

17. P. Freyd, D. Yetter, J. Hoste, W. Lickorish, K. Millet, and A. Ocneau.; A new polynomial invariant of knots and links., *Bull. Amer. Math. Soc.*, <u>12</u>, pp 239-246; (1985).

18. L. H. Kauffman, *On Knots*, Annals of Mathematics Studies, <u>3115</u>, Princeton University Press, Princeton, NJ, (1987).

19. P. J. Flory, *Principles of Polymer Chemistry*, Cornell University Press, Ithaca, N.Y. (1953).

20. *Scientific American*, Entire issue, **197**(3), (1957).

21. G. B. McKenna and D. J. Plazek, *Polym. Prepr. (Am. Chem. Soc., Div. Polym. Chem.)*, **30**(1), 75-6, (1989).

22. R.H. Crowell and R.H. Fox, *Introduction to Knot Theory*,New York, Springer-Verlag, 6,(1973)

23. J. Savoca, University of Akron Ph.D. Dissertation, (1982)

24. J. H. Conway, An enumeration of knots and links and some of their algebraic properties, *Computational Problems in Abstract Algebra*, Pergamon Press, N.Y., 329-358, (1970)

25. P. Collingwood, *The Makers Hand, Lark Books*, Asheville, North Carolina, Interweave Press, Loveland, Colorado, (1987).

26. I. Emery, *The Primary Structures of Fabrics an Illustrated Classification*The Textile Museum, Washington, D. C., (1966).

27. J. H. Hiatt, *The Principles of Knitting*, Simon & Schuster, New York, NY, (1988).

28. E.J.W. Barber, *Prehistoric Textiles*, 122 , Princeton University Press, Princeton, N.J.(1991).

29. C. M. Pancake and S. Baizerman, Guatemalan Gauze Textiles: A Description and Key to Identification, *Textile Museum Journal*, **19-20** (1980-81).

30. A. P. Rowe, J. B. Bird, Three Ancient Peruvian Gauze Looms, *Textile Museum Journal*, **19-20** (1980-81).

31. K. R.Markinson, *Shrinkage of Wool Fabric*, Science Series, Marcel Dekker, N.Y., (1979).

32. K. Reidemeister, *Knotentheorie*, Chelsea Pub. Co., N.Y.,(1948).

33. Hearle, J. W. S., Grossberg, P., and Backer, S.; *Structural Mechanics of Fibers, Yarns, and Fabrics, Vol 1*, 312-319, Wiley-Interscience, New York, (1969).

34. Bhowmick, A. K., Cho,J., MacArthur, A. and McIntyre,D., "Influence of Gel and Molecular Weight on the Properties of Natural Rubber." *Polymer*, <u>27</u>, 1889-1894, (1986).

Synthesis and Cutting "in Half" of a Molecular Möbius Strip—Applications of Low Dimensional Topology in Chemistry[1]

David M. Walba,* Timothy C. Homan, Rodney M. Richards and R. Curtis Haltiwanger[†]
Department of Chemistry and Biochemistry, University of Colorado, Boulder, Colorado 80309-0215

This paper appears with complete experimental procedures in a special issue of the New Journal of Chemistry devoted to "Topology in Molecular Chemistry, Catenanes and Knots", and is reprinted here by permission of the Publisher

ABSTRACT. — Topology has suggested several classic targets for chemical synthesis, including molecular Möbius strips. Details of the first synthesis of a molecular Möbius strip, possessing a three rung Möbius ladder molecular graph, are described. The structure is composed of crown ether rings fused by the tetrahydroxymethylethylene (THYME) unit—the rungs of the Möbius ladder are thus carbon-carbon double bonds and the single edge is a polyether chain of 60 atoms. A chemical realization of the classic topological trick of cutting a Möbius strip "in half" only to have it remain in one piece is accomplished by ozonolysis of the double bonds, and serves as an elegant corroboration of the topology of the structure. In addition, just as topology suggested targets for synthesis, the synthesis has stimulated new topology. A brief description of some of these new mathematical results is presented, including the discovery of a new and very "deep" kind of chirality. Thus the three rung THYME Möbius ladder molecular graph serves as the prototype of the class of intrinsically chiral graphs—possessing a kind of chirality which can never occur in knots or linked rings.

Introduction

In recent years a remarkable synergism between the disciplines of organic chemical synthesis and mathematical low-dimensional topology has resulted in substantive advances in both. In chemistry, the new field of scientific and mathematical inquiry resulting from this marriage—topological stereochemistry[2,3]—has seen demonstrated in the laboratory elegant examples of topological stereocontrol in organic synthesis, once the exclusive province of enzyme-directed biosynthesis. On the mathematics side, extrinsic graph theory, a relatively quite mathematical sub-discipline a decade ago, is now burgeoning. In this endeavor, known topology served to suggest novel targets for synthesis, the synthesis then inspired the invention of new mathematics, which then suggested new targets for synthesis, and the cycle is continuing.

This unusual burst of interrelated advancement in both chemistry and mathematics was, to a large extent, triggered by the synthesis and stereochemistry of one molecule: the 3-rung Möbius ladder where the rungs are carbon-carbon double bonds and the "uprights" are polyethers, the key structural feature being the tetrahydroxymethylethylene (THYME) unit.[4] Herein are described the details of the synthesis and properties of this topologically stimulating molecule and its cylindrical isomer. In addition, a chemical realization on the molecular scale of the classic topological trick of cutting a Möbius strip "in half", only to have it remain in one piece, is presented. The latter process serves as an elegant corroboration of the Möbius ladder structure. Finally, a discussion of new topology inspired by the synthesis is given, with some comments on future directions in topological stereochemistry.

Synthesis of the 3-rung THYME Prism and Möbius Ladder

THE THYME POLYETHER SYNTHETIC STRATEGY

Around 1960 Wasserman and van Gulick independently proposed Möbius ladders (they referred to the structures as Möbius bands or Möbius strips) as targets of organic synthesis. Wasserman's ideas were published in the now famous "Chemical Topology" papers,[5] which soon thereafter stimulated the birth of the important field of DNA topological chemistry.[6,7] In the first of that two paper series, published in 1961, Wasserman referred to "an extensive discussion of the possibilities of Möbius strips and braids in chemical systems" appearing in a preprint by van Gulick entitled "Theoretical Aspects of the Linked Ring Problem." Van Gulick's paper was never published, but his ideas received additional visibility when Schill included two of the Figures from that paper in his widely read 1971 monograph "Catenanes, Rotaxanes, and Knots."[8]

While both Wasserman and van Gulick considered the Möbius ladders chiefly as intermediates in the synthesis of catenanes and knots,[9] they both presented well conceived approaches to the synthesis of Möbius ladders. As shown in Figure 1, Wasserman envisioned "...conversion of a roughly rectangular surface into a Möbius strip by a half-twist about the long axis before the ends are joined to form a ring. Subsequent cleavage of the cross-links in such a two-stranded strip would lead to a large single ring...."[5b] Wasserman's key ladder-shaped intermediate I possesses a number of "rungs" between "uprights," where the ends of the uprights are functionalized to allow for the required connection processes. The uprights are identical and oriented antiparallel.

Connection of A to A and B to B results in only products with an odd number of half-twists (Möbius ladders), the simplest possessing one half twist. As shown in Figure 1, Wasserman's original depiction was of the 3-rung Möbius ladder with one half twist (II). Finally,

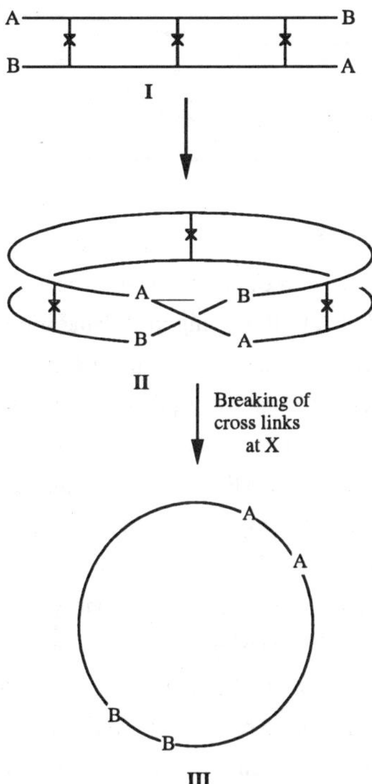

Figure 1. — Wasserman's three rung Möbius strip.

"breaking of cross links at X" results in a single macrocycle **III**. In the text, Wasserman also discusses the fact that a similar synthesis proceeding via a structure with 2 half twists would provide a catenane upon breaking of the cross links. In his 1962 paper Wasserman states "...separation of a two-stranded strip that has three half-twists will yield a trefoil," referring to the simplest nontrivial knot.

Van Gulick implicitly addresses a desire for additional regiocontrol in the bis-macrocyclization. Thus, in his "simplified" approach, which is even further simplified in the Figures used by Schill such that the original idea was not shown, a two rung ladder with four different functional groups at the ends of the uprights is proposed (Fig. 2). Such a ladder could lead selectively to products with an even or odd number of half twists depending upon the type of connection process.[10] In the Figures reproduced by Schill, twisted strips with zero to six half twists, and the products obtained upon breaking of the rungs, are shown.

While both of these approaches provide for entry into the Möbius ladder topology to the exclusion of prisms, they also both require that two different cyclization processes be utilized to

Figure 2. — van Gulick's two rung ladder intermediate.

achieve this regiocontrol. Thus, in Wasserman's case, the reaction connecting A to A cannot be the same as the reaction connecting B to B, otherwise an altogether different topological outcome would compete (i.e. connection of A to B, leading to a planar tricyclic). Van Gulick's implied scenario is similar. Finally, in either case, even with the high degree of regiocontrol afforded by these approaches, the topological stereocontrol required to obtain only products with greater than one half-twist, leading ultimately to knots or links, is still missing.

Our approach to the synthesis of a Möbius ladder is similar to that of Wasserman, but with the uprights oriented parallel instead of antiparallel. The functional groups affording the connection process are hydroxyl and tosylate groupings, and the synthesis gains added esthetic appeal since after the connection, the resulting functional array is identical to the rest of the "edge" of the ladder, affording a product with pleasingly high symmetry.

Thus, the strategy utilizes polyether uprights and double bond rungs, making the key structural features crown ether rings fused by the THYME unit[11] as illustrated by the 3-rung THYME ladder 1 (Fig. 3). This type of structure was dictated by our initial thinking on the problem, which was driven by several factors.

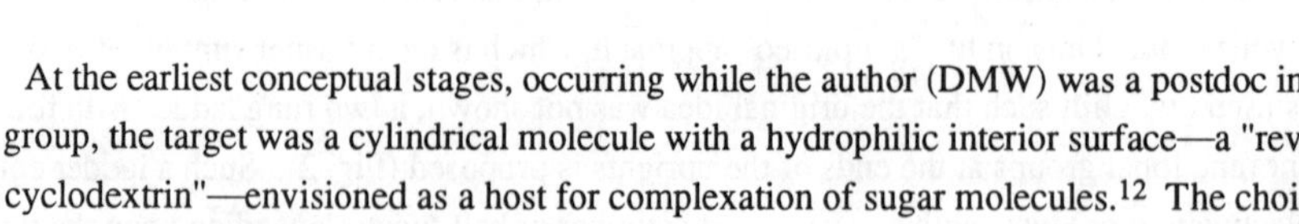

1

Figure 3. — The THYME diol-ditosylate Möbius ladder precursor.

At the earliest conceptual stages, occurring while the author (DMW) was a postdoc in the Cram group, the target was a cylindrical molecule with a hydrophilic interior surface—a "reversed-phase cyclodextrin"—envisioned as a host for complexation of sugar molecules.[12] The choice of the polyether framework and Williamson etherification macrocyclization reactions were natural in this context.

The THYME ring fusion seemed highly advantageous due to the lack of tetrahedral stereogenic centers in THYME, which if present could cause additional stereochemical problems in the synthesis. The choice of a key intermediate where the cyclization is completely intramolecular (as opposed to a tetra-ol ladder, for example, reacting with linear ditosylates) was made quite consciously in order to avoid oligomerization and additional regiocontrol problems. Finally, the choice of identical connecting reactions was made for the sake of synthetic simplicity, at the cost of some regiocontrol in the process.

Of course, the "undesired" regioisomer in the synthesis of the untwisted prism is the one half twist Möbius ladder, a fact recognized by the author immediately. Experimental work on the project in our laboratories was driven primarily by the prospects of realizing the Möbius strip synthesis envisioned by Wasserman and van Gulick.

We, too, were considering the Möbius ladders and prisms as intermediates for the synthesis of knots and links, following the tradition in the literature. From this perspective the one half twist Möbius ladder perhaps seems unimportant except as the first step towards more highly twisted products, leading as it does to the trivial unknot after breaking the rungs.

This attitude was due to ignorance of the truly novel topology of the one half-twist Möbius ladder itself, which was not discovered until after the synthesis was complete. In fact the one half-twist 3-rung Möbius ladder is perhaps the most topologically stimulating molecular structure synthesized to date, and is quite an interesting target in its own right. As an approach for synthesis of this target, the route involving structure 1 as key intermediate is clearly seen as an excellent compromise between synthetic simplicity and topological stereocontrol (see discussion below).

SYNTHESIS OF THE KEY INTERMEDIATE 3-RUNG LADDER 1.

The highly convergent and efficient synthesis of diol-ditosylate 1 relies on the use of tetrahydropyranyl (THP) groupings and the 3,4-bisalkoxymethylfuran grouping as orthogonal alcohol protecting groups. The key to this strategy is the use of the functionalized furan as a protected THYME diol-diether unit based upon the work of Magnusson,[13] and suggested to us by the elegant application of a similar approach (use of the furan ring as a protected maleic dialdehyde) for synthesis of an annulene analog of naphthalene by Sondheimer and Cresp.[14] The key step in the synthesis involves unmasking of the THYME diol functional array in the presence of 1° tosylates, affording key intermediate 1.

The chemistry of the THYME system has been discussed in considerable experimental detail in prior publications, specifically those outlining the synthesis and some complexation properties of the 2-rung THYME cylinder,[11] and describing the synthesis of the 4-rung THYME Prism and Möbius ladder.[15] In this report we start with the 22-crown-6 derivative furan-ditosylate 3 (available in a <u>total</u> of six chemical steps, five convergent, from the furan diTHP ether 2 in 39% overall yield) and the acyclic diol-ditosylate 4 (also used in the synthesis of 3), as shown in Figure 4.

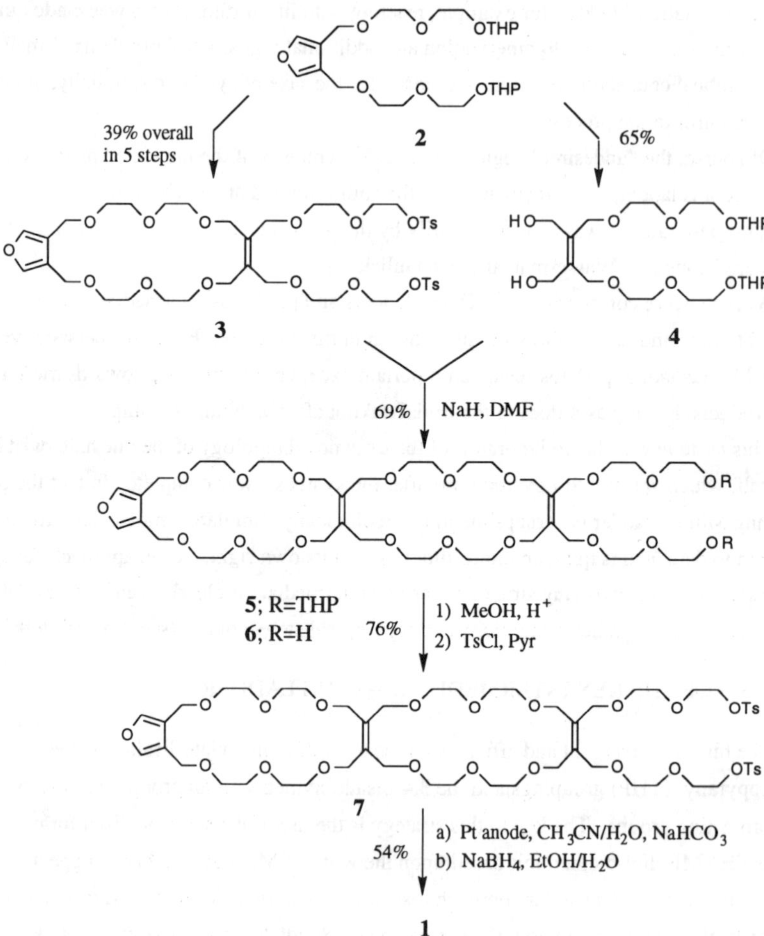

Figure 4. — Synthesis of THYME diol-ditosylate **1**.

Coupling of the ditosylate **3** with diol **4** in a 22-crown-6 forming reaction was accomplished by the standard procedure. In this system the crown formation is excellent, allowing isolation of a 69% yield of the desired 3-rung precursor **5**. The key process in the THYME synthesis, whereby the functional group array required for the cyclization is created, is illustrated in the conversion of furan-diTHP **5** to the target diol-ditosylate **1**.

Methanolysis of the THP groupings gives the crystalline furan-diol **6** in 79% yield. Tosylation of this diol under standard conditions then affords a 96% yield of the furan-ditosylate **7**. Conversion of the furan ring to the THYME diol unit involves anodic oxidation of the furan ring to a bis-hemiacetal, followed directly by borohydride reduction to the target THYME diol-ditosylate 1 without isolation of the intermediate bis-hemiacetal. Careful control of experimental variables[16] allows reproducible transformation of furan-ditosylate 7 into diol-ditosylate 1 in greater than 50% isolated yield after flash chromatography.

BIS-MACROCYCLIZATION OF DIOL-DITOSYLATE 1.

In general, crown ethers are created by base-promoted cyclization of a ditosylate or dihalide with a diol. This is also the strategy, applied twice in a bis-macrocyclization, utilized by Sauvage in his synthesis of the molecular trefoil knot.[17] In such cases, since the initial step is intermolecular, ultra-high dilution cannot be utilized to minimize undesired oligomer formation.

Cyclization of diol-ditosylate 1, however, is an intramolecular process, and ultra-high dilution affords a product mixture devoid of higher oligomers, as indicated in Figure 5. Experimentally, the required dilution is easily achieved by addition of a DMF solution of diol-ditosylate 1 (3.3 mM) to a suspension of NaH in DMF ("17 mM") over a period of 14 hours utilizing a syringe pump. The final concentration of polyether in the reaction mixture is 0.4 mM, but under these conditions it is assumed that the cyclization process occurs rapidly relative to the addition rate, and the actual concentration of diol-ditosylate (or "hemi-cyclized" product) is considerably less than this at any given time.

After a period of additional stirring (2-6 hrs), the reaction is quenched by addition of a small amount of water, and the solvent is removed utilizing a rotary evaporator fitted with a dry-ice condenser and attached to a vacuum pump. The solid residue is then partitioned between water and dichloromethane, and crude product isolated from the organic extract by rotary evaporation of the solvent.

Analysis of the resulting white waxy solid by TLC shows two components using either silica gel (THF eluent, R_f=0.60, 0.40) or alumina (THF eluent, R_f=0.65, 0.35). Preparative separation of these compounds by flash chromatography on either silica gel or alumina is easily achieved. The order of elution of the two components is the same on both adsorbents: Chromatography gives a high R_f clear, colorless viscous oil; the racemic Möbius ladder 10, and a low R_f crystalline solid; prism 11. This procedure has been repeated many times, and in general a slightly greater amount of prism 11 is isolated (average isolated yields of 22% high R_f material and 24% low R_f material, though in the best cases, 70% combined yields of the two materials have been obtained).

In order to examine the product ratio in the cyclization reaction more closely, a simple method for analysis of the crude reaction mixture was sought. As discussed below, ^{13}C NMR provides such a method. The olefinic carbon resonances for the two products are completely resolved at 22.5 MHz (FX 90Q), and at 62.9 MHz (WM 250) ($\Delta\delta$=13.5 Hz and 34.9 Hz,

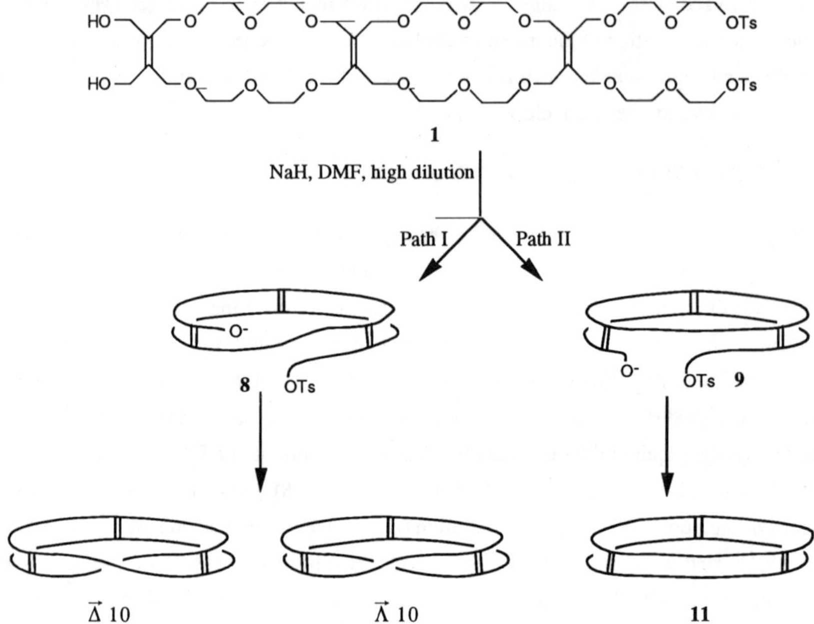

Figure 5. — Cyclization of diol-ditosylate **1**.

respectively). The carbon NMR spectrum of a mixture of the two polyethers easily allows integration of the olefinic carbon peaks, and the product ratio obtained in this manner is independent of differential T_1 relaxation effects (by using a long delay time between pulses, and observing that the ratio of integrated intensity is independent of delay time). When analyzed in this way (at both 25.5 MHz and 62.9 MHz), crude product mixtures obtained from three separate cyclization reactions run as described above gave observed product ratios of 41:59 ($\pm$ 3%), with the prism **11** predominating.

The formation of two polyether cycles in approximately equal amounts from cyclization of diol-ditosylate **1** is expected. As shown in Figure 5, it is assumed that the cyclization proceeds in stages, with diol-ditosylate **1** initially reacting to give one of two possible "hemi-cyclized" products **8** and **9** (in Figure 5, and throughout this paper, the structures of the cyclized THYME polyethers are captured by simplified graphs where the vertices represent olefinic carbons, the "double bonds" represent carbon-carbon double bonds, and simple arcs represent -$CH_2OCH_2CH_2OCH_2CH_2OCH_2$- units).

These products are constitutional isomers, and are of course formed via constitutionally isomeric transition states. Examination of space filling models, however, suggests that the effective activation energy for formation of alkoxy-tosylates **8** and **9** is very similar. Initial formation of an alkoxide (or bis-alkoxide) from diol **1** is followed by a 31-membered ring formation via path I to give "hemi-cycle" **8**, or by a 30-membered ring formation via path II to give "hemi-cycle" **9**. Both of these products are achiral, and both should be quite flexible (i.e. exhibit a high density of populated conformational states and similar entropies).

Once the initial cyclization event takes place, the topology of the final products is set. Initial cyclization by path I leads in principle to a family of products possessing an odd number of half twists, while path II leads to a family of products possessing zero or an even number of half twists. The members of the "zero or even number of half twists" (prism) family are all homeomorphic, as are the members of the "odd number of half twists" (Möbius ladder) family. Of course, every member of the family of prisms is non-homeomorphic to every member of the Möbius ladder family.

Models strongly suggest that for this system products with any number of half twists greater than one would be impossibly strained. While the rates of cyclization of alkoxy tosylates **8** and **9** are not expected to be the same, compound **8** has no choice but to afford the racemic Möbius ladder **10** upon cyclization, while compound **9** can only cyclize to give the tris-THYME prism **11**. Since compounds **8** and **9** should be formed in equal amounts, the production of two fully cyclized polyether products in equal amounts is expected, even though the twisted product **10** is expected to possess more strain and perhaps less entropy then the untwisted isomer **11**.

The formation of a larger amount of prism **11** relative to Möbius ladder **10** could be the result of formation of more hemicycle **9** relative to **8**. We feel this is unlikely, however, and interpret the product ratio as follows. Though difficult to isolate or characterize, byproducts must be formed in this reaction, since the combined isolated yield of the two polyether products is always much less than quantitative.

It is quite reasonable that some byproducts derive from intramolecular elimination competing with the desired nucleophilic substitution. It is thus attractive and reasonable to suggest that differing activation enthalpies and entropies for the substitution reactions **8** $\longrightarrow$ **10** and **9** $\longrightarrow$ **11**, and the corresponding elimination processes, are responsible for the observed product ratios. Specifically, one may propose that the activation entropy and/or enthalpy for cyclization of hemicycle **8** to Möbius ladder **10** is large relative to the intramolecular elimination pathway for hemicycle **8**, or the cyclization and elimination pathways for hemicycle **9**, leading to more elimination-derived byproducts from **8** than from **9**, and less product **10** than **11**.

Alternatively, it is possible that significant amounts of byproducts in the cyclizations are formed by intermolecular reaction with alkoxide or fortuitous hydroxide or dimethylamide in the reaction mixture. If byproducts are indeed produced by intermolecular reactions with unwanted nucleophiles/bases, then any retardation in rate of cyclization of **8** relative to **9** would allow excess byproduct formation from **8** relative to **9**, and ultimately afford less of the twisted product **10** than untwisted **11**.

Prism **11** is clearly achiral—the conformation suggested by the drawing in Figure 5 belongs to the D_{3h} point group. The three-rung Möbius ladder, on the other hand, possesses a topologically chiral molecular graph when the rungs are "colored" (see below).[18] Thus any molecular realization of this graph (e.g. structure **10**) must be "chemically chiral" (i.e. resolvable in principle) under conditions preserving constitution. Therefore, cyclization of the achiral diol-ditosylate **1** in the presence of only achiral reagents affords a racemic mixture of Möbius ladders **10** ($\vec{\Lambda}$) and **10** ($\vec{\Delta}$). Details of the proof of structure for the three-rung THYME polyethers is given below.

Physical Properties of the 3-Rung THYME Polyethers

MASS SPECTRAL AND GEL PERMEATION CHROMATOGRAPHIC ANALYSIS

While good arguments against the formation of higher oligomers in the cyclization of diol-ditosylate **1** are presented above, it is clearly important to rule out this possibility. While the two-rung THYME "prism" (actually a tetrahedral graph) homologue of compound **11** shows an abundant molecular ion under standard electron impact conditions, no molecular ions are detected for the products of cyclization of diol ditosylate **1** by electron impact mass spectrometry.

Using a standard chemical ionization technique, however, satisfactory mass spectral data for the products of the cyclization are obtained. Thus, in separate experiments, when equal concentrations of the high R_f oil and low R_f solid (2 mg/50 ml) in 1:1 acetonitrile/water are injected into the CI ion source of a mass spectrometer, similar, but not identical, mass spectra result. In each case, high ion currents are obtained, with the high R_f oil showing a protonated molecular ion at m/z=865 with relative intensity of 33% of the base peak; the latter being an ion at m/z=391; the low R_f solid shows a protonated molecular ion at m/z=865 which in this case <u>is</u> the base peak, and an abundant ion at m/z=391 (48%).

Additional evidence that both products of cyclization of diol-ditosylate **1** are monomeric derives from gel permeation chromatography. Both products of the cyclization exhibit identical retention times on Waters 100Å and 500Å μ-styragel columns (toluene eluent), and are retained longer than the bicyclic furan-ditosylate **7**. Both products exhibit well-resolved peaks (R_s=1.0) when co-injected with compound **7** on the 500Å column.

These results are in good agreement with expectation, given the larger molecular weight of compound **7**, and its expected considerably larger hydrodynamic radius relative to the tetracyclic compounds **10** and **11**. Analysis of the crude cyclization reaction mixture by gel permeation showed no products with retention times shorter than that of furan-ditosylate **7**.

Thus, three observations argue that both products of the cyclization of **1** are monomeric: The product ratio formed under ultra-high dilution conditions; mass spectra; and gel permeation chromatographic retention times. While none of these is absolutely unequivocal, the combined weight of all provides strong evidence for the molecular weights of the observed products.

MAGNETIC RESONANCE AND TOPOLOGICAL STEREOCHEMISTRY OF THE 3-RUNG THYME POLYETHERS

In achiral solvent

The ^{1}H and ^{13}C NMR spectra of polyethers **10** and **11** are beautiful in their simplicity yet contain interesting information. As reported in communication form earlier,[4] both Möbius ladder **10** and prism **11** show qualitatively similar spectra in chloroform-d solution: An AB quartet for the allylic protons and a broad singlet for the ethyleneoxy protons in the ^{1}H spectrum, and four singlets in the proton-decoupled ^{13}C spectrum. Interestingly, these observations are due to different effects in the two products. The AB pattern for prism **11** is due to molecular rigidity making the "inside-out" deformation of the "cylinder" slow on the NMR time scale at any accessible temperature, while the appearance of only a single AB pattern, and of only four carbon resonances in the spectra of Möbius ladder **10** is due to a rapid molecular motion which cannot be "frozen."

Thus, for the 3-rung THYME prism **11**, the expected average D$_{3h}$ symmetry in solution results in the four carbon resonances observed in the ^{13}C spectrum. This is the minimum number possible not allowing fortuitous isochrony since there are four heterotopic carbons in the structure, as indicated in Figure 6.

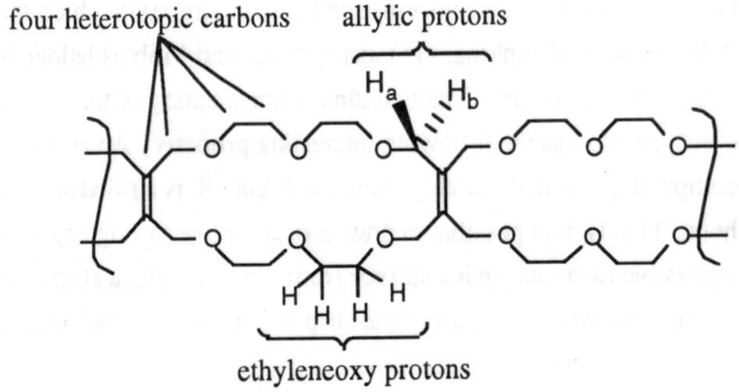

Figure 6. — Unique protons and carbons present in the THYME polyethers.

The AB quartet observed for the allylic protons (H_a and H_b) in the ^1H spectrum of THYME prism **11** results from molecular rigidity. Thus, the "inside out" motion which permutes allylic protons H_a to H_b as illustrated in Figure 7 must occur slowly on the NMR time scale. This is in contrast to the behavior of the 4-rung THYME prism, which exhibits a singlet for the allylic protons at room temperature, but affords an AB pattern similar to that shown for compound **11** when cooled to 213°K.[15] The data on the four-rung homologue suggest an effective barrier for the inside-out motion of about 12 Kcal/mol.

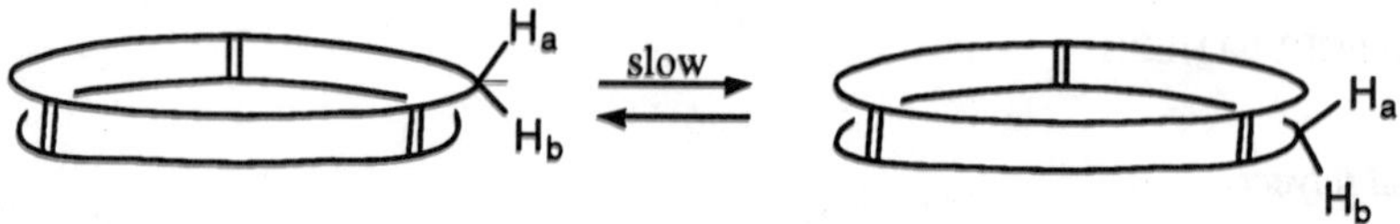

Figure 7. — Inside-out motion of a THYME prism.

A solution of three-rung prism **11** in DMSO d_6 was examined by ^1H NMR at 250 MHz at elevated temperatures to study the inside-out motion in this system. In fact, no change in the spectrum occurred up to 443°K, indicating an effective barrier to the inside-out motion of greater than 23 Kcal/mol. This is certainly not unreasonable based upon examination of CPK molecular models of prism **11**.

A Möbius strip, of course, cannot be turned inside-out since it possesses only one side—a property shared by the molecular Möbius ladder **10**. The observed ^1H and ^{13}C NMR spectra of racemic **10**, however, are consistent only with a dynamic structure. First consider the ^{13}C spectrum, showing only four peaks—again equal to the number of heterotopic carbons. Given the proposed structure, which possesses a topologically chiral graph, this observation proves the molecule possesses only one set of six homotopic olefinic carbon atoms, excluding fortuitous isochrony.

Is it possible to achieve a single conformation for structure **10** in which the olefinic carbons are all homotopic? In one way of thinking, the most symmetrical Möbius ladder is that wherein the twist is equally spread about the ladder. The two dimensional analog of this structure is termed an equilateral Möbius strip,[19] and has the following interesting property. An equilateral Möbius strip, illustrated by the computer-generated drawing shown in Figure 8, is a ruled surface with negative curvature everywhere. That is, it is possible to draw a straight line through each point on the surface, but it is impossible to create such a surface from, for example, a strip of paper. This is why paper Möbius strips always have a part of the strip which is "more twisted," and part which is "less twisted."

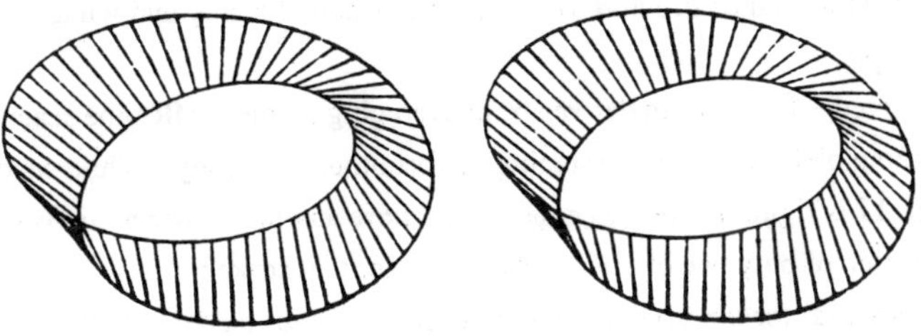

Figure 8. — Computer generated stereopair of an equilateral Möbius strip.

It is quite conceivable that compound **10** could achieve an "equilateral Möbius ladder" conformation, where the twist density about the ladder was the same everywhere. Such a conformation, however, still does not afford six homotopic olefinic carbons. Thus, even for the equilateral Möbius ladder there is one unique point on the axis of the ladder (the midline between the edges) where the edges pass through the plane containing the axis. This point might be called the "locus of twist." Clearly, there is no way to distribute the double bonds on the ladder in such a way that all six vertices are equivalent. The best that can be achieved for this type of conformation is illustrated in Figure 9. The equilateral Möbius ladder conformation (A) (the double bonds are represented here by bold lines), possesses three diastereotopic sets of homotopic pairs of olefinic carbons; black, gray and white.

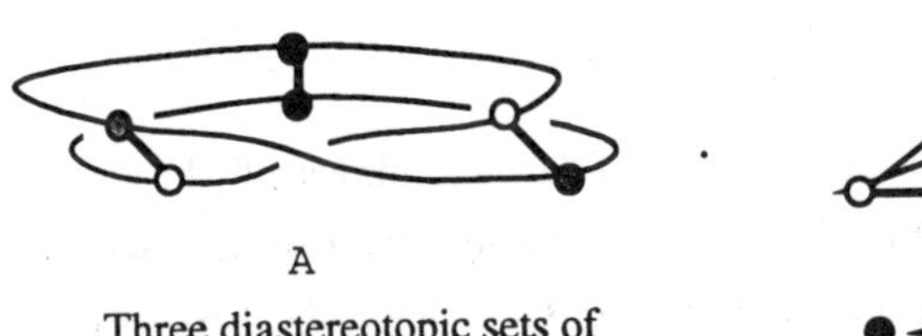

A

Three diastereotopic sets of
homotopic pairs of olefinic carbons

B

Two diastereotopic sets of
homotopic pairs of olefinic carbons

Figure 9. — Topicity of the 3-valent vertices of two symmetrical Möbius ladder presentations.

Interestingly, Simon has proven that no single rigid conformation of Möbius ladder **10**, even given infinite flexibility (the topological limit), possesses six homotopic olefinic carbons.[18] Thus, for a single conformation at least two sets of diastereotopic vinyl carbon atoms must be present. One conformation with this minimum number (albeit impossible from the Euclidean chemical standpoint) is indicated as structure B of Figure 9, where the gray and white carbons of structure A are homotopic.

Thus, the observed ^{13}C spectrum derives from averaging on the NMR time scale of conformations permuting olefinic carbons assuming no fortuitous isochrony. While we do not suggest that the prism **11** is rigid, it is not <u>necessary</u> to propose any molecular motions in order to rationalize the observation of only one olefinic carbon peak for that structure.

The obvious proposal for the structure of the dynamic Möbius ladder is illustrated in Figure 10. In order to rationalize the observed spectrum, it is proposed that a motion of the locus of twist about the ladder's axis by 60° is occurring rapidly on the NMR time scale. This interconverts conformations of type A and B of Figure 10. Note in conformation B one of the double bonds passes through the locus of twist. In this motion, all six carbons become homotopic on the time average.

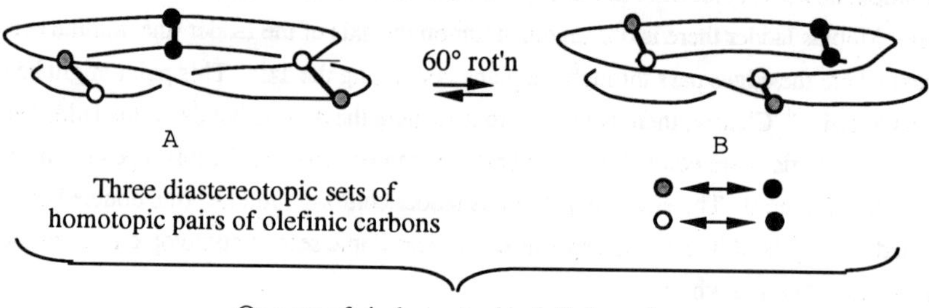

Figure 10. — The dynamic Möbius ladder.

This argument is not meant to imply that the edges of the molecule actually lie on an equilateral Möbius strip, but rather to illustrate the time average symmetry of a large number of conformations. Also, it should be noted that the spectrum does not rigorously prove a dynamic structure since fortuitous isochrony cannot be ruled out. Nevertheless, the dynamic structure is strongly suggested. In addition, the data suggest that the motions permuting carbons has a small activation energy, since even when cooled to -80°C in CD_2Cl_2, the olefinic resonance stays a sharp singlet, though some broadening of the other carbon resonances does occur.

The dynamic nature of the molecular Möbius ladder is especially interesting given the Fermion nature of the classical dynamic Möbius strip.[20] Thus, the dynamic Möbius ladder is a classical analog of a spin-1/2 system, since motion of the locus of twist about the axis by 360° gives rise to a physically identical state where the vertices are 180° "out of phase," as illustrated in Figure 11. This fact could in principle actually have chemical consequences relating to the quantum mechanical Berry's phase phenomenon.[21]

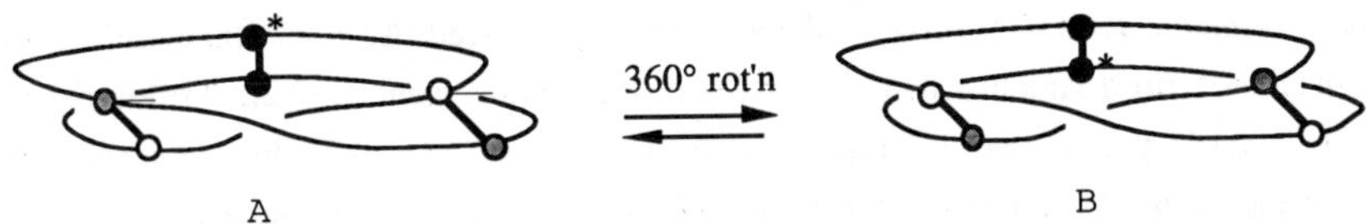

A   360° rot'n   B

Figure 11. — Illustration of the fact that a 360° motion of the locus of twist about the axis of the Möbius ladder results in a physically identical state which is out of phase with the starting state.

In contrast to the homotopic nature of the six olefinic carbons of compound **10**, where the topological symmetry of the system is realized in the actual chemical structure, the AB pattern observed in the proton spectrum has purely Euclidean origins. Thus, pairs of protons (one-valent vertices) connected to the same carbon must be topologically homotopic for any structure. In the actual molecular graph the allylic protons (see Figure 12) are pro-stereogenic[22] due to the Euclidean invariance of the tetrahedral bond angles. Thus, the two attached protons must be either enantiotopic or diastereotopic. This is true even though the molecule has only one edge, and the Möbius strip defined by the molecular graph has only one side. In the actual molecule it is impossible to realize the topological symmetry of the methylene groups.

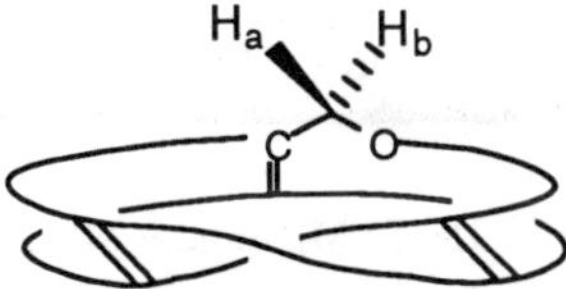

Figure 12. — Diastereotopic allylic protons on the edge of the Möbius ladder.

Since the structure is topologically chiral, the protons in question cannot be enantiotopic, and must therefore be diastereotopic. This, in combination with the conformational motion of the locus of twist, proves that a single AB pattern (assuming no fortuitous isochrony) is expected for the twelve allylic methylene groups of structure **10**. All the peaks in the spectrum broadened at low temperatures, but there was no evidence of any splitting of the single AB pattern, again suggesting a low activation energy for the twist motion.

In chiral solvent

As mentioned above, and discussed in detail below, structure **10** is topologically chiral and thus must be chemically chiral. As indicated in Figure 5, cyclization of the 3-rung diol-ditosylate **1** affords Möbius ladder **10** as a racemic mixture of two topologically enantiomeric structures. It was desirable to observe a physical manifestation of this chirality as a structure check. This was relatively easily achieved by NMR spectrometry in a chiral solvent. Specifically, ^{13}C NMR analysis of racemic **10** at 62.9 MHz exhibited two nicely resolved ($\Delta\upsilon$ = 1.7 Hz) olefinic carbon peaks in chloroform-d saturated with the Pirkle chiral solvating agent (S)-2,2,2-Trifluoro-1-(9-anthryl)ethanol (approximately 300 mg/ml), as shown in Figure 13. After the chiral solvating agent is removed by chromatography, the ^{13}C NMR spectrum of the racemate reverts to a single olefinic carbon signal.

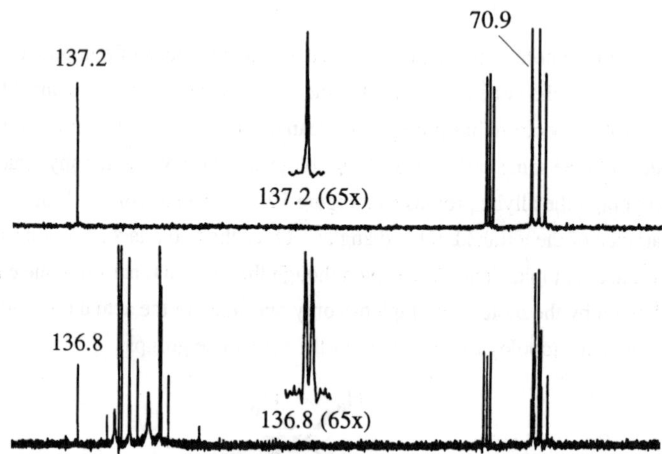

Figure 13. — ^{13}C NMR spectra (50-150 PPM), at 62.9 MHz, of 3-rung Möbius ladder **10**. The top spectrum was taken in CDCl$_3$ and the bottom spectrum was taken in CDCl$_3$ saturated with the Pirkle chiral solvating reagent. The enlarged peaks in the center of the spectra are the olefinic signals expaned about 65 times in the horizontal axis. The two peaks in the enlarged part of the bottom spectrum are from the two enantiomeric Möbius ladders, $\Delta\delta$ = 1.7 Hz.

As expected, analysis of the achiral Prism **11** by ^{13}C NMR using the same conditions (solvating agent, concentration of the polyether) exhibited only a single olefinic carbon resonance. Finally, the same examination of the olefinic carbon peak for a mixture of the two products in chloroform-d showed splitting for the Möbius ladder while the Prism remained a sharp singlet. These data show that the carbons producing the olefinic signal for the Möbius ladder come in two enantiotopic sets—consistent with the presence of two enantiomeric Möbius ladders. Of the data presented to this point, this splitting of enantiotopic carbons in the spectrum of **10** in chiral solvent is the only qualitative physical difference between the Möbius structure and the isomeric prism **11**.

CRYSTAL STRUCTURE OF THE 3-RUNG THYME PRISM 11

Based upon the analysis described above, the structures of both the Möbius ladder **10** and prism **11** are established. The assignment is confirmed by single crystal X-ray analysis of the crystalline isomer **11**. The crystal showed disorder, with two similar conformations of one of the crown rings present. One of these is illustrated by the orthogonal projections shown in Figure 14.

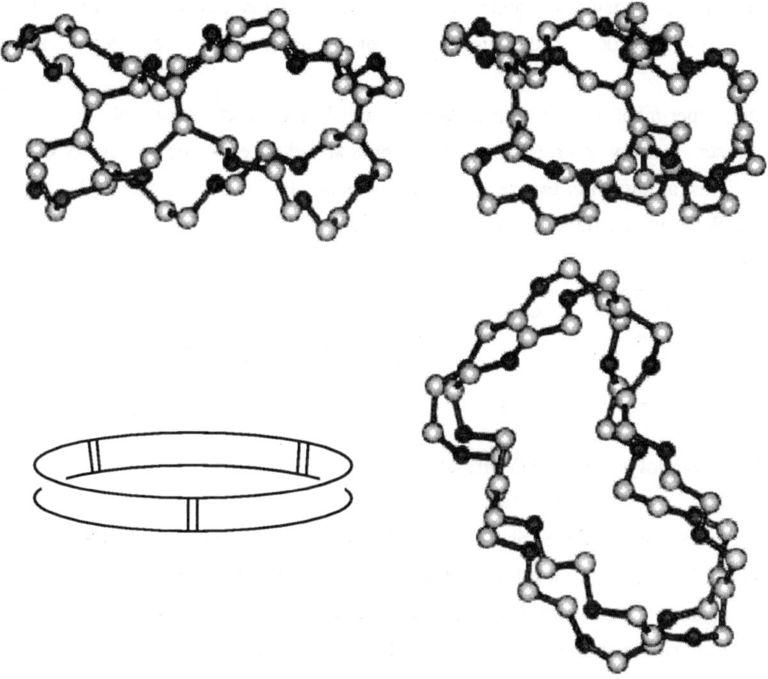

Figure 14. — Orthogonal views of one of the conformations occurring in the crystal of prism **11**.

Breaking the Rungs of the Möbius Ladder

The fact that when cut in half, a Möbius strip remains in one piece is perhaps it's most well known topological property. From the beginning of this project it was envisioned that a chemical equivalent of the famous parlor trick, as discussed by Wasserman in the first discussion of molecular Möbius strips in the literature, would serve as an interesting and elegant corroboration of the structure. Thus, chemically breaking the double bond rungs of the THYME Möbius ladder only to have it remain in one piece seemed an especially attractive reaction. Indeed, anticipation of the relative ease of such a process was considered one of the major advantages of the THYME system.

As we have reported in preliminary form in several communications, ozonolysis of compounds **10** and **11** does indeed afford the expected polyketones, but with some problems.[2b,15,23] Specifically, the ketone products are unstable, presumably with respect to intramolecular aldol condensation, and could not be characterized to our satisfaction. The solution to this problem has been described in preliminary form,[23] and is given here in detail for the first time.

Initial experiments on the simple model system **12** indicated that ozonolysis of the THYME tetraethers under carefully controlled conditions could afford the expected bis-alkoxy ketones efficiently. Thus, as shown in Figure 15, ketone **13** is obtained from tetraether **12** in 90% isolated yield after application of a standard ozonolysis protocol.

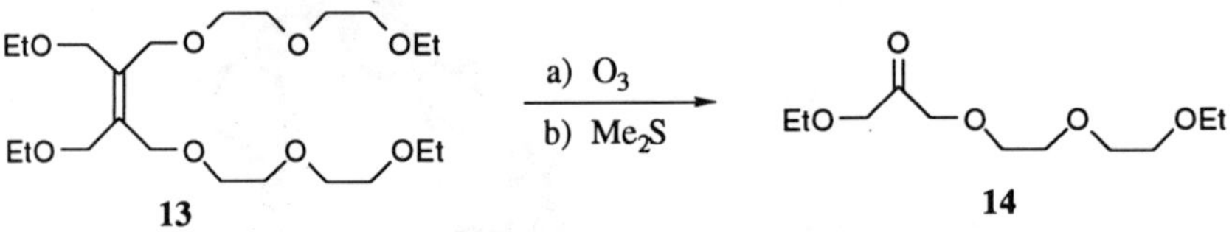

EtO OEt
EtO OEt
O O O O
13

a) O_3

b) Me_2S

EtO OEt
O O O
14

Figure 15. — Ozonolysis of a THYME tetraether model compound.

Similar ozonolysis of the tris-THYME Möbius ladder **10** and prism **11** also seemed straightforward. Treatment of a solution of Möbius ladder **10** and the indicator dye solvent red 23 in dichloromethane at -70°C with ozone generated by a commercial ozonator until the dye decolorized, followed by immediate quenching of the reaction by the addition of excess dimethyl sulfide (DMS), removal of solvent under reduced pressure and purification of the crude product on silica gel yielded the hexaketone **14**, as shown in Figure 16. Similar treatment of the prism **11** gave triketone **16**. Analysis of these materials by ^{1}H NMR at 90 MHz showed two sharp singlets, one for the protons alpha to carbonyl and the second for the ethyleneoxy protons. Carbon NMR showed four signals corresponding to the four heterotopic carbons, and the IR spectra were fully consistent with structures **14** and **16** as shown.

Analysis of the products by 500 Å ultrastyragel gel permeation chromatography showed that the product obtained upon cleavage of the Möbius ladder has a larger hydrodynamic radius than the product obtained from the prism. For triketone **16** the mass spectrum could be obtained by injecting a sample into the source of a chemical ionization mass spectrometer in acetonitrile/water solution (this solvent acts as the ionizing reagent), giving the expected protonated molecular ion. Several attempts to analyze hexaketone **14** under identical conditions gave no peaks

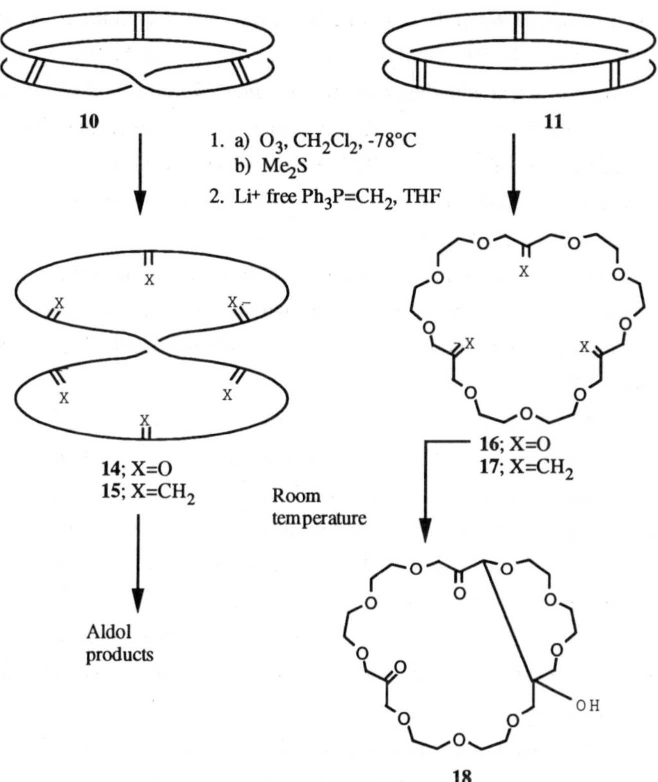

Figure 16. — Breaking the rungs of a Möbius ladder.

at all, apparently due to its low volatility. Good protonated molecular ions were easily obtained for both polyketones by fast atom bombardment (FAB) mass spectrometry using a glycerol matrix.

The first hint that problems might be encountered with this approach came upon examination of the ^{1}H NMR spectra of ketones **14** and **16** at higher field. It became apparent that the ethyleneoxy region of the proton spectrum for the ketones was always integrating high relative to the protons alpha to the carbonyl grouping (about 2.25:1 for the best samples).

The possibility that the integration error was due to a difference in relaxation times of the protons alpha to the carbonyl and the ethyleneoxy protons was ruled out by NMR experiments where the relaxation delay was increased between accumulations (up to 20 sec) with no change in the integration. The ketones **14** and **16** were purified by gel permeation chromatography to homogeneity using a 500 Å ultrastyragel column, and the ^{13}C NMR and ^{1}H NMR spectra showed no impurity peaks, yet the proton integrations were never acceptable.

Repeated chromatography of the triketone **16** on silica gel did not improve the integration either. In fact, a second spot by TLC analysis appeared. Analysis of the IR, ^{13}C and ^1H NMR of this impurity are most consistent with the product of an intramolecular aldol condensation (**18**), as shown in Figure 16. Even if the triketone is simply stored neat in an evacuated flask over a period of several days, the impurity appears in the TLC. Similar observations were made in the case of the hexaketone **14**. An integration ratio of 2.25:1 for the triketone (the best obtained) suggests that the aldol impurity exists as 12% of the mixture.

It is proposed that the intramolecular reactivity of these ketones is especially facile due to destabilization of the carbonyl groups by the electron-withdrawing alkoxyalkyl groupings, exacerbated by the intramolecular nature of the process (no intermolecular aldol products are observed for ketone **13**).

This inability to obtain correct proton integrations for either of the polyketones was deemed unacceptable, and thus a method was sought for blocking the carbonyl reactivity of the ozonized products to allow isolation and full characterization of stable materials.

An appealing aspect of the ozonolysis in this system was that the ^1H and ^{13}C NMR spectra of the products are particularly diagnostic of the structures due to their high symmetry. Maintaining this spectral simplicity was a key factor in the design of the carbonyl blocking group chosen. Thus, protecting groups which would complicate the spectra, or reductions leading to many stereoisomers, were avoided. In the end, Wittig methyleneation of the ketones was chosen as the optimum ketone protection scheme.

Initial experiments with model ketone **13** employing Ireland's[24] (BuLi, Ph3PCH3I in ether at -78°C) or Corey's[25] (NaH, Ph3PCH3I in DMSO at room temperature) procedures gave the corresponding olefin in high yield. However, application of Ireland's procedure to ketones **14** and **16** yielded no isolable product. The use of Corey's method under several different reaction conditions resulted in very poor yields of the corresponding olefins—2% and 8% for the hexamethyleneated and trimethyleneated products **15** and **17**, respectively.

The problem was nicely solved by methyleneation of the polyketones under "salt-free" conditions.[26] As indicated in Figure 16, treatment of crude hexaketone **14** with lithium free methylenetriphenylphosphorane in tetrahydrofuran (THF) gives the crown ether **15** in 67% isolated yield. Given that the reaction must occur six times without side reactions on a single molecule to obtain product, this is a remarkably efficient process, which works equally well on the triketone **16**, to give crown ether **17** (81% isolated yield).

Both of the polymethyleneated products are stable and can be obtained in very pure form and fully characterized. As desired for these polyene crown ethers, the ^1H and ^{13}C NMR spectra are nicely diagnostic of the structures. For each compound, the ^1H NMR shows singlets for the vinylic and allylic protons and multiplets for the ethyleneoxy protons with an integration ratio of 1:2:4 respectively, as shown in Figure 17. The ^{13}C spectra exhibited the expected five signals.

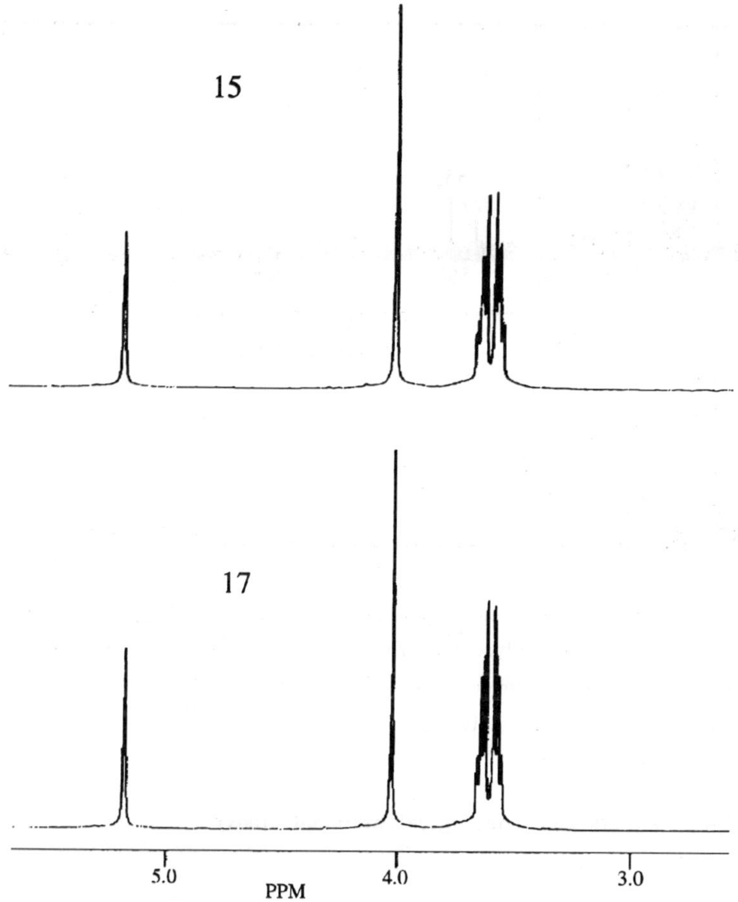

Figure 17. — Proton NMR spectra of the polymethylene crown ethers **15** (top) and **17**.

Analysis of the methyleneated products by gel permeation chromatography shows the product of the "cut" Möbius ladder to be larger than that from the prism. Finally, the FAB mass spectra of the cycles are fully consistent with the proposed structures, as indicated in Figure 18. Thus, as expected, cutting the prism **11** in half results in two pieces, while a similar cutting of the Möbius ladder **10** affords a single product of twice the size. This process completes at last a chemical realization of perhaps the most famous result in all of topology, as first suggested by Wasserman and shown in Figure 1.

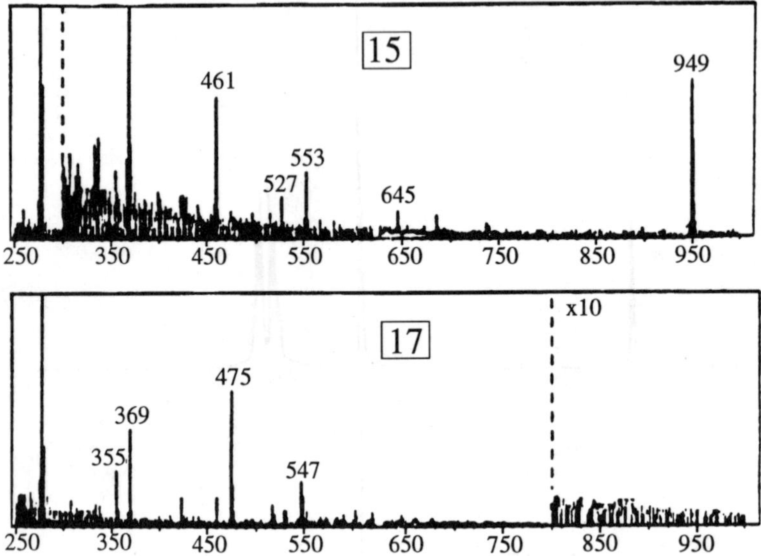

Figure 18. — FAB mass spectra of compounds **15** and **17**. The top spectrum is of compound **15** (m/z = 949 (M+1)$^+$) using a glycerol matrix with a trace of HCl. The bottom spectrum, taken under identical conditions, is from compound **17** (m/z = 475 (M+1)$^+$). Peaks at m/z = 645, 553, 461, and 369 are due to protonated glycerol oligomers.

The Möbius Ladders and Topological Stereochemistry

INTRODUCTION

This synthesis of the three rung THYME Möbius ladder served as a major stimulus for advances in modern topological stereochemistry. Topological stereochemistry, in turn, is an important part of the Möbius ladder story. A brief discussion of the topic is therefore presented here, though all of the key results described below have been published previously. This discussion is not meant to serve as a detailed review, however, and much interesting work touching on the subject is not presented. Apology is given in advance for these omissions.

THE INVENTION OF TOPOLOGICAL STEREOCHEMISTRY

Wasserman's Topological Isomers

In Wasserman's chemical topology papers he clearly considered "topological isomers" to be different from constitutional isomers and stereoisomers. Thus, while Wasserman considered the trefoil knot and the isomeric unknotted ring to be topological isomers, he also proposed the idea that linked rings and the unlinked pair may be considered as topological isomers where "...topological isomerism involves systems of molecules."

The idea that joined systems of molecules which are not covalently bonded, but joined by a "topological bond," should be considered as a single molecule, is certainly stimulating. Indeed, chemists are now very familiar with molecules wherein the constituents are not covalently bonded. No one would argue, for example, that Cram's cyclobutadiene carceplex[27] should not be considered as a single molecule even though the cyclobutadiene is not covalently bound to the carcerand.

Indeed, in this context Wasserman's discussion in a way presaged the advent of host-guest chemistry! Sokolov, in his 1973 paper "Topological Ideas in Stereochemistry,"[28] explicitly focused in detail on the topological nature of host-guest complexation ("molecules in a cage"), where "inside" and "outside" are distinguished, albeit by Euclidean constraints.

But the concept that a complex should be considered as an isomer (topological isomer) of the separated host and guest has not become a part of chemical thinking. The classical concept of chemical isomers as structures with the same number and kind of atoms is still extremely useful and universally accepted.

Topological Stereoisomers

Thus the link and separated rings are not isomers. The trefoil and unknot, however, are clearly stereoisomers, but of an unconventional type which had never been explicitly defined in the literature, though perhaps many chemists carried the idea in their heads.

After completion of the synthesis of the "first molecular Möbius strip" **10**, we began investigating what topological properties the molecular structure (the molecular graph, or the topologically similar "fattened" graph described by the van der Waals surface of the molecule) and a Möbius strip had in common. This led to the first statement of the now well known concept of topological stereoisomers as isomers possessing homeomorphic but non-homeotopic molecular graphs.[29] Conventional stereoisomerism, on the other hand, is a fundamentally Euclidean phenomenon; conventional stereoisomers possess homeomorphic and homeotopic molecular graphs.

Thus, while the link and unlinked rings are not topological stereoisomers, at the time of Wasserman's work the mathematical discipline of knot theory had long since proven that the trefoil (or any nontrivial knot) is non-homeotopic with the unknotted ring (this is the fundamental fact of knot theory).[30] When composed of atoms, these structures (trefoil knotted and unknotted circuit graphs[31]) are thus topological stereoisomers, and more specifically topological diastereomers, since they cannot achieve mirror image presentations. Also the mirror image trefoils were well known to be non-homeotopic; the molecular mirror image trefoils are thus topological enantiomers, each being topologically chiral.

Extrinsic Graph Theory

Regarding the question of what the Möbius ladder has in common with a Möbius strip, one important answer, immediately suggested by any chemist, is chirality. But the chirality of a Möbius strip is topological. That is, a Möbius strip is non-homeotopic with its mirror image. To our amazement and delight, at the time of the synthesis the topological chirality of the three rung Möbius ladder had never been proven, nor was it at all obvious to mathematicians used to the surprising possibilities introduced by infinite flexibility in 3 space!

Indeed, with regard to topological chirality, at the time of the synthesis of the three rung Möbius ladder the only mathematical results involving one dimensional objects embedded in 3 space dealt with knots and links. To our knowledge not a single mathematical treatment of the topological chirality or achirality of a graph more complex than a simple circuit or linked circuits (a straightforward extension of knot theory) had ever been published!

Topological chirality and achirality are extrinsic properties of one dimensional objects in 3 space, and the study of the topological chirality of graphs is a part of a discipline now termed extrinsic graph theory. Thus, while the study of the intrinsic properties of graphs was a thriving field by 1970,[31] very few extrinsic graph theoretical results had been developed by 1982. Kuratowski's Theorem regarding intrinsically nonplanar graphs[31] can be considered the first result of extrinsic graph theory since it deals with graphs embedded (or not embeddable) in a low dimensional space. Kinoshita and his students were also proving theorems dealing with nonplanarity of unknotted embeddings of the theta curve (a simple intrinsically planar graph) in 3 space.[32] But alas there was nothing regarding topological chirality.[33]

After completion of the synthesis of the three rung Möbius ladder we discovered, to our surprise, that a monochrome three rung Möbius ladder graph, the prototype of the class as first defined by Guy and Harary,[34] is in fact topologically achiral! The proof of this theorem consisted of finding an achiral presentation of the graph, which was published in 1983.[2a] This approach to proving topological achirality is similar to the exercise a chemist goes through in trying to establish resolvability (or lack thereof) of a new chemical structure, and was absolutely rigorous since a positive result had been obtained (that is, an achiral presentation had been found). To our knowledge, this represented the first result in extrinsic graph theory dealing with topological chirality.

Of course we, as chemists, were not equipped to prove the topological chirality of the actual molecular graph in question (structure 10), which involves edges of two colors (the THYME double bonds, and the polyether chains). Since a negative result (our inability to find an achiral presentation of the graph or any way of deforming the graph to its mirror image) is notoriously and rigorously nonrigorous in topology, the best we could do was conjecture that the colored three rung ladder was topologically chiral based upon our lack of success in proving it to be achiral.[2a] Note that the demonstrated chemical chirality of the material is not topologically compelling since many deformations of the graph are clearly impossible for the molecular structure (for example dragging a double bond and its appended chains through one of the crown ether rings, or rotating about double bonds).

Our work did stimulate Jonathan Simon, a topologist at the University of Iowa, to ultimately prove the topological chirality of the colored three rung ladder and monochrome ladders with any higher number of rungs in the standard embedding.[18,35] To our knowledge this was the first result of extrinsic graph theory proving the topological chirality of an object possessing no link and no knot, and helped popularize modern extrinsic graph theory.

The Most Chiral Molecules

The publication of the Möbius ladder molecule and our conjectures about the extrinsic topology of graphs also stimulated topologist Erica Flapan, who has since proven several very interesting theorems on chirality.[36] One such theorem, which serves to demonstrate the novelty of extrinsic graph theory relative to the topology known prior to the synthesis of compound 10, is highly relevant here.

In 1982 all known topologically chiral one dimensional objects were knots or links, since these were the structures being studied by knot theorists. All knots and links are homeomorphic to unknotted circles (nontrivial comment: this is the definition of knots and links). Thus, to put it in a somewhat fanciful yet elegant way, all known topologically chiral graphs could be moved out of 3 space, manipulated by continuous deformation, then projected back into 3 space as an achiral graph or graphs. Put in a more conventional way, all known topologically chiral graphs had achiral embeddings.

Stimulated by the synthesis of compound 10, Flapan proved that in fact this structure is topologically chiral in every embedding![36a,c,d,e] That is, every projection of 10 into 3 space is topologically chiral. We (ourselves and Flapan) call this property intrinsic chirality, by analogy with the well known intrinsic nonplanarity of the Kuratowski graphs, which cannot be projected into three space in any topologically planar embedding.

Thus, the THYME polyether 10 is the prototype of a very small class of "topologically most chiral" organic molecules possessing intrinsically chiral graphs.[37] Note that the topologically chiral trefoil knot and the oriented link[38] molecular graphs of Sauvage are clearly intrinsically achiral. It is also interesting to note that the four rung THYME Möbius ladder, whose synthesis was described by us in 1987,[15] is also intrinsically achiral, though it is topologically chiral in the embedding synthesized.[36] The existence of intrinsic chirality, together with topological rubber gloves,[2,36] led to the topological hierarchy of molecular chirality as published in 1991.[39] While we feel it unlikely that any unique physical manifestations of intrinsic chirality will be discovered, extrinsic graph theory certainly has important chemical consequences,[39] not the least important among them being the suggestion of new targets for synthesis!

TOPOLOGICAL STEREOCONTROL

Classic Synthetic Targets of Topological Stereochemistry

In the beginnings of topological stereochemistry, well known topology served to suggest novel targets for chemical synthesis. Thus, in the chemical topology papers[5] Wasserman provided a list of what might be called the classic targets of topological stereochemical synthesis—a list including several of the most famous knots and links, and of course the Möbius strips. As shown in Figure 19, these are the linked rings (**IV**), the trefoil knot (**V**), the figure-of-eight knot (**VI**), the link with a minimum of four crossings (**VII**), three rings linked in a chain (**VIII**), the achiral Borromean rings (**IX**), where all three rings are linked, but no two are linked, and the "Möbius strips" such as (**X**) and (**XI**), with two and three "edges," respectively (in addition Wasserman discussed two chiral isomers of three linked rings).

In the early work on topologically-inspired synthesis the Möbius strips were envisioned primarily as intermediates leading to knots and links. Thus, the Möbius ladder with two uprights and three half-twists leads to the trefoil, while the four half twist diastereomer (a prism) leads to the link **VII** upon cleavage of the rungs. However, no number of twists of a ladder with two uprights can afford the figure-of-eight knot (**VI**), chain (**VIII**), or Borromean rings (**IX**).

The Möbius ladder with three uprights (**XI**) provides for a chemical equivalent of cutting a Möbius strip into "thirds." The structure **XI**, with one half twist, gives two linked rings, one with twice the circumference of the other, upon cleavage of the rungs. And as van Gulick pointed out in his manuscript, many interesting structures, including the figure-of-eight knot (**VI**), chain (**VIII**) and Borromean rings (**IX**), can result from cleavage of the rungs of isomers of **XI** possessing the proper "braiding."

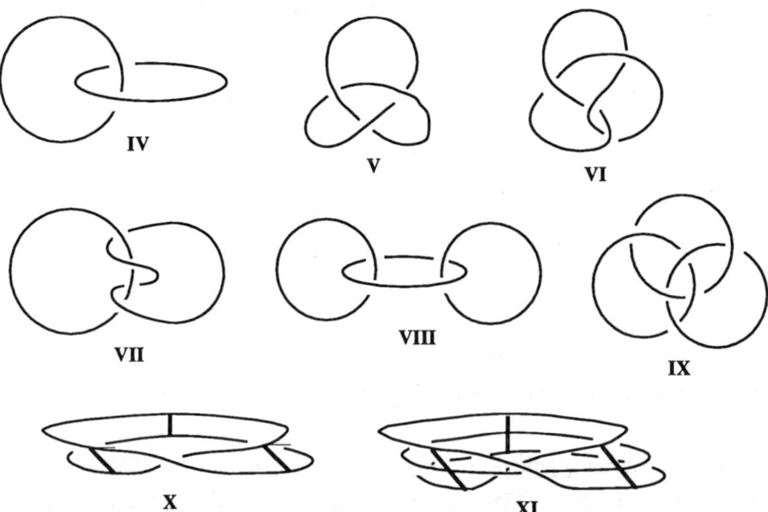

Figure 19. — Classic targets of synthetic topological stereochemistry.

Nonplanarity and Topological Stereocontrol

All of the structures given in Figure 19 possess a key topological property in common: All are extrinsically topologically nonplanar, and the Möbius ladders are also intrinsically nonplanar (the three rung Möbius ladder possesses a K_{33} nonplanar molecular graph[34]). In mathematical knot theory, the knot which can be presented in a plane is termed the unknot or the trivial knot (this being, of course, the simple unknotted circle). In fact, in the context of knot theory it makes good topological sense to consider any graph with a planar presentation to possess trivial topology.

To a large extent the key to synthesis of topological targets such as those listed in Figure 19 is the achievement of extrinsic topological nonplanarity. It is reasonable to refer to this goal as nontrivial topological stereocontrol. Thus, consider any simple cyclization reaction, as indicated in Figure 20. In fact, in all known examples of such reactions, the trivial knot **XII** is formed to the virtual exclusion of the trefoil **XIII** or any other nontrivially knotted product. This is, of course, an example of outstanding topological stereocontrol, since only one of an infinite number of possible topological stereoisomers is formed. Operationally and mathematically, however, this may justifiably be considered as an example of trivial topological stereocontrol. The same may be said of most of the myriad of reactions affording Euclidean stereocontrol. Indeed, from the topological perspective all of conventional stereocontrolled synthesis is trivial.

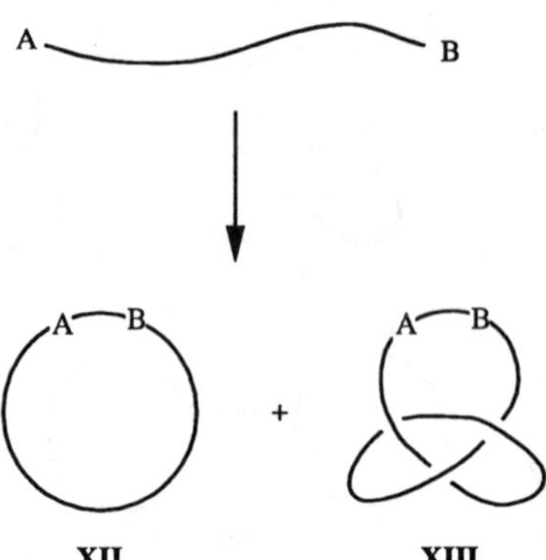

Figure 20. — Trivial and nontrivial topological stereocontrol in cyclization reactions.

The goal of synthetic topological stereochemistry, of course, is the synthesis of nontrivial targets. Thus, if the cyclization in Figure 20 could be accomplished such that the knot **XIII** were formed as the major product, then this would constitute nontrivial topological stereocontrol.

In fact, nontrivial topological stereocontrol is a central process in biochemical systems, involving the well known topoisomerases.[6] Excellent nontrivial topological stereocontrol in the synthesis of DNA macrocycles may be achieved using such enzymes. For example, the directed synthesis of a DNA figure-of-eight knot was recently reported.[40] In small molecule chemistry, however, examples of nontrivial topological stereocontrol are rare, the present synthesis being one. In the following section a brief discussion of the approaches developed to date for achieving this goal are described.

Topological Stereocontrol and the THYME Strategy

There seem to be two fundamental approaches for achieving nontrivial topological stereocontrol: A Euclidean approach and an intrinsic topological (regioselective) approach. Efficient Euclidean approaches involve using the metric geometrical properties of molecules to obtain nontrivial topology in a directed way, and are exemplified by the Luttringhouse-Schill strategy for synthesis of catenanes,[8,41] which has been extended by Sauvage[42,43] and Stoddart.[44]

In the Schill case, the Euclidean metrics of an organic covalent precursor provides a framework for forcing the formation of topologically nontrivial products. In the Sauvage approach a templating transition metal affords the desired Euclidean molecular shape, while Stoddart's method takes advantage of a strong non-covalent host-guest interaction for achieving the same end. Fundamentally, the Stoddart method is identical with Wasserman's original catenane synthesis[45] (often referred to as a "statistically random approach," where a macrocyclization reaction is accomplished in a macrocyclic solvent) except that the equilibrium favoring the desired (threaded) intermediate in Wasserman's case was much less favorable.

The Euclidean approach was also used to advantage in the synthesis of the first intrinsically nonplanar (K_5) molecule (in work stimulated by the topologically motivated ideas of Howard Simmons, Jr., and realized by Howard Simmons, III, working in the Woodward group at Harvard, and also in a non-topologically motivated synthesis by Paquette),[46,47] which unbeknownst to the authors, also provided the first topologically chiral organic molecule.[2,18] It is interesting to note that Woodward himself apparently considered molecular intrinsically nonplanar graphs to be unknown in the realm of natural products,[2,46] though the vitamin B_{12} structure (including the Co atom) is in fact a nonplanar graph. The statement regarding nonplanar graphs in nature is to our knowledge correct, however, if one excludes organometallic structures and polymers such as diamond[2] and the rare nonplanar proteins.[48]

The Möbius ladder strategy for synthesis of knots, as first suggested by Wasserman and van Gulick, is fundamentally different from the Euclidean approaches, relying instead on intrinsic topology (regioselectivity) to obtain nontrivial products. Indeed, both Wasserman and van Gulick envisioned such strong regiocontrol that in their Möbius ladder syntheses only nontrivial products would be produced. As mentioned above, the THYME strategy involves less intrinsic topological control in order that both cyclization steps may be the same transformation and accomplished in a single chemical step.

As an approach to the synthesis of nontrivial knots, without some Euclidean control thrown in, either the Wasserman/van Gulick or THYME strategies are inefficient and have been called "random" in the literature,[42,49] since the formation of the unknotted one half twist product competes with formation of the required three half twist (or higher) knotted targets. But as a strategy for synthesis of the one half twist Möbius ladder itself, mathematically perhaps the most stimulating nontrivial target, the THYME route is in fact highly efficient, competing favorably with all but the very best Euclidean approaches to nontrivial targets.

Thus, though cyclization of diol-ditosylate 1 can (and does) lead to both nontrivial and trivial products[50] (the Möbius ladder 10 and 3-rung prism 11, respectively), the intrinsic topology of the approach affords a relatively high degree of topological stereocontrol. The mixture of cyclized prisms and Möbius ladders are obtained in high yield due to the fully intramolecular nature of the cyclization, and due to the intrinsic topology of the ladder 1, almost half the product formed is nontrivial. Finally, and importantly, due to Euclidean constraints the only nontrivial product formed is that with a single half twist. The overall yield of Möbius ladder (about 30% isolated in the best runs) is lower than the best catenane-forming processes ($\cong 60\%$ yields) and the nonplanar graph synthesis (50% yield of nontrivial product), but is an order of magnitude higher than the yield of trefoil knot produced by Sauvage in his famous synthesis using a two rung "Möbius ladder" strategy with strong Euclidean stereocontrol (the beautiful biphenanthryl "double helix" templated by copper) but lacking the intrinsic topological control of the THYME strategy.[17]

Nouveau Targets of Topological Stereochemistry

As stated in the introduction, topology suggested targets for chemical synthesis (specifically those shown in Figure 19), the synthesis stimulated new topology (extrinsic graph theory), and finally, the new topology has suggested new targets for synthesis. A brief description of some of these targets, represented schematically in Figure 21, and the approaches to the topological stereocontrol being applied to their synthesis, seems a fitting conclusion to this paper.

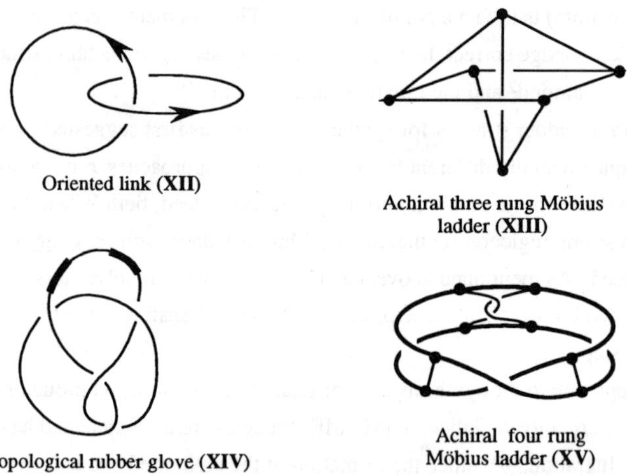

Figure 21. — Some new targets of topological stereochemistry.

It could be said that the first nouveau target of topological stereochemistry is the oriented link (**XII**). This is the prototype of a topologically chiral object, and a molecular realization of such a structure has been synthesized by Sauvage, stimulated by the work in topological stereochemistry.[38] This type of topologically chiral object was well known prior to the surge in extrinsic topology, however. The other targets shown in Figure 21 had not been suggested by mathematics until after the invention of topological stereochemistry and related mathematics.

Thus, to our knowledge no example of a topologically achiral three rung Möbius ladder (achiral K_{33} graph such as **XI**) has yet been prepared, though experiments directed towards realization of such a structure have recently been reported.[51] The topological stereocontrol involved in that work is Euclidean, and is somewhat reminiscent of that involved in the nonplanar K_5 graph syntheses of Simmons and Paquette.

While the figure-of-eight knot is a classic target of topological stereochemistry, the colored version illustrated by structure **XIV** has special properties making it an especially interesting nouveau target. Specifically, the knot **XIV** is topologically achiral but rigidly chiral in every presentation—a topological rubber glove (the monochrome knot **VI** is topologically achiral and possesses a rigidly achiral presentation).[1,2,23,36b,39] This target represents the only level on the topological hierarchy of molecular chirality not yet represented by a real molecule (assuming that a molecule is not real until it has been synthesized and characterized).[23]

While van Gulick has proposed an approach for the synthesis of such knots (the three braid Möbius ladder route mentioned above), the strategy seems impossible to realize at this primitive stage in the development of topologically stereocontrolled synthesis. Based upon discussions with Louis Kaufmann (a topologist at the University of Chicago, Chicago Circle) aimed at describing the simplest way to obtain to a figure-of-eight knot, we have been developing a strategy to the topological rubber glove which incorporates both the Euclidean and intrinsic topological approaches applied in successive steps—the hook and ladder route first described in 1987.[1,23]

As illustrated in Figure 22, the hook and ladder approach involves first creation of a functionalized catenane. Cyclization of the resulting hook and ladder graph **XVIII** then leads to several possible topological stereoisomeric graphs as shown. Since the starting material has nontrivial topology, all of the products do also.

Breaking the rungs of the 1/4 twist product **XIX** leads to colored unknot **XXII**, a similar process applied to 3/4 twist hook and ladder **XX** gives a trefoil (**XXIII**), finally breaking the rungs of the 1 1/4 twist hook and ladder **XXI** gives the topological rubber glove **XIV**.

We have recently reported the successful synthesis of the 1/4 twist product **XIX** using a combination of the Sauvage catenane synthesis for the creation of the hook and THYME rungs and uprights system.[1] Interestingly, while the product of breaking the rungs of this 1/4 twist isomer is topologically achiral, and the monochrome 1/4 twist hook and ladder graph is also topologically achiral, structure **XIX** itself is topologically chiral due to the coloring (this is similar to the situation with the three rung Möbius ladder).[52] Whether it will prove possible to achieve synthesis and proof of structure of the more twisted cyclized hook and ladders remains to be seen.

In order to avoid the obvious problems of requiring greater than the minimum number of twists involved in the hook and ladder strategy indicated in Figure 22, Sauvage has proposed in the literature a similar synthesis of a figure-of-eight knot where the "...knot is obtained by connecting two bis-chelate complexes in an orthogonal fashion."[49] This proposed route involves both strong Euclidean control and intrinsic topological control (two different cyclization processes are required—in a manner reminiscent of Wasserman's original Möbius ladder strategy), and in the ideal case could lead to the figure-of-eight as the least twisted possible product. However, the target as shown is not a topological rubber glove!

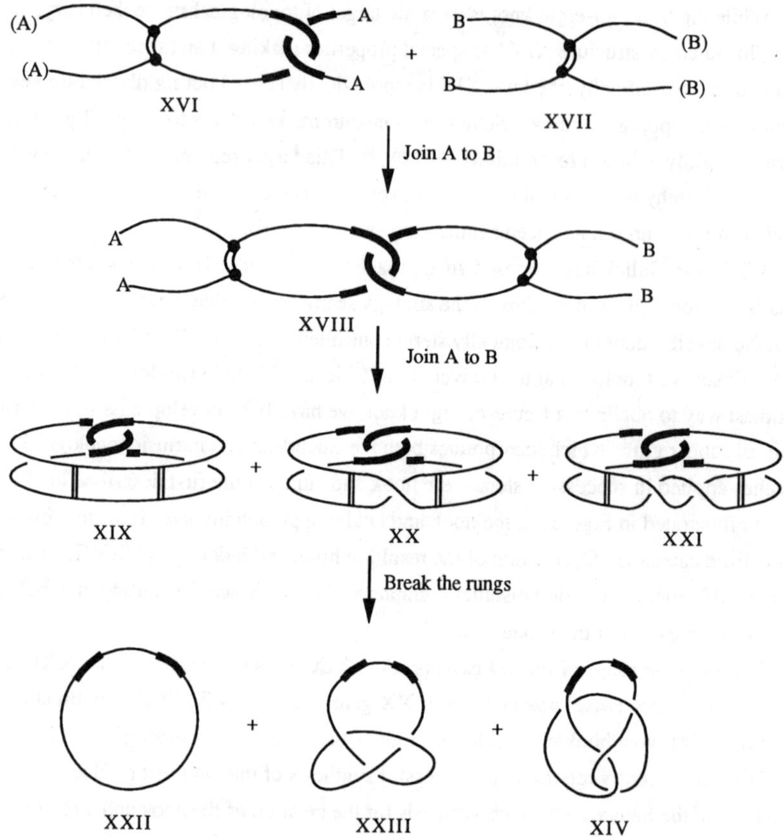

Figure 22. — The hook and ladder approach to a topological rubber glove.

Conclusion

While the synthesis of topologically inspired targets has no obvious current application in technology, if recent history is any indication synthetic studies on the new targets of topological stereochemistry will stimulate interesting new mathematics. Also, as mentioned above topological stereocontrol does play a crucially important role in biochemical systems. It is clear that improvements in small molecule topological stereocontrol will come with time (certainly one day molecular Borromean rings will be created by a directed approach!). It is conceivable that some time in the future this skill will enable important technological advances.

Acknowledgment

The authors wish to thank the Office of Naval Research and the National Science Foundation for financial support of this work.

References

†   University of Colorado Department of Chemistry and Biochemistry X-ray facility. Current address: SmithKline Beacham Laboratories, King of Prussia, PA.

1   Previous papers in the series: Walba, D. M.; Zheng, Q. Y.; Schilling, K. <u>J. Am. Chem. Soc.</u> **1992**, <u>114</u>, 6259-6260, and references therein.

2   (a) Walba, D. M. "Stereochemical Topology." In *Chemical Applications of Topology and Graph Theory*; King, R. B. Ed.; Elsevier: Amsterdam, 1983; Vol. pp 17-32. (b) Walba, D. M. <u>Tetrahedron</u> **1985**, <u>41</u>, 3161-3212.

3   Simon, J. "A Topological Approach to the Stereochemistry of Nonrigid Molecules." In *Graph Theory and Topology in Chemistry*; King, R. B. Ed.; Elsevier: Amsterdam, 1987; Vol. pp 43-75.

4   Walba, D. M.; Richards, R. M. <u>J. Am. Chem. Soc.</u> **1982**, <u>104</u>, 3219-3221.

5   (a) Frisch, H. L.; Wasserman, E. <u>J. Am. Chem. Soc.</u> **1961**, <u>83</u>, 3789-3795. (b) Wasserman, E. <u>Sci. Am.</u> **1962**, <u>207</u>, 94-100.

6   (a) Bauer, W. R.; Crick, F. H. C.; White, J. H. <u>Sci. Am.</u> **1980**, <u>243</u>, 118-133. (b) Gellert, M. <u>Ann. Rev. Biochem.</u> **1981**, <u>50</u>, 879-910. (c) Wang, J. C. <u>Sci. Am.</u> **1982**, <u>247</u>, 94-109. (d) Wasserman, S. A.; Cozzarelli, N. R. <u>Science</u> **1986**, <u>232</u>, 951-960.

7   Sumners, D. W. "Knots, Macromolecules and Chemical Dynamics." In *Graph Theory and Topology in Chemistry*; King, R. B. Ed.; Elsevier: Amsterdam, 1987; Vol. pp

8   Schill, G. <u>Catenanes, Rotaxanes, and Knots</u>; Academic Press: New York, 1971.

9   Twisted ladder precursors of catenanes, possessing necessarily an even number of half twists, are not Möbius ladders but rather prisms, a fact ignored by the early workers in the field.

[10] Interestingly in retrospect, van Gulick did not actually show any Möbius ladders in his paper, since as we have pointed out, the "strips" with an odd number of half twists and two rungs are not Möbius ladders, but rather embeddings of the tetrahedral graph.[2]

[11] (a) Walba, D. M.; Richards, R. M.; Sherwood, S. P.; Haltiwanger, R. C. J. Am. Chem. Soc. 1981, 103, 6213-6215. (b) Walba, D. M.; Richards, R. M.; Hermsmeier, M.; Haltiwanger, R. C. J. Am. Chem. Soc. 1987, 109, 7081-7087.

[12] The 2 rung THYME "prism" indeed provides an interesting example of such a hydrophilic cylinder.[9b]

[13] Froborg, J.; Magnusson, G.; Thoren, S. J. Org. Chem. 1975, 40, 122-123.

[14] Cresp, T. M.; Sondheimer, F. J. Am. Chem. Soc. 1975, 97, 4412-4413.

[15] Walba, D. M.; Armstrong, J. D., III; Perry, A. E.; Richards, R. M.; Homan, T. C.; Haltiwanger, R. C. Tetrahedron 1986, 42, 1883-1894.

[16] This interesting reaction is discussed in detail in reference 15 in the context of the four-rung THYME synthesis.

[17] Dietrich-Buchecker, C. O.; Sauvage, J.-P. Angew. Chem., Int. Ed. Engl. 1989, 28, 189-192.

[18] Simon, J. Topology 1986, 25, 229-235.

[19] We thank Professors Jonathan Simon and Walter Seaman, of the Department of Mathematics, University of Iowa, for pointing out to us the properties of the equilateral Möbius strip.

[20] This was pointed out to the authors by Professor Robert Whetten, Dept of Chem, UCLA.

[21] (a) Delacrétaz, G.; Grant, E. R.; Whetten, R. L.; Wöste, L.; Zwanziger, J. W. Phys. Rev. Lett. 1986, 56, 2598-2601. (b) Robinson, A. L. Science 1986, 234, 424-426.

[22] Mislow, K.; Siegel, J. J. Am. Chem. Soc. 1984, 106, 3319.

[23] Walba, D. M. "Topological Stereochemistry: Knot Theory of Molecular Graphs." In Graph Theory and Topology in Chemistry; King, R. B. Ed.; Elsevier: Amsterdam, 1987; Vol. pp 23-42.

24 Ireland, R. E.; Anderson, R. C.; Badoud, R.; Fitzsimmons, B. J.; McGarvey, G. J.; Thaisrivongs, S.; Wilcox, C. S. _J. Am. Chem. Soc._ **1983**, _105_, 1988.

25 Greenwald, R.; Chaykovsky, M.; Corey, E. J. _J. Org. Chem._ **1963**, _28_, 1128.

26 (a) Rietz, A. B.; Mutter, M. S.; Maryanoff, B. E. _J. Am. Chem. Soc._ **1984**, _106_, 1873. (b) Vedejs, E.; Meier, G. P.; Snoble, K. A. J. _J. Am. Chem. Soc._ **1981**, _103_, 2823.

27 Cram, D. J.; Tanner, M. E.; Thomas, R. _Angew. Chem., Int. Ed. Engl._ **1991**, _30_, 1024-1027.

28 Sokolov, V. I. _Russian Chemical Reviews_ **1973**, _42_, 452-463.

29 This idea is presented in reference 2a, and the mathematical term "isotopic" was used as the pairwise descriptor for structures interconvertable by continuous deformation in 3 space. In consultation with mathematicians, we later suggested the term homeotopic to mean the same thing (reference 23) since "isotopic" has a well known and altogether different meaning in chemistry. Homeotopic previously had no meaning in either mathematics or chemistry.

30 (a) Reidemeister, K. _Ergebnisse der Mathematik Vol. 1: Knotentheorie_; Springer-Verlag: Berlin, 1932. (b) Fox, R. H. "A quick trip through knot theory" In _Topology of 3-Manifolds_; Fort, M. K., Jr. Ed.; Prentice Hall: Englewood Cliffs NJ, 1962; pp 120-167.

31 Harary, F. _Graph Theory_; Addison-Wesley: Reading, 1969.

32 Kinoshita, S. _Osaka Math. J._ **1958**, _10_, 263-271.

33 The Kinoshita theta curve was latter shown to be topologically chiral. See Millett, K. C. "Algebraic Topological Indices of Molecular Chirality." In _New Developments in Molecular Chirality_; Mezey, P. G. Ed.; Kluwer Academic Publishers: Boston, 1991; Vol. 5, pp 165-207.

34 Guy, R. K.; Harary, F. _Can. Math. Bull._ **1967**, _10_, 493-496.

35 (a) Simon, J. "A Topological Approach to the Stereochemistry of Nonrigid Molecules." In _Graph Theory and Topology in Chemistry_; King, R. B. Ed.; Elsevier: Amsterdam, 1987; Vol. pp 43-75. (b) Simon, J., "Proceedings of Symposia in Applied Mathematics," Editor, 45, 97-130 (1992).

[36] (a) Flapan, E. "Chirality of Non-Standardly Embedded Möbius Ladders." In *Graph Theory and Topology in Chemistry*, King, R. B. Ed.; Elsevier: Amserdam, 1987; Vol. pp 76-81. (b) Flapan, E. Pacific J. Math. **1987**, <u>129</u>, 57-66. (c) Flapan, E. Mathematische Annalen **1989**, <u>283</u>, 271-283. (d) Flapan, E. "Topological Techniques to Detect Chirality." In *New Developments in Molecular Chirality*, Mezey, P. G. Ed.; Kluwer Academic Publishers: Boston, 1991; Vol. pp 209-239. (e) Flapan, E.; Weaver, N. Proc. AMS **1992**, <u>115</u>, 233-236.

[37] Flapan has recently proven that the Simmons-Paquette K_5 molecule also possesses an intrinsically chiral molecular graph (personal communication).

[38] Mitchell, D. K.; Sauvage, J. P. Angew. Chem., Int. Ed. Engl. **1988**, <u>27</u>, 930-931.

[39] Walba, D. M. "A Topological Hierarchy of Molecular Chirality and other Tidbits in Topological Stereochemistry." In *New Developments in Molecular Chirality*, Mezey, P. G. Ed.; Kluwer Academic Publishers: Boston, 1991; Vol. 5, pp 119-129.

[40] Seeman, N. C.; Du, S. M. J. Am. Chem. Soc. **1992**, <u>114</u>, 9652-9655.

[41] (a) Schill, G. Chem. Ber. **1967**, <u>100</u>, 2021. For the first synthesis of saturated hydrocarbon linked rings, see: Schill, G.; Schweickert, N.; Fritz, H.; Vetter, W. Angew. Chem., Int. Ed. Engl. **1983**, <u>22</u>, 889-891.

[42] Dietrich-Buchecker, C. O.; Sauvage, J.-P.; Kern, J.-M. J. Am. Chem. Soc. **1984**, <u>106</u>, 3043-3045.

[43] Dietrich-Buchecker, C. O.; Sauvage, J. P. Chem. Rev. **1987**, <u>87</u>, 795-810.

[44] (a) Anelli, P. L.; Ashton, P. R.; Ballardini, R.; Balzani, V.; Delgado, M.; Gandolfi, M. T.; Goodnow, T. T.; Kaifer, A. E.; Philip, D.; Pietraszkiewicz, M.; Prodi, L.; Reddington, M. V.; Slawin, A. M. Z.; Spencer, N.; Stoddart, J. F.; Vicent, C.; Williams, D. J. J. Am. Chem. Soc. **1992**, <u>114</u>, 193-218. (b) Ashtron, P. R.; Goodnow, T. T.; Kaifer, A. E.; Reddington, M. V.; Slawin, A. M. Z.; Spencer, N.; Stoddart, J. F.; Vincent, C.; Williams, D. J. Angew. Chem., Int. Ed. Engl. **1989**, <u>28</u>, 1396-1399.

[45] Wasserman, E. J. Am. Chem. Soc. **1960**, <u>82</u>, 4433-4434.

[46] (a) Simmons, H. E., III; Maggio, J. E. Tetrahedron Lett. **1981**, <u>22</u>, 287-290. (b) Benner, S. A.; Maggio, J. E.; Simmons, H. E., III J. Am. Chem. Soc. **1981**, <u>103</u>, 1581-1582.

[47] Paquette, L. A.; Vazeux, M. <u>Tetrahedron Lett.</u> **1981**, <u>22</u>, 291-294.

[48] Mao, B. <u>J. Am. Chem. Soc.</u> **1989**, <u>111</u>, 6132-6136.

[49] Dietrich-Buchecker, C.; Sauvage, J. P. <u>New J. Chem.</u> **1992**, <u>16</u>, 277-285.

[50] Here the terms trivial and non-trivial are taken from the field of knot theory, where the trivial knot is the unknot, the only embedding of a closed curve which may be placed entirely in a plane in 3-space. Knots which cannot be presented in a plane in 3-space (i.e. all knots which are not the unknot) are non-trivial embeddings of a closed curve. Extension of this terminology to graphs is straightforward: a trivial graph is "planar" in that it may be presented in a plane in 3-space. Nontrivial graphs cannot be presented in a plane. All nontrivial knots may be said to be extrinsically nonplanar. Nontrivial graphs may be intrinsically nonplanar (i.e. possessing the Kuratowski nonplanar graphs K_{33} or K_5 as a subgraph) or extrinsically nonplanar.

[51] Chen, C.-T.; Gantzel, P.; Ho, D. M.; Hardcastle, K.; Siegel, J. S., "Synthesis and Structure of the Novel Tetracyclic-Cyclophanes," Abstracts of the Division of Organic Chemistry, 205th ACS National Meeting, ORGN # 309 (1993).

[52] Simon, J., unpublished results.

Reprinted from JOURNAL OF MATHEMATICAL PSYCHOLOGY
All Rights Reserved by Academic Press, New York and London

Vol. 26, No. 3, December 1982
Printed in Belgium

Turning a Penrose Triangle Inside Out

THADDEUS M. COWAN

Kansas State University

Why do impossible figures, which cannot exist in three dimensions, appear to make three-dimensional sense? In order to shed some light on this question the limits may be tested to which three-dimensional operations on these figures can be performed. In this paper a particularly difficult operation, viz., torus eversion is attempted. Not only is an eversion found to be possible but an unfamiliar impossibility develops. The regular form of the eversion is shown to be unique.

The eversion of a torus is a well-known problem which appears now and then in recreational mathematics (Gardner, 1959). Its appeal comes from the challenge it provides for the visual imagination. The problem asks, "If a hole is cut in the side of, say, an inner tube, can the tube be pulled completely through the hole so that it everts or turns inside out?" You are now asked to compound the problem and imagine (or try to!) the eversion of an *impossible* torus such as Penrose's triangle shown in Fig. 1a.

The eversion of a Penrose triangle may have significance beyond its status as a curiosity. Figures like this cannot exist in our three-dimensional physical world, and we might deny their toroidal form on these grounds. Yet toruses they are; if they were not they would seem flat (literally) and uninteresting. While Fig. 1a is not a projection of a real three-dimensional object it is a representation, a "projection," of a *perceptual* object. The interesting question here is not why these figures appear impossible, but why do we make perceptual sense out of them at all.

As a beginning we might try to discover the formal properties of such a perceptual torus. Included in the list of formal properties are certain transformations we can accomplish with these figures. Since three dimensionality seems to be a particularly crucial property here we might ask if there exists any three-dimensional transformation that cannot be performed on a perceptual object. And while we cannot physically maneuver a Penrose triangle in three space, we might be able to discover what the results of a given maneuver would be.

Here we choose inversion as a transformation problem because it is sufficiently complex so as to make it nontrivial. Furthermore, there are reasons to believe that an eversion provides an impossible figure pattern that is different from the ones we have been accustomed to seeing. These reasons will become apparent later.

Address reprint requests to Thaddeus M. Cowan, Department of Psychology, Kansas State University, Manhattan, Kansas 66506.

252

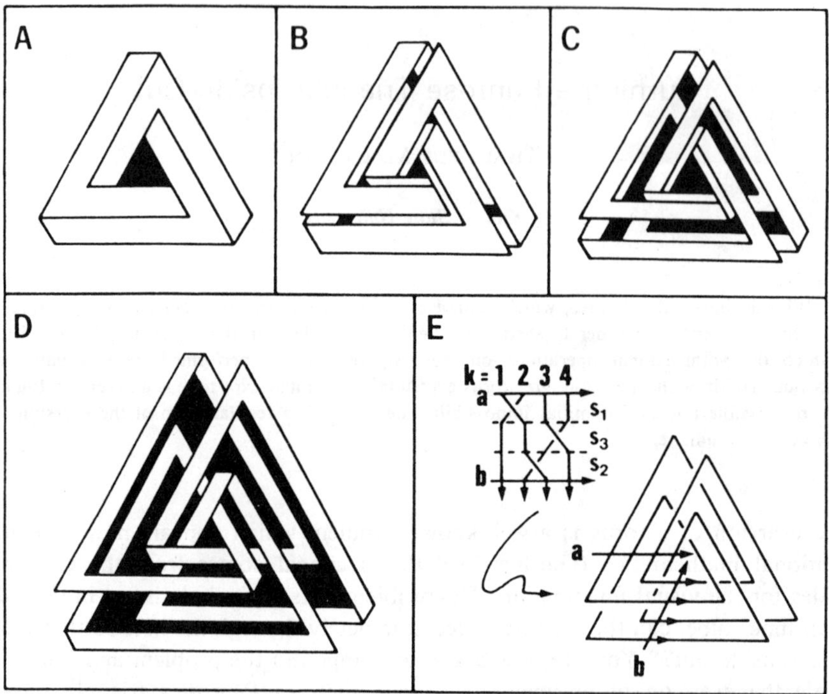

FIG. 1. (A)–(D) A Penrose triangle in knots and coming unglued; (E) an $s_1 s_3 s_2$ braided corner.

STIMULUS CUES FOR IMPOSSIBILITY

The Penrose triangle, like most impossible toruses, is edged and cornered, and this imposes a rigidity onto the figure which would seem to obviate eversion. It is evident that we must somehow free the torus by either relaxing the corners, the edges or both. However, it is necessary that those features which reflect the impossibility of the figure be retained during the eversion; we want to make sure that the impossibility is carried through the eversion process. Our first task then is to indicate which of the stimulus cues provides the perceptual confrontation.

We note that the corners appear right angled downward (or upward), and this leaves us with the impression that we are led out of the plane in a consistant direction at every turn, and yet we end where we begin. This means that every impossible torus figure has an impossible staircase counterpart. It is this aspect of the torus that creates its apparent contradiction, a fact that has been suggested a number of times (Harris, 1973; Cowan, 1977; Draper, 1978). Moreover, it has been empirically demonstrated that naive observers, although they are unaware of it, will rely on just this feature in judging the degree of impossibility of a wide variety of such toruses (Cowan and Pringle, 1978).

THADDEUS M. COWAN

It seems self-evident that to find the confrontation, the realization that one ends where one begins, we must scan the torus in a circular manner around the center following whatever stimulus cue that guides us in that direction. Therefore, the parts which best reveal this ascending–descending feature, the feature we want to preserve during eversion, are the *longitudes* or those edges which run parallel to the center hole.

In recent years there have appeared a number of papers promoting different analyses of figure impossibilities (Huffman, 1971, 1977; Cowan, 1974, 1977; Draper, 1978; Terouanne, 1980). Each of these should have its own thing to say about torus eversion. We choose the approach of knot topology and the related study of braids.

THE PENROSE TRIANGLE AS A BRAIDED STRUCTURE

One way to show how the edges are patterned in a Penrose figure is to show how the surface strips of the triangle are arranged. The edges define the boundaries of these strips, and as Fig. 1 shows the edge and strip patterns are the same. The strips in this figure have been separated. They are sometimes folded at the corners along lines parallel to the *meridians* (cross section lines) of the torus. We ignore these creases, because they are orthogonal to the longitude lines critical to impossibility.

The separated strips reveal a weaving pattern at each of the three corners. This weave can be considered a type of braid. The corners are well represented by braided lines, and the topology of braids (Artin, 1941) serves as a useful tool for the analysis of figures like this (Cowan, 1974). The braid for one of the corners is given in Fig. 1e. If we let the symbols s_k and s_k^{-1} represent the kth edge from the right (following the longitude direction) passing over (s_k) and under (s_k^{-1}) the $(k+1)$th edge from the right we find that each corner is $s_1 s_3 s_2$ braided. The corners of the Penrose triangle are all alike, and the braid of the whole Penrose triangle is

$$s_1 s_3 s_2 s_1 s_3 s_2 s_1 s_3 s_2 \quad \text{or} \quad (s_1 s_3 s_2)^3.$$

THE PENROSE TRIANGLE AS A KNOTTED STRUCTURE

Since we ignore the folds or creases in the braided strips (edges) we relax the corners and treat the edge pattern as a general knot which we call a Penrose knot (see Fig. 2a). The analysis of a knot is an extraordinary example of a figure–ground problem. A knot K is described, not by the course of the string entangled in it, but in terms of the way the space R^3 surrounding the knot $(R^3 - K)$ is curved. Space curves differently around different loops of the knot which are set apart from each other by certain nodes.

But how are we to describe the curvature of the various parts of the surrounding space? We can conceive of a given section of space as a collection of directed lines or

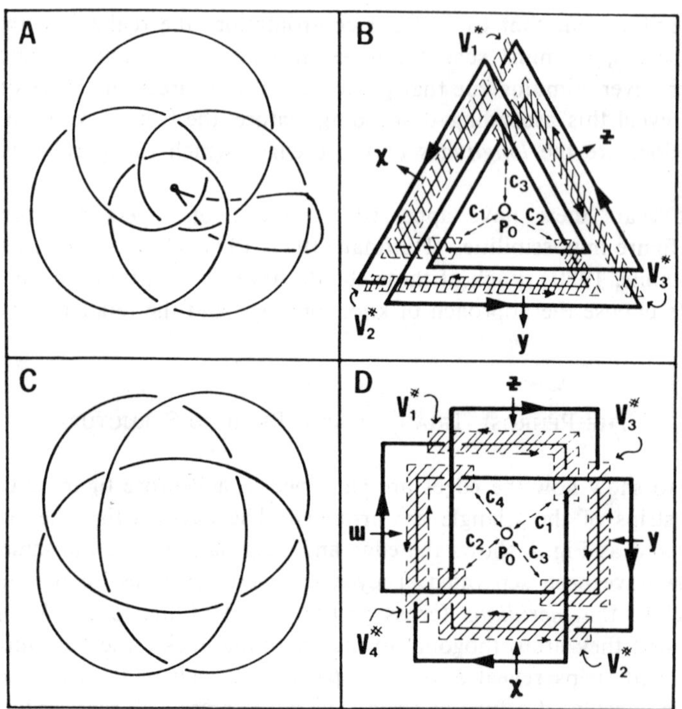

FIG. 2. (A)–(B) The Penrose knot and its over and under path group presentation. A single loop path (x, y, or z) is shown in (A); (C)–(D) the "everted" Penrose knot and the presentation of its fundamental path group. (A) and (C) are homotopies.

paths emanating from a single base point. One of these paths is shown in Fig. 2a. Any path is deformable into any other within the same section of space. Such paths are said to be *homotopic*. The equivalence classes of these paths with the basepoint exhibit a well-defined algebraic structure called the fundamental path group of the knot.

Since we use the algebraic structure of the knot as an analytic tool we must find the fundamental path group of the Penrose knot. The path group will point to a relationship between the Penrose knot and the common overhand knot or trefoil. We will use this knot form to show how an edged torus figure can be everted.

FUNDAMENTAL PATH GROUP

It is beyond the scope of this paper to give a detailed account of just how the path group is found, and we approach the problem in an algorithmic way by listing the

256   THADDEUS M. COWAN

steps in the process. The reader is referred to Crowell and Fox (1963) for their justification. An excellent introduction to knot theory can be found in a *Scientific American* article on knot theory by Neuwirth (1979).

In order to present a fundamental path group we must first define a set of elementary paths or generators of the group. Then we specify a set of group relations, an exhaustive list of equations each of which equals one or the identity of the path group. (The identity is the set of all paths that can be deformed, here retracted, into the constant path or base point. Such paths are homotopic to the constant path.) Each equation in the list of relations is reduced. That is, all contiguous inverse products in the set of terms (including the first and last terms) which can be reduced to one are eliminated.

The fundamental path group of any knot, in this case the Penrose knot, can be determined by following the steps listed below.

(1) Give each knot a polygonal form. This form comes naturally to the Penrose knot.

(2) Give a direction to the string of the knot (i.e., the edges of the Penrose figure).

(3) Mark the string segments which incorporate unbroken sequences of *overpasses*. These are shown by the heavy lines in Fig. 2b.

(4) Cross each of these segments with an arrow which indicates a left-hand screw (the segment it crosses comes in from the right). The individual arrows are actually segments of separate paths with a common base point which loop under and around the overpass lines. These loop segments, labeled x, y, and z in Fig. 2b, define the elements or generators of the algebraic group of the Penrose knot. The relations of the group, the set of reduced equations equal to one, are found in the following way:

(1) Enclose the segments which incorporate unbroken sequences of *underpasses* with a polygon.

(2) Draw a path c_i from each corner of the three polygons to the base point p_0.

(3) Give a counterclockwise direction to the edges of the polygon and a bidirectional label to c_i. The edges of the polygon and c_i form a path leading to the basepoint p_0.

We now have three loop paths $V_i^\#$, each of which are homotopic to the same constant path. That is, each path can be retracted back to the base point without becoming entangled in the knot. Each path, therefore, is a unit path. Notice that each of these loop paths crosses each overpass segment of the knot in the same way as the generator of the segment or its inverse (not surprisingly, this presentation is called an *over and under presentation*). Therefore, the unit paths $V_i^\#$ can be considered equal to a composite (product) of segments each homotopic to some generator.

For example, consider $V_1^\#$. Starting at p_0 we encounter c_1. Then the path crosses a knot segment in a direction congruent with z, then y, etc. Therefore, the equation for the $V_1^{\#}$ unit path is

$$V_1^{\#} = c_1 zyxzx^{-1}y^{-1}z^{-1}x^{-1}c_1^{-1} = 1.$$

Moving c_1 and c_1^{-1} to the other side we find that this bidirectional segment can be omitted from the equation without loss.

Continuing,

$$V_2^{\#} = xzyxy^{-1}z^{-1}x^{-1}y^{-1} = 1.$$

$V_3^{\#}$ can be determined from the other two. The formal presentation of the fundamental path group, $\pi(R^3 - K)$, of the Penrose knot then is

$$\pi(R^3 - K) = |x, y, z: xzyx = zyxz, xzyx = yxzy|. \tag{1}$$

THE PENROSE KNOT AND THE TREFOIL KNOT

We identify the three pairwise combinations of the three generators: $xz = a$, $yx = b$, and $zy = c$. Substituting these in Eq. 1 we find that $ab = ca$, and $ab = bc$. Some simple algebraic manipulations beginning with a solution for c leads us to $bab = aba$. Consider the presentation

$$\pi(R^3 - K) = |a, b: bab = aba|. \tag{2}$$

This is the presentation of the cloverleaf or trefoil knot (Crowell and Fox, 1963). Thus a relationship is established between the trefoil and the Penrose knots.

Multiply each side of the relation in Eq. 2 on the left (the algebraic group is noncommutative) by aba,

$$ababab = abaaba.$$

Let $u = ab$ and $v = aba$,

$$(u)^3 = (v)^2.$$

Equation 3 specifies two homotopic knot forms both of which belong to the same torus. For example, $(v)^2$ may be taken as a trefoil knot that follows a longitude path twice around the center of the torus (see Fig. 3a). That is, it winds itself two times around the center hole; any line from the outside to the center intersects two strings of the knot projection, and we say the *winding number* of the torus knot is two.

But what about $(u)^3$, and how do we find it? First note that if the exponent gives the winding number, then the number of times the string of the new knot wraps itself around the center hole should be three. Observe that there are three loops to the trefoil in Fig. 3a, and each loop can be seen to follow the meridian of the torus. If we

THADDEUS M. COWAN

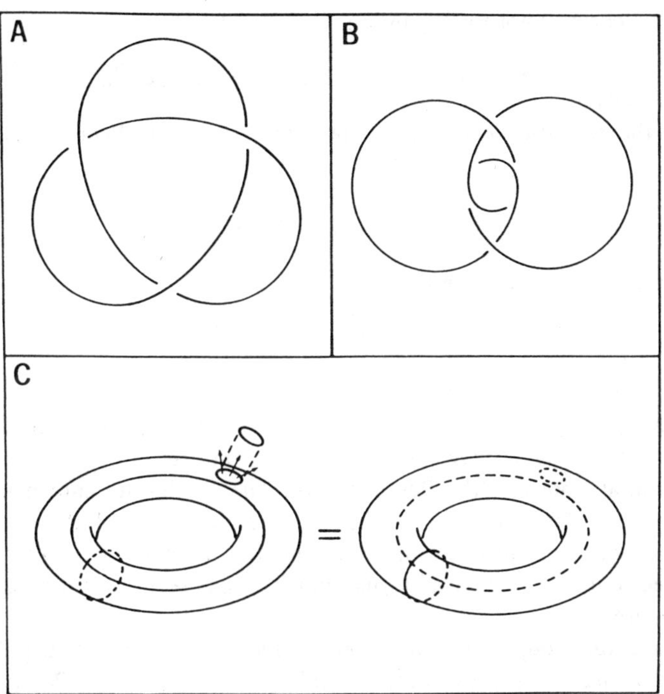

FIG. 3. (A)–(B) The common overhand knot (also called a trefoil or cloverleaf knot) and its homotopic form; (C) the eversion of a torus showing the exchange of the meridian and longitude.

carefully lift the right edge of each loop and, moving clockwise, thread each in turn with, say, a pencil, we create the new knot shown in Fig. 3b with the pencil in the center (an empirical verification of this with a real string has pedagogical value).

The new knot seems to be what we need; it has a winding number of three. But more importantly for our purposes, the meridian loops we threaded are now the longitude strings of the new torus. That is, the torus longitude line and meridian line have exchanged positions, and this is precisely what happens when the torus is everted (Fig. 3c). In short the eversion of the torus containing the knot $(v)^2$ will produce the knot $(u)^3$ and we can capitalize on this in everting the Penrose triangle.

EVERTING THE PENROSE TRIANGLE

Although the Penrose knot has the appearance of a trefoil knot $(v)^2$, it has a winding number of four not two; we might expect something like

$$(u)^3 = ((v)^2)^2 = (v)^4.$$

Return to the Penrose knot by expanding $(u)^3$ and $(v)^2$,

$$(ab)^3 = (aba)^2, \qquad (xzyx)^3 = (xzyxxz)^2.$$

We can use the relations in the group presentation of the Penrose knot (Eq. 1) to show that

$$(zyxz)^3 = (yxz)^4.$$

Setting $e = zyxz$ and $p = yxz$ we have

$$(e)^3 = (p)^4 \tag{4}$$

which is the desired result.

The Penrose triangle is evidently $(p)^4$, but how do we construct $(e)^3$? We can procede much in the same way that we did with the trefoil by physically threading the three meridian loops of the Penrose knot. The results are shown in Fig. 2c. Of course in creating this new knot we have found the edges of the everted Penrose triangle. Note that the winding number of the $(e)^3$ knot is three. This tells us that there are three edges to the everted torus, thus its cross section is prismatic. Prismatic impossible figures are unknown.

The fundamental group of this new knot is the same as that of the Penrose knot. The over and under group presentation of this new knot (see Fig. 2d) is

$\pi(R^3 - K) = |w, x, y, z: wzy = zyx, xwz = yxw, yxw = zyx|$. Again the fourth relation is determined from the other three. Solve for w in the third relation and substitute this value for w in the first and second relations,

$$\pi(R^3 - K) = |x, y, z: zyxzy = yxzyx, yzyx = zyxz|. \tag{5}$$

Using the relations in Eqs. 5 and simple algebraic manipulation we find

$$(zyxz)^3 = (yxz)^4 \qquad \text{or} \qquad (e)^3 = (p)^4.$$

We now transform this new knot into a cornered torus.

CORNERING THE EVERTED TORUS

We return to braid theory to help us form the corners. The braid representation of the everted Penrose knot is

$$s_1 s_2 s_1 s_2 s_1 s_2 s_1 s_2 = (s_1 s_2)^4 \tag{7}$$

as Fig. 4a shows. All of these braids are inverses but the inverse signs will be omitted for convenience; we will in effect be dealing with the mirror image of the figure. Note

THADDEUS M. COWAN

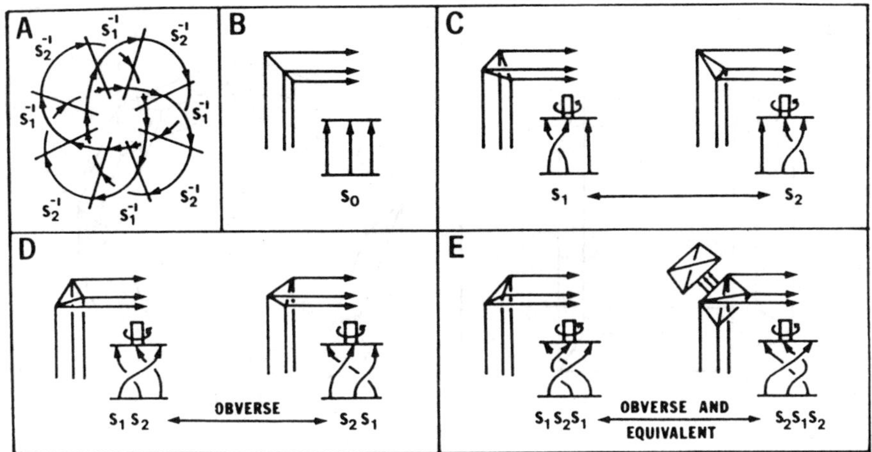

FIG. 4. (A) The homotopy of the Penrose knot divided into braided segments; (B)–(E) some prismatic corners possible with the Penrose homotopy. Each braid pair (not corner pair) are mutual obverses created by rotating the braid frame as indicated. (B) An identity prism corner which is not considered here; (C)–(D) two corners that are considered viable; (E) $s_1 s_2 s_1$ is considered viable but its obverse is not.

that since there are only three edges there are only two types of underpasses or overpasses (s_1 or s_2) in each braid.

Now we pay the price for relaxing the corner braids of the Penrose triangle prior to eversion, for here we do not know how to group the crossover nodes which will define the braids of the corners of the new figure. However, we can impose certain logical restrictions on the corners to aid us in our efforts. For example, the figure is prismatic, and we will accept only cornered crease patterns that match the cross section; i.e., the prism bends so that the creases form a triangle. This means that each edge bends once and only once when rounding the corner to give three points through which the triangular crease may form.

This has two consequences: First, there can be no more than three crossover nodes per corner, and second, no two edges may cross then recross each other. Therefore, corner braids with more than three nodes will not be allowed nor will braids with $s_1 s_1$ or $s_2 s_2$ node pairs. This leaves braids s_1, s_2, $(s_1 s_2)$, $(s_2 s_1)$, $(s_1 s_2 s_1)$, and $(s_2 s_1 s_2)$. These are shown in Fig. 4 with representative corners. The identity braid s_0 is also shown, but this corner will not be used here because of the reduced form of the knot relations.

The equivalence $s_k s_{k+1} s_k = s_{k+1} s_k s_{k+1}$ is one of two relations of the braid group presentation of which $s_1 s_2 s_1 = s_2 s_1 s_2$ is a particular. The other relation is $s_k s_{k+2} = s_{k+2} s_k$ which need not concern us since it requires a braid of at least four strings. One of the two $s_2 s_1 s_2$ corners is tetrahedral as Fig. 4e shows (the other is an obverse), and since the middle edge bends twice here this corner will not be allowed.

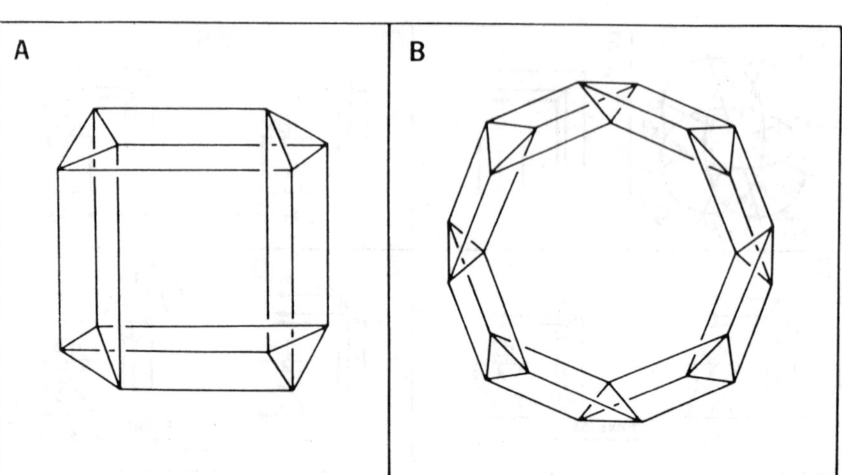

FIG. 5.   (A) The everted Penrose triangle in its regular form. Its braid equation is $(s_1 s_2)^4$. (B) An irregular form of the eversion. Its braid equation is $[(s_1)(s_2)]^4$.

Such a corner, when encountered, is easily transformed to the $s_1 s_2 s_1$ form.   The relevant braid relation allows us to write numerous equivalent forms of Eq. 7. If we take each form and define the torus corners by gathering the nodes by ones, twos, and threes, then include all their permutations and combinations, the number of torus figures becomes quite large. But the Penrose triangle is a regular figure. That is, all corners are the same. We might expect as much of its everse.

We reduce the set of possible everted toruses, then, to the set of regular forms. The possible membership of this set includes a form with all s_1 (or s_2) corners, a form with all $s_1 s_2$ (or $s_2 s_1$) corners, and a form with all $s_1 s_2 s_1$ corners. However, there is no way that we can use the equivalence relation to reduce all nodes to s_1 (or s_2), since both s_1 and s_2 are included in the relation. Furthermore, since there are eight nodes in Eq. 7 there cannot be a regular figure formed out of $s_1 s_2 s_1$ braided corners since the three nodes of $s_1 s_2 s_1$ do not divide eight evenly. The only possibility left is to partition Eq. 7 into four $s_1 s_2$ sets (or the obverse set $s_2 s_1$).

The everse of the Penrose torus is shown in Fig. 5. It has four $s_1 s_2$ corners, and it displays an interesting ascending–descending impossibility. An irregular variant is shown next to it for comparison. There is an esthetic symmetry here between the Penrose triangle and its regular everted form. The Penrose triangle has three corners and a cross section of four sides; its regular everse has four corners and a cross section of three sides. Whether this is true for all regular figures and their regular everses is not known.

RECEIVED: August 10, 1982

262   THADDEUS M. COWAN

REFERENCES

ARTIN, E. Theory of braids. *Annals of Mathematics*, 1941, **48**, 101–126.

COWAN, T. M. The theory of braids and the analysis of impossible figures. *Journal of Mathematical Psychology*, 1974, **11**, 190–212.

COWAN, T. M. Organizing the properties of impossible figures. *Perception*, 1977, **6**, 41–56.

COWAN, T. M., PRINGLE, R. An investigation of the cues responsible for figure impossibility. *Journal of Experimental Psychology: Human Performance and Perception*, 1978, **4**, 112–120.

CROWELL, R. H., & FOX, R. H. *Introduction to knot theory.* New York: Springer-Verlag, 1963.

DRAPER, S. W. The Penrose triangle and a family of related figures. *Perception*, 1978, **7**, 283–296.

GARDNER, M. Mathematical games. *Scientific American*, 1959 (January), **198** (1), 92–96.

HARRIS, W. F. Perceptual singularities in impossible pictures represent screw dislocations. *South African Journal of Science*, 1973, **69**, 10–13.

Huffman, D. A. Impossible objects as nonsense sentences. In B. Meltzer & D. Michie (Eds.), *Machine intelligence.* Vol. 6. New York: Halstead, 1971.

HUFFMAN, D. A. A duality concept for the analysis of polyhedral scenes. In E. W. Elcock and D. Michie (Eds.), *Machine intelligence.* Vol. 8. New York: Halsted, 1977, pp. 475–492.

HUFFMAN, D. A. Realizable configurations of lines in pictures of Polyhedra. In E. W. Elcock and D. Michie (Eds.), *Machine intelligence.* Vol. 8. New York: Halsted, 1977, pp. 493–509.

NEUWIRTH, L. The theory of Knots. *Scientific American*, 1979 (June), **240** (6), 110–124.

PENROSE, L. S., & PENROSE, R. Impossible objects: A special type of illusion. *British Journal of Psychology.* 1958, **49**, 31–33.

TEROUANNE, E. On a class of impossible figures: A new language for a new analysis. *Journal of Mathematical Psychology*, 1980, **22**, 24–48.

List of Authors

Thaddeus M. Cowan
Department of Psychology
Kansas State University
Manhattan, Kansas   66506

David Finkelstein
School of Physics
Georgia Institute of Technology
Atlanta, Georgia   30332

Herbert Jehle

Louis H. Kauffman
Dept. of Math. Stat. and Comp. Sci.
Univ. of Ill. at Chicago
851 South Morgan St.
Chicago, Ill.   60607-7045

Lord Kelvin, Sir William Thomson

Alison MacArthur
Department of Polymer Science
The University of Akron
Akron, Ohio   44325

Y.B. Magarshak
Biomathematical Sciences Department
Mount Sinai School of Medicine
City University fo New York
New York   10029

Maurizio Martellini
I.N.F.N., sezione di Roma
I-00185   Roma
Italy

Eckehard W. Mielke
Institute for Theoretical Physics
University of Cologne
D-50923
Koln, Germany

H. K. Moffatt
Dept. of Applied Mathematics and Theoretical Physics
Silver Street
Cambridge CB3 9EW
U.K.

Mario Rasetti
Dipartmento di Fisica
Politecnico di Torino
I-10129 Torino
Italy

Lev Rozansky
Theory Group
Department of Physics
University of Texas at Austin
Austin, Texas 78712-1081

David W. Walba
Department of Chemistry and Biochemistry
University of Colorado
Boulder, Colorado 80309-0215